THE SPARKS OF RANDOMNESS

Cultural Memory
in
the
Present

Mieke Bal and Hent de Vries, Editors

HENRI ATLAN

THE SPARKS OF RANDOMNESS

VOLUME I

SPERMATIC KNOWLEDGE

TRANSLATED BY Lenn J. Schramm

STANFORD UNIVERSITY PRESS · STANFORD, CALIFORNIA

Stanford University Press
Stanford, California

The Sparks of Randomness, Volume 1, was originally published in French under the title *Les Étincelles de hasard, Tome 1. Connaissance spermatique*, ©Éditions du Seuil, 1999. Collection de la Librairie du XXIe siècle, sous la direction de Maurice Olender.

This book has been published with the assistance of the French Ministry of Culture—
National Center for the Book and of the Human Futures Foundation.

This work, published as part of a program providing publication assistance, received financial support from the French Ministry of Foreign Affairs, the Cultural Services of the French Embassy in the United States, and FACE (French American Cultural Exchange, www.frenchbooknews.com).

Printed and bound by CPI Group (UK) Ltd, Croydon, CR0 4YY

Library of Congress Cataloging-in-Publication Data

Atlan, Henri.
[Étincelles de hasard. English]
The sparks of randomness / Henri Atlan ; translated by Lenn J. Schramm.
v. cm.--(Cultural memory in the present)
"Originally published in French under the title: Les Étincelles de hasard."
Includes bibliographical references and index.
Contents: v. 1. Spermatic knowledge
ISBN 978-0-8047-7357-7 (cloth : alk. paper)--ISBN 978-0-8047-6027-0 (pbk.)
1. Chance. 2. Philosophy and science. 3. Biology--Philosophy. I. Schramm, Lenn J. II. Title. III. Series: Cultural memory in the present.
BD595.A8513 2011
194--dc22 2010036068

Designed by Bruce Lundquist
Typeset at Stanford University Press in 10/15 Sabon

Mine is a dizzying country in which the Lottery is a major element of reality; until this day, I have thought as little about it as about the conduct of the indecipherable gods or of my heart. Now, far from Babylon and its beloved customs, I think with some bewilderment about the Lottery, and about the blasphemous conjectures that shrouded men whisper in the half-light of dawn or evening.

JORGE LUIS BORGES, "The Lottery in Babylon," in *Collected Fictions*, trans. Andrew Hurley (New York: Penguin Putnam, 1998), p. 101

One might say: "'I know' expresses comfortable certainty, not the certainty that is still struggling." Now I would like to regard this certainty, not as something akin to hastiness or superficiality, but as a form of life. (That is very badly expressed and probably badly thought as well.) But that means I want to conceive it as something that lies beyond being justified or unjustified; as it were, as something animal.

LUDWIG WITTGENSTEIN, *On Certainty*, §§357–359

The tree of life in the middle of the garden, and
the tree of knowledge, good and bad.

GENESIS 2:9

Life as such does not exist.

ALBERT SZENT-GYÖRGI, *The Nature of Life*

To Bela Rachel

CONTENTS

TRANSLATOR'S INTRODUCTION xiii
Lenn J. Schramm

INTRODUCTION 1

CHAPTER 1
THE SPARKS OF RANDOMNESS AND THE MANUFACTURE OF THE LIVING 21
1. Producing Life, Controlling Time 21
2. Seminal Reason and "It's All in the Genes" 23
3. Concept and Life 25
4. Golems 28
5. The Sparks of Randomness and Their Ambiguity 34
6. The Generations of the Flood, of the Tower of Babel, and of the Wilderness 37
7. Ashmedai and Solomon: Tricks of Nature and Tricks with Nature 39
8. The Tree of Knowledge, Good and Bad 43
9. Ambiguity and Polysemy 47
10. The Randomness of Birth: Evil at Its Source Is Not Evil 49
11. The "Mixed Multitude" in the Wilderness and the "Head That Is Not Known" 53
12. The Flood and the Tower of Babel 55
13. The Time of Magic and the Time of Technology 57
14. The Times of Memory 61
15. The Individual "Seal" in the Nature of Things 62

CHAPTER 2
NO ONE KNOWS WHAT THE BODY CAN DO:
THE TREE OF KNOWLEDGE AND GAMES OF ABSOLUTE DETERMINISM 66
1. The Shape of the Body and the Corporeality of God 66
2. Good and Evil: Moral Conscience or Knowledge of the True 72
3. The Tree of Knowledge and the Fall of the First Man, According to Rabbi Shlomo Eliaschow's *Book of Knowledge* 74
4. An Ambiguous Serpent 79
5. A Return to the Sparks of Randomness 82
6. The Yearning for Yearning: Eros and the Holy 84
7. Behemoth and Leviathan 86
8. What the Body Can Do 88
9. Mechanism and Responsibility 90
10. Nature's Ultimate Trick: The Parable of the Divine Intrigues (*'alilot*) 96
11. "Before Creation": The World of *Tohu* and the 974 Lost Generations 101
12. What Freedom? 109
13. The Call of Ethics 119
14. The Games of the Infinite—in Itself, by Itself, and for Itself 124
15. Incorporations 130
Appendix: The Dialectic of Absolute Determinism and Choice in the *Book of Knowledge* and in Ecclesiastes 132

CHAPTER 3
SPIRITS AND DEMONS: SUBJECT AND SHADOW 142
1. Hidden Causes 142
2. Spinoza and the Spirits 146
3. Causes in Nature and Animation of Subjects 149
4. "Natural Magic" and the Science of Bygone Times 149
5. "The Spirits Tell" 152
6. The Determining Causes of Nature and Their Representations in Classical Antiquity 153
7. The "Genetic" Causes of Scholasticism 155
8. Elements of Talmudic Demonology 156
9. Divination and Prophecy 158
10. The Shadow of a Shadow 162
11. Shadow and Shade, Inside and Outside 164
12. Subject and Shadow 166
13. Knowledge Without Conscience? 168
14. Wisdom of the Left Side and Wisdom of the Right Side 169
15. Knowledge "Below" and Unity "Above" 172
16. The Ego and the Subject 176

17. The Stakes of the Infinite 180
18. Angels and Demons in the Generation of Babel 182
19. "God" and the Names of the Name 184

CHAPTER 4
MYTH AND PHILOSOPHY, KABBALAH, "EXPANDED SPINOZISM" 189
1. Philosophical Questions, Mythical Answers 189
2. Criticism of Kant's *Critique:* Methodological Dogmatism and Empirical Skepticism in Salomon Maimon 200
3. Reconstruction and Active Thinking 217
4. East and West: New Philosophical Myths and Theological-Political Conflicts 210
5. The Indian East and Semitic East in the Contemporary History of Greek Philosophy 216
6. Spinoza's "Anti-Judaism" 221
7. Spinoza's "Pharisees" and the End of Prophecy 232
8. The Salvation of the Unlettered in the *Tractatus Theologicus-Politicus* 238
9. The Pharisees and the Birth of Judaism 242
10. The Book of Esther 246
11. Salvation for All and a Provisional Moral Code 251
12. Leibniz and Kabbalah: Mathematics and Theodicy 259
13. Leibniz and Spinoza: Science and Philosophy in the Europe of the Churches 268
14. Spinoza's "Christianity," Ancient Philosophy, and Eternal Wisdom 273
15. Expanded Spinozism and Limited Kabbalah 279

CHAPTER 5
THE DESACRALIZATION OF CHANCE:
FROM ORACULAR LOTTERY TO THE INDIFFERENCE OF THE RANDOM 289
1. Chance and Casting Lots 289
2. A Rereading of Joshua 7:1–19 294
3. Ruth and Boaz: Chance and Destiny 299
4. Desacralized Chance and the Paths of Knowledge 302
5. Throwing Dice 305
Appendix: Stéphane Mallarmé, "A Throw of the Dice Will Never Abolish Chance" 306

CHAPTER 6
HOW THE BIBLICAL GOD "GOES AT RANDOM" IN HEBREW, BUT NOT IN TRANSLATION 310

INDEX 317

TRANSLATOR'S INTRODUCTION
Lenn J. Schramm

It is a special privilege and challenge to render into another language a book whose germ is a problem of translation (see Chapter 6) and that concludes with the statement that "translation is always important." The responsibility and challenge are redoubled by the fact that I have worked with a living author (thus able to answer my questions and resolve my perplexities) who is fluent in the target language (and thus able to correct my misreadings and quibble with my lexical choices). And they are raised exponentially by the fact that the French original itself is to a large extent an exposition and exegesis of ancient texts in Hebrew and Aramaic, so that I have had to wrestle with nuances and semantic fields in multiple source languages and superimposed linguistic layers. Not everything that is in the ancient tongues made it into the French, and I confess my inability to render every subtlety of the French into English. As a translator, my advice is that you read Prof. Atlan in French, and the Jewish classics in Hebrew and Aramaic. For those not blessed with this capacity, the present effort will have to do.

This is not to say that there are no advantages to the English version. In particular, the close reading required to produce a good translation led me to discover many small errors in the French (especially source references, but also occasional slips by the author) that have been corrected here, with Prof. Atlan's gracious assistance and consent. In addition, I have inserted clarifications of various points where I hoped my knowledge of Hebrew and the Jewish sources could be of assistance to readers lacking it. Were the "glossed layout" of this volume carried to the end, these would appear in the margins of the margin, or in a different font. In the event, I have encased them in angle brackets.

A NOTE ON TERMINOLOGY

Several problems of terminology bear special attention.

Vie, vie, and *le vivant*

In the French text, Prof. Atlan distinguishes among *Vie* (uppercase), *vie* (lowercase), and *le vivant.* In English, all three are simply "life" in one sense or another.

To begin with, *Vie,* which I have rendered "Life," is (I quote Prof. Atlan's explanation to me) "a metaphysical notion that designates a reality that is essentially different from the mineral world"—the core of Vitalism. It was formerly invoked to explain the properties of living beings that could not be explained in any other way. Contemporary biology, based on the mechanistic model of physics and chemistry, has vanquished the vitalism that depended on this concept of Life. This is the sense of the word in one of the epigrams to this volume, Albert Szent-Györgi's otherwise paradoxical declaration that "Life as such does not exist."

By contrast, *vie*/life (lowercase) refers to life as we live it day to day, and in opposition to death.

Finally, *le vivant* 'the living', a coinage of several French thinkers cited by Prof. Atlan (Jacob, Canguilhem), is the object studied by the biological sciences. The term denotes the sum total of the properties of living organisms that, although they can be observed only in living organisms, must be explained by something (biochemical and biophysical processes) other than Life. In particular, it is not the same as "a living being" (*un vivant*). The fluidity of language makes this category somewhat slippery, however; sometimes *le vivant* does mean "living being" and has been rendered as such.

Sacré and *saint*

For Prof. Atlan, there is a sharp distinction between *le sacré* and *le saint* and their affiliated words. There is no doubt that the difference is not just linguistic: *le saint* denotes an intrinsic property, whereas *le sacré* refers to what is conferred by outside agency (think "consecration"). This distinction is tenable in French and Prof. Atlan holds to it religiously.

(To complicate matters, Hebrew has only one lexical family, *ḳadosh* and its relatives, to express both concepts. But Prof. Atlan always renders it as *le sacré.*)

In English, however, whose Germanic roots have provided a third family, that of *holy,* the simple binary distinction does not exist and idiomatic usage rides roughshod over the real conceptual difference. We may say "Holy Scripture" or "Sacred Scripture" interchangeably (though only the latter in this book, to respect the author's preference), but idiom prefers "holy matrimony" to "sacred matrimony" and we invariably refer to the "sacred precincts" of the "Sanctuary" or "Holy Temple" in Jerusalem and only to the "Holy of Holies."

The procedure adopted here has been to use "sacred" and its congeners whenever this does not offend English idiom. When quoting English works or published translations I have treated the author's or translator's usage as sacred. Readers are asked to bear in mind that in almost every case it is the extrinsic *le sacré* that is meant.

Péché and *faute*

Prof. Atlan never employs the French *le péché* 'sin', which he hears only in its primary, religious and theological sense of an offense against the deity.[1] For him, it is always "only" *une faute*, a word with a vast semantic field, ranging from "lapse," "mistake," or "error" all the way to "misdeed" and "sin." Although this breadth of meaning mitigates the culpability and, for him, washes away any religious connotation, it also means that most readers of the French do not realize that Prof. Atlan does not believe that Adam sinned.

The forthcoming Hebrew version,[2] bound by the biblical text and thousands of years of Hebrew usage, employs *ḥeṭ* 'sin'. This was amenable to the author for two reasons: first, the related verb *heḥṭi'* can mean "miss the mark"; second, according to the Talmud (B *Yoma* 36b), among the several words that designate "sin," *ḥeṭ* refers specifically to "inadvertent omissions."

Prof. Atlan proposed that "transgression" conveys his meaning. I am skeptical that English readers would hear the distinction without constant reminder; nor do I do want to be guilty of sinning against idiomatic usage. Later he suggested "fall" (in the sense of "stumble"), adding that both Spinoza and Maimonides describe the result of what Adam and Eve did as a "fall"—a fall in the quality of their knowledge, not in their moral essence (the strong Christian connotation that makes me uncomfortable with the term). Sometimes, taking my cue from Milton, I have written "disobedience"; sometimes, where the word was apt, "transgression"; sometimes "fall" (which also renders the French *chute*); and sometimes (mea culpa) I have stuck with "sin." Readers are asked to keep in mind that this word is not to be taken as bearing any theological weight.

Elohim

In the history of translation, the Hebrew word *'elohim* has been understood variously as "judges" (e.g., Targum on Exod. 21:6, 22:7–8), "angels" (pseudo-Jonathan and Ibn Ezra on Gen. 35:7), "gods" (e.g., Exod. 18:11), "god," and "God."[3] Although the word is formally a plural and occasionally construed as such (see Rashi on Gen. 20:13 and 35:7, who explains it as the "plural of majesty"), it almost always takes a singular verb.[4] In the biblical corpus, the singular form *'eloah* is found only in poetry and in books that postdate the Babylonian exile (Nehemiah, Daniel, and Chronicles), which strongly suggests that it is a back-formation or Aramaism. In *A Tort et à raison* Prof. Atlan often rendered *'elohim* as "dieux," no matter the context. Yielding to my (philological) discomfort with this, he allowed me to write "god(s)" in the English translation, *Enlightenment to Enlightenment*. In the present work, he adopted my approach and wrote "dieu(x)" in the French, so that I have been able to copy him faithfully in the English.

1. I do not know how he understands "J'estimais que le plus grand péché d'une femme est de n'être pas belle" (Anatole France)!

2. *Nizozot shel ḳeri*, trans. Ora Gringard, ed. Yoav Meirav and Henya Kolumbus (Jerusalem: The Bialik Institute, 2011).

3. Cf. "Elohim is a homonym, and denotes God, angels, judges, and the rulers of countries" (Maimonides, *Guide* I.2).

4. For Prof. Atlan's understanding of this issue, see especially Chapter 4, n. 18.

SOURCES

Classical Hebrew and Aramaic texts

Most translations of the Bible, Talmud, midrashim, etc., are my own, directly from the Hebrew or Aramaic, but guided by Prof. Atlan's French renderings thereof and following consultation with standard English versions (notably the New Jewish Publication Society and Revised Standard Version for the Bible and the Soncino translations of the Talmud and Midrash Rabbah). Bible references are to the Masoretic text; variant chapter/verse divisions are indicated in brackets.

For the *Zohar*, I have been able to consult Daniel C. Matt's ongoing translation as far as it goes (nearly to the end of the book of Exodus) and have given page references to this edition whenever relevant. Here too I have often had to (silently) modify the translation to suit Prof. Atlan's exposition of the text.

Spinoza

I have found it convenient to quote Spinoza in the venerable Elwes translation, which I have silently corrected on occasion (mainly in matters of word order and nineteenth-century commas). In difficult passages I consulted the original Latin. Where Elwes was quite unsuitable I have referenced the modern translation (Shirley or Curley) employed.

There are two numberings of Spinoza's letters: those taken from the *Opera Posthuma*, where they are grouped by correspondent; and those based on the Van Vloten–Land edition, which prints them chronologically. Because I have relied on Elwes, who follows the O.P., I give those numbers first, with the other reference in brackets.

Short titles for works cited

The following works are cited frequently and have been credited in the notes by short titles.

- **Works by Henri Atlan**

Atlan, *Enlightenment* = Henri Atlan, *Enlightenment to Enlightenment: Intercritique of Science and Myth*, trans. Lenn J. Schramm. Albany: State University of New York Press, 1993.

Atlan, *Entre le cristal et la fumée* = Henri Atlan, *Entre le cristal et la fumée*. Paris: Le Seuil, 1979.

Atlan, *La fin du "tout-génétique"?* = Henri Atlan, *La fin du "tout-génétique"? Vers de nouveaux paradigmes en biologie*. Paris: INRA, 1999.

Atlan, *L'Organisation biologique* = Henri Atlan, *L'Organisation biologique et la théorie de l'information*. Paris: Hermann, 1972, 1992, 2006.

Atlan, preface to *Le Golem* = Henri Atlan, preface to M. Idel, *Le Golem*, trans. C. Aslanoff. Paris: Le Cerf, 1992.

Atlan, *SR* II = Volume 2 of *The Sparks of Randomness*, forthcoming.

Atlan, *Tout, non, peut-être* = Henri Atlan, *Tout, non, peut-être: Education et vérité*. Paris: Le Seuil, 1991.

- **Works by Spinoza**

Curley = *Collected Works of Spinoza*, ed. and trans. Edwin Curley. Princeton: Princeton University Press, 1985.

Ethics = *Works of Spinoza*, trans. R. H. M. Elwes, vol. 2. London: George Bell & Sons, 1883; repr. New York: Dover, 1955. Cited by Part and Proposition; thus *Ethics* III 4 = Part III, Proposition 4.

Letters = Ibid.

Shirley = *Complete Works*, trans. Samuel Shirley, ed. Michael L. Morgan. Indianapolis: Hackett, 2002.

TTP = *Theologico-Political Treatise* [*Tractatus Theologico-Politicus*], in *Works of Spinoza*, trans. R. H. M. Elwes, vol. 1. Cited by chapter and page number.

- **Other authors**

Moses Maimonides, *The Guide for the Perplexed*, trans. M. Friedlander. 2nd ed. London: Routledge and Kegan Paul, 1904; repr. New York: Dover, 1956.

The Zohar: Pritzker Edition, trans. and comm. Daniel C. Matt. Stanford: Stanford University Press, 2004–.

TRANSLITERATION OF HEBREW AND ARAMAIC

א	’	ל	l
בּ	b	מ	m
ב	v	נ	n
ג	g	ס	s
ד	d	ע	‘
ה	h	פּ	p
ו	v, w	פ	f
ז	z	צ	ẓ
ח	ḥ	ק	ḳ
ט	ṭ	ר	r
י	y, i	שׁ	sh
כּ	k	שׂ	ś
כ	kh	ת	t

Note: Geminated consonants (except for שׁ) are indicated by doubling of the English equivalent.

THE SPARKS OF RANDOMNESS

INTRODUCTION

Adam knew his wife again, and she bore a son and named him Seth ["gotten"], because "God has got me another seed in place of Abel, for Cain killed him."

GENESIS 4:25

When Adam had lived 130 years, he begot in his likeness after his image, and he named him Seth.

GENESIS 5:3

The *sparks of randomness* are the drops of semen that Adam spilled, according to legend, during the 130 years that he was separated from Eve. The phrase is a literal translation of the Hebrew *niẓoẓot shel ḳeri*. When we analyze it closely in Chapters 5 and 6, we will try to show that the word *ḳeri* is derived both from *miḳreh*, a random or chance event, neither regular nor planned—an accidental occurrence—and from the involuntary emission of sperm that the Bible refers to as *miḳreh laylah* 'nocturnal event'.

A lapse of 130 years separated Seth's birth from that of Cain and Abel and the murder that followed, after Adam and Eve had eaten from the Tree of Knowledge and were banished from the Garden of Eden. According to a talmudic legend, developed at length in the midrashic literature and kabbalah, during these 130 years Adam lived apart from Eve; the sperm he spilled during that period created and nourished demons, the source of the "lost generations" of the Flood and the Tower of Babel.

This story of knowledge and sperm is difficult for us to understand today, after two thousand years during which the knowing subject has been separated from the body, considered to be one of the objects of its knowledge. But knowledge—whether the biblical knowledge of Genesis, the spermatic knowledge of the Midrash, or the seminal reason of the Stoics—is the forgotten source of our present intuition of a physical union between soul and body. It had to be forgotten so that an autonomous and wide-awake knowledge, removed from the fusion of dream and illusion, could emerge. But can forgetting fully play its part if we are not aware of it?

Today we must revive the knowledge acquired through sex and through the fruitfulness of this concept in order to understand what biology, the cognitive sciences, and psychoanalysis are trying to tell us, perhaps clumsily, *in addition to* what they may tell us explicitly.

Knowledge, sexuality, generation, concepts and conceptions, birth and abortion, angels and demons, aging, disease and death: science and technology constantly bring us back to these eternal problems, inherent in the human condition, while refashioning the terms in a way that is sometimes dramatic and unprecedented. Does a new science have a new morality? Who can decide this question? How, and with what tools? Using which concepts and conceptions of the world, of existence, of what is good and what is bad? What words can we use to talk about them? What style? Perhaps the empirical and logical mode of science and technology, whose terms are indispensable for posing the problems. Perhaps the narrative mode of literature. Or perhaps even the talking heads of television. Remember that today the standard response to the question of how virtue can be taught is Protagoras' rather than Socrates': good and evil are not taught only by means of scientific knowledge, but with the support of images taken from epic poetry, where moral problems are raised. In most cases, its heroes and antiheroes are the basis on which we accept or reject what we identify in our imagination as good or evil. Critical scientific and philosophical analysis makes it possible for us to delve ever deeper into technical and conceptual subtleties. But science and philosophy themselves are not the sources of universal norms. Nor can religious dogmas, despite the assistance their authority may furnish to those who hold to them, produce rules that are acceptable to everyone and suited to the complexity of specific situations.

In fact, myth has always taken hold of these questions and expressed them in its own synthetic and oneiric mode, built on visualization and association, which, perhaps better than science, can uncover the concealed threads of a hidden fabric woven in different registers of experience and knowledge, which analysis strives to distinguish: not only Prometheus and Oedipus, but also the biblical myths of the Tree of Life and Tree of Knowledge, the Flood, and the Tower of Babel. Speaking and writing about how we ought to live demands a style in which diverse languages—scientific and technical, legal and philosophical, poetic—can coexist without being confounded; where the perception of reality is always pregnant with, but never supplanted by, the contributions of the imagination; where the rationalities of science and of myth can subsist side by side, without being confused, and can criticize each other.

Thus the new reflections on ethics seem to require inventing a new form of discourse. Its birth was registered when we came to the realization that modern science, contrary to Condorcet's dream, not only fails to resolve all social and political problems but in fact creates new ones, because it spawns new possibilities without providing means for settling them. What is more, scientific discourse is not always free of dubious extrapolations. Sometimes, unknown to those who conduct it, the myth still manipulates it; and the ancient issues, thought to have been left behind long ago, return to the surface. The Big Bang restores creation, if not the Creator. As for the celebrated "human genome" with its poorly defined contours, referring to it as an "endowment" that is sacred and untouchable is no less imaginative than seeing the heart as seat of the passions and the bile as the medium of anger. In the new form of discourse that we must construct, we must burn whatever fuel is to hand. We must not hesitate to stoop to case-by-base legalism to argue, after hearing both sides, about what is permitted and what is forbidden. But we must place the analysis of the technical details, and, perhaps, an examination of more basic principles, alongside complicated plots, real or mythological, and interpret them on several levels, where the always-present not-said and not-thought can at least be rendered visible and thus become, even if only for the moment, partly said and partly thought.

But this form may not be as radically new as it seems. No doubt we can derive inspiration from the dialogues of the schools of antiquity, in which myth, science, and philosophy were not yet separated, neither from one another nor from the experience of right thinking in pursuit of right living—with oneself, with others, and with nature. Alongside the works of the ancient schools of philosophy, the inquiries conducted in the rabbinical academies of Palestine and Babylonia have come down to us in the unique style of the debates and narratives of the Talmud. The legal disputations aimed at establishing just laws on the basis of multilevel interpretations of biblical myths and statutes are interwoven there with new legends or *aggadot*. The midrashic and kabbalistic literature took up these accounts and developed them into new myths, re-energizing and amplifying their interpretive power. We too can be inspired by this form, without necessarily adhering, of course, to the literal sense, for at least two reasons. First, the social, scientific, technological, and philosophic context of two thousand years ago is incompatible with that of today, even if what we call "human nature" does not seem to have changed very much in the interim, at least in the biological sense. Above all, however, it is the nature of mythical narrative to be resumed again and again, generation after generation, in a recursion that amplifies it and in which the letter of the commentary, and of the commentary on the commentary, serves as a new text to be interpreted, as a pretext for new interpretations.

What is the status of the randomness of birth, of chance, of the ignorance of causes that we call "fate," in a world that we are increasingly able to control, where we can even plan for uncertainty by means of probabilistic estimates of risk? Isn't it the vocation—or destiny—of our species to use its inherent capacities, its large brain and its cognitive and linguistic abilities, to order and control the rest of nature? But does what applies to the rest of nature also apply to the human species? Is it humanity's destiny to suppress destiny by means of planning? Are human knowledge and technology violations of natural law, a rape of nature, on which they are imposed like some monstrous anomaly? Could they be a curse on humankind, generation after generation, massacre after massacre, always increasing in number and intensity, in proportion to humanity's control over everything that is not itself? Or are they merely one of the many products of that same nature?

Are we the children of Prometheus only? Are we not also the children of Adam, who was enjoined to fill up the earth, to occupy it and dominate it, to rule the fish in the sea, the birds in the sky, the terrestrial animals and every living thing that creeps on the earth? Clearly human domination of nature is not the *product* of the cultures that preserved this myth. On the contrary, the narrative merely expresses, in its own way, the dominion of the human species as it has always been experienced, in all latitudes and by all cultures.

In the same fashion, the narrative of Genesis is also the story of a curse. The narrative coils like a serpent around the two poles of the human anomaly, knowledge and morality, with *Homo faber* and *sapiens sapiens* at one end and the suffering they inflict on others and on themselves at the other. At one end is the evil that they do, suffer, imagine, and plan; at the other, the good, the happiness, and the bliss that they also imagine and try to plan—in short, the angels and demons with which they fill their universe, inside and outside themselves. The myths of Genesis and of the sparks of randomness make it possible to explore all this.

Curiously, as a matter of bourgeois morality and for law, the English term for an infant born out of wedlock is "natural child," as if the institution of marriage endowed children with the superior status of "artificial children"! (The same idiom is found in French.) Today, indeed, we are not far from the production of true artificial children, not through social institutions but through biotechnology. Will this be bad or good? For humanity today? For the men and women of the future, at least some of whom may be such children?

When we refer to the biblical account of Adam and Eve's transgression and the curse that followed it, we will do our best to forget the received ideas about "original sin." In the Western world, the history of Adam, Eve, and the serpent is generally associated with the Augustinian interpretation imposed on Christian orthodoxy since the fifth century: human beings are doomed to unhappiness and suffering because of the sin of the flesh committed by our first parents, Adam and Eve. Sex is fundamentally evil; holiness demands abstinence and celibacy. We inherit their sin and guilt at the moment of conception, which is produced by that very same sin; this explains all the unhappiness and suffering with which human beings are afflicted from the moment of birth, including infants who have not yet had time to do something bad of their own free will.

For some Augustinian theologians, this universal predestination to evil negates free will; the only way to escape it is divine grace and obedience to the authorities of City and Church. In this form, which centuries of catechism have made familiar to and inculcated in millions of children, the story has played and continues to play a decisive role in the moral and religious mind-set that is almost consubstantial with Western civilization. It has shaped notions that still hold meaning, even for those outside the Church, of male and especially female sexuality; of the family and the body; of birth and death; of guilt, holiness, and innocence. But the early Christian Church, subversive and persecuted, before the Christianization of the Roman Empire placed it in the saddle, did not always hold this interpretation. Augustine himself was able to impose it only after protracted theological and political debates. Elaine Pagels' *Adam, Eve and the Serpent*,[1] which offers a contemporary and critical Christian perspective on the history of primitive Christianity, clearly depicts the protagonists and issues of these controversies. It was a matter of divergent interpretations of the first few chapters of Genesis, where this foundational story is told. In the year 418, at the end of the Pelagian controversy, the Augustinian reading was proclaimed to be orthodox. All other views were condemned as heretical and their proponents were excommunicated. Their interpretations, less misogynistic and less appalling, did not highlight the sexual aspects of the story but rather its message of individual freedom and responsibility to obey or transgress divine law, in a world that is fundamentally good, in the image of its Creator. These interpretations, deemed heretical ever since (despite having been defended by bishops of the Church), had been preceded by other interpretations, of Gnostic inspiration, which were, if possible, even more heretical and had been condemned even more quickly. Today, especially since the discovery of the library at Nag Hammadi, we know that Gnostic interpretations of the Bible and Gospels were an integral part of early Christianity, at least during its first two centuries. One of their characteristics was the symbolic nature, cosmic rather than anthropological, they attributed to the biblical protagonists. For example, in some of these interpretations Eve is an icon of wisdom, the mother of

1. Elaine Pagels, *Adam, Eve, and the Serpent* (New York: Random House, 1988).

the universe, rather than a woman of flesh and blood. In a similar vein, the virginity of the mother of Christ was not understood literally.[2] In general, these extraordinary stories, including the first chapters of Genesis, were understood less as edifying tales with a moralizing bent, intended to nurture a particular social and religious doctrine upheld by the Church about the relative values of celibacy and marriage, for example, than as myths of origin, like the Greek or Egyptian traditions they supplanted. This difference must be underlined. For some Christian Gnostics,

> the story was never meant to be taken literally but should be understood as spiritual allegory—not so much *history with a moral* as *myth with meaning*. These gnostics took each line of the Scriptures as an enigma, a riddle pointing to deeper meaning. Read this way, the text became a shimmering surface of symbols, inviting the spiritually adventurous to explore its hidden depths, to draw upon their own inner experience—what artists call the creative imagination—to interpret the story. . . . Consequently, gnostic Christians neither sought nor found any consensus concerning what the story meant but regarded Genesis 1–3 rather like a fugal melody upon which they continually improvised new variations, all of which, Bishop Irenaeus said, were "full of blasphemy."[3]

But this was nothing new. Philo had employed and greatly expanded this method of allegorical interpretation. Similarly, for some philosophers, mainly Stoics, the Iliad and Odyssey were not to be understood according to their surface meaning as accounts of the rivalries and loves of the gods. Their surface meanings concealed deeper truths of natural philosophy that could be uncovered by a symbolic reading. The rabbis of the Talmud and the Midrash were raised in this type of interpretation, associated and superimposed on the plain meaning of the biblical text, to the extent that it allows itself to be grasped. Kabbalistic interpretations that uncover and develop the "hidden" meaning of the text merely amplified and systematized this tendency, already found in the Talmud. It is not astonishing that kabbalah sometimes demonstrates familiarity with Gnostic themes, as in its ideas about what preceded creation, to which we shall return later. Gershom Scholem saw this as reflecting the direct influence of Gnosis on some sources of the kabbalah.[4] Moshe Idel, on the contrary, suggests that it was Gnosis that drew on ancient Jewish influences, or at least that the influences were mutual.[5] The interpretations of the story of Adam and Eve that we will look at below, like the legend of the sparks of randomness that inspired the present work, are much closer in method, if not in content, to Gnostic construals than to those derived from Augustine.[6] They differ, nevertheless, in at least two points. First, the symbolic interpretation does not cancel out the literal meaning but is superimposed on it. Adam is at the same time the first man, an archetypal figure of the human nature in each person, "male and female," and *Adam Ḳadmon*, the primordial human being, a cosmic and divine figure that fills the universe, simultaneously creator and created. (In this respect, the kabbalistic readings are often less rigorously allegorical than Philo's, for

2. See eadem, *The Gnostic Gospels* (New York: Random House, 1979, 1989).

3. Pagels, *Adam, Eve, and the Serpent*, p. 64.

4. Gershom Scholem, *Origins of the Kabbalah*, ed. R. J. Zwi Werblowsky, trans. Allan Arkush (Philadelphia: Jewish Publication Society; Princeton: Princeton University Press, 1987); idem, *Kabbalah* (New York: Quadrangle, 1974), pp. 10–14, on "rabbinical gnosticism."

5. Moshe Idel, *Kabbalah, New Perspectives* (New Haven: Yale University Press, 1988), pp. 115–157.

6. See, for example, Michel Tardieu, *Trois mythes gnostiques: Adam, Éros et les animaux d'Egypte dans un écrit de Nag Hammadi (II,5)* (Paris: Institut d'études augustiniennes, 1974).

example.) A second and decisive difference is that, unlike the Christian gnostic interpretations, which were swiftly condemned by the Church and banished from its doctrine, the kabbalistic interpretations, with all their diversity, have remained an integral part of orthodox rabbinic Judaism. They are less well known than interpretations that are easier to understand and employ in religious instruction, whose goal is edification and enlightenment; but over the generations, until the beginning of the twentieth century, their authors were frequently prominent teachers of rabbinic orthodoxy, sometimes community rabbis. For all these reasons, we will use the biblical story of Adam and Eve and its rabbinic commentaries as a myth that is pregnant with multiple meanings, discarding the received notions of original sin and hereditary curse, the inevitable evil character of physical nature that derives from them, redemption through celibacy, and mortification of the flesh as the road to salvation. We will read it, not as a story with a moral, but as an album of images of diverse and contrasting aspects of the human condition, associated in particular with the protracted period of childhood and maturation that follows birth and with the long interval between sexual maturity and intellectual and emotional maturity: the Tree of Knowledge is assimilated before the Tree of Life (although we can easily imagine that the inverse chronology might have produced a happier outcome). If, all the same, there is a moral to be drawn from the story, it must involve the search for some sort of redress or reparation—that is, a way to ameliorate this condition by identifying the harmful effects, the sources of pain and suffering, in order to eliminate, attenuate, or transform them.

In the first chapter of this work, we will examine the theme of the Golem, the artificial humanoid of the talmudic and kabbalistic literature. We will consider how it relates to the multiple levels of knowledge and holiness that can be attained through study of the Torah, which the Talmud conceives of as a parallel search for the truths of nature and for the ethical and legal norms, both societal and individual, that make right living possible. In the light of the biblical and talmudic myths, the transformations of the human condition that the twentieth and twenty-first centuries seem to be producing may not be as extraordinary as they seem. More precisely, it is their move from the imaginary world of myth to concrete reality that seems to be quite new and unprecedented, given that they have always been present in narrative fiction, at least as possibilities associated with some image of human nature.

Little by little, science and technology seem to be liberating the children of Adam and Eve from the biblical curse of painful toil and painful childbirth. The era of machines, industrial and postindustrial, increasingly frees men and women from their sentence to life at hard labor. The amount of time people spend working has been decreasing for the last two centuries. Unlike the pessimistic and resigned interpretations that some catechisms give to the biblical narrative, this release from labor and pain corresponds to the highest and most basic vocation of the human race, which—certainly for the talmudic sages—is precisely the creative activity of knowledge and wisdom and by no means subjection to toil and pain.

"Every man is born for toil," stated Rabbi Eleazar, citing Job 5:7. However, the sage continues: "I do not know whether this means toil by mouth or the toil of physical labor. But when it says 'for his mouth compels him' (Prov. 16:26), I may deduce that toil by mouth is meant. Yet I still do not know whether [this means] toil in the Torah or in [secular] conversation. But when it says,

'This book of the Torah shall not depart from your mouth' (Josh. 1:8), I conclude that one was created to labor in the Torah."[7]

Work is no more than a regrettable necessity. Release from it would permit the true nature of human beings to blossom in the world of wisdom and of the words that express it. The curse imposed on the first man, condemned to earn his bread by the sweat of his brow, was not pronounced to be borne as inevitable. On the contrary, it is to be rectified along with the sin itself, generation after generation.

7. B *Sanhedrin* 99b.

The same applies to the curse that women will give birth in pain. Today childbirth is no longer the agonizing and dangerous ordeal that it was less than a century ago. We can even see, on the horizon, total release from its burden—at least for those women who perceive it as such. Symbolically, we can say that the pill and the automatic washing machine have set women free. The washing machine spares them a disagreeable daily domestic task and has also made men were more willing to pitch in and help. The pill permitted the revolution of manners that is called, perhaps not totally appropriately, "sexual liberation." Sex without procreation became relatively easy. It was the beginning of "family planning," which allows women to avoid unplanned pregnancies. The legalization of contraceptives and abortion has given them a large measure of control. For now this is merely negative—preventing the uncertainty of unwanted births. But the process of active planning, with the attendant risk, sooner or later, of achieving a total separation between reproduction and sex, is on the way. Children will then be produced from start (in vitro fertilization or cloning) to finish (artificial gestation) outside a woman's body. We are still rather far from this, especially ex vivo gestation; in principle, though, nothing prevents us from imagining the solution of the many technical problems associated with the invention of an artificial uterus. When that happens, the production of living beings—human and nonhuman—will accompany, more or less inevitably, the liberation of men and women from the existential curses that compel them to suffer simply in order to survive, feed themselves, and reproduce. The pains of labor—in both senses of the term—will have disappeared.

Human reproductive cloning would be another step in this direction. For now it seems fated to be outlawed by a broad international consensus; and this is a good thing, for reasons that (as we shall see) are social, rather than biological or metaphysical. But who knows what humanity will have become in a century or two? What "value" will be invoked to deny women the right to control their own bodies and to liberate themselves from the constraints of pregnancy? As in the legend of Jeremiah and the perfect golem he made, the only question is whether human societies can reach the moral level required to meet the challenge posed by a capacity to employ technology to fully control and streamline the lives of human beings. Would the individuals and institutions that wield this power be generous enough and good enough to avoid succumbing to the temptation of diverting it to their own profit? Would the "parents" of infants produced in this way be sufficiently generous to create an environment conducive to their physical and moral growth and felicitous freedom? For the moment, it would no doubt be overly optimistic and naïve to answer in the affirmative. But this might well change in the near or distant future. The most pessimistic scenarios are not necessarily the most probable ones. We can heed the prophets of disaster and avert their predictions. In the end, there is no reason moral progress cannot accompany technical progress. After all, slavery—a social norm in the ancient world, not to mention in recent centuries—has finally been abolished, almost everywhere, by virtue of the technological progress that diminished its utility—even if its surrogates survive in regions where humans are still exploited. For the first time in human history, the violence of war as a way to settle disputes—along with other related forms of brutality such as torture, genocide, and repression—is condemned by the international community. Certain measures, though still rather limited, are being taken; institutions like the International Court of Justice have been established to translate this condemnation into action, however timidly. Yet well into the twentieth century, not only was such violence the norm, it was the main source of glory in the ethos of nations. Pacifists were utopians even more than they were a minority.

So nothing prevents us from imagining an era when humanity, at peace and increasingly open to the refinements of the life of the mind, makes intelligent and constructive use of the results of technological progress, including those related to the production of life. Nothing is inherently evil, in this domain as in others. Everything depends on the intellectual and moral environment. We have no reason not to believe that practices that raise the specter of serious moral regression, in our present environment, might one day, in a different moral and intellectual context, be beneficial. Today we can only imagine such a context, in a more or less mad utopian vision of another world, or of a return to a lost paradise.

But reproduction without sex poses another problem. What will become of sexual knowledge if it is totally detached from procreation, when there is no longer a created child to support the family and social bond or just to serve as a pretext for this bond? Here, at least, the concept is not so far-fetched and we can already come up with some idea of the answer.

Every sexual union is intrinsically fertile. Even if it fails to produce a child, its effects are like spirits or souls that remain attached to those who produced them, angels or demons, depending on the spiritual state of the two partners at the moment of their union. Angels or demons, depending on whether the encounter went beyond self and was giving and open, or was merely use and abuse of the other party—it being understood moreover, that angels can become demons and vice versa. The Tree of Knowledge of Genesis, *'eẓ ha-da'at ṭov va-ra'*, is not only the "tree of knowledge *of* good and evil," as it is usually rendered. The Hebrew can also be rendered the "tree of knowledge, [which is both] good and bad," of life and death intermingled. It is contrasted to the Tree of Life, although the two trees are in fact only one, at their source, where, as we shall see, evil is not evil. This biblical knowledge is clearly nothing other than the libido, but the entire libido: an awakening of the senses and the mind to knowledge of sex through sex, an initiation in the reality of impulses and their stakes, in the power of desire, in happiness and unhappiness, and the good and evil that accompany their satisfaction; loss of the animal innocence of one who didn't know, good and bad knowledge, the source of joy, pleasure, happiness, and even of "blessedness," but also the origin of unhappiness, suffering, and death. In the wake of moral codes, law, and psychiatry, biology has now gained a hold on this ambivalence; we can—and perhaps should—use hormones to treat sexual deviants. The fact that, in these extreme circumstances, Eros has not been able to separate itself from Thanatos is the meaning of the warning to Adam and Eve, when their eyes were not yet open, in the (relative) innocence before they achieved puberty: "On the day that you eat of it you shall die" (Gen. 2:17). Not simply to die at the appointed time, in accordance with nature, but also not to live, unless it be a life that is entirely "for death."

In this story the Tree of Life is the remedy, the wisdom that is hard to reach but attainable. It is guarded by the ominous Cherubim who stand vigil over the gates of Eden after the expulsion; but the sap of the Tree of Life, the wisdom inscribed on the Tablets of the Law, waits to be incorporated and given loving voice, between the Cherubim who embraced in the Holy of Holies in the Temple. It makes it possible to invert the image, at least in principle, and to conceive of death as "for life." This evokes the notion of programmed cell death, what we now call "apoptosis," a physiological death that is part of the normal processes of embryonic development and tissue regeneration in organisms. As for the reality of our own death, which we can only imagine, wisdom tends to keep it within the bound of "death in its own time." While waiting, in a sort of mental apoptosis, we live it as a desire for and impending failure of the future and the fullness of possibilities.

In the Lurianic kabbalah, developed in Safed in the sixteenth century, the legend of the sparks of randomness served as the foundational myth of an arcane account, both moral and cosmic, whose crux, from the origin until the end of time, is precisely the career of these scattered sparks in the chain of generations and their role in a narrative of humanity as a series of begettings, the production of living beings by other living beings.

An analysis of this myth (an extension of the myth of the Tree of Knowledge) will serve us as the connecting thread for addressing several problems raised by the artificial fabrication of living things as it relates to the fundamental ambivalence of knowledge. Science is not neutral, as is often still said in an attempt to attach value judgments about its constructive or destructive effects to its applications only. But neither is it bad or good in itself; it is ambiguous, both at the same time. Science is simultaneously good and bad, open and closed, the carrier of both life and death, the source of truths and of illusions—just like life itself, which exists only by virtue of the thousand deaths that enable it to be renewed. The Tree of Knowledge and the Tree of Life are one. Good and evil, life and death: these are relative to the point of view, particular or universal, to the long-term or immediate interest. The myth of the Tree of Knowledge is not that of Prometheus; knowledge is not stolen, any more than life is. It is *given*, like life, through a sin without guilt that is an integral part of the system of giving and receiving, which can be recognized as a sin and can then set in motion the desire for atonement.

But the contemporary biological revolution is not merely technological. It has clear philosophical implications, inasmuch as the notion of Life has a new status. To paraphrase the biologist Albert Szent-Györgi, Life no longer exists as the object of scientific investigation. Life per se is no longer the object of those disciplines still called "the life sciences." But life, as the sum total of the lived experiences of our existence, clearly remains at the center of our social images and of our philosophical and moral questions. This is the source of the multiple misunderstandings and problems that accompany the progressive biologization of daily life. The objects of biology are increasingly remote and differentiated both from "life" and from the living in general, as this science focuses on the study of physical and chemical mechanisms that can provide an operational explanation of the structure and activity of organisms. The value of these explanations derives essentially from the fact that they make it possible to control the normal or pathological functioning of organisms and offer, for the first time, the possibility of producing biological artifacts (transgenetic organisms, chimeras, clones, etc.) in the way in which we are accustomed to manipulate inanimate objects using physics and chemistry. Our society's prevalent images of life and death; of the vegetable, animal, and human; of sexuality and illness; and of the normal and the pathological are superpositions of concepts derived from biology on top of outdated images that integrated the experience of daily life with traditional animist and vitalist ideas. This is why our most common images of the living (*le vivant*) are often at odds with contemporary biological ideas; the former are generally based on a confusion of the notion of Life, as the ostensible object of the biological sciences, with the subjective and intersubjective experience of daily life. Furthermore, these confused images depend on the accelerated pace of fundamental discoveries in biology and the upheavals that these discoveries have provoked and stimulate in our images of what we habitually refer to as "life."

As for the social sciences and humanities, their traditional object is human beings, of the same order as the animal and vegetable and disjoined from the inanimate mineral. Scholarly discourses about social images of the living generally maintain that the living, the object of the life sciences, is also what everyone automatically understands by the word "life." In fact, biology increasingly locates its objects at the cellular and molecular level, of the same order as the "inanimate" objects of chemistry and physics, severed in practice (if not in principle) from the living, of which human beings are supposed to be in some fashion the ultimate paradigm. This helps perpetuate the misunderstanding about the development of this science, which is still held to be, as in the past, the science of life or the science of the living. A similar situation exists with regard to psychology and social images of the mind. These are still permeated by ancient notions of the nature and activity of the soul, though today these belong to the realm of folk psychology. But the soul has long since ceased to be an object of scientific inquiry and psychological science no longer concerns itself with these questions. The same has happened, although much more recently, to "life," which used to be associated with the vegetable, animal, and intellectual souls and was held to explain the activity of "animate" (i.e., endowed with an *anima*, a soul) beings. This too is no longer an object of scientific inquiry, although it remains the province of what we might call "folk biology."

In Chapter 2, we will look at some consequences of this state of affairs. We will see that we have indeed returned to the age of seminal reason and spermatic knowledge. The absolute monism suggested by the current sciences of living and thinking bodies makes the ancient notions of the *logos spermatikos* or of the spermatic sparks of randomness, endowed with cognitive properties, less strange to us. The total integration of the physical and mental implied by these notions was difficult to assimilate into the vitalist and spiritualist (or Cartesian dualist, in the best case) context in which theology and then natural philosophy were propounded until the second half of the twentieth century. The materialist monism presupposed by contemporary biology, extended to human biology, is certainly quite different from the physicalist and animist monism of the ancients. But it leads us to rediscover the reality of a "union of soul and body" whose most remarkable properties we forgot while philosophy was busy with the Mind or with Life and the material sciences dealt only with the relatively simple physical objects of physics and chemistry. Until recently, these sciences were not able to extend their domain to the complex and compound bodies that are organisms. Three centuries later, Spinoza's prediction in the *Ethics* seems to have been realized: "We thus comprehend, not only that the human mind is united to the body, but also the nature of the union between mind and body. However, no one will be able to grasp this adequately or distinctly, unless he first has adequate knowledge of the nature of our body" (II 13, Note). This also entails that we deal seriously with the natural determinations we discover in our actions. They include, of course, biological causes that cannot be reduced to the effect of genes, because they include the effects of regulatory systems—nervous, endocrine, immune—that develop epigenetically in and through the life of each individual and guarantee his or her continuing identity. But there are also social and cultural causes, conscious or unconscious, that are integrated with the previous set. All of this reopens the question of free will and moral responsibility, this time on the basis of new foundations that cannot be reduced to those of Kantian or post-Kantian idealism. We shall see how the author of a kabbalistic treatise of the early twentieth century, *Sefer ha-De'ah* (The Book of Knowledge), interprets the biblical myths and the legends and parables of the Talmud by means of the metaphor of a divine intrigue or drama in which our experience of free will is an actor. The drama is that of an absolute determinism that exploits this experience while always proceeding according to the law of its timeless necessity, foreseen from all eternity.

In Chapter 3, we will tackle the question of the moral subject, whose actions are set in motion by hidden internal and external causes. This will lead us to elements of what might be called "talmudic demonology." More precisely, we will consider those elements of the myth of the sparks of randomness (and the many commentaries on it) that lead to intercourse with various natural forces that are held to be spirits, whether angels or demons. Such beings continued to infuse the science of classical antiquity and the Renaissance until the century of Descartes, with his "animal spirits." So it is not astonishing that they are to be found not only in the legendary narratives of talmudic *aggadah* and the Midrash, but also in the legal sections of the Talmud. Our reading of these texts will be guided by the hermeneutic method of ancient rabbinic thought, which postmodern literary criticism seems to have rediscovered in part. We will apply this method to a number of talmudic texts and commentaries that credit spirits and demons with causal agency in the natural course of events, beneficial or otherwise, and in the meaning of these events for the human beings affected by them. We will then employ our sketch of talmudic demonology to restore the use of causal explanations in the attempts to make the world intelligible, which has characterized human thought always and everywhere. The characteristics of this magical, prescientific causality may help us uncover the newer traps of scientific causality, into which those who would extend its effective operational control to serve as a source of meaning for human existence keep falling.

Through the lens of the paired myths of the Flood and the Tower of Babel, where we again encounter the demons of the sparks of randomness, we also see the shadow of subjects built from the scattered remnants of Adamic knowledge. Everywhere and always, ever since *Homo sapiens* invented articulate language—though without turning into a computer programmed by one of his descendants (that is, without emerging from his affective animality)—there has been a close link between the two modes of breakdown of the subject, represented by these myths. The alternatives are fusion into the undifferentiated whole, drowned by the flood, and solipsistic imprisonment in some language of Babel that is never truly understood without misunderstanding, except by each individual, alone, in the unique fortuity of the conditions in which he or she learned it. To join these two, for more than twenty-five hundred years human beings have been endeavoring to pave a path of linguistic intersubjectivity, called "reason," across the bridges built of "common" notions, which we recognize, well or poorly, as shared by many individuals endowed with articulate language. The other type of intersubjectivity, nonverbal and nonrational, neither rational or irrational—music, for example, or, more generally, aesthetic or amorous communion—does not diminish the tension between these opposing terms. On the contrary, it pulls on the separated beings, tugging on their nostalgia for an undifferentiated mystical fusion that today can only be both exciting and death-dealing.

Our hermeneutic ramblings in these first three chapters will expose us to various themes of classical philosophy, such as determinism and freedom, causality, and the mind-body problem. We will then discover a close resemblance, which may seem paradoxical and unexpected, to what is sometimes referred to as "absolute rationalism," as expressed in Spinoza's *Ethics*. This is why, in Chapter 4, we will attempt to untangle the threads of the ambiguous link between myth and philosophy in general, and, more particularly, between certain currents of kabbalah and Spinoza's thought. This chapter may appear to be a long digression from our main theme; readers in a hurry who decide not to linger there will not lose the thread of the book as a whole. Nevertheless, this is where the engine that propels our inquiry is taken apart, where the points of departure, angles of attack, and perspectives from which we speak, read, and write are enumerated and analyzed, to the extent possible. In that chapter we will try to clarify the mode of thought and method of analysis, based on association and dissociation, that I have referred to elsewhere as an "intercritique." Rather than an intercritique of science and myth, the subtitle of that earlier work, *Enlightenment to Enlightenment*, here we are dealing with myth and philosophy.

We shall see that even if the inquiry is, strictly speaking, philosophical, the answers that philosophers supply to their own questions are always permeated by elements of myth, ancient and modern. After three centuries during which the natural sciences and philosophy have been relatively independent of each other, the latter remains, as in the past, and despite the critiques of the Enlightenment, a jumble of philosophical questions and mythological answers. Even Spinoza in his key work, despite the rigor of the geometric method employed, proceeds in the tradition of ancient philosophy, for all that he radically refashions it in the mold of classical rationalism and the new science that was developing before his eyes and with which he was acquainted. In this context we shall analyze the relevance and foundations of several major critiques of Spinoza, including, among the most striking, Leibniz on Spinoza's "monstrous doctrine," "a combination of the Cabala and Cartesianism, corrupted to the extreme," and Salomon Maimon's apparently paradoxical concept of the kabbalah as "expanded Spinozism." But our concern will not be merely historical. Rather, we will embark on a process of construction, incorporating the great philosophers of the past, singly or in combination, into our thinking in the present, so that, transcending specific historical conditions, a *philosophia perennis* can emerge. In a certain sense, the scientific revolution of the seventeenth century did not reach its conclusion until the twentieth, because it is only for us—witnesses to the successes of contemporary biology, which is mechanistic, molecular, and physical-chemical—that the mechanical revolution that transformed first physics and then chemistry has had an impact on our knowledge of living beings.

Thus we find ourselves at the junction between modernism, which seems to be coming to an end, and a new era in the adventure of knowledge. Postmodernism, the deconstruction into caricature of established verities, is only a first step—a trial run?—a symptom of the transition between yesterday's modernity and tomorrow's. As for earlier transitions, we would do well to investigate its foundations and its links to the knowledge of the past. Who knows whether we will have need of new foundations that necessarily reuse older stones, though arranged in a different pattern? It is not useless to delve into the memorial debris of past eras and try to discover whatever is there that can still be of service. Spurring a dialogue among the ancient foundational texts, over the centuries and across languages, can help uncover some of these building blocks. The Talmud—produced at the interface of the Greco-Roman world and the Christian and Muslim Middle Ages, and the kabbalah, elaborated at the junction of the Middle Ages and the Renaissance—express a unique body of thought developed over centuries, in Hebrew and in Aramaic, in the crucible of the rabbinic academies; they also mirror the Greek and Latin philosophy (sometimes with a detour through classical Arabic) that parallels it. Later, at the watershed marked by the Scientific Revolution, we will of course encounter Pascal and Descartes, but also, perhaps more profoundly, Leibniz and especially Spinoza. We will see that the philosophy of the *Ethics* is an indispensable link between a natural philosophy, re-embedded in the present, and the ancient doctrines that understood ethics on the basis of knowledge of the determining causes of the undissociated body-mind.

We must pause here to avoid misunderstanding. It would be absurd and ridiculous to think that the references to Spinoza in this work, alongside analyses of texts from the ancient Jewish hermeneutic tradition, might justify an attempt to "recover" this philosophy for theology, rabbinic or other, from which it was always at pains to demarcate itself and against which it always fought. Although Spinoza's philosophy can serve as the consummate link between the seventeenth century and the Modern Era, its richness, originality, and coherence mean that it cannot be reduced to anything other than itself. His ideas have justly been referred to as absolute rationalism and have been tugged in various directions, sometimes toward modern materialism and sometimes toward the idealism of the philosophies of mind. Yet Martial Guéroult's "mystique without mystery" remains the most suggestive, if not the most appropriate, epithet for this philosophy, because of the strange way in which it seems to be paradoxical when it is in fact perfectly rigorous and in any case irreducible. Numerous contemporary studies, following on the heels of three centuries of fascinated ostracism and ignorance of the "accursed" philosopher, have begun to correct many misunderstandings. They make it possible to follow the proper motion of this thought in its timeless and eternal aspect, modeled on the mathematical truths to which its own format refers. Neither the critical philosophies of the eighteenth and nineteenth centuries nor positivism and the linguistic philosophies of the twentieth century have detracted from the relevance of this system, which seems from the outset to have shielded itself against blows from any source, not by retreating into its own shell but, on the contrary, by warding off in advance the successive assaults that these philosophies have made on their predecessors and on the very idea of a philosophical system.

In *The Atheism of Scripture*, Volume 2 of the present work, we shall see how the rabbinic hermeneutic tradition of midrash and kabbalah escapes the confines of a single and unique meaning that the presumed divine author of the text impressed upon it. The ultimate meaning, the secret or *sod* of kabbalistic interpretations, is more deeply "hidden" in the text because it is suggested by what is *not* there, by the infinite possibilities of the blank page, the margins, the white space between the words. Far from being deducible from the text, the formal rules for constructing interpretations project the latter onto it and often modify how the words are read, while leaving their written form untouched. Only scrupulous respect for the form of the written text and for the hermeneutic rules limits the apparent arbitrariness of these construals and guarantees that they retain some anchor in the text being interpreted.

Nothing could be more alien to this tradition of reading than the critical method of analysis of Scripture inaugurated by Spinoza in his *Tractatus Theologicus-Politicus*. It is clear, on the other hand, that in his youth he knew only the doctrinaire forms of rabbinic Judaism and that the maximum openness and rationality he could conceive of in this context was that of Maimonides and his followers. But he had to distinguish himself from them violently—while leaving them ample room—because he could find nothing in them "which was not a commonplace to those Gentile philosophers, . . . nothing but the reflections of Plato or Aristotle, or the like, which it would often be easier for an ignorant man to dream than for the most accomplished scholar to wrest out of the Bible" (*TTP*, chap. 13, p. 176). "Maimonides and others" are criticized for twisting the plain sense of the text merely "to extort from Scripture confirmations of Aristotelian quibbles and their own inventions" (ibid., chap. 1, p. 17)—in particular with regard to the incorporeality of God and the theoretical and abstract character of the divine word, thus clearly contradicting, time and again, the explicit meaning of the biblical narrative. But reproaching Maimonides for his Aristotelianism was also popular with kabbalistic writers, some of whom, like Naḥmanides in the thirteenth century, demonstrated how poorly this theology was suited to the plain sense of the biblical text. As for their own philosophy, linked to the ancient Jewish tradition of the Talmud and Midrash, and apparently closer to the Gnostics, Neoplatonists, Stoics, and even Epicureans, they based it on what the text does not say, rather than on its plain meaning (Hebrew *peshaṭ*), which they considered to be "simple" (Hebrew *pashuṭ*), meant only for the uneducated common folk (Hebrew *peshuṭei 'am*). Thus Spinoza's desire to detach philosophy from biblical exegesis, in order to guarantee his freedom to practice philosophy in a Christian society—the true objective of the *Tractatus*—led him to analyze the ancient phenomenon of prophecy and take it seriously. He ignored philosophical rabbinism, for which, nevertheless, "the sage is superior to the prophet"[8] and whose emblematic figure, as he was for Spinoza himself, is the philosopher-king Solomon, the "wisest of men" and author of Proverbs and Ecclesiastes. This mode of thought, which informs the debates and *aggadot* of the Talmud, replaced ancient prophecy in the Diaspora Jewish communities from the time of the Babylonian exile, several centuries before the advent of

8. B *Bava batra* 12a.

Christianity. It is also possible, as we shall see, that in his contemporary Amsterdam the author of the *Tractatus* was even more sensitive to the rabbis' "childish exercises" and the kabbalists' "madness" because he witnessed the false messianism of the Sabbatean movement (which, like present-day messianic movements in Israel and elsewhere, claimed to be based on kabbalistic speculations and other eschatological calculations of the End of Days).

But this did not prevent Spinoza from conducting a dialogue, *sub species aeternitatis*, with the ancient texts, even if he quotes them infrequently, simultaneously building on and breaking with the scholasticism of the Middle Ages and Renaissance, just as modern science established itself by developing and rejecting prescientific thought. For us, this dialogue is essential, not to pull Spinoza toward religion and theology but, on the contrary, because the tools and insights provided by reading the *Ethics* in our modern philosophical context can facilitate interpretation of the ancient texts. As we shall see repeatedly, his philosophy, more than any other, makes it possible to reread and understand the masters of that ancient tradition in a rational and open context that is particularly suited to the spirit of our own age. This spirit, whether we admire it or deplore it, is nourished by the "geometrical" sciences of natural causes, which demythify and render intelligible the world of living beings and of consciousness that were, until not so long ago, the last refuge of ignorance. On the other hand, despite the traces of prescientific thinking (ancient astrology, the natural magic of the Renaissance), the tradition of questioning these texts—the constant oscillation between myth and philosophy, between existential individual ethics and impersonal abstraction and shared reason—can be an inspiration for our own search for practical wisdom, going beyond science and technology while recognizing them as unrivaled means for achieving comprehension and control.

Once again, we are not talking about ignoring the differences. The same questions seem to be posed, even though the concepts are borrowed and transformed, and different and often contradictory answers are offered to them. Nevertheless, in Spinoza, as well as in some kabbalists, Stoics, and Neoplatonists, we find, more than anywhere else, elements of an epistemology that is associated with an original monistic ontology, neither idealist nor materialist, and that is particularly suited to a natural philosophy that is aware of what contemporary biology, as the physics and chemistry of organized beings, teaches us. The irreversible achievements of critical and positivist philosophy clearly cannot be ignored. But their proponents, whether Kant and the post-Kantians or the logical positivists (with the notable exception of Wittgenstein), frequently fall into the error of a critique that believes itself to be grounded absolutely and does not criticize itself.

According to Wittgenstein, "'We are quite sure of it' does not mean just that every single person is certain of it, but that we belong to a community which is bound together by science and education. We are satisfied that the earth is round."[9] We can apply our experience of rational thinking about the objects produced by scientific praxis to essay second-degree, critical thinking about our own thought itself. This is one of the tasks of the philosophy of science and where it differs from the sciences, although it can no longer claim to be their ground, as it once did. But there does not seem to be any reason to halt in midstream and believe that our thinking about our thoughts of things must not be thought in its

9. Ludwig Wittgenstein, *On Certainty*, ed. G. E. M. Anscombe and G. H. von Wright, trans. Denis Paul and G. E. M. Anscombe (Oxford: Blackwell, 1974), §§298–299.

own turn. To put it in Spinozist terms, we can believe that the idea of an object, conceived reflexively as the idea of an idea, should lead us to take this idea of an idea as an ideatum in its own right and to form an idea of an idea of an idea—and so on ad infinitum. But this infinite regress is meaningless and comes to an end with the first reflexive idea of an idea. The fact that we pose the question and answer it, even in the negative, seems to imply and constitute another level of reflexivity. But this second level is inextricably interwoven with the first one. A nonrelativist critical philosophy believes that it can consider a certain level of reflexivity as the source of the ultimate ground of its truth. A relativist critical philosophy knows it cannot have such a ground, because of our limited ability to hold multiple interleaved processes in the mind simultaneously and because of the limited capacity of language to express them. Furthermore, modern critical philosophies, both idealist and materialist, have proven to be out of step with the progress of the natural sciences, which they were unable to foresee. By contrast, we believe that by transposing and translating these older philosophies we can draw inspiration from them, without falling into a precritical regression, and try to produce or reconstruct a natural ontology, and perhaps an ethics, for our own time.

Another warning concerns the references to kabbalistic texts. Kabbalah, as the most important expression of the mystical currents of Judaism, is traditionally opposed to the "rationalist" legal and philosophical currents whose emblematic figure is Maimonides, the twelfth-century physician, codifier, and philosopher, a disciple of Aristotle and Averroes. Today, when irrationalism is reconquering lost territory and taking over the media in one form or another, mild or violent—mild in the astrology of the salon or broadcast studio, violent in the antics of certain God-crazed individuals and cults—the scientific mysticism of Córdoba[10] or the overt delirium of various millenarian sects—we should not be astonished that kabbalah is enjoying a spirited revival, in association with astrology, magic, tarot, and other occult pseudo-sciences, and that it is also popular among some fundamentalist and/or messianic currents of Orthodox Judaism. Hence we must strongly emphasize the classic distinction between speculative or philosophical kabbalah—which is what interests us here and is a scholarly tradition closely related to Neoplatonism, no less "rational" than the Aristotelianism of Maimonides that it supplanted in the Renaissance, at the dawn of modern science—and practical kabbalah, the contemporary substrate of superstition and eschatological frenzy. Today this distinction is even more germane than the difference between ecstatic or prophetic kabbalah and theurgic kabbalah propounded by some modern historians of religion. The latter distinction, part of the tradition of academic Jewish studies, is the product of an external comparative analysis more than a real difference of doctrine (it is, in fact, usually only a matter of emphasis). By contrast, the distinction between speculative kabbalah and practical kabbalah has grown wider since modern science and critical philosophy broke with the prescientific practices and lore of Renaissance natural magic. Here we are interested exclusively in speculative or philosophical kabbalah as a tradition of inquiry in which, as for the sages of the talmudic academies and the philosophers of antiquity, the pursuit of ethical values cannot be separated from the quest for rational intelligibility, which is always attentive to and dialoguing with the sciences.

10. At the 1979 colloquium in Córdoba, "Science and Consciousness," scientists, clerics, and Jungian analysts got together to demonstrate the supposed conceptual unity of modern physics and ancient esoteric traditions. For an analysis and critique of this strange enterprise, whose allure never seems to fade, see Atlan, *Enlightenment*, pp. 22–24.

By contrast, practical kabbalah as it exists today is merely an anachronistic throwback to the natural magic of the Renaissance, when—keeping company with alchemy, astrology, and Hermeticism—it was an integral part of the prehistory of modern science and technology. Today—indeed already by the time of the tragicomical messianic epic of Shabbetai Ẓevi in the seventeenth century—it is no more than another source of the superstitions that constitute the common underpinning of obscurantist and fundamentalist fads. Today, as then, we can and must mock and condemn these "kabbalists," who unblushingly employ an esoteric jargon without locating it in the context of its successive elaborations and of the philosophical and scientific concerns of the era in which it emerged. The allure of the mystical and the occult makes it possible to sell carloads of superstitions, talismans, astrological and numerological predictions, palmistry, and the like. There is third-rate kabbalah just as there is "cheap Buddhism," in which the exotic nature of the vocabulary vainly endeavors to camouflage the vacuity of the thought. Spinoza, in chapter 9 of the *Tractatus*, mocked the mystifying usages of esotericism as applied to extract the deep meaning of Scripture that was supposedly concealed in the formal literal meaning of the biblical text; marginal notes evoking different readings, numerology applied to the letters, peculiar punctuation, the shape of some letters—all were supposed to be a cipher for decoding the secrets of God. For Spinoza, these were nothing but "childish lucubrations. I have read and known certain Kabbalistic triflers, whose insanity provokes my unceasing astonishment" (*Tractatus*, chap. 9, p. 140).

We could say much the same about those who, generation after generation, continue to employ these exegetical and mnemotechnical methods as if they were techniques of divination. In this domain, too, the march of folly shows no sign of ending, and sophisticated computer programs for calculating probabilities are now being used to "prove" the divine origin of the biblical text exactly as we have it, by counting regular intervals between letters to extract "secret messages" that the divine author is supposed to have sent us across the millennia. These "Bible codes" (which it took a computer to discover) are supposed to offer Nostradamus-like predictions of events that only a supernatural author could have foreseen thousands of years before they took place. The acme of confusion and naïveté is the aura of scientific validity granted to these "childish lucubrations" (as Spinoza would have labeled them) by their publication in an international journal of statistics, before they were taken up and given broad circulation in a best-selling popularization. This is a confounding of form and substance, of formal structure and interpretive semantics, of a posteriori probabilities that apply when we know in advance what we are looking for (because it makes sense to us) and a priori probabilities (which would be conclusive but which apply only to a set of indefinite and unknowable possibilities, that is, to the infinite set of all meanings that could be constructed from regular combinations of letters). This is indeed "insanity," a psychosis far beyond the spectacular madness of Ben Zoma, who, among the four sages of the talmudic legend who "entered the 'Garden,' . . . looked and was stricken [with madness]."[11] He never returned from his adventure of initiation because he could not contain the torrent of what he perceived there, and his mind, once so brilliant, began wandering "outside" (i.e., he went "out of his mind"). For our "kabbalistic" prognosticators and eschatological calculators, it is most often merely a confusion of levels of interpretation and a belief in the objective reality of symbolic structures, characteristic of many exegetical deliriums.

11. B *Ḥagigah* 14b–15a.

Note that this confusion sometimes extends to scientific publications, including some of the most prestigious; when statistics and probability enter the picture, critical sense seems to vanish. For example, the perpetual debate about nature and nurture is constantly relaunched in ostensibly new forms, even though the methodological problems are regularly denounced.[12] We should long since have been aware that the problem itself has no meaning, as long as, for example, the additive hypotheses that underlie the computational techniques generally employed—such as the calculus of variance—remain unproven. But that does not seem to matter. The need to believe and feel reassured always wins the day.

12. See, for example, R. C. Lewontin, "The Analysis of Variants and the Analysis of Causes," *American Journal of Human Genetics* 26 (1974): 400–411; Ned Block, "Race, Genes, and IQ," *Boston Review* 20(6) (1996), pp. 30–35.

In the same vein, the sensational and accelerating discoveries about the "gene for this" and the "gene for that" are often based only on a correlation between a particular (and not always well-defined) condition or behavior—such as "homosexuality," "violence," and even "poverty"—and a genetic marker found with some frequency in the experimental subjects and members of their families. The same applies to the faulty interpretation of statistical tests in economics and the social sciences in general. We cannot avoid the conclusion that many editors and the specialists who referee articles for peer-reviewed journals (the norm of scientific publication) keep forgetting, not only that correlation is not causality but also and especially that the only value of statistical tests is to eliminate a hypothesis, not to establish its truth, even if merely as probable.

A hypothesis of correlation—not to speak of causation—is to be discarded as improbable if the outcome of the statistical test is not sufficiently "significant." On the other hand, a "significant" result in no way establishes the truth probability of the hypothesis being tested. It merely allows us to eliminate as unlikely some unspecified contrary hypothesis that would deny any correlation between the observed phenomena, including the trivial correlations of no interest that derive, for example, from the chance effects of a common partial cause. This is one of those cases where a double negative does not equal a positive. And we have not even mentioned the asymmetry between the necessary causal deduction that "the effect is contained in the cause," as one used to say, and the empirical induction of causes, inevitably incomplete, from supposed effects.

We might expect that all scientific publications would be vigilant to avoid falling into these traps, of which specialists are very well aware but which seem to be forgotten the moment socially and politically sensitive topics are involved. It certainly seems as if fashion and ideology can blind criticism, which is left only with the skeptical slogan that statistics can be made to say whatever you want them to, or, more precisely, whatever you want to believe. Erroneous interpretations do not detract from the value of statistics, of course. But given their proliferation and recurrence, we must never forget that the verisimilitude of a belief cannot rest solely on the positive result of a statistical test. More particularly, in the case of a scientific hypothesis about the reality of some causal relationship—the genetic or hormonal determination of behavior, the therapeutic effect of a drug, etc.—even when a positive statistical test makes it possible not to discard the hypothesis straightaway, it still cannot be accepted as likely without some additional knowledge about the mechanisms by which the relationship, ascribed to causation or to physical/mathematical law, is effected. In some cases, fortunately, we do have at least partial knowledge of these mechanisms;

the statistical test can then reinforce our belief in the validity of this knowledge, acquired by the application of empirical and logical methods that are suited to the object of investigation. In many other cases, however (notably for the partisans of astrology and parapsychology), the desire to believe and to take shelter under a comprehensive explanatory theory wins out over the critical suspension of judgment that elementary prudence ought to impose. This is where scientific causality, applied thoughtlessly under the auspices of misinterpreted probabilities and statistics—increasingly widespread today—flows into the deliriums of a magical causality that does not know how mad it is. Facing complex situations over which we have poor control, we seem to employ the result of statistical tests, whatever the cost, to simplify matters and eliminate uncertainty, while confirming a belief—forgetting once again that although the test does not eliminate the hypothesis in question, neither does it allow us to conclude anything further about the validity or nonvalidity of the hypothesis. That its validity has not been ruled out does not mean that it has been affirmed, or that its nonvalidity has been excluded. Such would be the case only if all our knowledge of the world were based on two-valued logic, in which it suffices that a proposition is not false in order for it to be true. Unfortunately—or fortunately?—our empirical knowledge is such that a proposition may be neither true nor false, simply because we do not know enough in order to reach a conclusion. The belief that statistics and the computation of probabilities can fully compensate for our deficit of knowledge is the new pseudoscientific version of the ancient belief in casting lots to make the oracle speak.[13]

13. See Atlan, *SR* II, chap. 8.

But "a throw of the dice will never abolish chance," as Mallarmé wrote in capital letters. In Chapter 5 we shall attempt to get a grip on the desacralization of chance observed when we move from the mythical context of the biblical texts to their talmudic interpretations. We shall see how this exegesis flows from a radical transformation, the replacement of the sacred chance of the oracle, pregnant with a hidden meaning that can be uncovered by casting lots as means of divination, by the randomness of indifference, which foreshadows the probabilistic chance to which we are accustomed today. This does not devalue the role of chance, although it is reduced to our ignorance of causes in a world that is held to be totally determined. Hence taking account of chance means being aware of this ignorance, that is, trying to know it as a lack of knowledge, rather than acting as if our incomplete and limited knowledge could overcome it. This is the modern function of the computation of probabilities, despite the confusion entailed by the method. It was explained clearly by Laplace, whose hypothesis of an intellect that has perfect knowledge of the future course of the universe ("Laplace's demon") is presented (we must not forget) in the introduction to a treatise on probability. This approach has long since led me to advocate respect for the aleatoric nature of birth as a way to save ourselves from the fantasies of omniscience and technical omnipotence associated with the artificial production of living beings.

Finally, in the last chapter we shall tackle the problem of translation, which is always implicit in anachronistic interpretations of ancient texts. We shall see how it is possible to follow the career of chance, starting in the Hebrew text, the *ḳeri* of birth and of nonbirth, of events and human actions, and the different ways of translating and interpreting the word, depending on whether we approach the text with a theological or naturalistic attitude. *Pace* Einstein's aphorism about the God of nature, it is up to human beings to decide whether or not the God of ethics plays dice. This, at least, is what the text of Leviticus tells us, when taken literally; but a theological reading makes it impossible even to conceive what it might mean. Hence translators have always been hard put to deal with this passage, although it is easy to understand when approached without theological preconceptions about the nature of the biblical God.

Just as we read certain texts with pen in hand, underlining and adding annotations in the margins, we can also write with a second pen in hand. In this case, however, the marginal notes we add are not really, or not only, a post-factum commentary, but a true second text from which the first emerged. In some sense the text is a commentary on the notes because the content of the latter usually reflects the reading and study that preceded or accompanied the process of writing. The written text offered here could just as well be a commentary on and annotation of the texts I read with pen in hand. So I am grateful to the publisher for agreeing to follow the ancient tradition of manuscripts and later of printed books, in which text and notes are juxtaposed by adjusting the margins to suit their respective lengths and layout on each page. Computerized page makeup makes it easy to jump over the centuries and revive the ancient glossed layout that preceded the learned practice of footnotes. Variable margins made it possible for the commentaries to be read as a second text, before or after the main text. Here, without giving up the modern convention of call numbers,[14] their association with marginal passages of widely divergent length allows the typography to express, to the extent possible, the sometimes tortuous path of thought, as commentary on commentary, as commentary on itself.

14. See Anthony Grafton, *The Footnote: A Curious History*, rev. ed. (Cambridge, MA: Harvard University Press, 1997) [first published in German translation as *Die tragischen Ursprünge der deutschen Fussnot* (Berlin, 1995)].

ACKNOWLEDGMENTS

Some of the questions addressed in the first two chapters were discussed at two conferences of Francophone Jewish scholars, whose proceedings were published as *Le temps désorienté*, ed. Jean Halpérin and Georges Lévitte (Paris: A. Michel, 1993), and as *Le corps*, ed. Jean Halpérin and Nicolas Weill (Paris: A. Michel, 1996).

At various stages of the process Bruno Besana, Philippe Blaizot, Martine Heissat, Françoise Rendu, Jean-Luc Simonin, Isabelle Stroweis, and Gilles Toublanc helped decipher the manuscript and make it legible, or were involved in the production of this work.

The editorial support and advice of Maurice Olender never failed me.

I would like to thank Hent de Vries of the Humanities Center at the Johns Hopkins University and Emily-Jane Cohen of Stanford University Press for taking the initiative to publish an English version of this book, despite the complexity of the task. My profound thanks go to Lenn Schramm, who produced the translation, for his deep commitment to conveying as precisely as possible the meaning not only of the original French but also of the Hebrew and Aramaic sources and of my own reading of these texts.

CHAPTER I

THE SPARKS OF RANDOMNESS AND THE MANUFACTURE OF THE LIVING

I. PRODUCING LIFE, CONTROLLING TIME

The (re)production of living beings has always been the chief business of humankind. It provides us with our awareness of passing time, the time of individuals and the time of generations. The prize it offers is control over death, individual or collective, a control that may be real but ephemeral, or eternal and more or less a fantasy. Ever since women and men discovered how babies are made, they have been trying, generally without great success, to exert control over their progeny, both by restricting their choice of mates (through laws that define permitted and forbidden unions) and by transmitting a culture that molds their children and turns them into adults who comply with the norms of their society. Concomitantly, human beings soon acquired control of the (re)production of plants and animals, much more effective than their control over their own offspring, through the techniques of agriculture, domestication, and stock-breeding.

Today, however, thanks to theoretical and technical discoveries in reproductive biology and heredity, this age-old quest for control has brought us up to and over a threshold never crossed before.

Of course this is not the first time such a threshold has been crossed; the invention of agriculture and stock-breeding, some ten thousand years ago, constituted a revolution by which Neolithic man acquired a real ability to transform and produce living beings. The current biological revolution is an event of the same magnitude, but this time with regard to human life.

Humanity's progress from hunting and gathering to agriculture and animal husbandry has left obvious traces in myth, in the form of opposed pairs of divine, semi-divine, or human figures. In the biblical account, the myth of the Tree of Life and the Tree of Knowledge in the Garden of Eden marks the origin of the nature and evolution of human societies. The very first conflict that follows, between Cain and Abel, is repeated and transformed in the later conflicts between the hunter Nimrod and the shepherd Abraham,[1] and between the hunter Esau and the shepherd Jacob.

These narratives have clear ethical and ideological implications for us, because they highlight contrasting ways in which human beings can relate to their environment, different and opposed human accounts of their relationship to nature—in brief, different *religions*. These conflicts or wars of religion, which seem to be deeply embedded in our nature, have proliferated in multiple forms that far transcend the earliest confrontation between nomadic hunter-gatherers and sedentary farmers. The transhumant shepherds and herdsmen soon came into conflict with the settled farmers, even though they lived alongside each other.

Like other mythical accounts, the biblical narrative expresses, in its own way, this divergence of human characters. It tells, among other things, of the birth of a nomadic race of shepherds, the Hebrews, in the fertile agricultural cradle of the Nile Valley and the creation of a new civilization that, although sedentary, was always marked by the nomadism of dispersion and exile, in the Land of Canaan, that is, the domain of commerce and trade.[2] The farmer was no longer the child of the land that nourished him, but the partner and husband(man)[3] who fertilized it, in an intercourse simultaneously amorous and contractual.

The memory of this evolution is engraved, quite "naturally," in the diverse emotional, linguistic, and political experiences that characterize human societies. These experiences always involve a link to truth and to the legitimacy of sovereign power: nature's sovereign power over the human beings it produces, or the power of human beings over the nature they fertilize, or the hegemony of some human beings over others. The different sorts of control mirror one another in the amorous, contractual, cognitive, nutritive, and fertilizing relationships in which the two sexes conjoin.

These battles for sovereignty and control generate the various levels of our psyche, in which we superimpose our multiple relationships on what we consider to be real or true and, in this context, on what we esteem and desire because it is "good," or on what we fear and want to reject as trash or excrement—although we may recover it later by reintegrating it indirectly, in greater or lesser spans of time and space. It is always a question of sovereignty and power arrangements, whether for the god-born kings of antiquity, heroes and warrior chiefs, the aristocracies of bygone times, or citizen democracies (which are in fact also based on an aristocracy, one of speech and appearance).

The discourses about what is worthwhile and what is true, originated by oracles, diviners, sorcerers, and other "masters of truth,"[4] and later by prophets (true and false), priests, judges, and sages, and finally by a new set of wise teachers and philosophers, compete to bestow legitimacy on sovereign power. Values and truths keep batting around the question of their own legitimacy—the legitimacy of values resting on their assumed truth, that of what is taken to be true being based on the presumed incontrovertible value of truth.

The new control of the living fostered by the current revolution in biological and computer technology is probably a

1. <E.g., *Gen. Rabbah* 38:13.>

2. The original meaning of *kena'ani* 'Canaanite' is "merchant," as in Isaiah 23:8. See also Rashi's commentary on Genesis 38:2.

3. See, for example, Isaiah 61:10–11.

4. See Marcel Detienne, *The Masters of Truth in Archaic Greece*, trans. Janet Lloyd (New York: Zone Books, 1996).

development of the same magnitude as the discovery of agriculture and animal husbandry, and later, in the Bronze and Iron Ages, of metal-working, with consequences at least as important for the evolution and molding of human nature. But today it is a matter of an unprecedented *direct* control of human life. As we shall see when we turn our attention to golems, such control was imagined in the past, and some mythical narratives dealt with it, in a certain fashion. For us today, the myth is being materialized in the concrete efficacy of biotechnology.

2. SEMINAL REASON AND "IT'S ALL IN THE GENES"

In the application of technology to transform nature (an enterprise that has always characterized the human race, to the point that we can say it is human nature to transform nature), awareness of passing time long seemed to be inseparable from awareness of life. The experience of the passing of generations, of the successive periods of birth, maturity, old age, and death, served as the point of departure and reference for an awareness of the flow of time.

When, very soon, the motions of the heavenly bodies came to be used to measure time, they were themselves perceived as processes of life and death, as a living macrocosm imaging our human microcosm. Every month the moon died and was reborn. When the lunisolar year of the Hebrew calendar is extended by a month, to keep the sequence of lunations synchronized with the Earth's orbital period, it is said to be *me'ubberet* or "pregnant"; the computations used to determine the relationship between the solar and lunar cycles—the intercalation of seven years with thirteen lunar months in every nineteen solar years—is called *sod ha-'ibbur* 'the secret of gestation'.

It was not until the mechanical revolution of the seventeenth century that human beings came to view the stars as inanimate beings and consequently to have an awareness of physical time that is not determined by the generations of living beings, by deaths and births.

Two centuries later, the contemporary biological revolution is extending our possibilities of control to the domain of the living, using physical and chemical techniques derived from this mechanistic conception of nature. The techniques of molecular genetics open utterly new perspectives, because they make possible, for the first time, the direct manipulation of the genetic material that carries heredity. This is why genes and genetics have come to dominate the practice and discourse of contemporary biology. However, this invasion by the idea that "it's all in the genes" is not without its ambiguities. It has a mythical if not indeed mystical aspect, as is often the case with scientific concepts whose theoretical content is less intelligible and more equivocal than their operational value.[5]

Today it is not so easy to provide a unique and general definition of the gene: the unit of transmission of a trait, the unit of developmental control, the unit of selection, a snippet of DNA encoding for a protein, etc. In general, several scattered fragments encode several proteins, so that the spatial and temporal identity of a gene, defined structurally, can vary over the course of the transcription and translation process. What is more, DNA molecules, left to their own devices, are inert. They have no activity and can do nothing at all—not even reproduce themselves—without the intervention of proteins, the regulatory enzymes that catalyze their replication or expression. Similarly, a functional definition of a gene implicitly includes a set of enzymatic reactions between DNA fragments and other

5. See Atlan, *Enlightenment*, pp. 191–193.

6. See: G. L. G. Miklos and G. M. Rubin, "The Role of the Genome Project in Determining Gene Function: Insights from Model Organisms," *Cell* 86 (1986), pp. 521–529; T. Kono, "Nuclear Transfer and Reprogramming," *Review of Reproduction* 2(2) (1997), pp. 74–80. For a review, see Atlan, *La fin du "tout-génétique"?*

The meaning of the word *gene* has evolved significantly from the early days of genetics, still permeated by vitalism, at the start of the twentieth century, to the information-sciences materialism of contemporary molecular biology. The gene, originally a notion used to formalize the Mendelian laws of heredity, became (rather paradoxically, precisely when its material substrate was discovered) more muddled and multiform, depending on the experimental and theoretical context in which the concept is used. The gene of the molecular biologist is not the gene of the population geneticist or the embryologist or the evolutionary biologist. The word has retained its ancient dynamic connotation, derived from Scholastic "genetics," in the sense of the producer of genesis and development, although it is now applied to static structures of DNA molecules. Since these are, in the words of the geneticist Richard Lewontin, as inert as can be imagined, they are quite incapable of doing anything by themselves, including replicating themselves, and all the more so of dictating all aspects of an individual's life (see R. C. Lewontin, "The Dream of the Human Genome," *New York Review of Books*, May 28, 1992, pp. 31–40; see also his *Biology and Ideology: The Doctrine of DNA* [New York: HarperCollins, 1991, 1995], chapter 3: "Causes and Their Effects"). Hence we can say, with no fear of paradox, that "genetics," as the totality of processes of an organism's genesis and development, "is not in the gene" (Henri Atlan, "ADN: programme ou données? ou, le génétique n'est pas dans le gène," *Bulletin of the European Society of Philosophy of Medicine* 3(3), special CD-ROM issue 1.01a and b [1995]; idem, "Le projet 'Génome humain': un exemple de transmission du savoir biologique," in *Le Génome et son double*, ed. G. Huber [Paris: Hermès, 1996], pp. 83–94). A gene, as the unit of the genetic *process* of development, necessarily includes, in addition to one or more sequences of DNA, several enzymes that are its only "active" part and that regulate expression of these sequences in the form of other proteins. The result is that genetic discourses are marked by ambiguity, especially those that misuse the metaphor of the "program" allegedly written in the DNA. More and more authors have realized this, thanks to the human genome project. See, for example, in addition to the earlier references: A. I. Tauber and S. Sarkar, "The Human Genome Project: Has Blind Reductionism Gone Too Far?" *Perspectives in Biology and Medicine* 35(2) (1992), pp. 220–235; Evelyn Fox-Keller, "Nature, Culture, and the Human Genome Project," in *The Code of Codes: Scientific and Social Issues in the Human Genome Project*, ed. D. J. Kevles and L. Hood (Cambridge, MA: Harvard University Press, 1992), pp. 281–299 and 355–357; eadem, "Language and Science: Genetics, Embryology, and the Discourse of Gene Action," in *Refiguring Life: Metaphors of Twentieth-Century Biology* (New York, Columbia University Press, 1995), chapter 1. Alongside the application of genetic techniques to agriculture and biomedicine, these discourses about the *action* of genes have erected an imaginary genome that occupies the place left vacant by the soul and demons as the hidden causes of what we are and what happens to us. This shift is now culminating in the fantasies of immortality entertained by candidates for human reproductive cloning, as if the identical reproduction of a genome really had something to do with an eternal essence (see below, §15, and Chapter 3).

molecules, mainly RNA and proteins, that determine the activity state of the "gene" so defined. The first comparative analyses of the genomes of different species revealed the complexity of the relationship between the structure of a sequence of DNA and its genetic function(s) in the organism. A single gene, taken as a fragment of DNA with a specific sequence of nucleotides, can have multiple functions, differing from one species to another or varying as a function of the stage of development. The results of these analyses, associated with the revolution worked by the successful attempts to clone mammals by nuclear transfer and to "reprogram" the genomes of already differentiated cells, strongly bolster the decades-old criticism of taking the genetic program metaphor literally. In fact, they show the extent to which an organism controls the activity of its genome at least as much as the genome controls the functioning of the organism.[6]

The resultant polysemy of the new concept of the gene as a material entity with variable limits and multiple properties brings us back to the ancient notion of *seminal reason*, long considered to be obscure and mythical if not indeed impossible to grasp—as if, skipping over twenty-five hundred years, molecular genetics imparts an operational content to the seminal reasons of Stoic physics.

A seminal reason was also a rational, if not indeed reasonable, seed (*semen* in Latin), a material object or "body" that determined an individual's strength and development. The physical manifestation in each individual of the shaping power of nature, simultaneously body and soul, fire, spark, light and logos, reason and language, the more or less mythical *logos spermatikos* of olden times, like the genes of today—though we know how to manipulate them and modify their structure and function—is a body that is simultaneously material and informing, in the sense of the "source of form." Here we find the same indissoluble union of matter and form, of passive and active, of structure and function, of inert and living, of inanimate and animate, that we now attribute to what we call, for want of a better term and not without ambiguity, "information-bearing macromolecules."

3. CONCEPT AND LIFE

Something of this union, hard for us to imagine, survives today in the word *concept*, which comes from "conception" (unless it is the other way around). It is interesting to trace, along with Georges Canguilhem,[7] the history of the dynamic and ambiguous relations between concept and life. The Aristotelian problematic, simultaneously logical and naturalist, has continued to resound in the life sciences for nearly twenty-five hundred years, whereas the Stoic theories were swept aside and the very idea of seminal reason became incomprehensible—inconceivable, one could say. Despite the mechanical revolution in the physical sciences, the essence of life—what distinguishes the "animate" from the "inanimate"—is still a soul, an *anima*. As Canguilhem puts it, this soul (which, for Aristotle, expresses the final cause of nature) "was not only the nature but also the form of the living [*le vivant*]. Soul was at once life's reality [*ousia*] and definition [*logos*]. Thus, the concept of the living was, in the end, the living itself."[8] But this also led to insurmountable difficulties concerning the nature of the relationship between the intellect, which knows life through concepts, and life, which produces that intellect; between the conception of concepts by thought and the concept of conception, that is, the concept of Life, which makes it possible to conceive of this conception as an activity that (re)produces the living.

These problems have reappeared in different forms and different philosophical contexts, notably in Kant, Hegel, and Bergson, whenever the word *life* designates an intelligible concept. In the Battle of Universals, William of Ockham and the nominalists tried to devitalize the concept by demonstrating its unreal (or, more precisely, artificial and conventional) character, associated with the use of vague and general words that are falsely believed to designate "universal" entities that exist in nature, or essences as modes of being. But they themselves kept running into the problem of reconciling the real individual with the form that it reproduces and that it has in common with other individuals of its species. If only individuals are real, the shared form is merely an artifact of the human mode of evaluating the resemblance between individuals. This evaluation, practiced by all men and women, is what makes it possible to speak about nature. Yet it is artificial, in that it is a conceptualization of natural things—of individual living beings—effected by means of generalization and abstraction. This

7. Georges Canguilhem, "Le concept et la vie," *Revue philosophique de Louvain* 64 (1966); repr. in *Etudes d'histoire et de philosophie des sciences*, 7th ed. (Paris: Vrin, 1994), pp. 335–364; translated (in part) as "The Concept of Life," in *A Vital Rationalist: Selected Writings from Georges Canguilhem*, ed. François Delaporte, trans. Arthur Goldhammer (New York: Zone Books, 1994), pp. 303–319.

8. Canguilhem, "The Concept of Life," p. 303. <Here and elsewhere, Goldhammer's "living thing" for the French *le vivant* has been replaced by "the living" (see Translator's Introduction).>

"introduces a cleavage in the system of the living, because the nature of one of those beings [human nature] is defined by an artifice, by the possibility of establishing a convention rather than expressing the order of nature."[9] The history of the relations between life and concept has oscillated between two sets of problems—those raised by the simple identification of the concept with the living and those associated with distinguishing them, such that the concept is produced by a human nature that differs from the nature of other living beings precisely in that it can construct the artificial and the conventional. It can be shown that Spinoza's ontology made it possible for his theory of the individual—his "physics"—to resolve the problems of nominalism by replacing the distinction between nature and artifice with that between adequate and inadequate ideas (or causes). As with the problem of the relations between body and mind, Spinoza's unique position in the history of philosophy jumps out at us, along with how he was misunderstood by his critics, who thought it possible to reduce his doctrine to some of his sources, in this case medieval nominalism and ancient Stoicism.

9. Ibid., p. 308.

Given the nominalist risk of devitalizing and denaturing thought and concept by reducing them to human artifacts, the idealism of the last three centuries, from Leibniz through Hegel and on to Bergson and Whitehead, passing by way of Kant and the proponents of *Lebensphilosophie*, attempted to tie things together by moving in the other direction and conceptualizing Life. Life, expressing the logic of the mind just as it expressed the geometry of the forms for Aristotle, once again became congruent and even identical with the concept. More precisely, the *thought* concept was again identical with itself as a concept *conceived* by Life. But these attempts ran into new difficulties, posed this time by the autonomous development of biology. When the latter was placed firmly on chemical foundations in the nineteenth century, it began to move away from the evocation of vitalist concepts—especially the notion of Life as some vital force—as a way to explain the results of experiments on organisms.

When biology and medicine became experimental sciences, the concept of Life began to lose its explanatory value; biological phenomena were now to be explained in chemical terms, alongside and supported by new concepts such as homeostasis, the interior milieu, internal secretions, etc. The heuristic value of these concepts, for all that they were often ambiguous and not always appropriate, made it possible for physiology to develop as a relatively autonomous science. Over the course of the twentieth century, and in association with embryology and genetics (the sciences of development and heredity), a new biochemical and biophysical biology emerged, one that had absolutely nothing to do with vitalist principles—at least in appearance. This development took place under the influence of neo-Darwinian theories of the evolution of species, which answered the question of origins, and as a result of and thanks to the massive infusion of new concepts from information science and linguistics (code, information, transcription, translation, program, reading, punctuation, sense and nonsense, etc.), which became the core of the language of molecular biology and seemed able to answer the questions about its mechanisms. These new concepts made it possible to believe, for a moment, in the reality of a "language of life" or a "logic of the living," thereby justifying, retroactively, Aristotle's and Hegel's intuition about the identity of concept and life.

In the 1970s, writing in the journal *Critique*,[10] François Jacob (the author of *The Logic of Life*[11]) considered the idea of a language of life written in the sixty-four triplets of the genetic code (as did the linguist Roman Jakobson). Jacob could not resist invoking the analogy with the sixty-four hexagrams of the *I Ching*, a Taoist practice of divination that fascinated Jung, among others. Fortunately, Jacob seems to have returned later to a more critical vision of biological theory, developing the idea of a sort of "tinkering" by nature, conducted within an extended field of possibilities and limited only by the physical and chemical constraints of the matter that composes living bodies.[12] A few years earlier, Canguilhem had fallen into the same trap of taking the notion of the language of Life seriously. He wondered, however, whether the concepts of information science were in fact imported metaphors, reminiscent of Claude Bernard's "repeated use of such converging metaphors [in] an attempt to pinpoint a biological reality for which no adequate concept had yet been formulated."[13] But he immediately rejected this possibility and considered biological heredity, as the communication of information, to be a sort of return to Aristotelianism, because "to say that heredity is the communication of information is, in a sense, to acknowledge that there is a *logos* inscribed, preserved and transmitted to the living."[14] Furthermore, "to define life as a meaning inscribed in matter is to acknowledge the existence of an a priori objective that is intrinsically material and not merely formal."[15] The only difference with Aristotelianism is that "contemporary biology . . . has dropped the vocabulary and concepts of classical mechanics, physics and chemistry, all more or less directly based on geometrical models, in favor of the vocabulary of linguists and communications theory. Messages, information, programs, code, instructions, decoding: these are the new concepts of the life sciences."[16]

In fact, the discovery in recent decades that the mechanisms of genetic regulation are more complex than had been imagined, along with the new metaphors inspired by dynamic systems and theories of self-organization, has shown that these concepts are imprecise metaphors. The genetic process has very few of the syntactic and especially semantic characteristics of a language.[17]

The genetic code itself seems to be an exception to this; but in fact it involves a static correspondence between the linear sequences of DNA and RNA and the structure of proteins, with none of the dynamism of human language and not even that of a computer program. In the 1960s, Canguilhem and most biologists—who believed in the nonmetaphorical aptness of the linguistic concepts they invoked—were in the situation that Canguilhem himself described with reference to Claude Bernard: "He did not perceive and could not perceive—all scientists are in the same situation—that the discovery that had led him to elaborate certain concepts blocked his path to further discoveries."[18] This situation continued in the evolutionary epistemology of Lorenz, Popper, and some of their successors, who were sometimes returned, by a literal application of the metaphors of information science, to a panvitalism and a panpsychism in which the same "cognitive"

10. *Critique* 322 (1974).

11. François Jacob, *The Logic of Life: A History of Heredity*, trans. Betty E. Spillmann (New York: Pantheon, 1973).

12. François Jacob, *The Possible and the Actual* (New York: Pantheon, 1982).

13. Canguilhem, "The Concept of Life," p. 315.

14. Ibid., pp. 316–317.

15. Ibid., p. 317.

16. Ibid., p. 316.

17. See Henri Atlan, "Du code génétique aux codes culturels," in *Encyclopédie philosophique*, ed. A. Jacob (Paris: PUF, 1989), pp. 419–430.

18. Canguilhem, "Le concept et la vie," p. 361 (a passage not included in the English version).

process was supposed to be at work from the amoeba to Einstein.[19] In fact, the "genetic language" or "genetic message" has no meaning outside the molecular structures that we observe; they are not signifiers, because they signify only themselves. The nature of organisms has no more to say than that of other chemical bodies. Molecular biology is simply the physics and chemistry of macromolecules, which, thanks to the metaphors of information science, can be described globally, with no knowledge of the mechanisms by which the integrated functions of an organism—the phenotype—are produced from the static structures of these macromolecules—the genotype. To put it another way, we are still, as Canguilhem wrote of Bernard, employing "metaphors [in] an attempt to pinpoint a biological reality for which no adequate concept [has] yet been formulated."

Thus there is a total disconnection, in our scientific representation of the world, between how we understand conceptualization as an activity of thought and how individuals conceive of other individuals. The very problem of the nature of their possible congruence has vanished. It is only with great difficulty that we can still understand the terms of the classical problem of joining, in life and in thought, the conception of concepts as a vital process with a concept of Life as a process of the rational or intuitive mind.

The *logos* is no longer *spermatikos*, reason is no longer seminal, and knowledge is no longer spermatic. Epistemology, which deals with knowledge about knowledge, inquires into the conditions in which concepts are produced and seeks to form some concept of concepts; the cognitive sciences attempt to elaborate plausible models of thinking that are compatible with the constraints of anatomy and physiology. As for Life, it is no longer, as such, the object of scientific inquiry. It has been replaced by the study of the physical and chemical mechanisms that produce and explain the diverse phenomena observed in living organisms, including those peculiar to them, such as sexual or asexual reproduction, growth and development, and metabolic and physiological functions at several levels of organization. Today, neither Life nor the soul is a concept in what we continue to misleadingly refer to as the "life sciences." Instead, they are conceived of and tested in the intimate lived experience of our own subjectivity. This is why it is important that we always keep in mind those ancient myths in which the congruence of the living body and the concept of life involved a tangible physical link: the brain produced knowledge, which, descending by the spinal cord and then implanted in a woman's uterus, like a seed in the fertile earth, germinated and developed into a new living body with a new brain.[20]

4. GOLEMS

These comparisons of present-day operational notions with ancient philosophical ideas that were still permeated by myth are not meaningless games of association. They are useful because they make it possible for us to step back from the current situation and enunciate the philosophical

19. See Henri Atlan, "Intentionality in Nature: Against an All-Encompassing Evolutionary Paradigm: Evolutionary and Cognitive Processes Are Not Instances of the Same Process," *Journal for the Theory of Social Behaviour* 24(1) (1994), pp. 67–87.

20. This image is found in one of the earliest works of medieval kabbalah, *Sefer Habahir*, §155 (see *The Bahir: An Ancient Kabbalistic Text attributed to Rabbi Nehuniah ben HaKana*, trans. Aryeh Kaplan [New York: S. Weiser, 1979; repr. Northvale, NJ: Jason Aronson, 1995]). It occurs there as a commentary on Kohelet's observation that "one generation goes, another comes" (Eccles. 1:4). In fact, it seems to have been extremely widespread in antiquity and can be traced back at least as far as Pythagoras (Diogenes Laertius, *Life of Pythagoras*, §19).

and sometimes mythological context, which would otherwise remain implicit and unanalyzed, in which these new biological concepts are employed.

The "spermatic knowledge" of certain kabbalistic texts, similar to but different from the seminal reason of the Stoics, can help us define, through the many myths it has inspired, an anthropological context in which we can analyze the theoretical and ethical questions raised by technologies for manufacturing life. As we shall see, the main difference between the roles played by semen in the myth of the sparks of randomness and in the Stoic notion has to do with the part assigned to chance in making this knowledge efficacious, which significantly alters their common experience of the absolute determinism of nature.

Elsewhere I have written about the ethical problems posed by present and future applications of biotechnology.[21] We must deal with these problems case by case, paying special attention to the consequences of a particular technology for our biological, emotional, social, and cultural life, to the extent that they can be foreseen. The exercise is more difficult—but also more necessary—because one technology may produce different effects, some of them desirable, others far from it, depending on the conditions in which it is employed.

Here I propose, instead, to analyze the origin myths, because their anachronistic juxtaposition with the new situations created by these technologies can yield fresh insights. It will allow us to project diverse (and perhaps contradictory) meanings onto the moral and legal positions we are tempted to take in a given situation. This does not make our decisions about what is and what is not appropriate any easier. But whatever the decision, and whatever the criteria applied—utilitarian, intentional, religious, or secular—these decisions acquire added anthropological scope and weight from the new meanings conferred on them by the ancestral myth into which they are set.

This is why, rather than describing yet again all of the possibilities cooked up by the untrammeled imagination of Frankenstein-like biologists, we shall begin by examining the myth of the fabrication of a man or woman, as recounted in the many narratives about the creation of a golem by wise and righteous rabbis who, as Rashi reports, employed the doctrines of *Sefer Yezirah*, the Book of Creation, from which "they learned to combine the letters of a [divine] name."[22]

Gershom Scholem devoted a long chapter to this question,[23] but now we have a comprehensive study (more than three hundred pages) by Moshe Idel, who reviews the many accounts of golems from the talmudic era through the nineteenth century.[24] These two authors eliminate several misunderstandings based on preconceptions: for example, the most famous golem of all, that of Rabbi Judah Loew ben Bezalel, known as the Maharal of Prague (ca. 1525–1609), is certainly apocryphal; the Maharal himself never refers to it in his copious and detailed writings, although, like many other authors,

21. H. Atlan and C. Bousquet, *Questions de vie* (Paris: Le Seuil, 1994). See also H. Atlan, "Le milieu naturel et la personne humain face aux biotechnologies," in *Le jaillissement des biotechnologies*, ed. P. Darbon and J. Robin (Paris: Fayard-Fondation Diderot, 1987), pp. 209–226; idem, "Personne, espèce, humanité," in *Vers un antidestin. Patrimoine génétique et droits de l'humanité*, ed. F. Gros and G. Huber (Paris: Odile Jacobs, 1992), pp. 52–63; idem, "Le projet 'Génome humain' et la transmission du savoir biologique," *Alliages* 18 (1993–1994), pp. 27–36; idem, "Transfert de noyau et clonage: aspects biologiques et éthiques," *L'aventure humaine* 8 (Dec. 1997), pp. 5–18.

22. Rashi on B *Sanhedrin* 65b, s.v. "Created a man."

23. Gershom Scholem, *On the Kabbalah and Its Symbolism*; trans. Ralph Manheim (New York: Schocken, 1965), chapter 5, "The Idea of the Golem."

24. Moshe Idel, *Golem: Jewish Magical and Mystical Traditions on the Artificial Anthropoid* (Albany: State University of New York Press, 1990).

he discusses the theoretical question of the manufacture of an artificial humanoid. The legend of the golem of Prague first appeared long after his death. What need was this posthumous attribution of a golem to the Maharal meant to fill?[25]

For many, these accounts are merely so many superstitions or popular legends of the sort to which fantasy tends to run, like the tales of Faust and Frankenstein. This may be; but the evocative power of myth is undeniably present, as indicated by its resonance with questions that the development of modern science has made increasingly weighty. In 1982, a U.S. presidential commission referred to the legend of the golem in its discussion of the possible benefits and dangers of genetic engineering as applied to human beings: ". . . each of these tales [the Maharal's golem, Dr. Frankenstein's monster] conveys a painful irony: in seeking to extend their control over the world, people may lessen it. The artifices they create to do their bidding may rebound destructively against them—the slave may become the master."[26]

Now we can appreciate the extent to which the traditional literature about the golem, which is much more extensive than the legend of the Maharal, comes into contact, sometimes in quite unexpected details, with the ambiguity of the quest for power through knowledge. When men and women discover that they can deploy their knowledge and technology to create a human being who is "artificial" but who is in every (or almost every) way identical to themselves, the essence of the human condition is revealed in its fullest complexity. Nowhere in the golem liter-

25. See Atlan, preface to *Le Golem.* One answer to this question may lie in the prescientific magic of the Renaissance and the Maharal's genuine and documented interest in what would prove to be the earliest accomplishments of the scientific revolution. André Neher, who studied the relations between the Maharal and his scientific milieu (*Jewish Thought and the Scientific Revolution of the Sixteenth Century: David Gans (1541–1613) and His Times*, trans. David Maisel [Oxford and New York: Oxford University Press, 1986]), proposed an analysis of the golem myth focused entirely on this legendary attribution in his *Faust et le Maharal de Prague: le mythe et le réel* (Paris: PUF, 1987). Drawing a striking parallel with the legend of Doctor Faustus, created a century earlier, he analyzed what he calls "the profound identity between the golemic structure and the Faustian soul of modern and postmodern man" (p. 198), where "Faust is the myth of the modern man and the (Maharal's) golem is the myth of postmodern man" (p. 9). Despite the evident similarities between these two myths, Neher underlines the differences, notably the element of demonic possession and the pact with the devil that is the core of Faust's career, quite unlike the case of the golem. The latter's adventure merely reflects the irreducible ambiguity of human nature, determined and determining, created and creative, bearer of both life and death. The subsequent centuries of the machine age have merely highlighted this ambiguity and brought it to the breaking point. It is in this way that the golem, a human machine in every sense of the term, is the creation of postmodern man, as suggested by the father of cybernetics, Norbert Wiener (*God and Golem, Inc.: A Comment on Certain Points Where Cybernetics Impinges on Religion* [Cambridge, MA: MIT Press, 1964]). As for why later legend credited the Maharal with the manufacture of a golem, we may offer the following hypothesis. The sixteenth and seventeenth centuries were seen in retrospect as a turning point when modernity struck deep roots, thanks to the scientific and technological revolution in Europe. How the alchemical, magical, and astrological traditions inspired by hermeticism and kabbalah influenced the origins of this revolution is well known. In this context, as Idel's book shows clearly, fabrication of a golem represented a zenith, evidence of the profundity and divine truth of the knowledge that makes it possible to build one. Better yet, to borrow from the sixteenth-century kabbalist Meir ben Ezekiel Ibn Gabbai, the rabbis' ability to make a golem was, for those who asserted it, an indication of the superiority of Hebrew science and kabbalah to pagan lore and magic. The Maharal of Prague was a major figure, and perhaps the most important, in the intellectual milieu at the intersection of ancient kabbalah and modern astronomy. It was thus natural that legend attributed to him the knowledge that leads to the ultimate *imitatio Dei*—creation of a human being—despite the absence of an explicit account in his writings. In this way, popular imagination preserved and recreated the memory of the preeminence of the sage of Prague over the other sages of the era that saw the beginnings of modern science, when, later, the formidable practical efficacy of the sciences had become the evidence of that modernity.

26. President's Commission for the Study of Ethical Problems in Medicine and Biomedical and Behavioral Research, *Splicing Life: A Report on the Social and Ethical Issues of Genetic Engineering with Human Beings* (Washington, DC, 1982), p. 58.

ature (at least originally) do we find a negative judgment about the science and creative activity of human beings, "in the image of God"; in this it differs starkly from the Faust legend.[27] Quite the contrary: it is in and through creative activity that human beings achieve their fullest humanity, in an *imitatio Dei* that associates them with God in an ongoing and ever-more-perfect process of creation. But like a two-edged sword, this activity—the essential humanity of human beings—jeopardizes the continuation of the process for which it is, nevertheless, indispensable.

The legend of the golem—a male or female robot, built by learned scholars—is no marginal or anecdotal theme, but rather a constant presence in traditional Jewish literature. It occupies a quasi-canonical place there because it has the sanction of the Talmud itself.[28] In a discussion of the various types of sorcery condemned by the Torah, we read, by way of contrast to these practices—which, though impure, are nevertheless considered to be efficacious—that the righteous have the power to create worlds and that only their transgressions deprive them of this ability. To illustrate this statement, we are told that the fourth-century sage Rava "created a man"! The talmudic account continues: "He sent him to R. Zeira. R. Zeira spoke to him, but received no answer"—because, Rashi explains in his commentary, the man in question was not endowed with the power of speech. Whereupon Rabbi Zeira told him, "You are a creature of the rabbis.[29] Return to your dust." And to drive the point home, the text continues, "R. Ḥanina and R. Oshaia spent every Sabbath eve [i.e. Friday] studying the *Book of Creation*, by means of which they created a fat calf and ate it" for their Sabbath dinner.

Given this narrative, reported by the Talmud and vouched for by Rashi as literal truth, the possibility of making a golem was an incontrovertible part of rabbinic thought. Some, such as Maimonides, did treat it as symbolic. But other leading rabbis, such as Isaiah Horowitz (1565?–1630),[30] one of the major halakhic authorities of his time, read it as a tradition about an efficacious and authentic lore in which the ancient sages were proficient, thanks to their great wisdom and holiness. Moshe Idel, in his book on the golem tradition, lists many kabbalistic accounts of the various techniques used to create one. The question, at least in some cases, is whether these authors believed in the material reality of the being

27. See Neher, *Faust et le Maharal de Prague*.

28. B *Sanhedrin* 65b.

29. <Thus Rashi, glossing the Aramaic *ḥavrayya*; others understand the word to mean sorcerers or Magi.>

30. Often referred to as the *ShLaH*, from the acronym of the title of his major work, *Shenei Luḥot Haberit* (The Two Tablets of the Law), first published in Amsterdam in 1649, after his death in Tiberias, following a long rabbinic career in Poland, Frankfurt am-Main, and Prague. (There is a recent Hebrew edition, *Shenei luḥot ha-berit* [Jerusalem, 1959], and an English translation, *Shney Luchot Habrit on the Written Torah*, trans. and annot. Eliyahu Munk, 2nd ed. [Jerusalem and New York: Lambda, 1999].) Horowitz's reference to the ancient traditions about the use of *Sefer Yeẓirah* (*Vayeshev*, p. 103; trans. Munk, p. 358) to create artificial human beings and animals is particularly fascinating, given the wide circulation of his commentary in the Orthodox rabbinic Judaism of the period. It occurs in an excursus that attempts to solve an exegetical crux. Commenting on Genesis 37:2, which states that Joseph slandered his brothers to their father, Jacob, he quotes the midrash that fills in the details of this report: Joseph asserted that his brothers were having illicit relations with local women and were eating meat that had been cut from live animals—a gross violation of the Noahide laws that were incumbent on them. Horowitz propounds a dilemma: either they were really guilty of these misdeeds, which is hard to accept of Jacob's sons, the ancestors of the people of Israel; or the accusation was false, in which case Joseph was a liar. As a way out, Horowitz cites the tradition that *Sefer Yeẓirah* was written by Abraham and handed down to Isaac and then to Jacob and his sons. As the talmudic account affirms, this would have made it possible for them to create artificial animals to which the prohibition of unslaughtered meat would not apply; similarly, by following its instructions they could have created artificial women. . . . But Joseph, not knowing this, believed that what he had seen were women and animals with two natural parents!

created in this fashion. For many, the golem was a purely mystical phenomenon produced by meditation on the letters of the divine Name, it being understood that combinations of letters are insufficient to produce a material being.[31] The thirteenth-century itinerant kabbalist Abraham Abulafia, himself a great adept of meditation on the letters, mocked those who took the talmudic account literally and believed it possible to create a calf by studying *Sefer Yeẓirah*: "Those who do so are themselves calves," he wrote.[32]

31. Thus the eighteenth-century kabbalist Ḥayyim Joseph David Azoulai, quoted by Scholem, *On the Kabbalah and Its Symbolism*, pp. 188–189.

32. Ibid. The word *golem* has several meanings, which support diverse interpretations as to the nature of this artificial being. In the talmudic narrative, the being created by Rava is a man (*gavra'*), not a golem. The first use of the word with the sense of an artificial human being is much later. In its only occurrence in the Bible (Ps. 139:16), *golem* is variously rendered as "embryo," destiny," etc., and taken to apply to Adam, the "natural" man. In the Midrash and Talmud, *golem* designates an intermediate stage in the creation of Adam from the dust of the earth, before God breathed a living soul into him (Gen. 2:7, Naḥmanides on this verse, *Gen. Rabbah* 14:8), when he was still a shapeless mass (B *Sanhedrin* 38b), or, in other versions, a lump of clay with human shape but not yet alive (*Lev. Rabbah* 29:1). By extension, a woman who has never had sexual relations with a man is called a *golem*, in the sense of the raw material used by a potter. According to Maimonides (commentary on M *Avot* 5:7: "Seven traits [are characteristic] of a *golem* and seven of a wise man"), who is following the lead of M *Kelim* 12:6 and Rashi's commentary thereon ("unfinished metal vessels"), this sense of something crude and unfinished is the most general meaning of the word. The same root appears in the modern Hebrew *ḥomer gelem* 'raw material'. In the introductory section of a liturgical poem that recounts the ritual performed by the High Priest in the Temple on Yom Kippur (part of the Additional Service for that day), the word refers to the raw material from which the first man was created.

Thus from the outset the word *golem* expressed the ambiguity found in all its later occurrences: both an "inanimate human body," like a statue or a corpse, which is in some sense the raw material without the imposed "form," in the Scholastic sense of that which provides the essence and life; and an "animate human form"—Idel uses the term "anthropoid"—that is a copy of a real man or woman. For the Spanish kabbalist Joseph Gikatilla (1248–ca. 1325), a golem is a human body animated only by the *nefesh*, which is the part of the soul that is closest to the body and common to human beings and animals (a sort of vegetative soul that provides the capacity for movement), but which still lacks the affective soul and the intelligent soul (*Ginnat 'egoz* [Jerusalem, 1989], p. 208). Following this line, the question is whether Rava's creation was a "real" man. As we have seen, the answer was obvious as soon as R. Zeira discovered that the creature lacked the capacity for speech and consequently recognized it for an artificial being. But this is not always the case, as we shall see; there remains the theoretical possibility that one human being could fabricate a "real" human being who in every respect resembles a person with a father and mother. Thus the manufacture of an artificial human being raises the question of the nature or essence of man. It is in this context that the word *golem* designates a human being made by another human being, by means of the lore derived from a natural science that is held to be "true" because it is identical with that employed by the Creator to create the world. Most often, nevertheless, the synthetic being is an unintelligent brute, a caricature of a man or woman. This is the sense in which the Mishnah contrasts the *golem* with the wise man or sage.

But we cannot exclude the possibility that, thanks to its maker's knowledge, virtue, and purity of intentions, a particularly successful golem might be endowed with speech and intelligence. What is more, because we all know that human beings without rational intelligence—golems in that sense—can be produced in the normal fashion by a man and a woman, the question of the human or nonhuman status of an artificial anthropoid remains open. We are left with the "form" of the human body and face—the *ẓelem* that the golem reproduces no less than the authentic man Adam—as the only criterion of the humanity of a living being.

This permits assimilation of golem and *ẓelem*, of the matter of the body being formed with the form of the human body. Thus the "man" produced by Rava can be interpreted not as a golem in the sense of an artificial man, but as a symbolic representation of the world in the shape of a human being. This is the theme of the first man in Genesis, whose dimensions, according to the midrash, were those of the entire universe—what Idel calls a *macranthropos*. This theme was later taken up by the Lurianic kabbalah as the primordial man, *Adam Ḳadmon*, both created and creator, the goal of creation as well as the formative intermediary between the Infinite (the *Ein Sof*) and the infinity of worlds and formed beings, to the point that he himself is sometimes referred to as the *Ein Sof*.

Even some kabbalists who hewed to a literal interpretation of the talmudic text saw the "man" created by means of combinations of letters from *Sefer Yeẓirah* as no more than a robot—endowed, perhaps, with animality, but not with true humanity, as indicated by its inability to speak. This seems to have been the case, in particular, of the philosopher-kabbalists Solomon Ibn Gabirol and Moses Cordovero. It is interesting to compare this attitude with what can be inferred from the stories (with their interpretations) about Albertus Magnus, theologian, philosopher, alchemist, teacher of Thomas Aquinas, and the inspiration for Meister Eckhart and medieval Neoplatonism. On his deathbed, Albertus Magnus defended himself against charges of having practiced black magic but stated his desire to "verify several elements of this art" in order to know "whether it contained even the tiniest grain of truth," in his effort "to study thoroughly everything that the human mind has attempted" (see A. de Libera, *Albert le Grand et la philosophie* [Paris: Vrin, 1990], p. 12). Albertus too was said by legend to have created an "android" during thirty years of labor, "with the favor of heaven and the planets, a veritable speaking statue, a Venus of Ille <a reference to the story by Mérimée> or a golem," which his pupil, Thomas Aquinas, is supposed to have discovered and, horrified, shattered to bits. Contrary to this legend, however, Albertus was interested only in the theoretical aspects of the operation of an automaton, in connection with the wooden goddess that Daedalus was supposed to have made (ibid., pp. 13–14).

Finally, as Yehuda Liebes remarked in a review of Idel's *Golem* (*Ha'aretz*, Aug. 2, 1996, p. D2), it is not always easy to distinguish between the material sense of a golem, whether a "true" artificial man or a robot that is more or less androidal, and the allegoric or symbolic sense in which "making a man" may mean healing the sick or reviving those who seem to be dead, as in the story of the son of the Shunammite matron, restored to life by the prophet Elisha (2 Kings 4:34). Although the *Zohar* does not explicitly mention the fabrication of a golem, it does explain that the prophet effected this miracle by writing the divine name of 216 letters on the child, as in certain techniques for making a golem (*Zohar*, Introduction, 7b [trans. Matt, vol. 1, p. 48]). "Making a man" may also mean teaching someone the road to salvation, as in the traditional understanding of Genesis 12:5, which refers to the persons whom Abraham and Sarah "made" in Haran. This allegorical interpretation with a pedagogical sense may remind us of the myth of Pygmalion and its twentieth-century literary avatar. In Ovid's *Metamorphoses*, Galatea is an ivory statue; the sculptor falls in love with his handiwork and begs Venus to bring it to life. In George Bernard Shaw's *Pygmalion*, she is a flower girl whose horizons are expanded by a professor of philology. Finally, in the context of the kabbalists' mystical experiences of the supernatural, the metaphorical sense can be assimilated to the intrinsic sense if one accepts that sages operate in a world different from that of our normal experiences. What we consider to be symbolic they take to be real, like the alchemists, whose experiments in fact focused on themselves through the symbolism of their project of transmutation.

In all cases, for the kabbalistic tradition the manufacture of a human being by a human being, whether taken literally or metaphorically, is part of the sage's imitation of the divine creation; study of the words of the Torah (perhaps by means of the letters of *Sefer Yeẓirah*) associates this imitation with the unfinished work of perpetual creation. This is stated explicitly in the *Zohar*, shortly before the interpretation quoted above about the child restored to life:

> Of this is written: *I have put My words in your mouth and covered you with the shadow of My hand, to plant heavens and establish earth* (Isa. 51:16). The verse does not read *the heavens*, but rather *heavens* [implying that the reference is not to the existing heavens, but to new heavens].
>
> . . . [The verse continues:] *To say to Zion: "You are* 'ammi *'My people'*. Do not read *'ammi* 'my people', but rather *You are 'immi* 'with me', becoming My partner! Just as I made heaven and earth by speaking, as is said: *By the word of YHWH the heavens were made* (Ps. 33:6), so do you [by devoting yourself to the study of Torah]. (*Zohar*, Introduction, 5b [trans. Matt, vol. 1, pp. 27–28])

<Note that *'ammi* and *'immi* are consonantal homographs, both written עמי.>

In the ellipsis above, the *Zohar* explains that the "shadow" refers to the fact that God covers over the words of the Torah when they are placed in the mouth of Moses and the righteous, to protect them from the "jealousy of the angels"; the angels, envious of the material nature of the human body, try to prove that the body makes men weak and unable to receive the divine words, which will necessarily incinerate them. Whence the need for God to hide both human beings and the words until "a new heaven and a new earth" have been established (see Chapter 3).

33. B *Yoma* 72b. It is "a potion of life" (*sam ḥayyim*) for one who learns with "fear of heaven" or is a craftsman of the Torah (*'uman*) and rejoices in his learning. But it is a potion of death, and even of double death, for those who do not. As Rashi explains, such a person inherits hell twice; he loses this world by renouncing the common pleasures of life, but also loses the world to come because his learning is sterile.

Whatever the case, the creation of a golem, far from being sacrilegious, was seen as the fulfillment, by means of the secrets of the Torah, of a long and difficult ascent toward wisdom and holiness. This is, no doubt, a perilous path; but, like the Torah itself, it can be "a potion of life or a potion of death,"[33] depending on the student's orientation and intentions. Of course the golem may be more or less perfect, depending on the eminence of its creator. We have seen that Rava's golem was still quite imperfect because it lacked the power of speech; since it was not really a man—perhaps only a robot—R. Zeira had no scruples about destroying it. But we also hear of other golems whose makers were evidently on a higher level and able to endow them with the power of speech and sexuality.

In texts of the thirteenth-century German Pietists, cited by Idel, we read that the prophet Jeremiah created a golem. This was not an act of rebellion against God, but the culmination of a long upward path toward holiness and knowledge, which are necessarily coupled in any *imitatio Dei*. Indeed, how else can we know that an initiate has truly deciphered and understood the principles of creation other than by seeing that his knowledge enables him, too, to create a world? The only way we can be sure that his knowledge of human nature is correct is that it permits him to create a human being. The truth criterion of the sage's knowledge—just like the criterion of scientific truth today—is empirical.

Unlike Rava's golem, Jeremiah's artificial man could speak and soon rebuked the prophet: Was he aware of the confusion he might introduce to the world? From that day forth, whenever people encountered a man or a woman, they would no longer know whether it was God's creature or Jeremiah's! The prophet, who does not seem to have thought of this before, asked the golem how he could amend the situation. The artificial man replied that just as Jeremiah had made him, he could unmake him: "Reverse the combinations of the letters."[34] The lesson would seem to be that we need not renounce the attempt to attain the total knowledge that would give us the capacity to create a human being; once we have achieved it, however, we must abstain from applying it. This is a great lesson that merits serious reflection.

34. See Idel, *Golem*, p. 64.

First we must wonder about this confusion that the prophet Jeremiah, despite his sanctity and wisdom, had not foreseen, so that his own creation had to enlighten him. If the man he made was as perfect as a man conceived and born in the normal fashion, what was Jeremiah's mistake? What was wrong with what he did? As we shall see later, the fault lies in the ambiguous nature of this project of creative knowledge, which is simultaneously good and bad.

Jeremiah was not aware of this problem because, as in the Promethean projects of modern science and technology, the ambiguity vanishes from the horizon of the learned scientist, and even the sage, the moment that his creative activity carries him to the heights where evil no longer exists and, consequently, there is neither good nor evil. The success of the knowledge-driven ascent toward holiness and beatitude carries him, as it were, into a world in which the Tree of Knowledge fuses with the Tree of Life, where Adam's transgression can be rectified. It is a world out of time, a world of eternity, "the world to come" (*'olam ha-ba'*) rather than simply a "future world." With regard to our existence, which is in time, this flaw, inherent in the perfection of eternity, is expressed in part by the relationships with time, death, and imperfection that are intrinsic to the golem narratives: you cannot *kill* a golem, because, in the abstract eternity of the world in which it is fabricated, the golem's time is reversible. Instead, it is *unmade* by means of the same formulas and operations used to fashion it, carried out in reverse order; or by removing the letter *alef* from the word *'emet* (אמת) 'truth' engraved on its forehead, leaving the word *met* (מת) 'dead'. In other words, golem time, unlike human time, is reversible, just like the fully deterministic time of mathematical physics. By contrast, when randomness is introduced into a machine, when the determining causes are no longer under full control, time becomes irreversible; and that is the kind of time we experience in our lives.

5. THE SPARKS OF RANDOMNESS AND THEIR AMBIGUITY

This general idea can be refined if we think about the role of chance in the conception of a child as the result of an encounter between a man and a woman. What appears to be negative—the random nature of fertilization, the lack of control—and thus a defect, if our goal is a creation that is planned by perfect knowledge, becomes positive when we realize that such perfect knowledge, because never possible, must be an illusion. Thus the randomness of birth becomes valuable, something to be preserved if we want to hold on to even the bare possibility of human autonomy.

The fantastic tales about the sparks of randomness and the commentaries on these stories can illustrate this thesis, at the same time as they project us into an unaccustomed world where one speaks of chance and determinism in terms of spermatic knowledge.

The first exposition is found in the Talmud:

R. Jeremiah b. Eleazar . . . stated: "In all those years during which Adam was under the ban [in punishment for eating the fruit of the Tree of Knowledge] he begot spirits and male demons and female demons, for it says: 'Adam lived 130 years and begot a son in his own likeness (*ẓelem*), after his own image (*demut*) (Gen. 5:3)'—from which it follows that until that time he did not beget in his own image."[35]

That is, as we also find in a midrash,[36] during those years he had produced offspring—demons and spirits—who were not in his own image. In fact, it is only with regard to Seth, the third son, that the biblical text invokes Adam's original divine image and form; they are not mentioned in connection with the birth of Cain and Abel. What is more, their birth is referred only to Eve—"*she* conceived and bore, . . . *she* bore" (Gen. 4:1–2)—as if the father, Adam, had nothing to do with it. By contrast, with regard to the birth of Seth, Genesis 5:3 says that "Adam . . . begot." Building on this, the midrash suggests that Cain and Abel, the children of that first disobedience relating to the Tree of Knowledge, were conceived from a mixture of Adam's semen with the serpent's; and that, like the demons just mentioned, their form was not really (though it appeared to be) that of the *ẓelem,* the original form in which Adam was created according to the first chapters of Genesis.

In another place[37] Rashi explains what these various sorts of creatures look like: spirits (*ruḥot*) have no form, male demons (*shedim*) look like human beings, and nocturnal female demons or succubae resemble women with wings—duplicates of their mother Lilith, mentioned by the Talmud.[38] We may add that, according to the midrash, the demons fathered by Adam during his 130-year separation from his wife were themselves the fruit of nocturnal encounters with succubae, which come "to heat the heart of men in their sleep"—just as, to keep things symmetrical, male demons come to heat the heart of women. For both men and women these encounters produce, often involuntarily, the "pleasures of human beings" (*ta'anugot benei ha-'adam*), mentioned in Ecclesiastes 2:8, again in the company of male and female demons, as we shall see below. These pleasures can be dangerous. They may turn into suffering and torment—that is, *'oneg* (ענג) 'pleasure' becomes its anagram *nega'* (נגע) 'torment, wound, blow'[39]—because they are accompanied by an emission of semen that does not find a human partner. This pleasure-pain in turn produces new spirits, male and female demons, that torment human beings and cause them serious injury. But, as we shall see, there is an ambiguity in all of this, both good and bad, part of a vast mixture of good and evil, like the Tree of Knowledge at the start of our story. Recall that the biblical phrase normally rendered "the tree of knowledge of good and evil" can also mean "the tree of knowledge, [which is both] good and bad." This latter reading is often evoked by the Midrash.

35. B *'Eruvin* 18b.
36. *Gen. Rabbah* 23:6 and 24:6.
37. On B *Sanhedrin* 109a.
38. B *Niddah* 24b.
39. *Sefer Yeẓirah* 2.4; see below, §10.

The talmudic text from tractate *'Eruvin*, quoted above, reflects this ambiguity by means of an objection raised in the name of Rabbi Meir, who comes to Adam's defense:

Adam was very pious [*ḥasid gadol*, with the connotation of "great ascetic"]. When he saw that on his account death had been ordained as a punishment he fasted for 130 years, separated from his wife for 130 years, and wore clothes of fig [leaves] on his body for 130 years. [There is no contradiction with R. Jeremiah ben Eleazar's statement quoted above.] That statement referred to the semen that he emitted under duress.[40]

The entire talmudic passage is developed by the Midrash and by the *Zohar*. In the Lurianic kabbalah this astonishing tale takes on the dimensions of a cosmic drama that runs through all of human history.

Let us come back to the transformation of *'oneg* into *nega'*, of pleasure into pain. In a different context, a midrash applies it explicitly to a verse in the book of Samuel, in which the Lord tells David about the destiny of his son Solomon in his old age: "I will chastise him with the rod of men and the afflictions of human beings (*nigge'ei benei 'adam*)" (2 Sam. 7:14).[41] This phrase is symmetrical with that in Ecclesiastes (2:8), which also relates to Solomon, its author, but where he mentions, instead, the pleasures he has experienced. Those *ta'anugot benei ha-'adam* are precisely the pleasures transformed into afflictions (*nigge'ei benei 'adam*) that the Midrash and *Zohar* materialize as the drops of semen, initially lost at random in the course of Adam's pleasure—drops of chance, or sparks of randomness, *niẓoẓot shel ḳeri*, as the text refers to them—and then transformed into demons, which never stop tormenting his children. King Solomon, the wisest of men, is a leading actor in this drama. His wisdom allowed him to perceive, more than anyone else, the fundamental ambiguity of knowledge, which is both good and bad, precisely the knowledge misappropriated by Adam and Eve, prematurely and under the influence of the serpent, and moved away from the Tree of Life, thereby reinforcing the bad side. This is why the same legend is referred directly to Solomon in a different context, when he is the prisoner of the king of the demons, Ashmedai (Asmodai, Asmodeus), in a fantastic talmudic narrative that Rashi cites in his commentary on 2 Samuel 7.[42] He explains that the "afflictions of human beings" is an allusion to "Ashmedai, who expelled Solomon from his kingdom," as recounted by the talmudic legend in which Ashmedai, king of the demons, overcame Solomon through a ruse and, assuming his appearance, usurped his throne. Rashi goes on to explain the legend further: "As for the demons, these are the sons of Adam; for during the 130 years that Adam was separated from his wife, after the death of Abel, the female spirits conceived and bore children by him."

We see how Rashi, in his characteristic brevity, combines the two main elements of this story of demons and pleasure, beginning with the drops of semen that Adam emitted after his failed experience of knowledge. The *Zohar*[43] employs the same material in a different fashion and explicitly refers to these drops of semen as random emissions, *ḳeri*, employing, as we shall see, the classic terminology of the Mishnah. We need to delve into the details of these commentaries on different aspects of the biblical story of the Tree of Knowledge, good and bad, on the fundamental ambiguity of our knowledge, which is both divine and demonic, both source of life and source of death. We shall return to the extraordinary adventures of Ashmedai

40. B *'Eruvin* 18b.
41. *Gen. Rabbah* 24:6.
42. B *Giṭṭin* 68.
43. *Zohar*, Genesis 19b [trans. Matt, vol. 1, p. 150].

and Solomon, which conclude—could it be otherwise?—ambiguously, with alternative endings: did Solomon regain his throne or not? Is the denouement the blessed reign of wisdom or the final triumph of cunning concealed beneath the robes of the sage? This will lead us on several long detours, following various interpretations, among the many the rabbinic literature offers, of the foundational myth of the Tree of Knowledge.

6. THE GENERATIONS OF THE FLOOD, OF THE TOWER OF BABEL, AND OF THE WILDERNESS

We begin with texts of the *Zohar* and Lurianic kabbalah that contribute a new dimension to this theme by adding new protagonists: the generation of the Flood, the generation of the tower of Babel, and finally Moses and the generation of the wilderness, also known as the "generation of knowledge" (*dor de'ah*).

Consider this passage from *Sha'ar ha-kavvanot* (Gate of Intentions), by the sixteenth-century mystic Rabbi Ḥayyim Vital, the leading disciple of Rabbi Isaac Luria in Safed and editor of the fundamental texts of the Lurianic kabbalah:

> Know that the Israelites of the generation of slavery in Egypt were of the nature of the sparks of randomness that Adam emitted during the 130 years that preceded the birth of Seth, as our rabbis have taught. These sparks were reincarnated in the generation of the Flood; this is why they too spilled their seed on the ground, like the root from which they were created and hewn, until they were blotted out [by the Flood]. As is written: "YHWH saw how great was the wickedness of man [*or* Adam (*ra'at ha-'adam*)] on the Earth" (Gen. 6:5); and, as is known, spilled semen (*ḳeri*) is called "evil" (*ra'*) and one who spills it is called "wicked" (*ra'*). . . . The generation of the Flood was literally *ra'at ha-'adam,* the wickedness of *the* man, meaning Adam the first man, as is written: "I will blot out the man (*ha-'adam*) whom I created" (Gen. 6:7); that is, Adam himself, who was created by God himself. And these [the generation of the Flood] are the sparks emitted by him and which God wanted to blot out. Then these same sparks returned and were reincarnated in the generation of the dispersion [from Babel]; of them it is written, "YHWH came down to see the city and the tower, which the sons of man (*benei ha-'adam*)"—specifically the sons of Adam—"had built" (Gen. 11:5). They are called the "sons of Adam" to imply that they were drops of semen. Consequently they are the sons of the man—the male without the female.[44] Then, too, they sinned further by rebelling against the Most-High.[45]

Other texts from the same corpus present variations on this idea, such as the reincarnation of these demons in the residents of Sodom and Gomorrah; or, before the Flood, in the generation of Enosh, the son of Seth and grandson of Adam, the generation that "began to call upon the name YHWH" (Gen. 4:26)—an initiative that is understood, somewhat curiously, in a pejorative sense (another example of the ambiguity of good and evil), as relating to thaumaturgy and idolatry that exploit the power of the divine name to work magic. Although the name invoked was that of the God of Moses and Israel—the Tetragrammaton—here their action is interpreted as injurious blasphemy.[46]

44. Later we shall look at the ambiguity of this property, which is concretely "bad" but also, at its source, the origin of some of the highest values of all.

45. Ḥayyim Vital, *Sha'ar ha-kavvanot* (Jerusalem: Peki'in, 5735), p. 79a.

46. <The interpretation is based on reading the Hebrew *huḥal* as if derived from *ḥallel* 'profane' rather than *haḥel* 'begin'.>

In all variants of this narrative, some elements remain constant. First of all, the word *ḳeri* 'randomness' (as "drops of randomness" or "sparks of randomness") is always applied to the drops of semen emitted by Adam while he was separated from Eve. The Mishnah[47] uses the word *ḳeri* to mean semen emitted outside a woman's body, whether voluntarily or involuntarily (see Maimonides' commentary on this mishnah). The word comes from the same root as *miḳreh* 'event', which may also be accidental or random.[48] The randomness of semen ejaculated without a women to receive it is a priori bad, as the commentators highlight in the case of Er, Judah's firstborn, who spilled his seed to avoid having a child by Tamar; a practice repeated by his brother Onan, whose name has come down through the centuries to designate this act. *'Er* (ער) is an anagram of *ra'* (רע) 'bad' or 'evil'; Genesis 38:7 declares that Er was *ra'*—wicked—in the sight of YHWH. In a similar vein, Vital understands the wickedness of the generation of the Flood, when "all flesh had corrupted its ways on earth" (Gen. 6:5, 12), to be their practice of random *ḳeri*, of seminal emission without a woman to receive it.[49] Here, though, we already see the ambiguity that will be evident later, because this randomness, bad in the cases of Er and of the generation of the Flood, is not the same as the *ḳeri* of Adam or *ḳeri* in general; it was voluntary and controlled, a conscious and deliberate emission of sperm outside a woman. Thus it is controlled randomness (if that is possible), as opposed to *ḳeri* per se, which is an uncontrolled, random, and unconscious emission during sleep, and also unlike the *ḳeri* of Adam, who, according to Rabbi Meir, quoted above, was a pious ascetic and an unwilling party to his encounters with the succubae.

By contrast, the recurrent theme of all of these narratives is that of a cosmic drama in which all human history is oriented toward the progressive development and repair (*tiḳḳun*) of these sparks, which were lost in demons and reincarnated in the wicked generations—notably those of the Flood and Babel—before being recovered and saved in their final incarnation as the generation of the Exodus (the generation of the wilderness). When that generation received the Torah, which is both the Tree of Life (Prov. 3:18) and the Tree of Knowledge, the first transgression (eating the fruit of the Tree of Knowledge, a mixture of good and bad) was rectified by being transformed into knowledge that is only good and in which death, as a curse, no longer exists.

But the sin of the Golden Calf relaunched the cycle and once again made it necessary to refine and concentrate these scattered sparks, generation after generation (just as gold is refined in a crucible, little by little, to remove the dross—another image employed by Vital), until its culmination in Solomon's construction of the Temple in Jerusalem. Then sin returned again, leading to the destruction of the First and then of the Second Temple. Each act in this drama, each failure, is followed by an attempt at repair, which meets with only partial success, after which the cycle is repeated again. Only in the future messianic and post-messianic ages, which are by definition outside historical time, can the *tiḳḳun* be accomplished fully.

47. M *Berakhot* 3:3; cf. B *Berakhot* 20b, 22b, 26a.

48. The word *ḳeri* is found in the Bible only in Leviticus 26, in a very different context. The standard translations, obscured by theological prejudices, render it there in various ways. See Chapter 6.

49. <This link is based on the verbal echoes between the two verses: "all flesh had corrupted (*hishḥit*) its ways on the earth (*'al ha-'areẓ*)," while Onan "destroyed (*shiḥet*) on the ground (*'arẓah*)" (Gen. 38:9). Because *hishḥit* and *shiḥet* are forms of the same root, when he conflated the verses Vital could write that the generation of the Flood were "corrupting" (*mashḥitim*) their seed on the earth/ground (*'al ha-'areẓ*).>

7. ASHMEDAI AND SOLOMON: TRICKS OF NATURE AND TRICKS WITH NATURE

Along the way, we again encounter the heroes of our adventure of knowledge, brought together this time by the wisdom of King Solomon. According to a legend (B *Giṭṭin* 68), Solomon, despite his great wisdom and vast knowledge, required the assistance of Ashmedai, the king of the demons, to build the House of Peace—the Temple of Jerusalem. Here we can only summarize the talmudic narrative, of which every detail is extremely suggestive. Solomon needed Ashmedai because the ashlars used in the House of Peace (just as his name *Shelomo* means "his peace") were not to be quarried or dressed using iron, the material from which weapons are made. The only alternative was to employ a certain mysterious element or worm, the *shamir*, which only the king of demons knew how to find.[50] Note that this Ashmedai, whose name, probably of Iranian origin, resonates in Hebrew of both destruction (*shemad*) and the divine name *Shaddai*, is a "good" demon because, according to a fundamental principle to which we shall return, evil at its source is not evil. Ashmedai himself, as *king* of the demons, cannot be bad, even though both maleficent and beneficent demons are produced by his power. He passes his time, incidentally, sitting in the talmudic academy engrossed in study, sometimes in heaven and sometimes on earth.

In this way Ashmedai establishes a bridge between the lore of the upper world and that of the lower world. Here too, however, there is ambiguity, depending on the direction of the traffic: it may make the experiences of life intelligible and lead to wisdom and liberty; or it may exploit wisdom as an instrument for trickery and control, as a way to cause others and oneself to dissolve into and be lost in the unending and unknown chain of causes. As I have shown elsewhere,[51] this "descent of the wisdom of the left side" is the kabbalistic equivalent of *metis*, the cunning intel-

50. The *shamir* was used to quarry and dress stone without the use of iron, in keeping with the biblical prohibition: "And if you make for Me an altar of stones, do not build it of hewn stones; for by wielding your sword on it you profane it" (Exod. 20:22 [25]). This warning is uttered after Israel has received the Ten Commandments, as if to say, "If you want to make this law reign over earth, do not use the sword." Only a house of peace (*shalom*) and completeness (*shelemut*) is appropriate. This accords with why King David was not allowed to build the Temple. He had been a warrior; but the House of Peace could be realized only by a king who ruled in peace, like his son Solomon (Shelomo). To build this house using stones that had to be cut precisely so they would fit together, the builder had to figure out a way to leave the stones whole, even while cutting them! This dilemma is even clearer in Moses' use of the *shamir*, alluded to in the same talmudic passage and expanded on elsewhere (B *Soṭah* 48b): engraving the names of the twelve tribes on the twelve gemstones of the breastplate of the High Priest. The gems had to be whole, even though the name was engraved on them. <This is learned from "in their mountings . . . engraved like seals" (Exod. 28:20–21), where *be-millu'otam* 'in their mountings' can also be understood as "in their fullness," i.e., with nothing removed.> Because ordinary engraving would have cut part of the stone away, "he wrote [the name] on them with ink and showed them [the written strokes] to the *shamir*, and they [the stones] split of their own accord (like a fig that splits open in summer and nothing is lost)." The High Priest used the gems on the breastplate to interrogate the oracle of the *Urim ve-Tummim*, a biblical practice that disappeared, like the *shamir* itself, with the destruction of the First Temple.

For both the ashlars of the Temple and the gemstones of the breastplate, the integrity of the whole had to be reconciled with the specific characteristics of the part. Peace means that the whole is entire in each of its parts, even though the latter have been cut and carved. The whole must be cut, yet left intact. This is what the *shamir* does. Only the king of the demons knew where it could be found, because cutting without dividing is possible only in the higher worlds, where division (the work of demons) does not separate. For the oracle to be prophecy and not devilish sorcery, it can be queried only by the High Priest in the House of Peace, making use of these stones that are both intact and engraved (see Chapters 3 and 4).

51. Atlan, *Enlightenment*, pp. 124–127 and p. 140 n. 103. We will meet this image again in several venues (see below, n. 68 and Chapter 3). For now, it suffices to apply it to our present theme, the ambiguity of knowledge. For human beings, in our concrete world of action and perception, knowing cannot

ligence of the Greeks.[52] On the arid mountain where he lives, Ashmedai has dug a well that he keeps sealed when he is in heaven, because it is only when he is on the Earth that he needs water to quench his thirst. This frailty is what makes it possible for Solomon's agent to capture him: he replaces the water with wine, the wisdom "of below," and binds the demon in chains when he falls asleep.

To find Ashmedai and force him to yield his secret, Solomon summons Benaiahu ben Jehoiada. His name (or rather his father's) is itself significant, combining a divine name (the first three letters of the Tetragrammaton) with *yada'*, from the verb "know."[53] Benaiahu employs a stratagem to fill Ashmedai's well with wine; the thirsty demon drinks from it and falls asleep. Then

take place in the shadow of the ambiguities of good and evil, of life and death, of happiness and suffering, of well-being and discomfort, both our own and others'; this is true even if we admit that some persons are motivated by the disinterested quest for truth. For knowing proceeds through and bears on the non-utopian reality of our senses and of the multiplicity of beings and species, whose coexistence in a finite world is necessarily conditioned by division. This is a form of violence, even if it is contained and attenuated, or sublimated in love and the fascination with others that may accompany the sense of solidarity and compromise.

This ambiguity of knowledge, which is the theme of the Tree of Knowledge, good and bad, is stated explicitly in the *Zohar*'s subtle distinction between motion "downward" and motion "upward," which are appropriate respectively to the wisdom "of the right side" and the wisdom "of the left side" (see, e.g., *Zohar*, Genesis 65a [trans. Matt, vol. 1, pp. 378–380], and especially the *Sullam* commentary by Rabbi Judah Ashlag, §§121ff., to be analyzed in Chapter 3).

Wisdom of the right side—which is the side of generosity and love, of free and unlimited giving—is an ideal form of knowledge in which the light of understanding spreads without restriction from its infinite source to an intelligence able to receive it and to retransmit it without distortion. To the extent that it can appear in our world of action and finitude, it does so from "above to below," that is, from the world of the possible to the world of the actual. This also implies a certain violence, jostling the actual out of its stasis and solidity; but this violence is in the service of love. The metaphor suggested is that of the "violent forces of rain" (*gevurot geshamim* [B *Ta'anit* 2a]), a discharge of water that holds the promise of life and abundant harvests in the future, but whose vehemence disrupts the tranquility of the present. This violence is how love as a gift, infinite but also undefined and undifferentiated—*mayim* 'water(s)' can be understood as the dual of *mah* 'what', as potential—bursts into the reality of the present. Put another way, this violence is good; it is good when the wisdom of the right side descends in this fashion from above to below and human beings strive to absorb it.

But this wisdom of the right side, which is luminous and straight, nevertheless lacks what it needs to penetrate and illuminate everything that exists and its finite details, differentiated by the uniqueness of each being. It fails to confront another form of violence, much more essential, the violence of multiplicity and separation that love can only attenuate but not eliminate. That is the task of the wisdom of the left side, for which nothing is simple or straight; all is ruse and stratagem, which can be opposed only by still greater deception. The first trap is set by this wisdom itself, which is adapted a priori to each unique entity or quality in its finitude, and thus able to elevate each finite being into an absolute, into perfect rigor, unaware of any opening to another, to generosity or love. But that is not so bad as long as this wisdom, thus multiplied to infinity, remains in its world "above," as an infinite power in which each individual being (like Leibniz's monad) is also coextensive with the infinite, so that the differences play only the role of *possible* (but not yet actualized) sources of violence. Put another way, wisdom of the left side, as the possible source of violence, already ambiguous, both good and bad, is dangerous only if one falls into the second trap: that of making it descend "from above to below" in order to actualize it. This is the trap into which Adam fell, initially motivated by a praiseworthy idea—permeating worldly objects and beings with the light of knowledge, knowing them as they really are so that they can be given individual names, precisely what he did when he assigned each animal its true name (Gen. 2:19–20)—but misdirected in the end by the serpent. As trickery-nudity and divination-oracle, both *'ormah* and *niḥush*, the naked serpent (*naḥash 'arum*) incarnates the wisdom of the left side that descends from above to below (see below, n. 68 and Chapter 3).

52. Marcel Detienne and Jean-Pierre Vernant, *Cunning Intelligence in Greek Culture and Society*, trans. Janet Lloyd (Hassocks, Sussex: Harvester Press, 1978). In the story of Ashmedai we find the trope of chaining analyzed by these authors, where, like trapping someone in a net, the trick is made even more effective (see also below, n. 63).

53. Benaiahu ben Jehoiada appears at the end of the book of Samuel and in the first chapters of 1 Kings. According to the simple and obvious meaning of these passages (the *peshat*; see Atlan, *SR* II, chap. 12), Benaiahu, one of David's corps of heroes, helped Solomon ascend his father's throne and was then appointed commander of the army. The Talmud (B *Berakhot* 18) reads the four verses (2 Sam. 23:20–23) that introduce him and his exploits in David's service at other levels of meaning, the allusive sense (*remez*) and the allegorical sense (*derash*). The *Zohar* (Introduction, 6b–7a [trans. Matt, vol. 1, pp. 38–43]) adds an interpretation according to the esoteric sense (*sod*), in which the meaning of his father's name, "God knows," is a key element. "The name prevails" (i.e., influences its bearer), states the *Zohar* by way of introduction, and it then proceeds to expound these verses at length. We learn that the name designates the "hidden wisdom" found

Benaiahu binds him in a chain that has the name of God, Ashmedai's master, engraved on it; this prevents the demon from employing his powers to release himself. En route to King Solomon's court, the demon gives impressive evidence of both his physical strength and his knowledge of hidden things, a result of his capacity for divination. He tells Solomon where the *shamir* is to be found, guarded by a fantastic bird.[54] When Benaiahu manages to make off with the *shamir* the bird kills itself in despair, because it had sworn to return the *shamir* to the "Lord of the Sea" from whom it had borrowed it. Finally, Ashmedai, still a prisoner, tricks Solomon into releasing him from his chains and swallows the king's seal. He assumes the monarch's appearance and usurps his throne, while Solomon is reduced to wandering from place to place and repeating, "I was king in Jerusalem"—a verse from Ecclesiastes (1:12), which the legend places in the mouth of the dethroned king whom no one recognizes. The ending is uncertain. In some versions, the counterfeit king is unmasked by his scandalous behavior with women, including Bathsheba, the mother of the true king, and Solomon returns to his throne; in others, he remains a beggar forever.

Like all talmudic legends, this intricate story, with many details omitted here, has spawned multiple interpretations that exceed the scope of the present discussion. What is relevant for us is that King Solomon, the wisest of all men, must engage in parlous commerce with demons in order to acquire knowledge of the esoteric and invisible, knowledge that is indispensable for building the Temple—that is, for repairing and re-establishing order in the world of human beings and inanimate objects.[55]

What is more, we have seen that one of the verses that mentions the "pleasures of the sons of man," frequently invoked in our tale of the sparks of randomness, is in Ecclesiastes, the book

at the apex of the *sefirot*, associated with the "the head that does not know and does not make itself known" (see below, §11).

This wisdom "does not make itself known"; the Tetragrammaton itself is present only in a hidden and implicit form, as its first three letters YHW. But, as the name indicates, it knows (*yada'*). The "head" from which it proceeds, on the contrary, "neither knows nor makes itself known." Infinite, it is neither thought nor desire (*Zohar*, Genesis 65a), which implies that it is concealed from what is, for us, desire and thought. The *Zohar* interprets this knowledge as sexual knowledge, the union of male and female (*zivvug*); in the realm of esoteric wisdom, the union of knowledge does not bear fruit, because the vessels or vehicles that could receive them in the world "below" are not ready. It is only secondarily, when this knowledge can be poured into beings of the world below, permitting them to unite and know in their own turn, that it emerges from its source in esoteric wisdom. In the latter, the union takes place without a receptacle and its effect is not always beneficent for beings of the world below, at least not at first. This passage attributes the destruction of the two Temples to this return of the esoteric wisdom into itself and its abandonment of the wisdom that "makes itself known."

As for the head that "neither knows nor makes itself known," we shall see below that its status and activity are defined by an "infinity of doubts" that contains the infinity of possibilities before any process of actualization begins.

54. A hoopoe, a brightly colored bird that will use just about anything to build its nest. In our story it employs the *shamir* to cut through a pane of glass that Benaiahu has placed on top of its nest.

55. Among other later interpretations with multiple ramifications, we encounter Ashmedai in another Lurianic work by Ḥayyim Vital, *'Ez ḥayyim* (Gate 49, chapter 6), where he is referred to as a "Jewish demon," a strange expression that derives from the *Zohar* (Deuteronomy 227a; see below, n. 78). There the term is applied to an aspect of the angel Metatron, who is sometimes an angel and sometimes a demon, "depending on the actions of human beings." The angel Metatron also constitutes the bridge between "upper" and "lower" knowledge. It is he, according to the talmudic legend of the Pardes or garden of knowledge, who misleads Elisha ben Abuya, the one of the four sages who came out of the garden and "cut down the saplings," into a sort of desperate nihilism (B *Ḥagigah* 15a). He is also the one who, according to the apocryphal book of Enoch, guides the initiates who ascend to the "palace" on high. Benaiahu and Solomon's adventures with Ashmedai and the guardian bird of the *shamir* represent the other side of an initiatory voyage into the garden of knowledge, where they encounter Ashmedai, the king of the demons, the other side of Metatron, the "prince of the face" or viceroy of the upper realm, "whose name is that of his master."

in which the aging Solomon rendered a balance sheet of his experiences and his wisdom. It is instructive to consider the wider context of this verse and the interpretations it has inspired. Solomon says:

And I applied my mind to know wisdom and to know madness and folly. I perceived that this too is running after wind. . . .

I also amassed silver and gold, the treasure of kings and provinces; I got singers, both men and women, and the pleasures of the sons of man, [female] demon and [female] demons (*shiddah ve-shiddot*).

So I became great and surpassed all who were before me in Jerusalem; yet my wisdom remained with me. Whatever my eyes desired I did not keep from them; I kept my heart from no pleasure, for my heart found pleasure in all my toil, and this was my reward for all my toil. Then I considered all that my hands had done and the toil I had spent in doing it, and behold, all was vanity and running after the wind, and there was nothing to be gained under the sun.

So I turned to consider wisdom and madness and folly; for what can the man do who comes after the king? Only what he has already done. Then I saw that wisdom excels folly as light excels darkness. The wise man has his eyes in his head, but the fool walks in darkness; and yet I perceived that one fate (*miḳreh ʾeḥad*) will befall everyone (*yiḳreh ʾet kullam*) [i.e., all will encounter a random destiny, with no particular meaning or intention]. Then I said to myself, "The fate (*miḳreh*) of the fool will befall me also; why then have I been so very wise?" And I said to myself that this also is vanity. (Eccles. 1:17, 2:8–15)

This passage from Ecclesiastes sets the context of wisdom, knowledge, folly, and absurdity in which its author refers to "the pleasures of the sons of man, demon and demons." But the word rendered here "demon/demons," *shiddah/shiddot*, is obscure, and many other translations have been proposed for it—in the Talmud, right before it recounts the legend of Solomon and Ashmedai,[56] and in the commentaries by Rashi and by Abraham Ibn Ezra. The Talmud understands *shiddah* in the rather bland sense of "box" or treasure chest; Rashi, building on this, suggests "carriage" or some kind of luxurious box-shaped equipage. Ibn Ezra proposes "beautiful women," in the sense of captives of war (related to *shod* 'spoils') chosen for their beauty; he justifies this rendering by the context because, in the list of pleasures that Solomon is reckoning, this form of gratification, although certainly among the most important of all, would be curiously absent were this term translated in some other fashion.

In the end, the Talmud proposes that we understand *shiddah* as the feminine of *shed* 'demon' and take it as an allusion to Solomon's adventures with the king of the demons (reviewed above), with its many details relating to various aspects of wisdom, knowledge, and intoxication.

This is also the sense assigned by the passage cited above from Vital's *Shaʿar ha-kavvanot* and by the *Zohar* when it uses the anagram of *taʿanugim* 'pleasures'–*negaʿim* 'afflictions' to link this verse to the demon spawn of Adam's sparks of randomness.[57]

56. B *Gittin* 68a.

57. After referring to the 130 years during which Adam engaged in sexual commerce with female spirits, the text continues:

When King Solomon descended to the depth of the nut, as is written, *I descended to the nut garden* (Song of Songs 6:11), he took the shell of a nut, contemplated all those shells, and realized that all the joy and delight of those spirits [which inspire sensual desires in human beings]—shells of the nut—consists solely in clinging to human beings and leading them astray, as is written: *The delights of the sons of men [are enjoyed by] male and female demons* (Eccles. 2:8). Further [as can be learned from the same verse], *the delights of the sons of men*, in which they indulge asleep at night, generate demons.

[God] had to create [all these things in order to repair] the world

8. THE TREE OF KNOWLEDGE, GOOD AND BAD

To catch a glimpse of even part of the light that these commentaries can shed on the biblical account of the Garden of Eden and its two trees, of life and of knowledge, we will have to make a long detour (in the next chapter) in pursuit of a better sense of how their authors understood Adam's disobedience concerning the Tree of Knowledge: not the Tree of Knowledge of Good and Evil but rather a tree that is itself good and evil, good and bad. The core of all these accounts is the fundamental ambiguity of knowledge. To complicate matters, it is extremely difficult to unravel the tangled levels, normally separated, that are mixed up so disconcertingly here, because we are always being referred from one register to another, from the abstract to the concrete, from intellectual knowledge to "biblical" (sexual) knowledge.

In the meantime we must make do with jumping from one domain of meaning to another, with strange confusions that are readily grasped only through metaphor. In this process we will be assisted—or perplexed—by the names of the main characters, because they are words that oscillate between different domains of meaning, with unexpected jumps between the concrete and the abstract, between the sexual and the intellectual, always playing in the dual register of *da'at*, biblical knowledge, which is simultaneously thought, emotion, and sex.[58] The Greeks, too, had this dual register

by means of these shells. For all consists of a kernel within, with several shells covering the kernel [*or* "of a brain with many membranes covering it"]. The entire world is like this, above and below, from the head of the mystery of the primordial point to the end of all rungs: all is this within that, that within this, so that one is the shell of the other, which itself is the shell of another. (*Zohar*, Genesis 19b–20a [trans. Matt, vol. 1, p. 151])

In this scheme we find the general arrangement that is referred to, probably under Neoplatonic influence, as "emanation." In the kabbalistic tradition, however, this word applies specifically to one of the worlds (*'azilut*), a divine world, the highest of all, where the shells or husks are not bad because, as we shall see, evil at its source is not evil (for the more precise meaning of *'azilut* see below, nn. 68 and 78).

As we have seen, both the Talmud and Rashi render *shiddah* in a more ordinary fashion, as "box," "chest," or "carriage." But the loop is closed in a fascinating allusion in Rashi's commentary on Ecclesiastes, referring not to beings but to objects associated with the various pleasures that King Solomon is supposed to have experimented with. This rendering of *shiddah/shiddot* as objects rather than as persons returns us, delightfully and unexpectedly, to the main theme of our story. We also encounter the word *shiddah* in two passages in tractate *Shabbat* (120a, 122b) that apparently have nothing to do with our subject, but which Rashi cites in his commentary on Ecclesiastes. There (and in several other talmudic passages) *shiddah* certainly has the sense of "chest" but is associated with two other objects, one appropriate to the Flood, the other to Babel. One of them is *tevah* 'box', which is also the biblical word for Noah's ark; the other is *migdal*, taken in the talmudic context as referring to some sort of cabinet, but whose normal sense is "tower." Here the link between the demons produced by Adam and the generations of the Flood and of the Tower of Babel appears symbolically at the level of objects; it is the box of pleasures—Pandora's box (which in Hesiod is actually an earthenware jar)—that is recapitulated by Noah's ark/box and reproduced in another fashion by the tower/cabinet of Babel.

58. The *Zohar* (Deuteronomy, *'Idra Zuṭa* 291a) takes the plural *de'ot* in the phrase "god of knowledge(s)" (*'el de'ot* [1 Sam. 2:3]) as indicative of three categories of *da'at* or knowledge: one that is hidden and unattainable; another that is veiled and relates only to "the head"; and a third, derived from the former, "that spreads through the body" and "fills the chambers," as in "the rooms are filled with knowledge" (Prov. 24:4). This is the key to the meaning, for ancient societies, of one of the negative assessments of women that can be found, alongside frequent praise for them, in the Talmud: *nashim da'atan ḳallah 'aleihen* (B *Shabbat* 33b and *Ḳiddushin* 80b), frequently (mis)understood to mean that women are weak-minded and of light (*ḳallah*) intelligence. In fact, this judgment does not refer to their intellectual faculties—elsewhere we are told that God endowed women with more understanding than men (B *Niddah* 45b)—but rather, as the immediate context of the two passages shows, their (supposed) lesser capacity to resist torture or temptation. Denominating that capacity as "knowledge" implies that this type of strength of character or moral resistance is considered to be both a cognitive trait and an emotional trait; the link between knowledge of what is good and the emotions is strong enough to empower resistance to whatever is contrary to that knowledge. The focus here is on feminine emotions and sexuality—as viewed by the male-dominated societies of past generations—and not on women's intellectual capacity. This lightness of knowledge is responsible for Eve's yielding to the seduction of the serpent and everything that ensued. The opacity of the human body, whose original integument of *'or* 'light' (אור) was replaced by its near homophone *'or* 'skin' (עור), corresponds to the replacement of consummate and glorious knowledge by shameful and "evil-smelling" sexuality (Isaiah

in mind when they spoke of the *logos spermatikos* or seminal reason. But Platonic idealism and Aristotelian intellectualism, the wellsprings of philosophy, made a powerful contribution to the disjoining of this association, to the point that it has become quite unintelligible and inconceivable for us except as a metaphor. No doubt the mythical Eros of cognition and knowledge has always permeated philosophy, which is the "love" of wisdom. But the relationship between sexual knowledge and intellectual knowledge quickly degenerated into a mere metaphor. The two modes of knowledge were sundered; their unity was thought about and experienced less and less, as the soul was progressively detached from the body. In the *Phaedrus* and the *Symposium* Plato praises homosexual love at length, evidently as a privileged form of philosophical knowledge. In the beauty of his beloved disciple the wise man finds the road that leads to knowledge of Beauty, which is itself the preferred route to knowledge of the True and the Good. For Plato, this love releases the soul from its corporeal prison so that it can contemplate the Ideas. Love, ideally, should remain "platonic"; that is, not only has the woman been eliminated from this arrangement, but the wise man will also overcome his carnal impulses and refrain from consummating his love. Later, the feminine side was reintroduced in the form of nature, viewed sometimes as a nurturing mother, sometimes as an object of desire, which the wise man, according to Bacon, must make his own, penetrating its secrets in order to master and dominate it. But all of these are only metaphors borrowed, in each case, from the then-current social images of gender differences and the division of labor.[59]

If we confine ourselves to a purely intellectual register, the relationship between knowledge and randomness is quite clear. Randomness is, first of all, a deficiency of knowledge. It is what we face when we do not know causes, or, at least, when we do not know them well enough to render a full explanation of an event. Our inadequate knowledge of causes in nature is our first experience of what we call "chance." Proceeding from there, we can imagine ontological chance, in which this inadequacy corresponds to events in nature that are not determined per se, independent of our capacity for knowledge. This latter species of chance, composed of gaps in the chain of causes, is what leaves room for human freedom. Elsewhere I have discussed this confused and quite mediocre

Horowitz, *Shenei luḥot ha-berit*, 3, *Vayeshev*, p. 96 [trans. Munk, p. 290]). This light or insubstantial (*ḳal*) character, contrasted to the gravitas of glory (*kavod*, related to *kaved* 'heavy'), had to be repaired. That this was the objective of the 130 years of Adam and Eve's separation is indicated, according to Horowitz, by the fact that the numerical value of *ḳal* is 130. But Adam emitted "drops of randomness" during this period, and these had to be remedied by the 130 years of exile in Egypt that preceded the birth of Moses (ibid.).

59. The masculine is "naturally" associated with action and control, whereas passive feminine Nature, sometimes stubborn and recalcitrant, will always be raped to give up its secrets. For Evelyn Fox-Keller (*Reflections on Gender and Science* [New Haven: Yale University Press, 1985]), the mechanical revolution of Bacon and Descartes in the seventeenth century modified and reinforced these images, thanks to the efficacy of the new science that emerged from it. The old hermetic alchemy of the Renaissance had preserved traces of an effective and unified bisexuality, in which the magician merged with nature, of which he was part. For the alchemists, God's immanence in nature, in the material world, and in the male and female human body was the warrant of this union, and science could still derive its power from the force of love and sexual knowledge. But this "natural magic," as it was known, soon faded from the scene, because its technical efficacy did not measure up to the objective rationality of the new science—both a cause and an effect of the dominant images that have reigned ever since. The male-dominated organization of society and division of labor have their correlative in a science that is based on a realist epistemology, on the principle of objectivity, and on the obligatory separation of subject and object. This produced the association of male, rational, and objective, on the one hand, and of female, emotional, and subjective (thus irrational and unscientific) on the other. These notions of complementary pairs—male/female, objective/subjective, scientific/fantastic, have reinforced one another in the triumphant scientific and industrial societies.

conception, with regard both to human freedom and to the very idea of a deterministic nature and our possible knowledge thereof.[60]

On the contrary, our experience of chance as ignorance implies that we can conceive of adequate knowledge that aims to be total and infinite (at least as an a priori possibility), even if the experience of our own finitude persuades us that we will never attain it. According to an interpretation common to Maimonides and Spinoza, it is this inadequacy that characterizes our first parents' fall.[61] We experience it every time we have inadequate knowledge—mere opinion or imagination—of something or someone: the experience of incomplete knowledge, always inadequate in comparison to the ideal of full knowledge, an ideal of which we can conceive even though we have never experienced it. Thus our knowledge is simultaneously good and bad, because it is both positive, affirming what it knows, and negative, handicapped by everything it does not know.

In a different register, one element of the experience of the sacred, and of the mystical experience in general, is access to what is supposed to be the perfect original knowledge, in which nothing is a matter of chance and everything has a meaning. These experiences are often accompanied by oracular rites in which some person (possessed or in a trance) sees or hears the hidden meaning of things present and past and their future ramifications, which are normally unknown. It is significant that casting lots, as a way to force the oracle to speak, to make the omens appear, to know what is normally hidden, has always played a central role in divinatory practices. As so often, the passage from normal consciousness to a modified state of consciousness of some "other reality" (qualified in various ways as a separate reality, the world of the sacred, prophecy, oracles, divination, etc.), is accompanied by the inversion of the negative into the positive. In these practices, the randomness of the lottery is turned into its opposite. What is usually inadequate knowledge—lotteries rely on the even odds entailed by our ignorance of causes—then becomes the means par excellence for uncovering the hidden, when the diviner enters the state that allows him or her to interpret whatever happens. In Chapter 5, we shall see how the experience of chance was progressively stripped of its sacred character as the prophetic experience disappeared from the cultural and social field of our civilizations. We shall also see how the talmudic interpretations of several biblical episodes reflect this evolution from an oracular chance that expresses divine knowledge of hidden things toward an indifferent chance that expresses only a lack of knowledge.

But it is the Eros of sexual knowledge that remains the most common and universal experience of the sacred. This is where the tension and ambiguity are greatest in the experience of a randomness that is, of course, ignorance, but also a sign of superior, complete, and infinite knowledge of causes that are normally hidden from us. We first perceive this ambiguity in the sexual act itself, which is both knowledge of another in the intimacy of his or her body as well as the lack of transparency, the opacity of the encounter, revealed by the naked skin.[62] But we also perceive it in the children created by this knowledge. Here everything is hidden and largely unconscious. First of all, the encounter between a man and a woman is an unforeseen event, quite unpredictable, especially in modern societies where marriages are not arranged as a function of prescribed and predefined kinship relations. Next, fertilization itself is the result of a process that

60. See Chapter 2; Henri Atlan, "Postulats métaphysiques et méthodes de recherche," in *La querelle du déterminisme* (Paris: Gallimard, 1990), pp. 113–120.

61. See below, n. 85.

62. See Chapter 2.

remained scarcely known until twentieth-century biology discovered most of its mechanisms. (Note, incidentally, that what we have learned from biology demonstrates the role of chance in the encounter between sperm and ovum that determines the child's genetic structure.) The Talmud makes God himself responsible for bringing men and women together so that new human beings will be conceived, in a chain of absolute determinism where "everything is foreseen" and in principle knowable by a perfect and infinite knowledge of the processes of nature.[63] A drop of semen reaches its goal by chance, but when it does so it creates the most significant carrier of meaning we can imagine: a new human being, capable of creating new meanings in its own turn, of giving sense to what happens, beginning with himself or herself as a first-person *I*. This is the origin of the clear idea that its source, like the source of every random event, is knowledge that is all the loftier and more valuable because it is normally hidden. Converting chance into a marker of this perfect and unitary knowledge, where everything is in its place, is thus equivalent to transforming the experience of sex into an experience of the sacred.[64] Conversely, accenting the fortuitous character of sex, tossed to the chance of the scattered and indeterminate, means the a priori denial of the reality of this knowledge and destroys its very possibility. (Later we shall see that, despite appearances, the relative control provided today by medically assisted techniques of human reproduction, which allow us, within limits, to overcome the random character of birth, is still not enough to give this chance a human meaning.) The sparks of randomness, when scattered, produce demons that merely highlight this dispersion. But they are also an indirect sign of the perfect and infinite divine knowledge to which Moses and his Torah are said to permit access. Whence the importance of repairing the ravages of these demons, until, in the generation of the wilderness, the "generation of knowledge" (*dor de'ah*), they were reunited at their source.

63. The Talmud compares the difficulty of bringing about the meeting and union of two persons with quite distinct personalities to that of splitting the waters of the Red Sea (B *Soṭah* 2b). Hence only God himself is capable of doing it (and, moreover, has devoted the best part of his efforts to it since he wound up Creation on the evening following the sixth day). After conception,

> the angel who is in charge of conception . . . takes up a drop [the fertilized ovum] and places it in the presence of the Holy One, blessed be He, saying, "Sovereign of the universe, what will be the fate of this drop? Will it produce a strong man or a weak man, a wise man or a fool, a rich man or a poor man?" But he does not say "wicked man" or "righteous man," in agreement with the view of R. Ḥanina. For R. Ḥanina stated: "Everything is in the hands of Heaven except the fear of Heaven." (B *Niddah* 16b)

This absolute determinism, maintained by the same Rabbi Ḥanina who affirms, in opposition to the other sages, that Israel is subject to astral determinism (see below, pp. 154–155), also applies to matchmaking: at the moment of conception a heavenly voice (a *bat ḳol*, *lit.* "daughter of a voice") proclaims who will be married to whom. This is the gist of the discussion in tractate *Soṭah* referred to above and of Rashi's commentary thereon. Here we also see how two sorts of determinism are superimposed: "natural" determinism (by the stars) and moral determinism. The latter would seem to require God's "miraculous" intervention but is in fact within the power of human beings, in that it is steered in one direction or another by their good or bad deeds. As we shall see, this theme of the intermeshing of human responsibility with the structure and activity of the divine component of the world is frequent in kabbalistic literature. One of its most radical expressions is found in *Nefesh ha-ḥayyim* (the Soul of Life), by Rabbi Ḥayyim Volozhiner (1824). On this see, inter alia, Charles Mopsik, *Les grands textes de la Cabale. Les rites qui font Dieu, pratiques religieuses et efficacité théurgique dans la Cabale des origines au milieu du XVIII*[e] *siècle* (Paris: Verdier, 1993). In Chapter 2 we will see that this notion does not prevent recognition of an absolute and timeless determinism in which "all is foreseen," including the choices made by human beings, in a game where "everything happens as if" we made the decisions about our own actions. In brief, it is our temporal and relative awareness of things, along with knowledge of the causes themselves, including those that direct our choices, that makes us responsible for our actions, and not some disembodied free will that eludes the determinism of "all is foreseen."

64. See Rashi on Leviticus 19:2: "in every passage where you find a restriction on sexual relations you also find [a reference to] the sacred."

This oscillation between registers, implicit in *da'at*, biblical knowledge, is also expressed in other key words, which conflate several usually different senses, in the midrashic narratives and their talmudic and kabbalistic commentaries.

9. AMBIGUITY AND POLYSEMY

Sefer Yeẓirah associates the Hebrew *millah* 'word' with its homonym *mīlah* 'circumcision' (which the Bible applies exclusively to males).[65] Both, in fact, go back to a root that means "cut." In language, cutting off strings of letters or sounds to produce words is what generates meaning. The cut of circumcision, emblematic of the difference between the two sexes, makes it possible for the two to unite in an alliance that expresses something other than the fusion of the same. In Hebrew, one *cuts* (i.e., concludes) a treaty or pact (though the verb employed is different).

The word *naḥash* 'serpent', with its sexual symbolism, itself divided into straight and crooked, also means "augury" or "divination" (Num. 23:23)—an intuitive guess, a form of knowledge that is approximate but sometimes dazzling, and that is perverted in the soothsaying of the *menaḥesh* or "oracle."

The Tree of Knowledge, as a tree (*'eẓ*), is simultaneously a source of nourishment and of the counsel (*'eẓah*) provided by the serpent. The Tree of Life, originally forbidden, is in fact the Torah itself (Prov. 3:18), the wisdom that existed before the created universe, set aside to be given to Moses at Sinai; it is unsullied knowledge that can be only good. The two trees, the Tree of Knowledge and the Tree of Life, were originally only one.[66] Similarly, the serpent is both the brazen serpent that works miracles, the *neḥash neḥoshet* of Numbers 21:9, and Moses' staff,[67] with which he led the generation of the wilderness, the *dor de'ah* or "generation of knowledge," because it received the Torah in a face-to-face encounter composed of signs and portents—themselves ambiguous because they also led to the errors and sins of the Golden Calf. The serpent of Genesis, by contrast, is *'arum*, which means "naked" as well as "shrewd" or "crafty."[68]

Finally, as we have seen, the scattered drops of semen are

65. Chapter 1, mishnah 3.

66. See, inter alia, Rabbenu Baḥya's commentary on Genesis and his reading of Rashi on the trees, *both* of which are said to be located at the very center of the garden (Gen. 2:9).

67. See below, n. 78.

68. Genesis 3:1. A possible connection between cunning and nudity, in the context of knowledge, intelligence, and wisdom, is suggested by the two possible readings of Proverbs 8:12, in which wisdom-intelligence is speaking of itself: "I, wisdom, dwell in craftiness/nudity (*'ormah*)" (Prov. 8:12). The Talmud (B *Sotah* 21b) proposes both meanings of *'ormah* (from the same root as *'arum*), corresponding to two aspects of wisdom. On one level, it observes, "when wisdom enters a man cunning enters with it." When associated with cunning, wisdom is opposed to the innocence of the naïve and simple and serves as a means to control objects and beings through knowledge of the causes that affect them and of how to redirect them, perhaps even against themselves if necessary. Or, as Rashi puts it in his commentary on the verse, "when a man studies Torah, he absorbs every sort of cunning" (*lit.* "the cunning of every thing"). When associated with nudity, on the other hand, wisdom is the fullness that fills the void, or the clothing that adorns and embellishes the body. Thus, on the other level, the Talmud considers prior nakedness—the absence of preconceptions—to be a precondition for acquiring this wisdom, which is the words of the Torah: "The words of the Torah remain only with him who makes himself naked for them." These two aspects of wisdom and knowledge correspond to the two types of wisdom—those, respectively, of the left side and of the right side (see above, n. 51); as perceived by the latter, the former is not necessarily and in all circumstances bad and to be condemned.

Rather, as we have noted, everything depends on their goal and orientation. Wisdom of the right side overflows "from above to below" and dictates the law that applies to all, whereas wisdom of the left side runs the risk of being applied in the service of particular and potentially destructive interests. Whence the rule of subordinating it to the wisdom of the right side, directing it from "below to above"; that is, from the particular, with its limitations, toward the general and universal, which is potentially infinite (see Chapter 3, §14). The double sense of *'arum* allows the ambiguity of knowledge to

also sparks of randomness, elements of disorder. Here randomness seems to be contrary to every attempt at intelligibility and order, as if absurdity and the radical negation of all meaning were being introduced into the most concrete fabric of human history, that of their generations.[69]

A priori, then, there would seem to be individualized knowledge, an intentional ordering of things, a humanization of nature—"good knowledge"—and, contrasted to it, a bad knowledge that takes possession of us, in which knowers are depersonalized and enslaved by what they know. On the one hand, there is the transparency of adequate knowledge for which, at its limit as divine knowledge, "the knower, the known, and knowledge are one," to use the Aristotelian phrase taken over by Maimonides and later by the kabbalistic tradition and by Spinoza.[70] On the other hand, there is opaque knowledge that is confused and inadequate, the knowledge of dreams, hallucinations, and delirium.

But things are not so simple. Good and evil, what is good and what is bad, are always inextricably mixed up. Transparency, too, can give rise to uncertainty when there is a sort of mystical fusion between subject and object. On the other hand, the distance between subject and object may be a source of clarity, of the sort afforded by the rigor of objective knowledge.

In the Midrash we find this interweaving of the good and the bad with regard to different types of demons or spirits,[71] in association with the "sparks of knowledge" that, after being scattered randomly by the disobedience of Adam, Eve, and the serpent, serve as their souls and animate them. A distinction is drawn between familiar spirits, those of the house, and alien spirits, those of the field. Which are benevolent and which are malevolent?

> [Some hold] that house spirits are benevolent, because they grow up with [man]; the opinion that they are harmful is based on the fact that they understand man's evil inclinations [from the inside]. He who

be superimposed on that of the serpent, which is both cunning and naked, its nudity understood both literally, as sexual, and metaphorically, as thought stripped to its bare essentials.

It is also useful to analyze the relationship between wisdom and craftiness/nudity that is suggested by the verb *shakhanti* 'I dwell', in the same verse from Proverbs cited at the beginning of this note. The Hebrew text has no preposition and is, literally, "I wisdom dwell craftiness/nudity"; to avoid a possible misreading, Rashi supplies one: *'eẓel* 'in' or 'alongside'. This is the same preposition that does appear in the biblical text later in the same chapter, where Wisdom declares, "I was alongside him (*'eẓlo*), like a master workman; and I was daily his delight, rejoicing before him always, rejoicing in his inhabited world and delighting in the sons of men" (vv. 30–31). This same word *'eẓel* is the source of *'aẓilut*, which in the Neoplatonic kabbalistic tradition is usually taken to mean "emanation." It is the process whereby multiplicity emerges from the One, such that the product, not yet separate, is still contained in the process itself, as if it dwelt there, but can nevertheless be distinguished from it, and thus can be considered to be "in" it or "alongside" it.

This reading of Proverbs suggests that Wisdom, the first of the emanated *sefirot*, dwells in craftiness/nudity and alongside its creator, serving as the blueprint for the rest of creation. This brings us to another midrashic conceit, to be developed in Chapter 2, in which the creator of a world that is totally determined and in which everything is foreseen toys with human beings who exercise (or think they are exercising) their free will and holds them responsible for their choices, even though these choices are inevitable and foreseen from all eternity. It is precisely in this world of *'aẓilut*, the world "above," that "evil does not dwell" (Ps. 5:5), because evil, at its source, is not evil.

69. We should remark, briefly for now, that casting lots is a traditional form of divination—for example, when the lot (*goral*) falls on Jonah as the cause of the tempest (Jon. 1:7). Similarly, Haman casts lots to determine the date for the massacre of the Jews (Esth. 3:7). In these cases, as in traditional techniques of divination, the random nature of casting lots is transformed into destiny (also *goral*), which is revealed through the devious knowledge of the oracle's serpentine tricks (by the *niḥush* of the *menaḥesh*). But it can also be a tangible substitute for a divine choice, as, for example, in the incrimination of Achan (Josh. 7:16–18) and the election of Saul (1 Sam. 10:20–21). See Chapter 5.

70. "God, God's intellect, and the things understood by God are identical" (*Ethics* II 7, Note). See Chapter 4 for a discussion of the meaning of this expression in the various philosophical contexts in which it appears.

71. See Chapter 3.

maintains that the spirits of the field are benevolent does so because they do not grow up with him, while as for the view that they are harmful, the reason is that they do not comprehend his evil inclinations.[72]

These demons or spirits, sparks of randomness scattered to every wind, were originally the opposite of any ideal of liberation by reason and by organized, clear and distinct knowledge; they are the hazards of existence that appear unexpectedly, sudden passions, sources of pleasure and pain, of hope and fear, of alienation. Their origin may be external (demons of the field, objective events, both foreign and strange) or internal (household demons, ideas, sentiments, the fantasies of imagination and dreams, both familiar and threatening). The *Zohar* portrays these demons, sources of pleasure for the sons of man, as players in the world-game, using the sons of man for their own pleasure.[73]

10. THE RANDOMNESS OF BIRTH: EVIL AT ITS SOURCE IS NOT EVIL

The semantic richness and mythical themes of this wordplay are inexhaustible. Here I want to analyze only the ambiguous status, good and bad, that these narratives attach to the randomness of birth. Why is the salvation of these sparks of randomness so important, generation after generation—the generations of the Flood, of the tower of Babel, of the wilderness, etc.? The writings of Ḥayyim Vital (*Sha'ar ha-kavvanot, Sha'ar ha-pesukim, 'Ez ḥayyim*, and other works, from which we have quoted only an important excerpt[74]), tell us explicitly.

At first—before they were carried off by demons—the souls that existed in potential in these "random drops" were of surpassing excellence. They derived from the same level as Moses himself, from the world of *da'at 'elyon*, the "knowledge of the upper regions." They were superior to the ordinary souls whose course is directed by a woman's common desire to receive one of them in order to become a mother, determined in some fashion by her desire to have a child. These souls had a loftier origin because they came from what the *Zohar* refers to as the "arousal above,"[75] a higher and in some fashion more disinterested desire. But how are we to understand it? On the one hand, chance, randomness, objects and events that are produced accidentally or unintentionally, are a priori evil, because they represent disorder and the negation of meaning; they partake of the "mixed multitude" (*'erev rav*) in which everything is undifferentiated and violent. Evil at its source, though, is never evil.

The kabbalists are constantly dealing with this question of evil in nature, whether because of their intimate experience with it as part of their people's tormented history or because, as for Gnostics and Buddhists, any profound reflection on being and existence leads inevitably to the problem of evil. For the kabbalists, and indeed the entire biblical and rabbinic tradition, however—and unlike the Gnostics—there could be only a single and unique source from which all that exists (including evil) proceeds. This is precisely what the prophet has the Creator of the universe state explicitly: "I make peace and create evil" (Isa. 45:7). Nevertheless, evil qua evil does not come from its source: "Nothing evil descends from above."[76] At its source, evil is not evil.[77]

72. *Gen. Rabbah* 20:11 and 24:6.

73. *Zohar*, Genesis 19b (see above, n. 57).

74. See above, n. 45.

75. <E.g., *Zohar*, Genesis 35a (trans. Matt, vol. 1, p. 219).>

76. *Gen. Rabbah* 51:3. Here the Midrash, commenting on the fire and brimstone that rained down on Sodom and Gomorrah, considers the possibility

of exceptions to this principle, such as the Psalmist's declaration that God "will rain coals of fire and brimstone on the wicked" (Ps. 11:6, and, implicitly, Micah 1:12: "evil has come down from YHWH to the gate of Jerusalem"). Isaiah Horowitz (see above, n. 30) takes over and refines this expression: "Nothing evil descends from above; rather, we produced the evil by being bad" (*Sha'ar ha-shamayim*, in *Siddur Ha-shlah* [Jerusalem, 1987], commentary on the Yom Kippur liturgy, p. 614). This is easily reconciled with God as creator of evil, as stated by Isaiah; we produce evil only because our nature allows us to do so. As Rashi states in another context (see below, p. 100), the Creator fashioned us to include our evil impulse. Thus he indirectly makes us do the evil we do. Later we shall see how this radical thesis is supported and amplified by the Talmud and some rabbinic commentators.

77. This idea, so hard to grasp, can be approached from different directions. We find it in Spinoza, expressed, of course, in the context of his philosophy: "The knowledge of evil is an inadequate knowledge" (*Ethics* IV 64). Evil is experienced and suffered, but it is never the object of adequate knowledge. There is no adequate knowledge of evil qua evil, because evil has no essence; and adequate knowledge, meaning knowledge of things as they really are and not as we imagine or perceive them, is knowledge of their essence:

> God is absolutely and really the cause of all things which have an essence, whatsoever they may be. If you can demonstrate that evil, error, crime, etc., have any positive existence, which expresses essence, I will fully grant you that God is the cause of crime, evil, error, etc. I believe myself to have sufficiently shown, that that which constitutes the reality of evil, error, crime, etc, does not consist in anything, which expresses essence, and therefore we cannot say that God is its cause. . . . None of these qualities express aught of essence, therefore God was not the cause of them, though he was the cause of [the criminal's] act and intention. (Letter 36 [23], to William Blyenbergh)

This is even clearer in the context of kabbalistic philosophy, when we adopt the perspective of an absolute natural determinism and of the divine *'alilot* or intrigues (see Chapter 2). From this angle, all human actions are part of the infinite chain of causes, foreseen from all eternity. This means that the God of ethics and religion holds us responsible for our actions, although the God of nature is their true author, working through us. As a human being I am bound to render account and assume responsibility for the evil wrought through me, as if I were the cause—for example, if my ox damages my neighbor's field or if my underage children get away from me. When I repair this evil to the extent possible, always separating myself from the *I* that did it—by means of regret or repentance, which in Hebrew is *teshuvah* 'returning'—I return it to its true source and author, where it is no longer evil because it is reinserted in its place in the infinite and eternal plot of the absolute and timeless determinism of nature—or in the causal sequence of "essences," to use Spinoza's terminology. This is how *teshuvah* makes forgiveness possible: not by causing the transgressions or crimes to be forgotten, nor by erasing them as if they had never existed, but, in even more radical fashion, by transforming them into merits, that is, into good (B *Yoma* 86b).

The entire ritual of Yom Kippur, *teshuvah* and *kapparah*, "return" and "expiation," revolves around this notion of sending human misdeeds back to their true author, to their source where they are not bad, but without annulling human responsibility. In the next chapter we will see how the expression *nora' 'alilah* 'terrible in deeds', a central and recurrent theme of this ritual and understood on the first level as praise of God, actually expresses the ambiguity of the dialectic, when we take *'alilah* in the sense of "plot"—thus "terrible in intrigue"—and as the pretext that holds human beings responsible for deeds they are destined to perform from all eternity. (The expression *nora' 'alilah*, recurs, hauntingly, in the passage that begins the final hour of the Yom Kippur service, called *Ne'ilah* or "closure." But the night before, at the start of the fast, the invocation that introduces the confession of sins, both grave and minor, begins with this same word *'alilah* in its pejorative sense, applying it to human deeds; the *yezer*, the totality of the impulses that inform and produce the passions, is referred to as *'alilot be-resha' lehit'olel*, "working deeds of wickedness." Later, though, the text applies the same word to the Creator, with the ambiguous sense of *nora' 'alilah*.)

This theme of God's responsibility for the transgressions and crimes for which we must nevertheless render account and repent is stated explicitly in the great kabbalistic poem *Keter malkhut* (Royal Crown), by the eleventh-century Spanish Jewish philosopher Solomon Ibn Gabirol (known in Latin Europe as the Neoplatonist Avicebron, author of *Fons vitae* [Source of life]). There we read:

> I will flee from Thee to Thyself,
> And I will shelter myself from Thy wrath in Thy shadow,
> . . .
> Remember, I pray Thee, that of slime Thou hast made me,
> And by all these hardships tried me,
> Therefore visit me not according to my wanton dealings [*ma'alalai*, from the same root as *'alilah*],
> Nor feed me on the fruit of my deeds.
> (*Selected Religious Poems of Solomon Ibn Gabirol*, trans. Israel Zangwill [Philadelphia: Jewish Publication Society, 1923; repr. New York: Arno Press, 1973], "The Royal Crown," XXXVIII, p. 118)

It would be difficult to express more paradoxically, yet at the same time more expressively, the tension that pits our knowledge of God-Nature, the cause of the determinism from which nothing and no one escape, against the human experience of good and evil and of our total responsibility for the fact that one or the other—good or evil—dominates all beings and objects, as if we ourselves produce what is done through us.

Finally, we must underscore the fundamental opposition between this attitude and some theological byways that would minimize the reality of evil and of the suffering of the innocent in order to justify, at any price, a personal God who is held to be both infinitely powerful and infinitely good. That "evil at its source" is not evil does not eliminate the evil reality of its effects. The ultimate experience that we call "Auschwitz" will not allow us to forget this. One of the recurrent themes of kabbalistic thought is the integration of evil into a system of creation that

For example, according to the *Zohar,* "above, in the Tree of Life, there are no bad [*lit.* foreign] husks, as it is written, 'evil does not dwell with you' (Ps. 5:5)."[78] In particular, this is what *Sefer Yeẓirah* says about the transformation of *'oneg* into *nega'*: "There is no good higher than delight and there is no evil lower than plague."[79] To put it

is compatible with the "I make peace and create evil" of Isaiah's God. But this has nothing to do with a theodical "explanation" of evil and of suffering. Such an explanation, utterly scandalous from a moral and even a religious point of view, would justify evil in some fashion by embedding it in a divine system of reward and punishment, in which God is by definition innocent and the victims are guilty. If there is an explanation, it can be found only "at the source" and does not diminish in the slightest our experience of evil and of the suffering that nature (not to mention human beings) can inflict on us. The "banality of evil," invoked by Hannah Arendt in her book on the Eichmann trial, does not eliminate its horror. The same thing could be said of the most egregiously perverted crimes. If one insists on posing the question in theological terms—though that is certainly not necessary—Hans Jonas's attempt remains, perhaps, the least offensive. See "The Concept of God After Auschwitz" (1984), in Hans Jonas, *Mortality and Morality: A Search for the Good After Auschwitz*, ed. Lawrence Vogel (Evanston, IL: Northwestern University Press, 1996), pp. 131–144.

78. *Zohar*, Genesis 27a. This theme is developed in other contexts, using different terms that always shed new light on it, and related to the ambiguity of the knowledge that engages us here. For example, the river of Genesis 2:10, which "issued from Eden to water the garden," is interpreted as analytical intelligence (*binah*), which understands one thing from another and thus, by means of definitions, distinctions, and divisions, waters the garden of knowledge (*da'at*) because it issues from an intuitive wisdom (*ḥokhmah*) that is synthetic and free of conflict (Eden). Isaiah Horowitz invokes this interpretation in his explanation of the thirteen divine attributes of mercy in the Rosh Hashanah liturgy (*Sha'ar ha-shamayim*, p. 400b), adding this remark, also borrowed from the *Zohar*:

> That river, although the source of strict judgment [understood as the root of suffering], is itself compassion (*raḥamim*). With regard to the same verse, the passage from the *Zohar* we have just quoted (Genesis 27a) identifies the river with the ministering angel Metatron, who himself plays an ambiguous role in the talmudic account of the four sages who entered the garden of knowledge (B *Ḥagigah* 15a) and again in the kabbalistic tradition: "The 'river' is Metatron; 'issued from Eden' means that it issues from his delight [*'iddun*, from the same root as *Eden*]; 'to water the garden' refers to his garden, which was entered by Ben Azzai, Ben Zoma, and Elisha ben Avuya [the three sages who did not successfully navigate their experience of initiation in the garden and, unlike R. Akiva, did not leave it in peace]. Its husks are good from one aspect and bad from the other aspect. And this is forbidden and permitted, proper (*kasher*) and improper (*pasul*), impurity and purity. . . ."

The text continues, now with reference to Moses' staff, called the "rod of God" (Exod. 17:5 and 9):

> This rod is Metatron, from one side of whom comes life and from the other death. When the rod remains a rod, it is a help from the side of good, and when it is turned into a serpent it is against him [an allusion to Gen. 2:18: "I will make him a 'counterpart' helper (Heb. *'ezer ke-negdo*)"; which Rashi explains, "if he merits, she will be his helper; *if not she will [act] counter [to] him*"]. God delivered it [Metatron/the rod] into his hand, [where] it typifies the Oral Law which prescribes what is permitted and what is forbidden.

According to Yehuda Ashlag's commentary on the *Zohar*, the *Sullam*, "permitted" and "forbidden" mean the aspect of good and the aspect of evil, as opposed to the Torah, which is *'aẓilut* 'emanation', composed entirely of the names of God, and about which it is written, "evil does not dwell with you" (Ps. 5:5; see Naḥmanides' preface to his commentary on the Pentateuch).

In another passage we read:

> It may happen that a man is subjugated by the evil impulse [i.e., the burden of his crimes in a previous avatar] and his soul can find no rest, because this impulse pursues him to penetrate the body to which he is subject. That is a "Jewish demon" (*shed yehudi*). The soul, a "Hebrew maidservant" [referring to the biblical law on Hebrew slaves, whose bondage was temporary], serves "for 10 years" [represented by the Hebrew letter *yod*], by the addition of which *shed* [i.e., "demon"] becomes *shaddai* [a name of Divine power], because it was able to guard that soul so that it could do penance. . . . Thus what was a demon becomes his angel, of the category of Metatron, and is transformed into *shaddai*, for the numerical value of *metatron* is the same as that of *shaddai* [314]. . . . And that demon is like Moses' rod, which turns from rod to serpent and from serpent to staff; so too this demon turns from a demon into an angel and from an angel into a demon, depending on a man's deeds. . . . And with regard to the demons that come from him, the Sages of the Mishnah learned that some of them are like the ministering angels—they are the disciples of the wise, who know what was and what will be. In their form on earth they are masters of philosophy, the astrologers of Israel who know what was and what will be from the signs of the sun and moon and their eclipses and every star and constellation and what is shown in the world. And some of them are like animals, procreating and multiplying like animals. Their appearance [in the world] below is that of the unlearned (*'ammei ha-'areẓ*). And the sages of the Mishnah learned that they are an abomination and their daughters are swarming creatures, and of their daughters it was written, "cursed be he who lies with an animal" (Deut. 27:21). They detest those who look like sages of the Mishnah, who really are ministering angels. This is why the sages of the Mishnah applied the verse to a man, namely, that if he is an angel of the Lord of Hosts, they should seek Torah from his mouth [after Mal. 2:7]; and if not, they should not seek Torah from his mouth. (*Zohar*, Deuteronomy, *Ra'aya Meheimana* 277a])

79. *Sefer Yeẓirah* 2:4, trans. A. Kaplan (York Beach, ME: S. Weiser, 1997).

another way, what is pleasure above becomes plague and torment below, by transposition of its letters; but conversely, what is torment below becomes pleasure above, at its source, where it is softened and transformed by its context.

When it comes to biblical knowledge and prophecy, the Balaam pericope (Numbers 23) offers just such an example of the transformation of random evil into good. The midrash on this text, cited by Rashi, contrasts the prophetic modes of Moses and Balaam. Balaam delivers his oracles as the result of a random encounter (*be-mikreh*), without a prior personal summons, a sort of inversion of Moses' prophecy, just as the serpent's knowledge inverts that of the Tree of Life. Whereas Balaam's prophecy comes upon him randomly (*be-'ar'ai*), Moses' is intentional and personal. In Hebrew the contrast is expressed by the difference of one letter between *vayyikkar* (ויקר) *'Elohim 'el Bile'am*, "God met Balaam" (in the sense of coming upon him by chance), and *vayyikra'* (ויקרא) *'el Mosheh*, "[God] called Moses."[80] Nevertheless, although Balaam employs various theurgic and divinatory practices condemned by the Torah, he "knows the knowledge of the Most High" (*or* "of the upper regions").[81] This is why it is not astonishing that the evil of Balaam's random prophecy is ultimately transformed into good. Balaam wants to exploit his prophecy to curse and bring down evil on Israel. But his bad intentions count for nothing and the content of this prophecy transforms evil into good, curse into blessing.

So evil at its source is not bad. Indeed, that is the original mechanism of *teshuvah* or "repentance" (from the verb *shuv* 'return'), the return to self by which evil can be transformed into good, sin into merit.[82] What is more, it represents a very large wager—double or nothing, as it were—in a redemptive strategy that is infinitely superior to the innocent and angelic stagnation of a frozen good that is given once and for all time.[83] From this perspective, evil at its source may be even more precious than is good in reality.

It is our partial and passive knowledge that makes us perceive things as good or bad, by projection and extrapolation from our perceptions of pleasure and pain.[84] In proportion as our knowledge is more extensive and active, it is less conditioned by these perceptions. Gradually it replaces "good and evil" with "true and false," with "real and imaginary." The essence of this process is a "return" to the mythic prelapsarian state, when Adamic knowledge was not of good and evil but of true and false.[85] Thus

80. Respectively Numbers 23:4 and Leviticus 1:1. See Rashi on Leviticus 1:1. See also Chapter 3.

81. Numbers 24:16. For the talmudic commentaries on his supposed carnal relations with his ass and spirits as part of these practices, see B *Sanhedrin* 105a–106a.

82. See above, n. 77.

83. This is one of the most important themes of the thought of Rabbi Abraham Isaac Hacohen Kook, a twentieth-century kabbalist, for whom the biological theory of evolution was in perfect harmony with the notion of *tikkun* or the repair of the world, the engine of an infinite progress of being toward unity and perfection through the multiplicity of its expressions. See, e.g., "Absolute Good and the Good That Elevates Itself," in *'Orot ha-kodesh* (Lights of Holiness) (Jerusalem: Mosad Harav Kook, 1964), 2:466 (Hebrew).

84. Atlan, *Tout, non, peut-être*, chap. 7; idem, "Le plaisir, la douleur, et les niveaux de l'éthique," *Journal international de bioéthique* 6 (1995), pp. 53–64.

85. It is remarkable that Spinoza and Maimonides, often violently opposed on essential points, nevertheless seem to agree about this interpretation of the transgression by the first man, although in different contexts.

For Maimonides, "through the intellect man distinguishes between the true and the false. . . . It is the function of the intellect to discriminate between the true and the false—a distinction which is applicable to all objects of intellectual perception" (*Guide* 1.2, p. 15). Good and bad, however, like beautiful and ugly, are not matters of intellectual perception and necessity, but of opinion and probability. When Adam had perfect and complete knowledge of all things as objects of intellectual perception "and was guided solely by reflec-

the doctrine of salvation through knowledge, which has taken diverse forms from the Stoics to Spinoza's quest for blessedness, seems to be a doctrine of clarification and disillusionment about the absolute character of the opposition between good and evil. This opposition tends to dissolve when our concept of evil as such disappears (although this does not, of course, eliminate our perception of pain) and is replaced by knowledge, to some extent amoral, of the paths by which things happen.

As we have seen (and will appreciate better in subsequent chapters), this knowledge of causes eliminates neither our perception of painful events and of bad actions nor the horror they may inspire in us. What is more, it is the precondition for achieving true freedom, in which the absence of free will does not absolve us of our responsibility for acquiescing in what is effected through us.

tion and reason, . . . he was not at all able to follow or to understand the principles of apparent truths" (ibid.). Good and evil did not exist for him as such, because only what reason made him understand to be necessary existed. But the loss of this all-embracing intellectual knowledge reduced him to a status in which he knew things other than by reason, in a contingent fashion, as a matter of opinion—such as beautiful or ugly, good or bad.

Spinoza, too, refers to the biblical myth: "If men were born free, they would, so long as they remained free, form no conception of good and evil. I call free him who is led solely by reason; he, therefore, who is born free, and who remains free, has only adequate ideas; therefore he has no conception of evil, or consequently (good and evil being correlative) of good" (Ethics IV 68 and Proof). But Spinoza goes on to observe that the hypothesis that human beings are born free is untenable, because they are part of nature and as such are subject to causes outside themselves. This impossible hypothesis is "seen to have been signified by Moses in the history of the first man." In the perspective of that hypothetical (but impossible) original state of humanity, Moses wrote "that God forbade man, being free, to eat of the tree of the knowledge of good and evil, and that, as soon as man should have eaten of it, he would straightway fear death rather than desire to live" (ibid., Note).

In any case, the defining condition of humanity after the fall is knowledge of good and of evil, an approximate and disputable knowledge that replaces knowledge grounded on clear and distinct rational ideas, the only possible source of human freedom. Hence for these philosophers knowledge of good and evil characterizes the enslavement of human beings as they are born in the real world, in contrast to the world of freedom and happiness that rational knowledge of the true would give us.

The role of science and of philosophy is to provide us with a path, assisted by reason, to a knowledge of good and evil that is true, because derived from the "true" nature of human beings. Thus true knowledge of good and evil is conceivable.

On Spinoza and Maimonides see also Shlomo Pines, "On Spinoza's Conception of Human Freedom and of Good and Evil," in *Spinoza: His Thought and Work*, ed. N. Rotenstreich (Jerusalem: Israel Academy of Sciences and Humanities, 1983), pp. 147–159; repr. in Shlomo Pines, *Collected Works* (Jerusalem: Magnes Press, 1997), 5:735–747.

11. THE "MIXED MULTITUDE" IN THE WILDERNESS AND THE "HEAD THAT IS NOT KNOWN"

For the Midrash, the "mixed multitude" (*'erev rav*) that left Egypt with the Israelites (Exod. 12:38) is a token of this liberation by and through knowledge. This mob of Egyptians and slaves of various ethnic origins, who joined with the Hebrew slaves to share their liberation, remained pagans in spirit and are blamed by some commentators for the sin of the Golden Calf. But the same mixed multitude is also referred to as "the people of Moses,"[86] in contrast to the tribes of Israel, the people of their god YHWH. The challenge for Moses, the master of knowledge, was to save this mixed multitude as well as the tribes of Israel. Just as evil at its source is the essence of salvation, randomness at its

86. See Exodus 32:7—"And YHWH said to Moses, 'Go down; for your people, whom you brought up out of the land of Egypt, have corrupted themselves'"—and Rashi's comment: "It does not say, 'The people have acted corruptly,' but 'your people.' These are the mixed multitude whom you accepted on your own initiative."

source has inestimable value, because this source is infinite knowledge itself, the knowledge that only Moses could attain (and even he only partially), and to which Moses' Torah, despite its finitude in time and space, can serve as the introduction.[87] To represent this, we can use the *Zohar*'s image of "the head that is not known" (*resha' delo' 'ityadda'*)[88] at the pinnacle of the *sefirot* and "faces."[89] Unlike the other parts of this structure, which are fully defined in all their details, this head is unknowable because, moving constantly, it produces, in the words of Ḥayyim Vital, an "infinity of doubts" about its effective state. Nevertheless this is the source of the knowledge of above, from which descends directly, albeit metaphorically, through the brain and spinal column, the fertilizing seed that knows and is known by the woman. This infinity of doubts is precisely the infinite world of possibilities. For us it the source of randomness, when one of these possibilities becomes real and we cannot know the infinite causes of its realization. Thus Moses' superior knowledge is the infinite knowledge that, for a necessarily finite being, can be only of an infinity of possibilities; that is, knowledge that knows its own ignorance.

We find this theme in the midrash that recounts how Moses cannot understand what Rabbi Akiva is teaching his students, even though the sage insists that it is "the law given to Moses on Sinai."[90] It is not meant as praise of ignorance or religious obscurantism that takes refuge behind ignorance of the impenetrable designs of God. We should bear in mind that, according to the legend, Jeremiah's construction of a golem was motivated by a refusal to give up the effort to know the structure of the universe and an attempt to gain power over it comparable to the Creator's.

In retrospect this irreducible ignorance, which carries with it future knowledge, can be viewed only negatively, as uncertainty, chance, randomness. But the experience of this randomness keeps knowledge from withering up into some lore or form of being that are given once and for all time, closed and definitive. That is why this *ḳeri* or chance, at its source, is the bearer of the greatest hopes, even though we do not know how to plan them and doing so is in any case quite out of the question.

87. In the final analysis, the Mosaic Torah is limited by the historical and geographical conditions of Sinai. According to the Maharal of Prague (introduction to his commentary on Tractate Avot), although Mt. Sinai is designated "the mountain of God" (*har ha-'elohim*) and is the counterpart of Moses, "the man of God" (*'ish ha-'elohim*), the Mosaic Torah that resulted from their encounter is only the necessarily finite introduction to the infinite Torah that no human being can receive directly from the infinite God. This is the meaning of the statement, at the beginning of Tractate Avot, that "Moses received the Torah from Sinai" and not "from God at Sinai," as is generally held. According to tradition, Moses could not advance beyond the forty-ninth of the fifty gates of understanding: "Rav and Samuel both said, 'Fifty gates of understanding were created in the world, and all but one were given to Moses, for it is said, "You made him <in the original context, "man" and "son of man" in the previous verse; but this homily takes the reference to be to Moses> little less than God" (Ps. 8:6 [5])'" (B *Nedarim* 38a). The fiftieth gate, "Jubilee," which contains all the others and opens onto another dimension, another "world"—that of infinite wisdom itself (see above, n. 53)—remained closed to him. This is the meaning of Moses' request that he be allowed to behold the glory of God or the divine Presence (Exod. 33:18), which was granted only in part.

Sefer Habahir interprets this deficiency of Moses' knowledge as related to how the divine mind exerts its power in each of its parts: "Moses said: 'I know the ways of the Powers, but I do not know how Thought spreads through them. I know that Truth is in Thought, but I do not know its parts.' He wanted to know, but they would not tell him" (trans. Kaplan, §194, p. 53).

88. <Alternatively, the head that "is not made known" or even "that does not make itself known": although the primary use of the Aramaic *etpa'al* is to express the passive voice, generations of readers, influenced by the cognate Hebrew verb pattern *hitpa'el*, whose primary sense is reflexive, have understood the text in the last sense. Prof. Atlan believes that it can and should be understood both ways.>

89. *Zohar*, Deuteronomy, *'Idra Zuṭa* 288.

90. *Yalḳut Shimoni*, Exodus, §173. See Atlan, *SR* II, chap. 12.

12. THE FLOOD AND THE TOWER OF BABEL

But this randomness becomes evil when it is fossilized in specific realizations that mistake themselves for their own final causes. Two extreme representations are the archetypal generations of the Flood and of Babel.[91] In extremely schematic fashion, they represent modes of being that embody two extremes in the control of the randomness of birth and begetting.

In the age that preceded the Flood, "all flesh had corrupted its way" (Gen. 6:12). This too was a mixed multitude of individuals and species, prey to random encounters, desires, and events, to the *miḳreh* of what happens without being named: events without meaning and the contingency of what might not have been, the unmeaning of the undifferentiated in which everything is always possible but nothing that exists carries any meaning, because everything is interchangeable. Far from being, ipso facto, a bearer of death, this proliferation of the undifferentiated leads to a constant renewal of life, with its fundamental property of fertilization and procreation, of "being fruitful and multiplying." But this luxuriance of life soon becomes self-destructive and lethal, because its excesses preclude stability; life, in the sense of a path or individual development by which a singular being can fulfill itself uniquely and thereby acquire the meaning its name should endow it with—the species name for plants and animals, the family name and given name for human beings, is no longer possible. Being alive is a necessary condition, but not a sufficient one, for individuality and the emergence of a self that can say "I." The living, as long as it is nameless, jeopardizes the emergence of the "I"—in Freud's sense that wherever there is an id, the ego (an "I") must emerge—unless it is channeled into and stabilized by processes of gestation and maturation that make it possible to distinguish and name families and individuals. The sparks of randomness are quickly extinguished if they are not absorbed by receptacles that can catch fire and burn without being consumed too quickly. Better, perhaps, than the multiplicity of scattered sparks is the image of the candle flame or, better still, the bush that "was aflame but was not consumed" (Exod. 3:2).

The Flood drowned all this undifferentiated life. A new start soon produced the people of Babel, who essayed the antithetical mode of controlling randomness. They tried to plan everything, to strip the sparks of life of all possibilities for self-expression by double-locking them within stones and mortar that subordinated them to a project aiming at total control of the future:[92] forging a single name for all humanity and waging war against heaven, the source of possible renewal, because it is also the source of the uncontrolled rains and storms that might interfere with construction—that is, creating a unique source of meaning that can be imposed on and determine the entire universe, on earth and in heaven. They said, "Come, let us build ourselves a city, and a tower with its top in the heavens [which the Midrash interprets as meaning "wage war against heaven"], to make a name for ourselves, lest we be scattered all over the earth" (Gen. 11:4). This is the pursuit of control over objects and control over gods, the planned and coordinated self-creation of the human being, of a one-dimensional collective Adam. There too, individuals and subjects are sacrificed in the end; not, this time, in the undifferentiated whole and fleeting multiplicity of too much chance but, on the contrary, in a deluded attempt to impose order and eliminate it.

91. As we have seen, it is the function of a third generation, that of the wilderness, to repair the damage and lead these conflicting sparks of randomness to an existence that is finally reconciled—even if there are relapses later.

92. Gen. 11:1–9.

In practice, these contingencies of existence can be handled in two ways: one, deeper, in which random events are transformed into singularities that bear various meanings, which appear in original and unique constructions that are always being started afresh; the other, superficial and deceptive, in which random events are hemmed in by the law of large numbers in the homogeneity of statistical averages. This makes it possible to control and master randomness by replacing it with a statistical calculation, but only on condition of wiping away all differences and eliminating every individuality. Here chance is not eliminated, but congealed into a homogenous random whole in which causal relations are collective and all individuals are interchangeable, like molecules in statistical thermodynamics; any uniqueness is excluded, along with any spontaneity and any novelty. Opposed to this death by dilution in the ineffable multiplicity of chaos is death by the rigidity of imposed order, directed by the predetermined collective name and exclusive meaning.

The first manifestation of this homogenization of the operational is the way in which language, the key symbol of the story of Babel, is used. Collective planning can work only where there is a shared formal and objective language (as in mathematics and the information sciences) from which everything that is unique, subjective, or, in brief, animated by what may be called an inner life is excluded as incommunicable. At this stage, salvation can come only from the fuzziness and relative non-understanding, the creative misunderstandings, that are reintroduced by the messy "vagueness"[93] of natural languages. The confused and scattered tongues of Babel are superior to the one and univocal language of Paradise.[94] At least they do not leave us with the illusion that the transparency and universality of a technical and functional language, in which there is no distance between sign and act, could recover the full reality of individual and relatively autonomous human beings.

Between the courses charted by these two "bad" generations there is another, narrow path, where the sparks of randomness are organized, but in an unplanned fashion, and can be assigned individual names that endow them with both life and meaning—meaning, through what the name says or designates; and life, in that the meaning is not overdetermined a priori but is on the contrary emergent, constructed, ordered, and organized a posteriori.

Calling (*ḳeri'ah*) an individual by her own name, is the inverse of *miḳreh* and *ḳeri*, of whatever transpires impersonally, without a name; the haphazard prophet Balaam is the foil of Moses, the prophet who is called. It is true that the inhabitants of Babel wanted to "make themselves a name." But that was a collective name, not one that could be spoken; it could only be made to exist, like a well-built tower, a unique sign visible to all and dominating everything. As a source of identification for individuals, the collective name—like an ideology or a religion—threatens to stifle them with an overdetermined meaning and prevent them from being called, each one, by his or her own name.

In this perspective, the subsequent history of humankind is the attempt to amend the catastrophes caused by these two con-

93. This idea is suggested by, among others, the Wittgenstein of the *Philosophical Investigations*. See Atlan, *Enlightenment*, pp. 370–371.

94. It is fascinating to follow, with Maurice Olender (*The Languages of Paradise: Aryans and Semites, A Match Made in Heaven*, trans. Arthur Goldhammer, rev. and augm. ed. [New York: Other Press, 2002]), the attempts, simultaneously grandiose and ridiculous, both "scientific" and mystical, to base the supremacy of a religion that lays claim to exclusive truth and universality on the identification (utterly serious) of what language was "really" spoken in Eden.

trasting methods of managing the randomness of the succession of generations. The story culminates in the generation of the Exodus and the wandering in the wilderness, who experienced both the collective birth of the nation and the individual release from slavery, followed by the gift of the Torah, which is simultaneously life and the middle path between death from too much living and death by suffocation in a predefined rut or a sealed tower.

This is where the talmudic sages, the disciples of Moses, delving deep into the study and practice of this path, brought order to the sparks of randomness without extinguishing them. For every sage, as for Jeremiah and all those said to have made a golem, the goal of the project is clearly himself, in his own existence. The golem he seeks to fashion and animate is just himself. But how can one know whether what one finds and creates in this fashion is indeed the Tree of Life, the Torah, and not one of their many caricatures? How can we know that we are not being deceived by the numerous mirages of the Tree of Knowledge, necessarily ambiguous when separated from the Tree of Life, even though it cannot be reached without the help of knowledge? We can judge only on the basis of the objective product of this creative activity, which is just as objective as what makes the world exist and spawns the generations.

We see, then, that the undecided and uncertain development of these possibilities, of these sparks of randomness with their potential for destruction or for gain, is the crux of a story whose meaning is not written in advance—with the attendant risk that it may have no meaning at all, if no one wishes to or is able to construct one post factum, or if the attempts to forge a meaning prove vain, because too limited and blind about themselves; with the risk, in other words, that the story, for those who chance upon it, may itself be just a matter of chance.[95] The sin—or rather the failing—of the first man and woman did not lie in their ineluctable confrontation with this challenge, for that is in fact their highest calling and endows them with the ability to "create worlds";[96] it lay, rather, in how it was done, by piling ambiguity on ambiguity, until it was revealed in retrospect to have been even more fertile. What ensues is human beings' desire to create human beings, to master the creation of life through mastery of the Name and of names, an ambition that has a devilish resemblance to the project of the people of Babel. But the same project was pursued, as we have seen, by Jeremiah, who did not recoil from fabricating a perfect human being—only to learn, from his own handiwork, that there is a limit not to be crossed, on pain of seeing this creation destroy itself and amplify thereby the confusion it was intended to diminish.

95. See Leviticus 26:1, 27, 40–41, where the word *ḳeri* is usually rendered as "contrary," "disobedient," or "hostile." See also Chapter 6.

96. "Rava said: If the righteous desired it, they could create a world" (B *Sanhedrin* 65b); see above, §4.

13. THE TIME OF MAGIC AND THE TIME OF TECHNOLOGY

And what of us today? We no longer dispose of magic or practical kabbalah to produce a golem. Instead, we have science and technology, on the one hand, confronting the irrational and religious obscurantism fed by the fears they inspire, on the other.

When it comes to prescientific modes of knowing and controlling nature, there is no great difference between the practices of pagan magicians and *'avodat ha-ḳodesh*, the "Great Work of the sacred" of the teachers of the esoteric lore of Israel. One expression of this tension is a dramatic dialogue between Rabbi Simeon bar Yoḥai and his son, Rabbi Eleazar, reported in *Tikkunei ha-Zohar*.[97] The latter asks how it is that idolaters can successfully manipulate the forces of nature. Rabbi Simeon, referring explicitly to the Tree of Knowledge from which the first man ate, explains that they know the *shem ha-meforash*, "the explicit name"[98] (and the various divine names derived from it), and employ combinations of their letters in formulas that give them knowledge of and control over these forces. Rabbi Eleazar is astonished that pagans can wield these names effectively. His father replies that it is a consequence of the knowledge of good and bad shared by all of Adam's descendants. This knowledge allows them to enlist the forces of nature in their own service, to satisfy their desires, for acts of death as well as of life; this, in fact, is the import of the Torah's ban on pronouncing the divine name in vain. A long dialogue ensues on the theme of the mixture of good and evil, of good and bad, of life and death, of the unified and the dispersed—the essence of these pagan practices, all of which are derived from their exploitation of the Tree of Knowledge and which, retroactively, reinvigorate the tree.

The dialogue seems to be over when Rabbi Eleazar concludes that the Jews, too, should be forbidden to employ angels and names, because such a mixture is inevitable and is necessarily accompanied by the idolatrous destruction of unity. But Rabbi Simeon reassures him, noting the difference between exploitation of the Tree of Knowledge, which is necessarily impure, and that of all the other trees in Eden, including the Tree of Life, which is the Torah. The latter, for those who know its secrets, is not a compound of good and bad, but is like the Tree of Life in Eden before the fall, where "evil does not dwell." In other words, all these rites of control based on names rest on the same underlying principle; but there is all the difference in the world between the practices of pagan magicians, like Balaam and Nebuchadnezzar, and those of Rabbi Simeon and his colleagues, because of their purposes and intentions. The former are moved by the crafty drive for selfish, reductive, and utilitarian ends; the latter, by the aspiration to repair the worlds and restore their unity—or, according to the kabbalistic formula, to unite *Ha-kadosh barukh hu* with the *Shekhinah*, that is, God in his post-biblical name, "The sacred-who-is-blessed," with his Presence. One group combines the letters of the Name in holiness, the others in impurity; one group in unity, the other in multiplicity and dispersion; one group in an exclusive perspective of life and construction, the others in the mixture of good and bad, joy and suffering, that is characteristic of the serpent's tree.[99]

In such a context, when the problem of the possible fabrication of made-to-order human beings is again on the agenda, we reach a limit where the scale of values seems to be inverted and randomness is better than planning and order.

97. *Tikkunei ha-Zohar*, end of chap. 6 (Vilna, 1876, pp. 100–101; Jerusalem, 1963, pp. 271–273). See the Gaon of Vilna's commentary on this passage, ibid.

98. <The Hebrew term is frequently rendered as "the Ineffable Name"; but in fact the *shem ha-meforash* may or may not be the same as the Tetragrammaton YHWH, which is "ineffable" because never pronounced as written since the destruction of the Second Temple.>

99. In his commentary on Exodus 20:3, Naḥmanides distinguishes three types of idolatry from which magical practices derive their power: that of the agent intellects, that of the stars and constellations, and that of spirits and demons that are active on earth (including through powerful deified monarchs like Pharaoh and Nebuchadnezzar). Nations submitted to these kings because of their identification with a particular god, that is, with a star or constellation that gave them their strength and success. Naḥmanides attributes the magical ability to manipulate the

If the issue is not simply curing disease or preventing suffering, but exercising the power of an ostensibly complete knowledge that can dominate and control nature, we come to the barrier that must not be crossed: the boundary where respect for the random element in the birth of a future being wins out over the order that could be contributed by knowledge. This unexpected inversion, which seems to have astonished Jeremiah himself, derives from the fact that the knowledge that plans and controls, however perfect it may be, is nevertheless that of a finite mind and will. To avoid being the slave of the mind and will of another, the future human individual must not be subject to them; if possible, its being must be defined only by an infinite understanding to which no human being can have full access. It is this origin beyond the limits of a finite understanding that we necessarily perceive as indeterminate, as a random set of possibilities.[100]

This does not call into question the truth, efficacy, or legitimacy of the knowledge whose bounds we are always seeking to extend, even to the point of the ability to create a hypothetically "perfect" human being. It is, rather, taking the next step and acting on this knowledge that, in this particular case, turns the scale of values on its head. This total determination by knowledge that is efficacious and true in its own legitimate domain

forces associated with the stars, the "hosts of heaven," to the dispersion of the nations after Babel and their subsequent subjection to various tutelary deities, each reigning in its own place. By contrast, the initial project of humankind at Babel was to "to make *a name* for ourselves, lest we be scattered all over the earth" (Gen. 11:4); that is, to employ for its own profit all the possible advantages of control of the Name. In this sense, pagan magic—exploiting the Name by fragmenting its unity into its different letters—is the normal condition of the scattered tribes of humanity after Babel. Only Israel, with its special (philosophical?) vocation to achieve unity at any price, is forbidden to employ it. This is the meaning of the curious statement in Moses' admonition to the people: "[Beware] lest lifting up your eyes to heaven, and seeing the sun and the moon and the stars, all the host of heaven, you are led astray to bow down and worship them, [objects] that YHWH your god has allotted to all the [other] peoples everywhere under heaven" (Deut. 4:19).

The medieval poet-philosopher Judah Halevi, who preceded Naḥmanides and the publication of the *Zohar*, seems to draw on the same source of inspiration in a poem that has been incorporated into the Yom Kippur service, "The name of your divinity is the One-Name," which can be construed as "the name expressed by your divinity is [for us] the One-Name." The letters of this same name (the Tetragrammaton) can be multiplied to produce what acts like a company of multiple divinities. In Chapter 3 we will examine Abraham Ibn Ezra's construal of the first of the Ten Commandments, "I am YHWH your god(s)" (Exod. 20:2), as meaning that it is the Tetragrammaton, "the glorious name that is written but not read," that all must accept as their only god. Drawing on the *Zohar*'s reading of Zechariah 14:19 (see below, Chapter 3, §§18–19), we will see how the unity of YHWH depends on the oneness of his name, the name of the name in some sense, which is simply, for example, "God," i.e., all of the divinities that human beings name in their languages.

100. This relationship between an infinite mind and a finite mind that is part of it recalls Spinoza's distinction between God, "insofar as he is infinite," on the one hand, and "insofar as he constitutes the essence of the human mind," on the other: "Hence it follows, that the human mind is part of the infinite intellect of God; thus when we say, that the human mind perceives this or that, we make the assertion, that God has this or that idea, not insofar as he is infinite, but insofar as he is displayed in the nature of the human mind, or insofar as he constitutes the essence of the human mind; and when we say that God has this or that idea, not only insofar as he constitutes the essence of the human mind, but also insofar as he, simultaneously with the human mind, has the further idea of another thing, we assert that the human mind perceives a thing in part or inadequately" (*Ethics* II 11, Corollary; see also III 1, Proof). These two *insofar*s refer only to God "insofar as he is a thinking thing" or "insofar as he is considered as a thinking thing, not insofar as he is unfolded in [*or* "explained by"] any other attribute" (II 5). The same distinction can be made concerning God insofar as he is explained by some other attribute, notably insofar as he is extended substance; except that it would no longer be a case of ideas but of body, and the very notion of an idea that is inadequate because incomplete is inconceivable. The question of inadequacy as a character of the finite can be asked only with regard to modes of thought, because of their reflexive character ("ideas of ideas") and their cognitive role. This relative superiority of the body over the mind reverses the superiority that reflexivity seems to accord to the mode of thought; "knowledge" of the body by the body is necessarily adequate. It is the reflexivity of knowledge through ideas that can render it inadequate (see Chapter 2, n. 22).

(and thus "adequate" in Spinoza's sense), but that is nevertheless finite (and therefore applies only to common notions, which are "equally in a part and in the whole"[101]), totally rules out the possibility that this human individual could enjoy a (relatively) autonomous existence and locks her up in a future that is already inscribed as her fate.

Thus something valuable remains of the prescientific traditions of knowledge we have evoked, on condition that we not mine them for a pseudo-science that would compete with modern science.

Natural magic, in the strict sense of the Renaissance, no longer exists; Paracelsian alchemy and medicine have been supplanted by much more effective sciences and technologies. As for practical kabbalah, most people regard it with great suspicion; we seem to have taken Jeremiah's lesson to heart. Creative action applied to objects has been replaced by internalization of this knowledge[102] and by work on oneself, whether this means the popular and emotional context of the Hasidim of the Ba'al Shem Tov or the elitist individualism of their opponents, the Mitnaggedim.[103] In any case, so far as its contemporary status is concerned, practical kabbalah has joined natural magic, alchemy, and astrology as sources of the superstitious regressions of those who insist on seeing them as natural science.

Nevertheless, in our modern age, when science has been divorced from magic and kabbalah has been internalized, the stakes remain the same, although the starting conditions are different. What will be the nature of the time of those generations, the time in which humanity produces itself? The time of the golem, the time of magic, is radically different from the time of technology—all the more so after the concept of the golem has been internalized in ritual and meditation. The time of technology is the objective time of transforming objects. The time of magic is the time of internal experience, the time of the shaman, for example, in which the relationship with objects—through names and spells—is just as immediate as one's relationship to oneself. This is why technology is incomparably more effective at transforming objects. Today we know that language does not act on objects directly. You cannot bring objects into being or modify them by reciting spells. Some investment of matter itself, objective and independent of the mind that employs it, is indispensable. This is one of the main achievements of what we call the experimental method, by which modern science made its final break with the natural magic of the Renaissance.

A magician who had achieved great wisdom and purity needed only to pronounce the formula while concentrating on it in order to act directly on objects. For science and the technology based on it, we know that this is not enough; spells exert no direct effect outside the person who utters them. Despite the efficacy of the abstract formulas of mathematical physics—which is in fact quite astonishing—you can recite Maxwell's equations thousands of times and never pro-

101. This is how Spinoza defines the object of "reason" or "knowledge of the second kind," which is certainly adequate but, unlike the "third kind of knowledge, . . . intuition," cannot know things and beings in their essence (*Ethics* II, 37–39, 40, Note 2; V 25–28).

102. The appropriation by both Hasidim and Mitnaggedim of the notion of making a golem through knowledge of names is illustrated by a story much closer in time to us—the end of the eighteenth century—about the Gaon of Vilna and his disciple, Ḥayyim Volozhiner, who seem to have taken Jeremiah's lesson to heart. According to the disciple's introduction to the Gaon's commentary on *Sefer Yeẓirah*, from the age of thirteen the latter possessed the ability to create a golem but refrained from doing so.

103. In the early twentieth century, Rabbi Abraham Isaac Hacohen Kook, both a major political figure and a great mystical poet, made an original attempt to merge these two traditions.

duce electromagnetic radiation; constant repetition of $E=mc^2$ has never caused anything to blow up. There has to be an objective arrangement of matter outside the subject for the abstract equations of physics (and, today, of molecular biology) to be effective.

Technology is a mediated form of magic, whereas magic wanted to be a technique of an immediate relationship between the abstract and concrete. Technology succeeds where magic fails. But this is true only when we are talking about things in their objective reality, distinct from any internal experience. Technology's effectiveness in manipulating objects—the most recent extension of the ambiguity of the Tree of Knowledge—comes at the cost of having to establish a space and time of things that are totally separate from the interior space and individual progress by which a subject can be constructed.

14. THE TIMES OF MEMORY

It bears note that the first domain to which magic was applied was the human mind itself—chiefly memory—which was to be transformed and improved along with objects. This is the source of the "magical memory," based on the use of mnemotechnical formulas, of which traces can be found in the Talmud and Stoic writings and of which Giordano Bruno, the great magus of the Renaissance, was a master.[104] It may in fact be our most valuable relic of the prescientific traditions: the experience of a time of human mind and life informed by action on memory, in which ritual and myth play a privileged role.

Thus giving up the idea of manufacturing a human being as an object, just when one has acquired the knowledge, wisdom, and holiness that make it possible to do so, means converting the technical goal of fabricating another body as an object into an educational goal of using memory to create human time, with all its components—memory of facts, of course, but also memory of the texts and formulas that convey the myths and rituals.

What remains effective in the time of the golem (which we have decided not to make) is the use of the knowledge and wisdom acquired to create a human time, employing an internalized magic that acts chiefly on memory. We can understand how such action on memory could be dangerous, like every human action, because it can be put in the service of various goals, just as education can serve noble ends or disastrous ideologies. Its goal cannot be posited once and for all as something that can be known in advance. On the contrary, it must be tested by the creation of meaning in the daily routine—or in eternity, which comes down to the same thing—and not in a planned future that can be only a repetition, or at best an extrapolation, of the past.

Thus the reality of the historical time of the generations depends on how the generations are produced. If they are created by planned processes, even those aiming at the noblest of utilitarian objectives,[105] this planned manufacture will eradicate the reality of the time of the generations and stagnate in repetition of the same thing, a fossilized reproduction of the Tower of Babel.

104. See, for example, F. A. Yates, *Giordano Bruno and the Hermetic Tradition* (Chicago: University of Chicago Press, 1964), esp. pp. 190–201.

105. For example, producing the fantasy of a perfect humanity, free of crime and pain, a sort of improved *ẓelem*. Some still speak of improving the human race by means of genetics, as if past experiences with breeding human beings and radical eugenics had not transformed these sometimes naïve and angelic intentions into a catastrophic and satanic reality.

15. THE INDIVIDUAL "SEAL" IN THE NATURE OF THINGS

In this context, a note of relative optimism emerges when we observe that the full realization of such planning is probably impossible. The *ẓelem*,[106] the image and formula that express nature in human beings with the imprint of their divinity, implies by definition the uniqueness of every individual; hence any replica, produced by cloning or some other means, would be no better than a caricature. As the Talmud notes, all human beings are "impressed with the seal (*ḥotam*) of the first man, yet none of them resembles his fellow."[107] This is how the natural object produced by the Creator differs from the artificial product of human beings, who, employing the same seal, can make only identical repetitions of the same thing—clones. But the talmudic text goes on to ask why it is important for human beings to be non-identical; or, if you wish, what are the social and moral obligations that result from human differentiation.[108] When considering this rhetorical question ("Why was it necessary that no two persons resemble each other?"), we again encounter the dialectic of common universality and individual differentiation represented by the generations of the Flood and of Babel.

This passage begins by noting the value of the unity of the human species. Its origin from a single couple rules out any pretensions to innate superiority by one family or group over another, with the violence that always accompanies such pretensions. But why do individual members of this single species have to be unique? The response is, "lest a man see a beautiful dwelling or

106. See Atlan, *SR* II, chap. 5.

107. M *Sanhedrin* 4:5 (B *Sanhedrin* 37a and 38a). The verb *ṭava'*, here used for the action of impressing a seal onto clay, has an interesting history that may be relevant to our discussion. Medieval Hebrew translators coined the word *ṭeva'* 'nature', not found in the Bible or Talmud, from the same root. (In the Bible [Exod. 15:4] the verb signifies "sink, drown.") This means that the philosophic notion of "nature," with its scholastic forms *natura naturans* and *natura naturata*, taken over later by Spinoza, is expressed in Hebrew by the same word that the Talmud employs here to indicate the impression in each individual of his "stamp" or "seal," that is, his "nature," in its human dimension that is simultaneously universal and particular. (See, for example, the usage of these terms by Joseph Gikatilla, a thirteenth-century kabbalist whom we will meet in Chapter 6.)

108. The format of such questions should not mislead us. For the authors and readers of these texts, it does not necessarily imply some pre-established harmony in which a particular effect of nature is to be explained by its utility, in an overall system of redemption planned by the Creator. In the sixteenth century, for example, the Maharal of Prague used a question of the same type to explain that things could not have been created other than they are. Tractate *Avot* (5:1) asks why the world was created by means of ten divine utterances, when a single fiat would have sufficed, and provides the teleological answer that it was "in order to punish the wicked who destroy the world that was created by means of ten utterances, and to reward the righteous who preserve a world that was created by means of ten utterances." In his commentary on this text, the Maharal underscores that the possibility the world might have been created by means of a single divine utterance is evoked only rhetorically; it was in fact necessary, given the nature of the universe, which is a unified multiplicity, that the smallest numerical unity specify such a whole. (This unity is represented by ten, no doubt because of its meaning in decimal notation, which, in this context, is held to be "natural.") In other words, such questions are not asking "Why are things this way?" as if they could have been otherwise, had not some specific intention guided the Creator's will. The real question, to which the text supplies an answer, is, "What can we learn, what moral can we draw, from the fact that the world was created by means of ten utterances (that is, from what is dictated by the nature of things and is expressed by the symbolism of ten as a unit within multiplicity)?" A possible answer is then given, which assesses blame or credit for destroying or preserving this world as a function of the number of utterances, that is, of the value and complexity of what is destroyed or preserved. The form of the question asks about deliberate ends, but this is merely a rhetorical device divorced from the true reasons, which belong to the order of necessity. The same applies to the passage from *Sanhedrin* cited here: Why was it necessary that people not all look alike? The answer actually addresses a different question: "What moral can we draw from this state of affairs?" It does not deal with the fact of physical differences or reasons for it, but with the value of human individuality expressed in this fashion. The true reason derives from the unique character of the sum total of all the determinations that constitute an individual.

a beautiful woman and say, 'that one is mine!'" This answer is based on Rabbi Meir's dictum that individuals differ in their faces, their voices, and their minds (or knowledge). These differences save them from a sameness that could be the author of even greater violence.

Once again we encounter the association between the mixed multitude and the rampant theft and violence of all against all that characterized the generation of the Flood, applied to two domains: that of being, represented by the encounter between the sexes (a beautiful woman), and that of having (a beautiful dwelling), in which one's relationship to another person involves rules of property and ownership. People have to look different and not be like clones of one another so that, in the presence of a beautiful woman or a beautiful dwelling, no one can exploit the general lack of differentiation and pretend that she or it is exclusively his.

People are recognized by their face and voice—by sight during the day, by sound at night—eliminating the total confusion in which, among other problems, every man would be indifferently the same for every woman, and vice versa. The Other is not an abstraction, but the perception of a specific appearance and voice.[109] As for what each person knows deep inside about the objects that are his own, the fact that this knowledge is hidden from others limits the undifferentiated claim of all to everything, along with the rampant theft and violence of all against all that accompany it.

We can take this as grounds for optimism—relative, of course—if we observe that this irreducible singularity of the individual is an essential element of the nature of human beings, in their conjoined body and mind. In the worst case, even reproductive cloning will never totally eliminate the differences. Only in the fantasy that "it's all in the genes," in which we might be fully "programmed by our genome," can one imagine otherwise. Indeed, reproductive cloning or the splitting of embryos, which seems to be technically feasible—that is, the production of multiple individuals with one and the same genome, armies of identical twins—would not eliminate every singularity and produce individuals who are "totally" identical. For, whatever the world may think, biological causes do not

109. The other encountered concretely is not the Other, but another and yet another and still another—each one different, but each a different other. How can there be so many others? Ancient philosophers, following Plotinus, never stopped sounding this unfathomable abyss, which millions of pages about monads cannot fill up. For it involves not just an intellectual experience but also the emotions, or rather the indefinite sum total of emotions, as well as memory, imagination, and intellection, which constitute the experience of the *I*, projected by the imagination on another *I*.

This is a mystical experience that can no more be put into words than spiritual illumination can, and with the same paradox of the millions of pages in which those individuals who speak cannot keep themselves from trying to say the unsayable.

The rationalism of philosopher-mystics may be the least-bad manifestation, pedagogically, of this simultaneously futile and indispensable enterprise. For these writers, philosophers in their method, mystics by their ineffable experiences of the individual (in their own *I*, in the *I* of others, in imagination, vision, the illumination of the All, or of the Infinite, or of the Other), the formal exposition of the Neoplatonic progression, as found, for example in Proclus' *Elements of Theology* and its medieval version, the *Liber de Causis*, may support this impossible (because always paradoxical) bridge, which is indispensable for teaching purposes.

These attempts at formal and scholastic rationality, incomprehensible in isolation from the axiological role that these philosophers' works played in their own lives (see Chapter 2, n. 127), serve an undeniable educational function when associated with other mental and physical experiences—artistic, athletic, or psychedelic—in short, with "knowledge."

Closer to our day, two philosophers introduce us, by very different paths, to this rational doctrine of the experience of the one and the many: Spinoza, with his idea of a single infinitely infinite substance present in each of its modes in its *conatus*; and Leibniz, with his monads, an infinity of infinite substances each containing, in its own fashion, the totality of all the others. We shall return to this in Chapter 4.

operate so simply, like a computer program "stored" in our genome. The epigenetic level introduces other causal factors and a certain play among them.[110]

At most, as in the case of real twins, they would look alike. But even this is enough to justify banning the technique and gives a new and unexpected meaning to the talmudic passage discussed here. We should oppose all attempts at human reproductive cloning for social reasons, associated with physical resemblance and bizarre ideas of planned manufacture, rather than for reasons of a biological or metaphysical order. There is no need to invoke some blasphemous rivalry with God ("playing God") or some assault on a human "essence" supposedly contained in the genome. It is enough to refer to the disruptions that identical appearances would produce in the social images of who is who and who owns what.[111]

But just as projects for psychosocial conditioning or subjugation of the masses by brainwashing have always been defeated by our ability to say "no," to revolt and free ourselves from slavery, we may assume that the same fate would befall a project of repression based on genetic uniformity. Projects that aim at total planning are always frustrated by unexpected accidents or results (the human community at Babel was scattered in all directions and the tower was never completed) because there can never be "perfect" control of the fabrication of human beings. Ultimately, as at the conclusion of Jeremiah's experiment, it is better not to tempt the devil—Ashmedai and his demons—and not even to try it.

110. See above, n. 6, and Chapter 3. See also R. C. Lewontin, Steven Rose, and Leon J. Kamin, *Not in Our Genes: Biology, Ideology, and Human Nature* (New York: Pantheon Books, 1984).

111. See Atlan, "Transfert de noyau et clonage." For the present, we may admit that a clone would have the resemblance of a natural twin to the being that was cloned, with perhaps a one-generation time shift. But we cannot be certain of this, because the biological differences between the two clones would be greater than that between identical twins, not only because of the age difference but also because they would be produced from different ova, with different cytoplasm. If the physical resemblance produced by the identity of chromosomic genes is not as great as thought, one of the arguments against human reproductive cloning would be removed, at least when only one clone is produced. For now, however, several arguments underline the dangers involved in applying this technique to human beings. Even were we certain that the technique is biologically safe—which is not yet the case—eliminating the randomness of birth would be one more step toward transforming individuals into manufactured objects. As underlined by the report of the [French] National Ethics Advisory Committee for the Life Sciences and Health [CCNE], "Reply to the President of the French Republic on the subject of Reproductive Cloning," Opinion 54 (Apr. 22, 1997), http://www.ccne-ethique.fr/docs/en/aviso54.pdf, it would totally disrupt society, involving, in part, an instrumentalization of the individual that would eliminate genetic indeterminacy, that is, much of the random element in the mechanisms of sexual reproduction. This indeterminacy is normally the fruit of the chance combinations of parental genes—the so-called genetic lottery—that presides at the unique production of the genome of each individual at the time of fertilization.

> Identical twins are . . . limited to brothers or sisters born at the same time. . . . One can well imagine the kind of social reality brought about by a production of clones, no longer the fruit of chance and exception, and no longer necessarily coexisting in time. These human beings, individuals in terms of their psyche in spite of their genetic similitude, would be seen in both the literal and the figurative senses of the word as identical copies of each other and of the cloned individual of which they would truly be a copy. In this way the symbolic value of the human face and body as the substrate of the person's uniqueness would be undermined. . . . Human clones would know they are clones and would know that others see them as clones. One cannot be blind to the intolerable lowering of a person to the status of an object that would ensue. Who could guarantee that such a destabilising effect on cardinal social representations would not pave the way for attempts at utilitarian creation of human varieties, which means the creation of new kinds of slavery, to which some well-known scientists dare to allude with carefree partiality? (p. 18)

However, slavery, like torture and other crimes against humanity, is a moral regression that flouts but does not eliminate the humanity of the men, women, and children who are its victims. (The case might be different for the creation of hybrids, part-human part-ape, for example, like the sheep-goat chimeras produced by the early fusion of embryos.) As noted by the National Bioethics Advisory Commission in the United States, "Even for absolute opponents, the process of cloning humans only violates human dignity; it does not diminish human dignity" (*Cloning Human Beings*, Report and Recommendations of the National Bioethics Advisory Commission [Rockville, MD, 1997]).

In the absence of unforeseen novelties, time is reversible and disappears as such, reduced to a spatial coordinate with no future. If the time of the generations is to continue, the sparks of randomness must retain their unexpected character, integrated into a process that is simultaneously created and creator, where they may perhaps receive (albeit after the fact) new meaning.

CHAPTER 2

NO ONE KNOWS WHAT THE BODY CAN DO: THE TREE OF KNOWLEDGE AND GAMES OF ABSOLUTE DETERMINISM

1. THE SHAPE OF THE BODY AND THE CORPOREALITY OF GOD

"No one has hitherto laid down the limits to the powers of the body, that is, no one has as yet been taught by experience what the body can accomplish solely by the laws of nature." This is Spinoza's note on the second proposition of *Ethics* III, where he establishes that "body cannot determine mind to think, neither can mind determine body to motion or rest or any state different from these, if such there be."

No one knows what the body can do. So it is in terms of knowledge, or rather of ignorance, that the question of what the body does is raised. One of the paradoxes of knowledge has to do with the body's knowledge of the body. We know only through our body, but we do not know how our body knows, or who knows through our body.

Nor do we know how we know, as long as we do not know what knowledge our body has. Even if we allow that the body does not cause the mind to think,[1] we can know the world—not to mention ourselves—only by means of our perceptions, our sensations, our actions, without which we can have no idea of anything and can know nothing. Our very identity cannot be dissoci-

1. We must not forget that, for Spinoza, the reason "the body cannot determine mind to think" is not that mind and body are separate entities; on the contrary, they are one and the same thing conceived under different attributes (*Ethics* III 2). (This is why Spinoza's monism, which does not ignore our lack of knowledge about the physical nature of our body—far greater in his time than today—led him to observe that we do not *really* understand the unity of body and mind, because "no one will be able to grasp this [unity] adequately or distinctly, unless he first has adequate knowledge of the nature of our body" [*Ethics* II 13, Note].) Furthermore, "men . . . are . . . convinced that it is merely at the bidding of the mind that the body is set in motion or at rest, or performs a variety of actions depending solely on the mind's will or the exercise of thought. However, no one has hitherto laid down the limits to the powers of the body, that is, no one has as yet been taught by experience what the body can accomplish solely by the laws of nature, insofar as she is regarded as extension. No one hitherto has gained such an accurate knowledge of the bodily mechanism, that he can explain all its functions; nor need I call attention to the fact that many actions are observed in the lower animals, which far transcend human sagacity, and that somnambulists do many things in their sleep, which they would not venture to do when awake: these instances are enough to show, that the body can by the sole laws of its nature do many things which the mind wonders at" (*Ethics* III 2, Note).

ated from our body; we are our body. In Hebrew, the grammatical notion of first/second/third person is *guf rishon/sheni/shelishi* (first/second/third *body*). Yet most of the time we do not know how our body knows and how something or someone comes to be thought by us—including the idea of our own "first person." Usually we have only a vague and inadequate idea, confused and mutilated (as Spinoza puts it), even of our own body.

Some fifty years ago, one of the pioneers of the cognitive sciences, Warren McCulloch, reviewed this question in a book entitled *Embodiments of Mind*, asking (the title of the first chapter), "What is a number, that a man may know it, and a man, that he may know a number?" This recalls another question, suggested by the most ancient treatise of kabbalistic cosmology and cosmogony, *Sefer Yeẓirah* (The Book of Creation), with regard to the nature of the ten *sefirot*:[2] Do we count by tens because we have ten fingers, or do we have ten fingers because there are ten *sefirot*—that is, because of some essential property of the number ten?[3]

This paradoxical ignorance of the body, which is nevertheless involved in all aspects of our knowledge, probably explains why we speak about it in so many ways. The discourses about the body proliferate without end; this text is merely one more. For there are several bodies—ultimately, as many as there are individuals. The body as conceived of by the natural scientist is not the same as that of the phenomenologist, the anthropologist, the psychoanalyst, or the lawyer. Very different, too, are the bodies of the artisan or the athlete, not to mention what the scholar refers to as a "body" of knowledge. Finally, we all perceive our own bodies from the inside, in a singular and irreducible fashion, as a unique complex of sensations, thoughts, and fantasies. It is this body-subject that is sometimes expressed, when talent is present, in the singularity of a work of art.

Here we shall limit ourselves to two problems that have obscured and continue to obscure the role of the body in what constitutes the humanity of human beings. One of these problems is very old and has troubled theology and philosophy for centuries: the incorporeality of God. The other, of much more recent vintage, emerged along with contemporary biology. This is the problem raised by the

2. *Sefirot* almost defies translation. It is the plural of *sefirah*, which, by association with other words derived from the same root, has multiple senses: the arithmetical meaning "count" or "counting" or the geometrical meaning "sphere"; but also the letters in which a book (*sefer*) is written, their count or numerical value (*sefar*), and the story (*sippur*) they tell. This triple nature of written language and its animation in a book (*sefer, sefar, sippur*) is invoked in the very first paragraph of *Sefer Yeẓirah* (1:1). The first state, that of the written text itself, corresponds only to the twenty-two letters of the Hebrew alphabet, which are silent because there are no vowels. In the commentary ascribed to R. Asher ben David they are compared to a golem, in that they are the inanimate raw material of language and are animated and energized only when read with the vowels. Counting, in which the letters represent digits, is an intermediate state; here they express only the abstract order of numeration. Starting in the second paragraph, *Sefer Yeẓirah* speaks of "sefirot of nothingness," *sefirot belimah*. The latter word is generally understood as *beli-mah*, literally "without anything," as in Job 26:7, where the earth is suspended from *belimah*, that is, from nothingness, which has no existence in itself, no foundation. In this sense, the *sefirot belimah* are abstractions with no separate real existence, no ipseity. But *belimah*, written and read as a single word, can be taken as a nominal form of the verb *balam* 'brake, bridle, restrain' (as in Ps. 32:9). The *sefirot belimah* can then be understood in the context of *Sefer Yeẓirah* itself, which enjoins, shortly afterwards, "Bridle (*belom*) your mouth from speaking and your heart from thinking" (1:8). See, inter alia, Saadia Gaon's commentary on *Sefer Yeẓirah*, and Moses de Leon's *Sheḳel ha-ḳodesh* (see below, n. 18); also the commentaries by Rashi, Gersonides, and Ibn Ezra on Job 26:7.

3. See Moses Cordovero, *Pardes rimmonim*, new edition (Jerusalem: M. Attia, 1962), chapter 1. Here the ten *sefirot*, explicitly related to the ten digits of the two hands (and of the two feet) by *Sefer Yeẓirah*, are explained by reference to a verse in Psalms: "When I behold Your heavens, the work of Your fingers" (Ps. 8:4 [3]). In the imagery of Psalms, God's digits correspond to the arrangement of the *sefirot* as the vehicles for the shaping activity of the heavens. *Sefer Yeẓirah* describes this arrangement by referring to the physical model of the human fingers.

mechanistic and nonvitalist character of our current knowledge about the structure and function of living bodies—human beings, animals, plants—all of them composed, with greater or lesser degrees of complexity, of physical bodies with which they share the same material nature.

There are some very fine and learned remarks about the corporeality of God in the kabbalistic literature (which we are probably wrong to continue to describe as mystical, as if its traditional contrast to theology gave it an obscure, incomprehensible, irrational character). I do not have much to add here, other than to refer to earlier texts,[4] especially to *Nefesh ha-ḥayyim* (The Soul of Life) by Rabbi Ḥayyim Volozhiner (1749–1821),[5] where this problem is expounded in its full complexity and can be readily understood if we rid ourselves of all prior notions about the theological character of these works.[6] Nevertheless, a few words on the subject are essential, because the two problems we are going to tackle are not independent. In fact, the traditional theology in which God is incorporeal and pure spirit was supported over the centuries by a vitalist science and philosophy in which the human body was considered to be animated (in the double sense of living and conscious) by the soul (Latin *anima*), an immaterial essence that directed its movements and actions.

This is an old tradition, expressed in various ways—from the idealism of Plato to the dualism of Descartes[7]—but with which contemporary biology is increasingly at odds. Today we picture the life of our body, and thus *our* life, as the product of physical and chemical causes that, as a matter of principle, leave no room for the suprasensible, for a soul in which free will might reside, with the primary responsibility that derives from it and is a property of ourselves as moral subjects. What is more, these two problems have clear moral implications. Hence we shall examine several aspects of the foundation myth we know as the biblical narrative of the Tree of Knowledge and the fall of the first man.

With regard to the theological problem, Maimonides, following Aristotle, and like Averroes and Thomas Aquinas, posited the incorporeal nature of God as a truth that need not be discussed, because already proven, like God's existence and oneness. For him, these were not articles of faith specific to one religion or another, but universal truths incumbent on any rational mind educated in logic and science, that is, the content of Aristotle's *Physics* and *Metaphysics*, considered at the time to be definitively established scientific truth. *'Ein lo demut ha-guf ve-'eino guf*—"He has no semblance of a body and he is not a body"—is one of Maimonides' Thirteen Principles of Faith. But there is a fundamental misapprehension here; the status of these propositions about God's existence, unity, and incorporeality changed irrevocably when the Renaissance challenged Aristotelian physics. *Pace* Maimonides, they can no longer be accepted as demonstrated scientific truths but only as articles of faith, as fundamental tenets of religious dogma.

As for the truth of these propositions, they were soon contested not only by philosophers but also by adherents of the tradition. Spinoza's is probably the best-known philo-

4. See: Atlan, *Enlightenment*, chap. 3; *SR* II, chap. 11.

5. R. Ḥayyim Volozhiner, *Sefer Nefesh ha-ḥayyim* (1824) (Jerusalem, 5743, and various other editions).

6. Henri Atlan, "Une rationalité symbolique," *Les Nouveaux cahiers* 86 (1986), pp. 43–51.

7. The image of a team of horses is used in both cases, though differently. In the *Phaedrus*, Plato compares the soul to a pair of winged horses that pull the body, guided by a charioteer; but one of the horses moves in the right direction, while the other goes hither and thither, under the influence of the body that imprisons it and prevents it from following the coachman. For Descartes, man is an

sophical challenge to God's incorporeal nature. But there are explicit challenges to Maimonides' thesis in the kabbalistic literature, too, including those by Naḥmanides (thirteenth century) in his commentary on the Torah;[8] by the Maharal of Prague (sixteenth century),[9] in *Gevurot Hashem* and in *Derekh ḥayyim*, his commentary on tractate *Avot*; and by Moses Cordovero,[10] especially in his commentary on *Shi'ur ḳomah* (The Measure of the Stature). As Cordovero writes, "all material form is related to God";[11] thus there is no organ of the human body that the *Zohar* refrains from attributing to God. Although we should not understand these attributions in a simplistic and literal fashion, neither should we understand them symbolically, as referring to spiritual qualities, because the forms of the spirit are equally anthropomorphic and limited when compared to the formless infinite.[12]

Without going into the details of these texts, we can say that they are all variations on the theme of *mi-beśari 'eḥezeh 'eloah*, "I will see [*or* have a vision of] God from my flesh" (Job 19:26). Of course, none of them asserts, in so many words, that God's nature or essence is that of the human body. On the contrary, they keep up their guard against a simplistic reading of the Prophets and the *Zohar*, which might induce the error that anthropomorphisms that attribute all the organs of the human body to God actually express his nature or essence.

But the question is not posed in this way. The divine that is made visible by the form and structure of the body is clearly not the formless infinite. The divine *world*, that of the *sefirot*, seems

earthly carriage drawn by non-earthly horses.

8. Commentary on Genesis 46:1. See Atlan, *Enlightenment*, chap. 3.

9. Judah Loew ben Bezalel, *Gevurot Hashem*, second introduction (Tel Aviv: Pardes, 1955); idem, *Derekh ḥayyim*, chaps. 5 and 6 (Tel Aviv: Pardes, 1956). In the latter text, the Maharal explicitly takes issue with Maimonides and, siding with the kabbalists against the philosophers, affirms that divine knowledge can be considered only as one of the realizations of divine power. "Knowledge is not his [God's] essence" is the formula he employs repeatedly; it leads him to observe that we can know nothing of the essence of the one and undifferentiated Infinite, because we cannot say it is matter and cannot say it is spirit. It is precisely for this reason, he says, that the kabbalists could speak freely of the multiplicity of attributes that express the effects and actualizations of his power through the ten *sefirot*. This multiplicity, which is itself an object of knowledge according to a model that is both spatial (the shape of the human body) and temporal (the alterations that constitute its history), in no way affects the unity and simplicity of the Infinite, for, in any case, "His essence is neither matter nor intellect" (ibid., pp. 177–179.)

10. Moses Cordovero, *Shi'ur ḳomah* (Warsaw, 1883; repr. Jerusalem, 1994).

11. Ibid., chap. 61.

12. A number of kabbalists borrowed several of Maimonides' formulations but gave them quite different meanings. A striking example with regard to the issue of corporeality is a passage by Rabbi Isaiah Horowitz (sixteenth century) in his *Shenei luḥot ha-berit*; he comments on Maimonides' thirteen principles of faith, one by one, and integrates them into his doctrine, even though his system is inspired by kabbalists like Moses Cordovero and Moses Ibn Gabbai. When he reaches the Third Principle ("He has no semblance of a body and He is not a body"; *Shenei luḥot ha-berit* [Amsterdam, 1649; new ed., Jerusalem, 1959], Part 1, Third Principle, p. 79), he cites an interpretation by Menaḥem Azariah da Fano (who expounded the works of Cordovero and Isaac Luria) that argues for the superiority of matter over form—clearly paradoxical in the context of the standard interpretations of Maimonides. Here we are dealing with a well-known kabbalistic theme, in which the material world, which comes last in the scale of being, has a loftier origin than the others, because its order and repair are the ultimate objective and goal of creation. Similarly, the Sabbath-eve hymn *Lekhah dodi* (by the sixteenth-century Safed kabbalist Solomon Alkabetz) describes the lofty status of the Sabbath, the last day of the creation, "the last item in [the process] of creation comes at the start of the thought [that conceives it]." Similarly, for the *Zohar*, the material nature of the human body is "envied" by the angels (Introduction, pp. 4b–5a; see above, Chapter 1, n. 32). This inversion, introduced by da Fano, is extracted from Maimonides as follows: "Why is it necessary to affirm that He is not a body, after we have said that He does not have the semblance of a body? Because matter has a higher origin than form, so that one might think that the attribution of a 'formless' body to God might be appropriate" (David Cohen, *Ḳol ha-nevu'ah: Ha-higgayon ha-'ivri ha-shim'i* [Jerusalem: Mosad Harav Kook, 1970], p. 263). As for the lack of bodily form, this makes perfect sense in the kabbalistic context when it is applied to the "simple infinite" (*'ein sof ha-pashuṭ*) and not to the divinity of the Chain of Being and *sefirot*, which has form and structure.

to play the role of intermediate essences, the *havvayot* referred to by Joseph Gikatilla in his *Ginnat 'egoz*. This is almost to say that that world is not an intermediary and does not have its own existence; and this is how idolatry could insinuate itself through the eye of a needle or through the tip of the letter *yod*, which connects the Name of this world, the Tetragrammaton, to the infinite. That divine world is indeed the same as our own, but in the light and transparency of its own intelligibility. The divine world is thus the divine immanence in this world, that is, created nature (or, we might say, *natura naturata*) in indissociable union with its creator.

To put it another way, what we have here is a sort of flattening of transcendence into immanence, in a way that may be different from what Spinoza envisioned in his expression; but this is not certain, because we actually find the scholastic expressions *natura naturans* and *natura naturata* in Gikatilla's *Ginnat 'egoz*, written in the thirteenth century (though not printed until 1615).[13] The *Zohar* compares another type of union to that of soul and body: the union of the vowels and letters. In Hebrew, the vowels are "invisible," because they are normally not written. (They are frequently indicated by dots placed above, below, or alongside the letters, whence their common designation *neḳuddot* 'points'.) Despite their invisibility, they are nevertheless united with the letters, animating them and setting them in motion. Spinoza uses just this analogy in his treatise on Hebrew grammar.[14]

In any case, the authors of all these texts reject the thesis of God's nonmaterial or noncorporeal nature, because this thesis, widespread in Greek antiquity, also asserts a theological preeminence of mind over matter. In other words, like Spinoza, although in a different way, they reject the thesis of a purely intellectual deity, as well as the Maimonidean thesis of the creative intellect or understanding. As the Maharal put it, God's essence is "neither matter nor intellect."[15]

Of course the infinite cannot be compared to the body, because the nature of the body is finite and limited.[16] But a mind or soul is just as finite and limited as a body, in comparison to the indefinable infinite. The metaphor proposed by this tradition is that of an infinite light, undifferentiated and uniform, neither body nor mind, which, limiting itself, produces bodies and minds. This image of unbounded uniform light suggests an infinite luminous *extension* no less than an illuminating *thought*, it too infinite.[17]

13. Joseph Gikatilla, *Ginnat 'egoz* (Jerusalem, 1989), p. 34. Here we should note that, even for Spinoza, although God is extended substance as well as thinking substance he is not corporeal in the sense of a finite being with a particular magnitude and form: "Some assert [erroneously] that God, like a man, consists of body and mind, and is susceptible of passions" (*Ethics* I 15, Note). God is no more body than mind, because those are merely *modes*, that is, manners of being that God produces in himself. In other words, using Spinoza's ideas and terminology, we cannot say, along with the Stoics, that God is a body or the totality of all bodies. This is because bodies are modes of infinitely infinite substance and not substance itself. On the other hand, we can indeed say that God/nature is present and active, as *natura naturans*, in all bodies, and in each body as *natura naturata*. Furthermore, he is present and active as *natura naturans* in the totality of all ideas ("infinite understanding") and as *natura naturata* in each idea.

14. ". . . among the Hebrews vowels are called [the] *souls of [the] letters*, and letters without vowels are *bodies without souls*" (Benedictus de Spinoza, *Hebrew Grammar [Compendium grammatices linguae hebraeae]*, ed. and trans. Maurice J. Bloom [New York: Philosophical Library, 1962], chap. 1 [in *Complete Works*, p. 588]). In the eighteenth and nineteenth centuries the properties of the Hebrew vowel-points nourished theological and political controversies about the original language of humankind, in the context of a nascent historical thought that wanted to see itself as rational and scientific. See Maurice Olender, *The Languages of Paradise: Aryans and Semites, A Match Made in Heaven*, trans. Arthur Goldhammer, rev. and augm. ed. (New York: Other Press, 2002), chap. 2, "Divine Vowels."

15. See above, n. 9.

16. As Isaiah asks rhetorically, "to *whom*, then, can you liken God, with *what* form compare Him?" (Isa. 40:18): neither to a person or a thing.

As for the form of the body, which, in spiritualist theology, does not apply to God, for these authors it is on the contrary the perfect form.[18] It is through this form that the infinite is revealed in creation, simultaneously body and mind, in rung after rung on the ladder of being and in relative fashion, so that one rung of the ladder can be sometimes mind and sometimes body in relation to another rung: mind, vis-à-vis a lower and more specific degree, or body, vis-à-vis a higher and more general degree.

In *Nefesh ha-ḥayyim*, Volozhiner offers a clear explanation of the necessary distinctions between the infinite, which, as we have seen, cannot be likened to any object or any person, and the image of a body animated by its soul, which is used to describe the relationship between God and the worlds that he fills and animates.[19] Here the intention is to guard against the error of comparing the essence of the Creator with the essence of the soul, despite the image of God as the world-soul, which, like the soul of a body, animates it although it remains unseen. What can be compared, however, is the nature of the union of soul and body, the connection between the two, which is the same as the nature of the union between God and the world. The consequence of this union is that we can know the properties and actions of the soul only from the perspective of its relationship with the body, to which it is attached; and this is true even though the soul is itself created, just like the body. A fortiori, the uncreated Creator can be grasped and conceived only from the perspective of his union with the worlds. It is these divine worlds that are structured like a human body. This is not only a nominal identity, metaphors taken from anthropomorphic

17. The comparison of the union of God and the world with that of mind and body is analyzed by Volozhiner (*Nefesh ha-ḥayyim* 2.5). He begins with the image of the soul, itself invisible and unknowable, which expands and extends in the body it animates; in some sense the soul becomes material by sharing the body's property of extension (*hitpashṭut*).

18. We can already find this distinction between the formless infinite and the system of the *sefirot* in an exposition of the "system of divinity" (*ma'arekhet 'elohut*) by Moses de Leon (thirteenth century), who was probably the editor of most of the *Zohar*, in his *Sheḳel ha-ḳodesh* (1292): "He, may he be blessed, is excluded from every idea and thought, because none is capable of grasping [him], and he, may he be blessed, has neither image nor form" (Moses de Leon, *Sheḳel-Haḳ-ḳōdesh*, ed. A. W. Greenup [London, 1911; repr. Jerusalem, 1969], p. 19). But this is a warning for readers and students of this system of the *sefirot*, which is indeed presented in the form of the human body. Hence de Leon stipulates that this bodily structure is to be understood as a *hierarchy*, with no material nature or ideational content, that is, as an arrangement of and set of relationships among its elements, which makes it possible for them, like the organs of the human body, to be unified within themselves by the One that is manifested through them. This distinction between a purely relational definition of the parts of this form and an image or representation can be understood as proceeding from the nature of the *sefirot* as presented in *Sefer Yeẓirah*, the ten *sefirot belimah*, which can be understood as "ten spheres *or* numbers, without an object," reading *beli mah* as "with nothing." As we saw above (n. 2), the other reading, suggested by *Sefer Yeẓirah* itself, is "ten *sefirot* of restraint," as in "Bridle (*belom*) your mouth from speaking and your heart from thinking" (Isa. 1:8). Ultimately this leads to two aspects of the same idea, namely, that the *sefirot* are like digits or numbers, or geometric entities, with no content that can be spoken or imagined. Here we may be meeting with a current of ancient Platonism, taken up later by Proclus, in which mathematical or logical knowledge (*dianoia*) has an intermediate status between the divine intelligence and mere sensory opinion. Mathematical entities, numbers with no content that can be represented other than "intelligible substance," are endowed with a nature that is intermediate between the "indivisible substances" of "the intellect" and the objects that can be divided and known by the senses. They are pure images of the intelligible, not cloaking the material nature of sensible things, and it is the totality of their relations, produced directly by the principles of the finite and the infinite, that constitutes their substance. Without object, neither material nor ideal, each number is an irreducible unity, at the same time as they are all in relationship with one another. As such, they express these first principles that derive directly "from the cause of the One," although we can neither say nor conceive anything about an "object," an ipseity or a quiddity, that would make it possible to define its content. (See the first few pages of the prologue in Proclus, *A Commentary on the First Book of Euclid's Elements*, trans. Glenn R. Morrow [Princeton, NJ: Princeton University Press, 1992].)

19. E.g., *Nefesh ha-ḥayyim*, 2.5. But Volozhiner also develops these distinctions at length in the first section, where he explains the *ẓelem 'elohim* or divine image.

language, but a structural identity between the human body and these worlds.[20]

In other words, this bodily form characterizes the structure of the divinity that allows itself to be known in the world; it is in this way that man is said to be in the divine image. By observing the human body, with the material nature of its organs and the relationships among them, we can have a notion of ("see") the divine, as Job intimated.

These theological and metaphysical differences about the question of corporeality are important, because they come to be expressed in radically different value judgments about the properties and experiences of our body. It is enough to compare, for example, the attitude toward sexuality: a shameful activity, intended solely for procreation, according to Maimonides, or a privileged ascesis for attaining holiness, according to the author of the *Epistle on Holiness* (*'Iggeret ha-ḳodesh*).[21]

20. See ibid., n. 2.

21. See *The Holy Letter: A Study in Medieval Jewish Sexual Morality, ascribed to Nahmanides*, trans. Seymour J. Cohen (New York: Ktav 1976); *Lettre sur la sainteté, Le secret de la relation entre l'homme et la femme dans la Cabale*, trans. and annot. Charles Mopsik (Paris: Verdier, 1993). <This work has been variously attributed to Naḥmanides, Joseph Ibn Gikatilla, Joseph of Hamdan, and Ezra b. Solomon of Gerona.>

2. GOOD AND EVIL: MORAL CONSCIENCE OR KNOWLEDGE OF THE TRUE

In general, the fact that we consider the origin of the moral problem of good and evil to be a foundation of our humanity is itself strongly influenced by the paradoxical experiences of our knowledge, which does not know itself—that is, the knowledge of objects and of our body by our body.[22] In fact, our experiences of physical pleasure and pain are the first source of our ideas of good and evil. Good is, first of all, what makes us feel good; evil, what makes us suffer. It is only secondarily, when processed by our memory and imagination, that we project the possibility of pain or of pleasure onto the future and onto others. Only then do we make judgments about what is good and evil, drawing on our memories and imaginative projections and employing strategies in which we conceive how a present good can produce a greater future evil, or vice versa, on the basis of similar experiences we may have had in the past.[23]

But this imaginative knowledge of moral good and evil, extrapolated from our immediate bodily knowledge of pleasure and pain, is itself ambiguous, both good and bad. It is good in that it extends to other persons and transcends the present moment and immediate sensation of pleasure and pain; but it is bad in that it is always approximate and incomplete knowledge produced by experiences that are necessarily limited but that nevertheless institute an absolute good and evil, held to be universal and binding—when

22. Placing mind and body on the same ontological plane, as Spinoza does, offers an original and useful way of tackling this paradox. It can be shown that Spinoza's monism, in which "the order and connection of ideas is the same as the order and connection of things" (*Ethics* II 7), fits very well with the current physical and mechanistic development of biology. Of the two attributes of the single substance, thought seems to be privileged, in that it produces not only its modes, that is, ideas, but also the ideas of the modes of all the attributes, including itself—ideas of things and ideas of ideas. But this special status of reflexive thought morphs into the body's special status as the object of knowledge that is indispensable for going beyond the spontaneous knowledge of our mind, which is imaginative and inadequate. In practice, it follows from the fact that "the object of the idea constituting the human mind is the body, and the body as it actually exists" (*Ethics* II 13, Proof); not only that "mind and body are one and the same thing"—a mode of substance—"conceived first under the attribute of thought, secondly, under the attribute of extension" (*Ethics* III 2, Note), but also that, to know the nature of the mind and its union with the body, we must have adequate knowledge of the nature of the body of which the mind is the idea: "No one will be able to grasp

clearly they are always conditioned by particular experiences and stories. Knowledge of good and evil, the end of the child's innocence in which good and evil exist only in the moment of pleasure and pain, is, of course, the opening to duration, to the possibility of constructing a future based on the experiences of the past. "Their eyes were opened," as Genesis (3:7) tells us, when Adam and Eve acquired this knowledge. At the same time, however, it is a closure that freezes past experiences of pain into absolute evil, clearly imaginary, because projected onto the future, onto another place and another person that we cannot know.

Such is the ambiguity of the Tree of Knowledge of Good and Evil, itself both good and bad, because this knowledge makes one believe in the reality of some absolute evil that is opposed to some absolute good. As such, Adam and Eve's "sin" is indeed a fall, because it defines the human condition vis-à-vis morality instead of truth. It is a fall from an original state in Eden (mythical, of course), where the relationship with the world was mediated by knowledge of the true and not by morality. In Eden, as in the upper worlds of which the Midrash and the *Zohar* speak, evil does not exist.[24]

This is how Maimonides,[25] like Spinoza (for once the two are almost in agreement), interpreted Adam's fall: as the acquisition of knowledge of good and evil to supplant knowledge of true and false. Knowledge, gauged by morality, can only be approximate and enslaving, because it is based on fear of evil and hope

this [union of mind and body] adequately or distinctly, unless he first has adequate knowledge of the nature of our body" (*Ethics* II 13, Note). This formulation is among those that are sometimes wrongly understood to express Spinoza's "materialism." Similarly, the apparent special ontological status of the attribute "thought" does not indicate idealism. As has been clearly shown by Martial Guéroult, the realist explanation of ideas and the idealist explanation of their objects are refuted simultaneously. Spinoza's ostensible materialism and idealism are erroneous interpretations (see, inter alia, Martial Guéroult, *Spinoza* [Paris: Aubier, 1968], vol. 1, p. 223; vol. 2, pp. 62–73). Spinoza's doctrine is based "on going beyond traditional doctrines that, in their antagonism, seem to exhaust the entire sphere of possibilities" (vol. 2, p. 63 n. 37). The reflexive reduplication of an idea as the idea of an idea in thought distinguishes thought from the other attributes. Thought's apparent privileged status makes it possible to know, by means of the ideas it produces, not only things but also the ideas of things. But this special dispensation is also the source of its inadequacy, when it knows in an incomplete, confused, and mutilated fashion, which is necessarily the case when "man thinks" (*Ethics* II, Axiom 2)—at least as long as it does not go beyond its first type of knowledge. Compared to this inadequacy, the body, which cannot know anything by itself, because it does not form ideas of its affections and even less of these ideas, cannot be deceived, in the sense of having partial and inadequate knowledge. On the contrary, the body is that in which and by which we exist; joined with our mind, it constitutes our essence as a human being. Nor can we know this essence—whatever it is that constitutes the union of mind and body—as long as we do not know it adequately. The inadequacy of our mind derives from the inadequate knowledge it gives us of objects, of our body, and of itself. The body has no cognitive inadequacy, because it knows nothing consciously, that is, through ideas. But it is the only real object of knowledge of which we have immediate experience. For our mind, knowledge of our essence, as the union of our body and mind, cannot be complete as long as knowledge of the body is not complete. This is also true of the mind's knowledge of the mind, of course. But this is self-evident and does not need to be said, because the mind must know itself by the ideas of its ideas; this is one source of the errors and inadequacies of imaginative knowledge (see above, Chapter 1, n. 100).

23. See Atlan, *SR* II, chap. 2.

24. Many texts of the Midrash and *Zohar* invite us to reconquer or reconstruct an Edenic nature thanks to the Tree of Life, which the *Zohar* compares to the Torah of *'azilut* (the Torah of the world of emanation), formed of the same letters as the Torah we know in our world of action but arranged in a different order. This Torah of *'azilut* does not contain commandments and prohibitions related to good and evil, because it consists only of combinations of the letters of the names of God, as they exist in the upper world, the source of all things, including what causes us pain/evil, but where evil as such has no reality. The Torah of the permitted and forbidden, of good and evil, is seen as a propaedeutic, a textbook, the antechamber to this Torah of liberty, the Tree of Life, above good and evil and at the center of the Tree of Knowledge of Good and Evil that is its husk (see above, Chapter 1, §10, and below, n. 43).

25. *Guide*, 1.2. Spinoza borrows this interpretation in his *Ethics* (IV 68, Note), with some modification (see Chapter 1, n. 85).

for good, rather than on contemplation of the true, which would be self-sufficient and leave no room, says Maimonides, for judging whether it is good or bad. In other words, moral knowledge (conscience), being wretched knowledge that is dominated by fear and hope, is an inferior form of knowledge, a fall from knowledge of truth.

3. THE TREE OF KNOWLEDGE AND THE FALL OF THE FIRST MAN, ACCORDING TO RABBI SHLOMO ELIASCHOW'S *BOOK OF KNOWLEDGE*

Now I would like to discuss a different reading of the biblical narrative, one that does not, incidentally, exclude the former, because it bears directly on our subject. In practice it introduces us to the knowledge of good and evil as described by Genesis, literally as the ingestion or "incorporation" of the fruit of a tree, a plant, by the mouth that eats it. This knowledge is also, according to the Midrash, the serpent's penetration of and "incorporation" by Eve. Our guide is Rabbi Shlomo Eliaschow, the author of one of the last great works of Lurianic kabbalah, written at the start of the twentieth century, *Sefer ha-De'ah* or the Book of Knowledge, part of a longer work entitled *Leshem shevo ve-'aḥlamah* (Opal [*or* Jacinth], Agate, and Amethyst—the third row of precious stones on the breastplate of the High Priest [Exod. 28:19, 39:12]); specifically, the section headed "Explanations of the World of *Tohu.*" Eliaschow employs the metaphor of "the sparks of randomness" that were scattered after Adam disobeyed and ate of the Tree of Knowledge.[26] It is in this context that we must try to understand his sin as a myth of origins, rooted in the experience of spermatic knowledge and its relations with the sexual organs, the body, and the mind.

According to Eliaschow, the sin of the Tree of Knowledge has three aspects.[27] The first is the desire or intention to know, an intellectual curiosity that is not bad in itself but that, through impatience and distraction, holds the danger of one's getting lost in the things one wants to know—a danger that even King Solomon, the wisest of men, did not escape. It was in part this curiosity that, as we have seen,[28] caused his worst problems and grave misadventure with Ashmedai, the king of the demons.

The second aspect of the sin has to do with knowledge as nourishment, ingestion, appropriation, and assimilation. They *ate* of the tree and "their eyes were opened." It was an overpowering experience of revelation that was provoked, triggered automatically—chemically, we might say—by the body's appropriation of another body. The latter is, at first, utterly external and strange. It is the body of a tree, a plant that produces subtle and invisible essences that can act directly on the mind. It is an experience of direct knowledge, of the incorporation of another body, a union of souls and bodies, such that man's body-soul is filled—possessed—by the body-soul of the plant, and the appropriator is at the same time appropriated. The thought and speech of the man and woman whose eyes are opened are no longer their own. This is the advice (Hebrew *'eẓah*, which has the form of the feminine of *'eẓ* 'tree') that the serpent hisses into the ear. These are the words of the "little gods who speak,"

26. See above, Introduction and Chapter 1.

27. S. Eliaschow, *Leshem shevo ve-'aḥlamah. Sefer ha-De'ah* (Piotrkow, 1912; repr. Jerusalem, 1976), "Explanation of the Tree of Knowledge" (*derush 'eẓ ha-da'at*), §3, 2:290. This "book of knowledge" (*de'ah*) must not be confused with Maimonides' "Book of Knowledge" (*Sefer ha-Madda'*), which might be more aptly rendered today as the "Book of Science." In chapters 5 and 6 of the latter we encounter the Maimonidean theory of human free will and divine omniscience, which, as we shall see, is quite different from that expounded by Eliaschow.

28. See Chapters 1 and 3.

as the Mexican shaman Maria Sabina called the hallucinogenic mushrooms of her ritual.[29] In that case, too, danger lurks in direct knowledge by the body, which is transported to heights of ecstasy and prophetic vision, but more quickly and more intensely than it can assimilate; the vessels are shattered by a light that is too strong and that they were not prepared to receive. This is the story of Nadav and Abihu (among others), the two sons of Aaron who died for having burned, in excessive quantity and with an "alien fire," the incense (*ḳeṭoret sammim*, literally "vapor of drugs") that may be burned only in the sanctuary.[30]

The third aspect of the sin, which will hold us longer, is knowledge by means of sex. This began badly, in the union between Eve and the serpent, which discharged its semen in her, as a sort of sexual venom. Adam was not the first to "know" Eve. Consequently the primordial sexual knowledge of the man and woman was tainted by the mixture inside her of Adam's sperm and the serpent's. This redoubling and reinforcement of the sin produced the first two children conceived and born, Cain and Abel, the dual human nature represented by the murderer Cain and his double-and-victim Abel, in which the original divine form, the *ẓelem* or creative image, can no longer be recognized.[31] The bodily form and face that constituted the humanity of Adam and Eve were replaced, according to the Midrash, by those of the serpent and the murderer. This last aspect of the sin is the inclusion of the serpent in the sexual relationship, as a third party that comes between Adam and Eve. This too opens their eyes and makes them like gods. What is this knowledge of good and evil, or rather this knowledge that is simultaneously good and bad, which they acquired in this way, after the serpent insinuated itself as the medium of their sexual knowledge?

To unravel the intricacies of these interlocking narratives we must turn to the main actor in this drama: the serpent. Who is this serpent before the sin, not yet cursed and thus not yet creeping in the dirt, but on the contrary walking upright on its legs,[32] capable of understanding the divine prohibition and of interpreting it for Eve? How is it able to tempt her, predicting—quite accurately—that her eyes and Adam's will be opened, making them like gods, after they have eaten from the tree? How does it know this? Has it already tried that fruit itself? Where did the serpent acquire its knowledge of what knowledge is? What is the relationship between knowledge and the human sexuality that it can pervert by means of what the Midrash calls its "pollution," *zohamat ha-naḥash*, the pollution of the serpent and by

29. Alvaro Estrado, *Maria Sabina, Her Life and Chants*, trans. Henry Munn (Santa Barbara, CA: Ross-Erikson, 1981). See Atlan, *Enlightenment*, pp. 339–343.

30. Lev. 10:1–2; Exod. 30:7–9.

31. *Zohar*, Genesis 55a; B *'Eruvin* 18b (see Chapter 1). The fact that the murderer and the victim are represented as a couple, both of them denatured and dehumanized by the mixture of semen that produced them, does not mean, of course, that they are identical or on the same plane. In the inextricable mixture of good and bad, the immediate *experience* of the good remains that of well-being; even more so, the experience of the bad remains that of pain. Whether the angel plays the beast or the beast plays the angel, angelic *behavior* remains supremely angelic. It is not because we can conceive of life, at a certain level of temporal development, as death overmastered, and of death as a process of life, that the experience of death is confounded with that of life. The same midrash presents Moses, the great amender of a later period, as the *gilgul* (revenant or reincarnation) of Abel and not of Cain, even though what he had to amend was the source of Cain's murder.

32. In certain texts, such as *Pirḳe de-Rabbi Eliezer* (chapter 13) and *Sefer Habahir* (§200), the serpent is described as a camel ridden by Sammael, the angel of death and destruction, who is often assimilated to the Satan of the first chapter of Job. This raises and resolves the problem of the origin of evil or of the first responsibility in this story. The blame no longer attaches to the serpent, no more than it does to Adam and Eve. According to this midrash, the serpent's actions in this were similar to those of "a man in whom there was an evil spirit. All the deeds which he does, or all the

the serpent, as today we might refer to pollution of the environment and by the environment?

The Hebrew of Genesis offers several keys for answering these questions. It tells us simply, at the start of the story, that the serpent was most cunning of the animals. It possessed a form of knowledge that was both animal and cunning; in this respect it was the most developed of all. As an animal it was one rung above the tree and all plants. It possessed a form of knowledge that could communicate with human beings by virtue of their animality, because a man or woman is also an animal, a "living soul" (*nefesh ḥayyah*) like all the other animals, and thus has direct experience of this animal knowledge. Relative to an ideal of specifically human knowledge, to which we shall return, the knowledge possessed by the serpent (and by human beings) is not only animal, but also cunning—indirect, devious, approximate. Nevertheless, sometimes it may be more successful than transparent and naïve knowledge, in those intricate and complex circumstances where the knowledge possessed by an "innocent" human being would be insufficient. The Hebrew word for serpent, *naḥash*, expresses this perfectly, with the same ambiguity of good and bad, because cunning knowledge is not necessarily and intrinsically bad. The root *n.ḥ.sh* means "guess" or "suppose," in the sense of intuition of something hidden. It is by this activity, which can resolve enigmas and interpret signs, that Joseph, the great interpreter of dreams, later defines himself for his brothers, when, after springing the trap he set for them, he pretends to have divined the whereabouts of the cup he accuses them of having stolen. To indicate that nothing can be hidden from him, Joseph asks them, rhetorically, "Do you not know that a man like me 'serpentizes' (*naḥesh yenaḥesh*)?" (Gen. 44:15); that is, "I am able to uncover hidden things." This cunning knowledge, which characterizes the serpent in Eden, reappears at the end of the story when Adam and Eve, having eaten the fruit, hide because they "knew that they were naked." The very same word *'arum*, which means "cunning" in the case of the serpent, denotes nakedness when Adam and Eve, having lost their presexual innocence, perceive themselves and know that they are naked. What is the relationship between the serpent's cunning and Adam and Eve's nakedness? Clearly this is one of the hinges, in this myth, of the intricate links among serpent, sex, knowledge, and

words which he utters, does he speak by his own intentions? Does he not act only according to the idea of the evil spirit, which (rules) over him? So (was it with) the serpent. All the deeds which it did, and all the words which it spake, it did not speak except by the intention of Sammael. Concerning him, the Scripture says, 'the wicked is thrust down in his evil-doing' (Prov. 14:32)" (*Pirḳe de-Rabbi Eliezer: The Chapters of Rabbi Eliezer the Great*, trans. Gerald Friedlander ([New York: Sepher-Hermon Press, 1981; first pub. London, 1916], pp. 92–93). Like the other angels, Sammael, a great prince of the celestial host, was "envious" of Adam and Eve (see above, n. 12) and decided to make them fall. But like Satan in the book of Job, his project is ultimately that of the Master of the Universe, on whom he depends, who "makes peace and creates evil" (Isa. 45:7). These texts offer a radical answer (to which we shall return) to the question of the origin of human misdeeds. This answer involves a game nature plays with man and with itself. The midrash suggests something similar with regard to the Creator's reaction to seeing Sammael astride the serpent, referring to God a verse from Job that in the original context applies to a large bird (often rendered as an ostrich): "it laughs at the horse and its rider" (Job 39:18). As *Sefer Habahir* puts it: "If not for [man's] sins, there would not be any differentiation between [man] and [God]. . . . [Man] sins, while the Blessed Holy One does not. . . . He has no sins. But the [Evil] Urge comes from Him." The text continues by recommending the proper course of action for us: "[The Evil Urge] originated from Him until David came and killed it. . . . [Thus he could say] 'evil cannot abide with You' (Ps. 5:5). How was David able to overcome it? Through his study, since he never stopped [studying] day or night" (*The Bahir: An Ancient Kabbalistic Text attributed to Rabbi Nehuniah ben HaKana*, trans. Aryeh Kaplan [New York: S. Weiser, 1979; repr. Northvale, NJ: Jason Aronson, 1995], §196). This is an allusion to the talmudic tale about the end of David's life, which has him studying without stop in order to keep the Angel of Death at bay; for the latter cannot take a man's soul when he is studying the Torah.

sin. Adam and Eve have literally incorporated the serpent's naked cunning into their mode of knowledge, and this is what allows them to know, through cunning knowledge, that they are naked. But what indeed does it mean "to know"—to *dis-cover*—that one is naked?[33]

Knowing that you are naked implies a certain relationship to your own body, a form of knowledge of your body that can be extended to the body of another. What relationship does this knowledge have to knowledge of self and knowledge of the other? What is more, in our story knowledge as incorporation, through the mouth or through the genitalia, has two forms: the idealized form of before the fall, and another form, *defined* by this fall and by the long process of repair that follows. Prelapsarian knowledge is immediate, transparent, and direct. Self-consciousness is evidence of the interior of one's own body, which is simultaneously, in the celebrated formula, knower, known, and knowledge,[34] the self's presence for the self. Prelapsarian knowledge is also an encounter with the other of the opposite sex, in which knowledge by the body and knowledge by the mind, inseparable, proceed through a skin made of translucent light, without opacity, according to the well-known midrash in which Adam and Eve's skin, their *'or* (עור, with initial *ayin*), did not yet cover their internal light, *'or* (אור, with initial *alef*).[35] The model of immediate knowledge of self is projected onto the other, in this biblical ideal in which the man and woman know themselves/each other, when they become a single body—*baśar 'eḥad* 'one flesh' (Gen. 2:24).

33. The various midrashic readings of this verse, quoted in particular by Rashi and by Rabbi Ḥayyim Ben Attar in his commentary *Or ha-ḥayyim*, interpret this knowledge as an experience of the sadness that followed the transgression. This is compared to the renunciation of the garment of the Law, with its obligations and prohibitions. The immediate satisfaction of desire provoked, as it were, an emptiness, naked destitution, like death, which, according to Genesis, must follow the transgression of the only prohibition with which Adam and Eve were "clothed" in Eden. As we saw with regard to the interpretation by Maimonides and Spinoza, the object of this ban was to preserve them from the fall that impended when knowledge of good and evil replaced knowledge of truth. Another significant aspect of these commentaries is their reference to a dialectic of the satisfaction and repression of desire in which the relationship to the Law is necessarily death-bearing (it makes no difference whether desire is satisfied too quickly or repressed too long) when it proceeds only from knowledge of good and evil, which is unavoidably approximate and relative. As Solomon knew, "Hope deferred makes the heart sick, but desire fulfilled is a tree of life" (Prov. 13:12). In this sense, one can say that Spinoza's blessedness, like the Tree of Life that should have triumphed over the Tree of Knowledge of Good and Evil, constructs a prelapsarian Eden, where "no one rejoices in blessedness because he has controlled his lusts, but, contrariwise, his power of controlling his lusts arises from this blessedness itself" (*Ethics* V 42, Proof).

Nevertheless, these commentaries do not deal explicitly with the relationship between the literal sense of physical nudity, in the biblical text, and the metaphors they develop. I do not know whether the interpretation I have dis-covered (because the text invites us to do so) derives from hermeneutical cleverness or merely from the immediate evidence of words used in a wide multiplicity of meanings, that is to say, from the multiple and distinct things they designate. But perhaps it all comes down to the same thing after all. Is not every translation, which necessarily diminishes the multiplicity of meaning inherent in the original, a form of indirect knowledge and a cunning interpretation?

34. This triplicity is customarily ascribed to God, notably by Maimonides, by which he means (unlike Spinoza, who also uses it) that God's knowledge, as the Active and Creative Intellect, is qualitatively—and not just quantitatively—different from human intellectual knowledge. In the kabbalistic thought that borrows this formula, this unity and immediacy of knowledge also relates to the self-consciousness of all those who can say of themselves "I" (*'ani*). This is why the name of God can be yoked to "I"—"I am (*'ani*) YHWH" (or, better still, "'I' [is] YHWH)—a recurrent biblical expression that standard theological readings do not explain very well (see below, n. 98; see also *SR* II, chap. 5).

35. See, for example, Isaiah Horowitz, *Shenei luḥot ha-berit* (Jerusalem, 1959) Part 3, *Vayeshev*, p. 96; English translation: *Shney Luchot Habrit on the written Torah*, trans. and annot. Eliyahu Munk, 2nd ed. (Jerusalem and New York: Lambda, 1999), p. 290. In this play on words, where *alef* is supplanted by *ayin*, the shift from luminous transparency to opacity is isomorphic with the passage from the unity of *alef* (whose numerical value is 1) to the multiplicity of *ayin* (whose numerical value is 70). What is more, a second play on words is applied to the origin of clothing, which covers nakedness (of skin) under garments of animal skins. As in a double negative that would hide what obscures, the primitive light of the body, now hidden by the opaque nakedness of its own skin, is suggested by the skins that cover it in turn.

In this picture of Eden before the fall, knowledge of bodies is knowledge of the other from the inside, as if the other's body were one's own body. This is why the bodies involved, transparent to themselves and to one another, are not considered to be hidden by the opacity of skin. The serpent's knowledge is clearly quite different. As its name indicates, the serpent, the *naḥash*, knows by *niḥush*, by "hypothesis," indirectly, through trial and error, through successive attempts, though interpretation and through oracular cunning that is devious to a greater or lesser extent. The fall triggered by the sexual incorporation of the serpent is that the transparent and immediate knowledge of self by self and of self by other, characteristic of sexual union, is replaced by a devious and approximate knowledge that involves trickery and seduction.

The first thing that Adam and Eve discover after they incorporate this serpentine knowledge, beyond the fact that it is pleasurable, is their own body, but now seen from the outside, by the indirection of looking. The transparency of self to self (immediate knowledge of the undivided mind-body by itself) is replaced by objective knowledge of one's own body, as if it were an external object or even an instrument, as if it were the body and skin of someone else who can be looked at. Because that immediate knowledge of self no longer exists, it cannot be projected on the body-mind of the other. Sexual knowledge is changed; instead of being immediate knowledge of the interior of the other, as if it were one's own body, it is knowledge of self that becomes knowledge of someone else, other than the other, not known from the inside, as knowledge of self, but with the eye of a stranger, with the eye and perceiving consciousness that distinguish and latch onto the visible aspect of skin and sex by which the body knows itself to be naked.[36]

It is this "serpentine pollution" that is the sin itself: knowing oneself, as well as the other, sexually, through the knowledge that one is naked—an external and objective form of knowledge. The consciousness externalized here is no longer immediate knowledge of self and other, but indirect and iterative knowledge, by successive approximations, of the concealed that is unveiled and of the opaque nakedness that is veiled in its turn—the cunning and possessive knowledge of seduction, simultaneously knowledge *of* and knowledge *how*,[37] in which aesthetic pleasure itself is yoked to the service of power. Physical beauty and elegance can then be severed from those of the mind and, by seduction of the body (but also of the mind), can be turned into weapons of war or instruments of sorcery and oppression. This separation, this objectification of the knowledge of

36. The ambiguity of the outside view we can have of our own body may be the origin of a curious expression that the Bible applies to prohibited sexual relations. Incestuous unions with close relatives, as well as what Françoise Héritier calls "secondary incest" (*Two Sisters and Their Mother: The Anthropology of Incest*, trans. Jeanine Herman [New York: Zone Books, 1999]), are first referred to (Lev. 18:6–17), euphemistically, as a ban on "uncovering nakedness"—using a word, *'ervah* (ערוה), whose root is close to *'or* 'skin' (עוֹר) and *'ivver* 'blind' (עִוֵּר). Only later (Lev. 20:16–22) do we find explicit language about forbidden sexual unions. It is forbidden to look at a relative's uncovered and objectified nakedness. At the first level, this objectification repeats in some fashion Adam and Eve's discovery that they were naked, now projected onto relatives and replacing the more subjective—more intimate, in the sense of internal—relationship that is established by family closeness. Of course, the simplest reading of these texts understands the prohibitions as related to incest, or at least its preliminaries. It is seen as the expression of a "horror of the same" (Héritier, *Two Sisters*) and of the mixing of relations. But this horror is relative(!), because the differences are reunited in the infinite essence of the One. This is why, in the mythical world of origins, incest among siblings, unavoidable because they are all the children of the first and only couple, is considered to be *ḥesed* (Lev. 20:17), in this context generally rendered "disgrace" or "shameful act," but a word that otherwise means "favor, loyalty, kindness, steadfast love" and that denotes the unlimited love that makes it possible to build the world. This is an alternative reading of Psalm 89:3 [2]: "A world will be built by *ḥesed*" or "a world of *ḥesed* will be built." When it is no longer a tale

our body, produces multiple catastrophes, of which the most important is the association, previously nonexistent and unknown, of Eros with Thanatos, of love with violence and death. "Make love, not war," was the motto of the American hippies during the Vietnam War. They did not know, in the dizzy heights of their drug-induced trips, that in the realm of practical knowledge the two go hand in hand; and, even worse, that the children of this union would be like Cain and Abel, both murderers and victims.

4. AN AMBIGUOUS SERPENT

Fortunately for us, we are not only the products of the fall of the first man and woman and of their murderous and murdered descendants. In fact, according to the biblical narrative, we are descended neither from Cain nor from Abel, but from Seth, the third son of Adam and Eve, conceived only after they had been separated for 130 years—an interlude necessary, according to the Midrash, for them to rid themselves of the serpent's poison. Seth reclaimed the divine form, the *ẓelem 'elohim* of his parents, which is the hallmark of the humanity of human beings and which Cain and Abel, fathered in part by the serpent, had lost.[38] The repair begins with Seth's humanity but is never completed. The pollution of the serpent, although expelled from the bodies of Adam and Eve, can still

of origins, however, but a matter of our day-to-day existence, *ḥesed* in the pejorative sense is used only for incest between brother and sister.

We should examine the wording and context of the passage more closely, noting the expressions used, in order to suggest what may be a deeper source of meaning of what is called, in the broad sense, the "incest taboo." The series of prohibition begins with "None of you shall come near anyone of his own flesh to uncover nakedness" (Lev. 18:6). After this, various relatives are specified: "You shall not uncover the nakedness of your father—the nakedness of your mother; she is your mother, you shall not uncover her nakedness. You shall not uncover the nakedness of your father's wife; it is your father's nakedness" (vv. 7–8). The text continues with "your sister—your father's daughter or your mother's daughter," and, after other relatives, "your father's sister" and "your mother's sister," who are your father's and mother's "flesh." As Naḥmanides underscores (commentary on Lev. 18:7), the first prohibition applies to one's mother, and the others are derived from it directly (the father) or indirectly (through the father). But the prohibition of incest with one's granddaughter is expressed in a curious fashion, climbing back up this chain of relationships that began with the mother: "for their nakedness is your own nakedness" (v. 10). To put it another way, suddenly the text reminds us that there is a naked skin even closer than one's mother—namely, one's own body.

This must be the source of all the prohibitions, at least insofar as they concern looking at the naked body, the "uncovering of nakedness," short of sexual union per se. The text seems to be telling us that it is the naked skin, the opacity of your own body, that you are looking at from the outside when you uncover the nakedness of a close relative. Ultimately, of course, this "uncovering" is not forbidden (there is no prohibition on disrobing and looking at one's own body), perhaps because it does not cancel out the immediate experience of internal knowledge of self. But this is the source of meaning of the subtler prohibitions on looking at the naked body, beyond the forbidden incestuous unions per se that are based on those between mother and son, with their possible consequences of confused relationships, where a child may be, for example, simultaneously his father's son and brother, etc.

37. There are several modes of knowing. We customarily distinguish knowing *that* from knowing *how*—that is, we distinguish the objects of knowledge, the content of knowledge that can be said, described in words, as something that is known, from that which is done, the result of knowing the technique for doing something, which does not always have to be stated in so many words once it is present as know-how. But knowledge of self cannot be reduced to either of these. Knowledge of self is more than knowledge of something, even if one can say many things about oneself; and it is more than knowing how, even if it clearly involves many skills. More precisely, knowledge of self implies knowing how to live, in the sense that what is known is the lived experience of the being that knows; it is immediate and transparent knowledge in which, as we have seen, the knower is also what is known as well as the knowledge relationship itself. Thus the prelapsarian knowledge of self, and of the other, was knowledge of being that may be confounded (perhaps as infants or animals do) with life itself. As in the *Zohar*'s image of evil at its source (Genesis 27a; see above, Chapter 1), the Tree of Knowledge is not separate from the Tree of Life.

38. See above, Chapter 1. We read in the *Zohar* (Genesis 55a [trans. Matt, vol. 1, p. 308]):

[Adam's] other sons derived from the clinging of the slime [*or* filth, pollution] of the serpent to its rider [Sammael], so they were not in the image of Adam. . . . R. Yose said: "Look at what is written: *Adam knew Eve his wife, and she conceived and bore Cain* (Gen.

be found in the external aspect of our world, in our experiences of objective knowledge, more or less confused, more or less adequate, by the senses and by the mind. These are the male and female demons of the legend, scattered all around us, which we find as the partial and partially fantastic causes of things, the random sparks of spermatic knowledge. We occupy an intermediate state in which the cunning knowledge of the serpent has become ambiguous—not only bad, but both good and bad and sometimes even indispensable, if only to foil the serpent's own tricks of exteriority. Several passages in the Bible express this ambiguity.

Thus, as we have seen, Joseph, son of Jacob and viceroy of Egypt, traditionally referred to as "Joseph the righteous,"[39] calls himself a *menaḥesh*, a diviner, from the same root as *naḥash* 'serpent'. After he gains a powerful position at Pharaoh's court, thanks to his capacity to uncover hidden things and interpret dreams, and employs a ruse to manipulate his brothers, he says to them, "Don't you know that a man like me *naḥesh yenaḥesh*"—literally, "to divine, will divine" (Gen. 44:15).

Elsewhere, too, the episode of the brazen serpent, the *neḥash neḥoshet* (Num. 21:6–9), employs the repetition of forms of the root *n.ḥ.sh* to express its simultaneously destructive and restorative character, intrinsically both good and bad, in the context created by the fall and the mechanism of its repair. When the Israelites are being plagued by the lethal bite of real serpents, which kill many of those who are complaining, once again, about the travails of their life in the desert, Moses, at God's behest, fashions a brazen serpent and mounts it on a pole so that, "if a serpent bit any man, he would look at the bronze serpent and live" (v. 9). This brazen serpent cures the fiery bites of the real serpents, the *neḥashim śerafim* or "seraphic serpents," as the text refers to them. (*Śaraf*, from a root meaning "to burn," is the singular both of seraphim, the angels of fire that, curiously, are generally depicted as a sweet and gentle, and of this species of serpents with their lethal fiery bites.) The pole on which the serpent is mounted is called a *nes*, a word that can mean both "miracle" and a sign to indicate a direction. For, as Rashi points out, quoting the talmudic sages: "Could the serpent slay or the serpent keep alive? It is, rather, to teach that when the Israelites directed their thoughts on high and kept their hearts in subjection to their Father in heaven, they were healed" (M *Rosh hashanah* 3:8). Now this brazen serpent, which healed those bitten by the fiery serpents by making them look up and "direct their thoughts on high," is a sort of serpent squared, a *neḥash neḥoshet*, because *neḥoshet* 'copper, brass' derives from the same root as *naḥash* 'serpent'. God instructed Moses to make a *śaraf*. Moses, to be sure of its effectiveness, made a *neḥash*

4:1). Certainly so! It is not written: *and he engendered Cain.* Even of Abel, it is not written: *He engendered*, but rather: *She continued bearing—his brother, Abel* (v. 2). . . . But of this one [Seth], what is written? *He engendered in his likeness, according to his image* (5:3). Furthermore, remarks the *Zohar*, the name Seth (שת), written with the last two letters of the alphabet, "the secret of the end of the letters of the alphabet," refers to the secret of the lost breath, Abel (Abel, *hevel*, means "breath"), which reclothed itself in the world in another body, just as Eve said when Seth was born: "God has provided me with another offspring in place of Abel" (Gen. 4:25). In other words, this is where it may be possible to unify the two sides, the inner and outer, of this fantastic story of serpentine pollution and understand how the offspring of the serpent in Abel and Cain can still be found, after their death, in Adamic humanity, in Seth, in the form of the spirits, breaths, and demons expelled from Adam and Eve but subsequently embodied in the corrupted generations of the Flood and of Babel, and then in the redeemed generation of the wilderness, which received the Torah at Sinai.

39. In the kabbalistic typology Joseph is associated with the *sefirah* of *Yesod* (foundation), focused in the sexual organs; this is the particular *sefirah* of the righteous, in one reading of Proverbs 10:25: "The righteous is an everlasting foundation."

neḥoshet so as to have, through the power of language, a truly serpentine serpent; as the Midrash notes, the "terms correspond."[40] Much later, according to 2 Kings 18:4, the Israelites in their land descended into idolatry and created a cult of this brazen serpent, preserved since the time of Moses, until the pious king Hezekiah, who uprooted idolatry from the land, shattered it into pieces.

Nor, of course, was this Moses' first encounter with a serpent. As we know, Moses' staff, with which he worked miracles, was itself sometimes a coiled serpent and sometimes a straight rod, the "rod of Elohim" that he raised heavenward to bolster Israel in their battle against the Amalekites (Exod. 17:8–15)—in a procedure analogous (as the same mishnah notes) to the later effect of the brazen serpent. The altar that commemorates that victory over the Amalekites is also designated by Moses a *nes*, a miracle, which he attributes to the Tetragrammaton, calling it *nissi*—"my miracle" but also "my rod." The *Zohar* carries these multiple ambiguities associated with the nature of the serpent, an approximation of true knowledge just as brass is an approximation of gold, to an extreme: "'The staff of Elohim was in his hand.' This rod (*maṭṭeh*) is Metatron, from one side of whom comes life and from the other death."[41] "Metatron" is the name of an angel that is also the primordial man, a sort of magistrate who rules over the world after Adam's transgression with the Tree of Knowledge, applying the Torah as the law of the permitted and forbidden, the law of laborious and bitter knowledge, a drug of life for those with merit, a drug of death for those without it. The Tree of Life, by contrast, designates the Torah of the world before the sin and after the amendment, without the ambiguity of punishment and reward and of life and of death. For the *Zohar* and those influenced by it, the rod-Metatron, like the Tree of Knowledge, is good and bad, life on one side, death on the other. As Isaiah Horowitz put it,

> when the rod remains a rod, it is a help from the side of good, and when it is turned into a serpent it is against him. God delivered it [Metatron/the rod] into his hand, [where] it typifies the Oral Law which prescribes what is permitted and what is forbidden.[42]

Thus it is the side of good and the side of bad, like Metatron himself, the secret of the Torah of the Created World (*beri'ah*) as opposed to the Torah of the Emanated World (*'aẓilut*),[43] formed completely of divine names, in which "evil cannot dwell."[44]

40. *Gen. Rabbah* 31:8; cited by Rashi on Gen. 2:23. <The Hebrew idiom *lashon nofel 'al lashon* covers all forms of parechesis (plays on sound—assonance, consonance, alliteration, etc.), including paronomasia (plays on meaning). Taking each word in its primary sense, the literal meaning of the expression is "tongue falling on tongue"; but the appropriate senses of noun and verb yield "word matching word." Prof. Atlan's rendering here, "la langue entraîne la langue" ("the tongue/language leads the tongue/language"), is an excellent image but inexact.>

41. *Zohar*, Genesis 27a; B *Ḥagigah* 15a (see above, Chapter 1, n. 78). See also: *Le livre hébreu d'Hénoch ou Livre des palais*, trans. and annot. C. Mopsik (Paris: Verdier, 1989); Moshe Idel, "Enoch Is Metatron," *Jerusalem Studies in Jewish Thought* 6 (1987), pp. 151–170.

42. This alludes to Genesis 2:18 and the creation of Eve: "I will make him a 'counterpart' helper (Heb. *'ezer kenegdo*)"; which Rashi explains as meaning "if he merits, she will be his helper; if not, she will [act] counter [to] him."

43. This is also the Torah of the world before sin (see above, n. 24). On the concept of the Torah of *'aẓilut*, see below, §14; and Naḥmanides' preface to his commentary on the Pentateuch. See also: Gershom Scholem, *On the Kabbalah and Its Symbolism*, trans. Ralph Manheim (New York: Schocken, 1965), chapter 2; and, on the interesting use that Spinoza would have made of this idea, J. Askenazi, "La parole de l'écriture selon Spinoza," *Les Nouveaux Cahiers* 26 (1971). See also below, Chapter 4.

44. *Zohar*, Genesis 27a (see above, Chapter 1, n. 78). See also Yehuda Ashlag, *Zohar with the Sullam Commentary* (Jerusalem, 1945) (Hebrew), ad loc.

5. A RETURN TO THE SPARKS OF RANDOMNESS

The ambiguity of postlapsarian serpentine knowledge appears in striking fashion in the myth of the sparks of randomness, where it no longer involves the serpent itself but rather the future career of its "pollution" in post-Edenic human nature, as determined by the fall and the ongoing repair. Recall that these sparks or drops of randomness, *niẓoẓot shel ḳeri* or *ṭippot ḳeri*, were emitted by Adam (but also by Eve) during the period when the first couple were separated, between the birth of Cain and Abel—offspring of the serpent's pollution and incapable of reproducing the original human form or image (*ẓelem*) that existed before the fall—and that of Seth, who regained this form. By the time of his conception, all traces of the serpent had been eliminated. But this does not mean that we, as Seth's progeny, have recovered the lost unity and transparency of the original Eden. It does mean, however, that this "new" human nature that is ours is at least open to the possibility. In other words, we find ourselves in an intermediate state, where the Edenic nature is not a given but is nevertheless possible through our cognitive experiences. These are always marked by their fundamental ambiguity, which for us is precisely the ambiguity of the possible: "Everything is foreseen but permission [*or* possibility; Hebrew *reshut*] is granted." We shall see how this dictum of Tractate *Avot* expresses the interplay between an absolute timeless determinism and our temporal experiences of what is possible. The "possibility" that is granted us implies that reality contains within itself, from all eternity, a thing and its contrary, and that each moment of our existence is pregnant with a future that is good and bad, both good and evil.[45]

45. According to Joseph Gikatilla (*Ginnat 'egoz*, p. 185), for a thing to change its natural behavior in an extraordinary fashion, as in a miracle or prodigy, the nature imprinted in it (*muṭba'*, which can mean both "imprinted" and "naturata") must contain, in potential, both contraries.

More precisely, in the context of the myth, this ambiguity derives from the fact that the emission or expulsion of the sparks of randomness has two different effects. On the one hand, Adam and Eve eliminated the serpentine pollution that had possessed them, thereby regaining something of their previous nature. On the other hand, the sparks they expelled engendered male and female demons. Thus these demons acquired an objective existence in the outside world, beyond the subjectivity of individual human beings. At the same time that something of the human image was recovered, concrete evil acquired objective existence. Now existing on the outside, it could penetrate and possess, in a reverse movement, the generations of collective violence and nondifferentiation. Thus the elimination of the serpent's spoor from Adam and Eve is both good and bad. The result is Seth, our human nature, which is no longer limited to Cain and Abel, that is, to the sadomasochistic pair of the inevitable murderer and the victim whose action is taken as a provocation. But it must confront the tension between an inner life that is potentially "innocent" and a dangerous outside from which it remains separated by the opacity of its covering skin.

A reserved domain, that of our private and inner life, can subsist in the lee of this disjunction. This means, however, that the inside is no longer "like the outside." Having "an inside like the outside"[46] is reduced to an ideal of sincere and

46. Just as the Ark of the Covenant, which held the Two Tablets of the Law, was made of wood (and thus opaque), but covered by gold both inside and out, so too "any scholar [whose study of the Torah contributes to its existence] whose inside is not like his outside is no scholar" (B *Yoma* 72b). This is one of the properties that make it possible to identify the true scholar, the *talmid ḥakham*. Elsewhere (B *Bava meẓi'a* 23b)

true conduct. Standing in opposition to this ideal is the reality of the relationship to what is not self, where love cannot eliminate misunderstanding, not to speak of dissimulation, hypocrisy, and mendacity, which courtesy, modesty, and amicable benevolence sometimes make essential. To put it another way, the expulsion of the serpent's pollution purifies the subjectivity from which it has been excluded, but at the price of enclosure in the bounds of an inner life that runs the constant risk of being severed from reality and is thus ultimately an illusion. When Adam and Eve finally expelled the serpent's animal knowledge, which had invaded them, their capacity for inner knowledge, knowledge of being—knowledge and life at the same time—was restored. But it is only a possibility, an ideal, like some yearning that struggles with the outside world, where the expelled demons can now give free reign to their actions. The sparks of randomness, now externalized, give life to the objective and collective evil of the generations of violence and sameness. The prize of this combat is the truth of each individual, as he or she faces the external world and the other. But this truth is no longer the same—or not yet the same—as that to which objective knowledge aspires, and notably the particular impersonal, theoretical, and operational truth to which the sciences have accustomed us. It assumes the existence of the subject, including the moral subject, at least in potential. But this subject is torn between its innermost being and knowledge of things. It can maintain itself only through a sort of bad faith, a game that it plays, both actively and passively, with the world around it—and with itself as the ball. For it can only oscillate between its angelic innocence, grappling with the demonic aggression of what is not itself, and the guilt of being identified with its assailant to the point of assimilating it, reincorporating the violence of the evil on the outside. In Hebrew, *réa*ʿ (רֵעַ), our fellow human being, the other whom we are told to love, is a homograph of *ra*ʿ (רַע) 'evil'.

In this game, the angelic innocence of inner life is necessarily an illusion. We are governed by the "knowledge that we are naked" and, even more so, by the objective knowledge of things that we have acquired. This tension, which pits the ideal of unity and transparency against the reality of separation and opacity, at first works in favor of division; separation and dissociation are transplanted into the very core of our knowledge. The combat between ideal and real is unequal. The ideal can no longer be posited even as a goal or objective. When we are confined within an ineffable subjectivity, no longer able to be object or objective, this ideal seems to be unattainable, except for certain privileged experiences to which we shall return. What remains is only a yearning that is increasingly divorced from reality, increasingly unreal, a yearning for a yearning, whose only surviving use is to nourish a compound of obscure angelic subjectivity and restrained (to a greater or lesser extent) objective violence—what we call "religion."

he is characterized by his conduct with regard to lies. By definition a seeker of truth, because he serves the Torah (both "teaching" and "law") of truth, he knows the conditions in which truth itself needs to be covered by the prevarication of modesty to avoid becoming lethal (see Atlan, *Enlightenment*, chap. 9). In other words, an inside that resembles the outside, despite the opacity, is posited as the ideal of unity for someone whose objective is to construct unity in a world whose discords and pitfalls he knows. For those who do not follow this path—the unlettered "people of the land" (*'ammei ha-'arez*) whom the talmudic tradition opposes to the scholars—this ideal can no longer be embodied in the task of expounding and bringing into being, on earth and in society, through patient construction, a Torah of justice and truth that takes account of the discord of reality and the disappointments of knowledge. As we shall see below, in the story of Behemoth and Leviathan, this ideal then becomes "spirituality." Far from establishing the transparency and unity it asserts, it cannot avoid fighting against the reality of the bodies that surround it.

6. THE YEARNING FOR YEARNING: EROS AND THE HOLY

What we retain from this mixture of good and bad, of life and death, is a yearning for a place like that mythical Eden where our experience of knowledge and desire would no longer be divided. This condition, described by a large body of sacred erotica, of which the Song of Songs is an eminent representative, is the yearning for a physical Eros that is not sundered from the sacred aspect of the mind, and, conversely, for a mode of the sacred that is not exclusively spiritual.

Our daily lot, by contrast, strikes us as an experience of disjointed knowledge. Today objective knowledge, which is appropriate for objects, fills and shapes all reality.[47] As such, this type of knowledge makes it possible to resist, albeit imperfectly, the violence of ecstasy and love that would embrace everything. But this objective knowledge may sometimes generate another form of violence, colder and more systematic, which excludes and crushes those who have no part in it, replacing what Georges Bataille called knowledge of the "intimate." In his *Theory of Religion*,[48] Bataille evokes this original knowledge in "innocence," where what is conceived is neither object nor thing but a self-immanent world of unashamed intimacy, rooted in animality and childhood. But that disappears when shame comes on the scene, when sex becomes the occasion for an objectified knowledge of the outside, in the opacity of fascination.

When this happens, everything falls apart. The body is known as an object and the mind is disembodied; the sacred as immanence, as presence of the true, the beautiful, and the good, is stripped of its reality. A transcendence of mind and gods sets itself up as the rival of another transcendence, that of objective knowledge and control of things. As a consequence, what remains of the sacred is sundered from the reality of things; or, as Bataille says:

> Various existences, all opposed to the *things* that pure objects are, resolves into a hierarchy of *spirits*. Men and the supreme being, but also, in a first representation, animals, plants, meteors [etc.], are spirits. . . . [T]he hierarchy of spirits tends to be based on a fundamental distinction between spirits that depend on a body, like those of men, and the autonomous spirits of the "Supreme Being," of animals, of dead people, and so on, which tend to form a homogeneous world, a mythical world, within which the hierarchical differences are usually slight. The "Supreme Being,"

47. Objective knowledge of bodies has become scientific knowledge. The sense of one's own self—proprioception, which physiology opposes to exteroception—has become an object of neurophysiological research. The history of this transformation has now reached what we call the cognitive sciences. For an account of this story, which is not yet over, see Marc Jeannerod, *De la physiologie mentale. Histoire des relations entre biologie et psychologie* (Paris: Odile Jacob, 1996).

In this context, the capacity to see and picture ourselves from the outside is one of the traits that distinguish the human species from the other animals. In practice, the reflexive character of human knowledge, the fact that we know that we know, as well as our ability to use language, seems to be the qualitative distinction between it and animal or plant knowledge. This distinction itself, according to André Pichot ("Pour une approche naturaliste de la connaissance," *Lekton* 4[2] [1994], pp. 199–241), is the result of an exteroceptive image of one's own body that makes it possible for human beings to perceive their own movements as those of another object that moves in the same space. It is this externalization of the human body's knowledge of itself, the result of the sensorimotor coordination realized by our brain, that gives it such reflexive knowledge. This reflexivity distinguishes knowledge, as a psychological faculty, from the simple interaction between an organized system and its environment, even when, as in the case of an animal, this is more than a simple inorganic physical-chemical interaction because it has an organizing effect on that environment, which is transformed into what Pichot calls the "external milieu." If we follow this author's hypothesis, it is the human body's capacity to perceive itself from the outside and thus to know, among other things, "that it is naked," that enabled it to develop the psychological and cognitive capacities that distinguish it from a nonhuman body.

48. Georges Bataille, *Theory of Religion*, trans. Robert Hurley (New York: Zone Books, 1989).

the sovereign deity [*le souverain des dieux*], the god of heaven, is generally only a more powerful god of the same nature as the others.

The gods are simply mythical spirits, without any substratum of reality. The spirit that is not subordinated to the reality of a mortal body is a god, is purely *divine* (sacred). Insofar as he is himself a spirit, man is divine (sacred), but he is not supremely so [*souverainement*], since he is real.[49]

On the contrary, in the talmudic context of Adam and Eve before their transgression (and later as well, in the perspective of the possible repair that begins with Seth, their third son), man is divine, not "inasmuch as he is mind," as in the vision that "religion" has of him, but in that he bears the imprint of the divine seal on his form (*ẓelem*),[50] in which his body and mind are inseparably united. It is after the transgression and before the redemption, with Cain and Abel, the children of the Serpent, who have lost this form, and with the descendants of Seth, who have recovered it but imperfectly, because the repair is not complete, that this disconnection between the mind—which is assuredly "divine," but nevertheless ambiguous because divided, good or bad, angelic or diabolic—and the body, which can only be murderer or victim, exists.

This is why what corresponds to the violence of this form of the sacred—fallen, enchanting, and cunning, that of the serpent and the spirits, of the angel made beast or the beast made angel—is a longing for something "original" of which we can have only limited experience in the real world in which we live. This experience is that of necessarily incomplete and imperfect platonic love, what Walter Benjamin sees as the experience of the aura as a manifestation of the original, through the name of the beloved, the form of the divine seal of Adam in the formal and creative language of the prelapsarian world. Benjamin finds this longing for something divine that is no longer ambiguous, for an angel that is not a demon, in the search for beauty; not the beauty of a body-object, but rather that of the truth, whose "representational impulse . . . is the refuge of beauty itself,"[51] in philosophical discourse and poetry. This is where Benjamin hopes to find the Adamic language of truth[52] of a prelapsarian Eros in which words

49. Ibid., pp. 36–37.

50. M *Sanhedrin* 5:4 (B *Sanhedrin* 37a–38b); see above, Chapter 1, n. 107.

51. Walter Benjamin, *The Origin of German Tragic Drama*, trans. John Osborne (London: Verso, 1998), p. 31.

52. In practice, the same scheme applies to language, for it too is divided. From within its rupture, however, it still hints at that original time when it was unified. This is one of the messages of our story of the random sparks that spawned the generation of Babel (see Chapters 1 and 3). The restoration of the original unity is deferred until the end of time, until the timelessness of the "world to come." While we wait, this unity is *spoken* by the language of myth and of ritual, as in the myth of the random sparks and the Yom Kippur rites of atonement. This unity can also be spoken in the formal language of a logic or ontology constructed "geometrically" in which the union of the sundered body and mind is directly *named* and thought, even while remaining hermetic and incapable of explicit expression in the detailed analyses of our empirical knowledge (see, for example, Atlan, *SR* II, chap. 6). Like the simple unity of the infinite—or of Spinoza's substance—this unity "is thought but not perceived" (see below, n. 60). What is more, here we run up against the problem of using words to analyze and explain myth. We find it difficult to represent to ourselves in coherent fashion, other than through metaphors or hallucination, this union of inside and outside, of sex and thought, of knowledge and semen. What do the ancient Hebrew words that express this union mean for us, in the light of the divided experiences through which we are aware of our world, interior and exterior?

The myth itself tells us the origin of this divided language. The random sparks produced by the expulsion of the serpent's pollution were embodied first in the generation of the Flood and then in the citizens of Babel. The one language of that generation was used to express the violence and falsehood

are totally adequate. "Truth is beautiful: not so much in itself, as for Eros."[53]

This longing for a platonic love that would in some sense be the genuine Eros recalls the unconsummated betrothal of the prophet Hosea, after the savagery of his beloved's prostitution and the betrayal of her adultery;[54] or again, the ambiguous image of the sister-wife, as an antidote to sexual violence, attached to the founding couples, Abraham-Sarah and Isaac-Rebecca.[55]

But, in a constant balancing act, Benjamin's longing for a disembodiment in which the beautiful survives in philosophical truth, in the mechanical work of art, seems to correspond to the agonies of Bataille (to whom we fortuitously owe the preservation of Benjamin's own writings), agonies over the severing of angel and beast, to which he does not resign himself. The immanence of Eros explodes in two opposing movements of fall and disembodiment, and we cannot know which of them is the source of the greater evil: its fall into the violence of objects or its disembodied dilution in the spiritual and in religion.

of their own origins. In a different context, the prophet Isaiah refers to this union of warped language and spilled semen in the formula "seed of falsehood" (*zera' sheker*), which evokes seminal reason gone astray: ". . . sons of a sorceress, seed of an adulterer and a harlot. Of whom are you making sport? Against whom do you open your mouth wide and stick out your tongue? Are you not children of sin, the seed of falsehood?" (Isa. 57:3–4). In addition to the open mouth and protruding tongue found in this passage, biblical Hebrew employs a number of metaphors for the perversion of the organs of speech that serve as instruments of power and protective demons: "smooth lips" and a "tongue that speaks arrogance" (Ps. 12:3–4); lofty words and a harsh mouth (1 Sam. 2:3). To these are opposed "pure words, refined silver" (Ps. 12:7) and "the God of knowledge" (1 Sam. 2:3). The generation of Babel exploited its common language in its project to achieve total mastery of beings and things, doubly corrupting the original language of Adam, who first named all creatures and objects. First it reduced language to a means of communication in an enterprise of group control (we might call it "constructivist," to keep up the notion of the builders of the tower). This in turn led to the breakdown of this single language into mutually incomprehensible languages, leaving communication itself impossible from that time forth. But this second corruption bears within itself the possibility of atonement for the first, in that it demonstrates, albeit in negative fashion, the yearning for something that lies beyond communication: an encounter that is not verbal, but that the richness of the ambiguities and misunderstandings of language—what Wittgenstein saw as the pervasive vagueness of natural language (*Investigations*, §§72–73, §§77–81, §§104–108)—makes it possible, paradoxically and approximately, to verbalize.

53. Benjamin, *The Origin of German Tragic Drama*, p. 31.

54. Hosea 1.

55. Genesis 12 and 16.

7. BEHEMOTH AND LEVIATHAN

The conflicts and contrasts between these two effects of the fall are represented in Rabbi A. I. Kook's interpretation of the talmudic legend of Behemoth and Leviathan.[56] These two mythological animals, mentioned at the end of the book of Job (chapters 40 and 41), appear in the talmudic account of the celebrated hunt or contest that will be witnessed only by the righteous, like a great cosmic game.[57] *Behemoth*, the plural of the noun *behemah*, which normally designates large herbivorous land animals (often domesticated ones), represents the muscular strength of large bodies, some sort of imaginary elephant or hippopotamus. *Leviathan*, which refers to no real animal, is the name of a mythological marine reptile (in some translations it is rendered "crocodile") and modern Hebrew for "whale." In the midrash

56. A. I. Kook, *'Orot ha-kodesh* (Jerusalem: Mosad Harav Kook, 1964), vol. 2, pp. 317–318.

57. B *Bava batra* 74. <This aggadah of the End of Days makes several appearances in the rabbinic literature, notably *Leviticus Rabbah* 13:3 and B *Bava batra* 75a. The righteous will join the Heavenly Host in hunting the leviathan and the behemoth. When the archangel Gabriel cannot kill them, the two monsters will fight to the death and the righteous will watch the spectacle. In some versions Leviathan and Behemoth kill each other; in others, God slays them.>

it refers to the two *tanninim* or "sea monsters" (Gen. 1:21), created on the fifth day; but God "killed the female and preserved it in salt for [the benefit of] the righteous in the Time to Come; for had they been permitted to be fruitful and multiply the world could not have survived them."[58]

According to the legend the male leviathan, which continued to haunt the rivers and oceans, has the form of a serpent, sometimes straight, sometimes coiled.[59] In Rabbi Kook's interpretation, it represents the subtle form of a fluid "spirit," which can extend from one end of the universe to the other. The talmudic combat between Behemoth and Leviathan pits body against mind in the arena where inner life and social life contend. Their combat represents the contests between irreconcilable but equally legitimate ideologies or impulses. The righteous watch them fight, perhaps within themselves, as in a theater, with just enough distance from the spectacle to enjoy it.

By definition, the righteous realize the perfect unity in which evil is not evil; this is why they can be entertained by watching these battles between body and mind. The outcome can only be good, in their eyes, just as in the playhouse where everything turns out well in the end and even the corpses stand up for their curtain call; or in the gladiatorial arena where losers and victors contribute alike, like "good sports," to the triumph of the games.

But in the divided reality of our experiences, dominated by our objective knowledge of things, subjective and interior knowledge, to which aesthetics and religion assign the highest value, is shunted aside into the unreal arena of the "spiritual." There it allows us, in a pinch, to conceive of the perfect and transparent knowledge of Eden, which is simultaneously knowledge of being and knowledge of life, where the Tree of Knowledge is no longer separate from the Tree of Life. But this is something that, as Eliaschow says, we can think of but never attain.[60] Any experience of this knowledge is deferred to a timeless future, to the "end of time." Until then, only the timelessness of

58. Rashi on Genesis 1:21.

59. "For that light of the higher knowledge is the Leviathan–straight serpent and coiled serpent, which will be disclosed at the time of the final *tiḳḳun*, and its disclosure will begin with the advent of the Messiah" (Eliaschow, *Sefer ha-De'ah*, vol. 2, p. 19 ["Explanation of the World of *Tohu*" II, §1.7.6]. Eliaschow employs the same image, applied to the white space between words, in another of his books, which serves as an introduction to *Sefer ha-De'ah:* namely, *Haḳdamot u-she'arim* (Piotrkow, 1908; repr. Jerusalem, 1970), *Liḳḳutim* 1.4, pp. 193–198. (On white space and the infinite potential of the unsaid it suggests, see Atlan, *SR* II, chap. 12).

The expressions "straight serpent" and "coiled serpent" associated with the Leviathan can be traced to Isaiah 27:1 and Job 26:13. The image of these two serpents as two forms of "spirituality" is also suggested by their aspect as celestial bodies; some read these verses as applying to the constellation Draco (e.g., Ibn Ezra on Job 26:13). The matter of which celestial bodies are made is supposed to be more luminous and "spiritual," less gross than that of terrestrial bodies.

The text of Isaiah is often interpreted, notably by Rashi, as referring to the idolatrous civilizations of Egypt and Assyria, which are compared to leviathans and to straight and coiled serpents. The identification of these two serpents with the constellation Draco is also understood as an allusion to the idolatrous rites of these cultures, which perceived and celebrated the celestial bodies as the divine sources of their power. Another possible reference is to the alchemical significance of the serpent Ouroboros, which swallows its own tail (see Chapter 3, §4 and n. 25).

In Eliaschow's conception of the world of *tohu* these two serpents are described at their "source," which is that of the "higher knowledge," where they express the very first divisions produced by the "carving" of "straight and coiled" lines, which demarcate and circumscribe the letters of the text and distinguish them from the surrounding white space, like "light from shadow" (Eliaschow, ibid.).

60. <Or "perceive"—the Hebrew word has both meanings.> Eliaschow used this phrase with reference to the unity of the infinite essence and the name. "How its hidden truth itself is uncovered in that name—for us all of this is only an object of knowledge and not of perception" (*Haḳdamot u-she'arim* 1.6, p. 12). We may compare this idea to Spinoza's distinction between clear and distinct knowledge of the unity of substance and our inability to know how "each part of nature agrees with its whole, and is associated with the remaining parts." We know "the reasons which induce us to believe" this, but not "the means whereby the parts are really associated, and each part agrees with its whole," because "to answer such a

logic or ritual, notably that of the real/unreal in-between of the conclusion of the Day of Atonement, can let us imagine it. And this idealized experience of "the intimate" soon decays into illusion, a prevarication composed of prophetic dream and hallucination, if we believe that the ideal has already arrived, fully realized; the erotic degenerates into pornography and undifferentiated love produces war, in the quasi-intoxicated and ecstatic experiences we can have of it.[61]

8. WHAT THE BODY CAN DO

The result is that in our actual world, postlapsarian but in a process of repair, what we normally experience is the cunning animal knowledge that makes love and war at the same time. For it is only retrospectively, and by contrast, that we can imagine prelapsarian sexuality, which never really existed—but which could have existed, according to the Midrash, had Adam and Eve waited to grow up and be ready for it.[62] It might also exist in the future End of Days, in an eschatological world-transcendence that has at least the merit of an organizing utopia. In the meantime, the ambiguity of which we are composed—good and evil, the good aspect and the bad aspect of knowledge—cannot be unraveled. But in this reality of ours, not only are cunning and seduction carriers of death; they are also forms of life, inevitably the most agreeable and most alluring, that nature offers us as an indirect, convoluted, and opaque means to make the truth, light, and transparency of the future emerge. What is more, we may wonder what life would be like without these games of beauty and seduction; or, rather, we can imagine it only too well as a form of death caused by boredom and repetition. Although it may be true that, in this context of trickery and seduction, beauty of the body and even of the face may harbor a certain danger, ugliness for its part is no less aggressive, especially when displayed in full nakedness. The modest and seductive trickery of clothing miti-

question, we should have to know the whole of nature and its several parts" (Letter 15 [32], to Oldenburg). Compare this with *Sefer Habahir* on the "fiftieth gate of knowledge," which remained closed even to Moses (see Chapter 1, n. 87).

61. We should note in passing that this interpretation of the combat between Behemoth and Leviathan illuminates the last chapter of the book of Job. God answers Job by invoking the precedence of nonhuman nature over the childish morality of human beings who judge good and evil only in terms of what affects them as good and bad. We shall see that the "great works" (*'alilot*) of Elohim in nature are necessarily perceived as "pretexts" (also *'alilot*) of bad faith when they are attributed to a personal God who works evil only to punish the guilty. Among these effects of God-Nature, "terrifying [in his] great works" (*nora' 'alilah*), the overwhelming and untamable power of Behemoth and Leviathan, interpreted respectively as "material" and "spiritual," acts simultaneously on the exterior and interior of human beings. This is what finally persuades Job of the absurdity of his questions and rebellion, for they can be expressed only in the limited context of the subjectivity of an individual moral consciousness, restricted to the categories of obedience and piety. Like a child, who is ignorant of the world and the interlinking of its forces, the moral theologian can conceive of misfortune only in terms of punishment of the guilty by the supreme judge; in the absence of sin it is accordingly scandalous and inconceivable. For those who can perceive the sequence of causes (*'illot, 'alilot*), however, sin and guilt, even when instantly recognizable, can always be seen as pretexts employed by bad-faithed Nature. These pretexts make it possible—though only locally and temporarily—to relate seriously to the moral and legal responsibility of ostensibly free agents, who are in fact determined by a series of events of which they are largely unaware. In these conditions, the deity who distributes reward and punishment appears indeed to be *'el nora' 'alilah*, "a terrible [awesome *or* terrifying] god of intrigue" (see below), when he is identified with God-Nature, that power which, as the book of Job has it (38:4ff.), "laid the earth's foundations . . . [and] fixed its dimensions, . . . closed the sea behind doors," and "bound the chains" of the constellations, which gives life to wild animals, each according to its might and intelligence, and which, finally, created Behemoth and Leviathan. But it is precisely that "terrible god" who is invoked in the concluding service on the Day of Atonement, in the last hour of the ten *yamim nora'im* or Days of Awe (*lit.* "terrible days"); sins can be forgiven precisely because the guilt has no ontological reality and is only a pretext for assigning responsibility.

62. See Eliaschow, *Sefer ha-De'ah*, vol. 2, p. 294 ("Explanation of the Tree of Knowledge," §6). See also Joseph Gikatilla, *Sha'arei 'orah* (Warsaw, 1883; repr. Jerusalem, 1960), Gate 10, pp. 209–210.

gates this aggressiveness and contributes, instead, to peace. The cunning and seductive knowledge of the serpent—not in the conditions of an ideal paradise where there is neither ugliness nor suffering, nor the externality of knowledge, but in the real world of our actions and passions—is not bad. It is like the Tree of Knowledge, simultaneously good and bad, because its lies are not necessarily true lies but can be the tricks of a modesty that conceals the nakedness of the truth to make it better understood.[63]

What is more, even in the time of our real existence, outside Eden, we still (or already) have some special and fragmentary experiences of immediate knowledge of the self by the self, which never fail to astonish us because of our normal ignorance of what our body can do. A brief review of the content of these special experiences is an interesting exercise. It is as if there lay, buried within ourselves, some remnants of this capacity for that original mythical knowledge, transparent and immediate, by which the body-mind knows itself even though the mind itself does not know how it knows. There is an entire gamut of activities in which we perceive and sense our body both acting and knowing, while at the same time, as if in some form of split existence, we observe what it knows how to do without knowing how it knows it. I am referring here to those extraordinary experiences in which we are the simultaneously active and passive witnesses of a bravura gesture that our body carries off and that attests to some knowledge that is the fruit of learning and memory. The action is all the more astonishing because we really do not know how we mastered it. Every athlete, for example (and every dancer, too), experiences these quasi-miraculous successes, in which the flawless move seems to execute itself, effortlessly. Nevertheless, it was preceded by a strenuous apprenticeship in practices that seem to be quite counter to nature and far removed from what we would do spontaneously. What could be more unnatural than the movements of a tyro skier, or the techniques involved in learning how to serve in tennis, or the precision of shooting a basketball or heading a soccer ball? Nevertheless, after the motion is mastered it is assimilated into the life of the body, which thereafter knows how to do it far better than we can explain it and which can retain this knowledge for years, even though we are not necessarily conscious of it. Physical exercise makes us experience, at the deepest level of self, an unsuspected unity with mind; it is without doubt the quest for this extraordinary experience that has been developed by Far Eastern traditions through ritualized practices, such as archery and the martial arts.

Closer to us and in a more worldly vein, a team of Italian researchers who specialize in training competitive athletes has attempted to understand and rationalize this knowledge that the body seems able to acquire only if it has no conscious knowledge of how it does what it does when it does it.[64] They have elaborated a theory and practice of training in which the appropriate movement is learned not, as is normally the case, by consciously and progressively correcting errors, but through the initial exaggeration of these errors. The only way the body can learn what the mind wants it to do is by indirection, through the conscious negation of what we want the body to do at the end of the process, without thinking about it. This experience of the mind by the body informs the life of the musician, the painter, and the sculptor. For them, too, the gesture can be carried off, in its aesthetic and especially

63. See Atlan, *Enlightenment*, chap. 9.

64. "[A method of error amplification]," in W. Bragagnolo, P. Cesari, G. Facci, and P. Olivato, *Apprendimento e ridimensionato motorio* (Rome: Società Stampa Sportiva, 1993).

its mystical aspect, only when the analytical hesitations of conscious learning are over and it becomes an immediate expression of the internalization, in the body and by the body, of knowing how to be. In her charming meditation on the piano and martial arts, *The Beauty of Gesture: The Invisible Keyboard of Piano and T'ai Chi*, Catherine David writes of the "surprising . . . embodiment of thought in a sensation, this mutation of a mental image into solid flesh," and, as in the case of the ritual gesture, of "remembering something one isn't aware of knowing."[65]

In another genre, Robert Pirsig describes what he calls the experience of quality, that is, what the subtlest speculations of ancient philosophy push us to aspire for, but which we can attain only by going beyond these speculations themselves, by walking or, better yet, by riding a motorcycle, which he elevates to the rank of an initiatory practice.[66]

65. Catherine David, *The Beauty of Gesture: The Invisible Keyboard of Piano and T'ai Chi* (Berkeley, CA: North Atlantic Books, 1996), pp. 21, 20.

66. Robert Pirsig, *Zen and the Art of Motorcycle Maintenance* (New York: William Morrow, 1974).

9. MECHANISM AND RESPONSIBILITY

These ultimate experiences of what the body can know of itself are important because they reinforce the "longing for the original" through the life of our extended body, which moves about in the space of things, and not only in the interior space of our emotions or intellect. At the same time, however, this merely reinforces the contrast with objective and shared knowledge of things, including our body, as we perceive them from the outside.

Hence we must now turn our attention to the development of this objective knowledge that clearly does not stop with the knowledge that "we are naked." This will lead us to recognize one last effect of the cosmic serpent of our myth, in the ultimate trick of nature that makes human beings believe they are guilty when, as a matter of principle, they are not, but without derogating in the slightest from their responsibility for what they do and who they are. This dialectic of the nonculpability of human beings who sin is developed in a subtle and fascinating way by Eliaschow in *Sefer ha-De'ah*, drawing on talmudic and midrashic texts that confront the question of free will and responsibility in a universe that is *totally determined* since "before the creation of the world."

Before we get there, however, we should take a brief look at our attempts to know ourselves in our bodies by means of the objective science that has led to contemporary biology. The objective science of bodies has always been the engine of the medical sciences. Today, taking over the baton from the traditional forms of medical science, it is biology that teaches us, with unsurpassed effectiveness (even if the lessons are always incomplete and subject to revision), how natural laws govern the birth, development, and death of living bodies. It reveals sequences of causes and effects, of principles and consequences; in brief, mechanisms by which bodies are formed and act. Although this form of knowledge is indirect, linked to the vagaries of the experimental method, and increasingly abstract and even formal—and thus far removed from the immediate and transparent knowledge of the self by the self—it is much more effective than anything ever known in the past. It has produced, among other things, a technological mastery of the processes of life and death that makes it possible to control, to some extent, not only birth and death but also, and especially, illness and pain. There have been many debates about the ambiguity of biotechnology, with its inherent risk of hubris and excess. It is to be feared all the more because it is effective and can confer the power of life, as well as death, on those who master it.[67] Here, though, I want to speak of a different ambiguity, perhaps even more profound, which has to do with the moral consequences of the discovery of the mechanisms that produce us, as body and mind, along with our actions. The old problem of absolute determinism versus freedom is inevitably posed in new terms because contemporary biological knowledge is fiercely monistic and mechanistic, allowing less and less room for a mind that, separate from the body, could somehow escape natural causes and thereby constitute, as formerly, a separate domain of the suprasensible, the intentional, and freedom.

Until the start of the twentieth century, the life sciences were distinguished from the physical sciences by the evident impossibility of purging them of the role of organizing forces that guide organisms in the development of their structures and accomplishment of their functions. Even Claude Bernard, who sought to reduce physiology to the quest for strictly defined physical and chemical mechanisms, nevertheless invoked a sort of invisible hand to explain how every biological phenomenon could be directed in its relations with others so that all of them together could constitute an organized functional whole. This led other biologists to wave the flag of a stubborn vitalism and to take these vital forces, sometimes compared to Aristotle's entelechies and supposed to explain the animate character of living beings, as opposed to the inanimate—soulless—character of physical objects, as their research object. They set themselves to refuting the mechanists who wanted to eliminate these holdovers of classical metaphysics, in which the immaterial soul organizes and directs the body. In pursuit of this end, the latter looked for and discovered more and more physical and chemical mechanisms that could explain the properties of living things without the need to invoke vital forces or even Claude Bernard's invisible hand and could thereby short-circuit the philosophical debates about the material or immaterial nature of life. Today, the debate between vitalists and mechanists seems to have been conclusively decided in favor of the latter. It is no longer necessary to invoke "Life" in order to explain the properties of organisms. Physical and chemical mechanisms determined by the chemistry of carbon and the catalytic properties of macromolecules have replaced it.[68]

67. See, for example: Henri Atlan, "Personne, espèce, humanité," in *Vers un anti-destin? Patrimoine génétique et droits de l'humanité*, ed. F. Gros and G. Huber (Paris: Odile Jacob, 1992), pp. 52–63; idem and C. Bousquet, *Questions de vie. Entre le savoir et l'opinion* (Paris: Le Seuil, 1994).

68. This situation is, of course, the result of the breakthroughs made by molecular biology, which has reduced what were formerly the most mysterious phenomena associated with living beings—reproduction and heredity—to chemical reactions. But the reductionist method, classically employed in physics, probably cannot explain the overall properties of organisms observed as integrated functional wholes. This result has been achieved through the concomitant development of a new physics, the physics of complex or disordered (or nonlinear) systems, which has expanded the concept of mechanism. Because they could not be explained by simple physical mechanisms, the only ones known and studied then, "emergent" or "self-organizing" *phenomena*, in which we observe a whole producing more than the sum of its parts, structures and functions unknown at the elementary level that appear at the global level and have every appearance of a goal that steered their development, were formerly deemed to be specific to Life and its teleological character. Now we know about emergent and self-organizing *mechanisms* that are common to inanimate physical and chemical systems and to organisms. Contrary to the misunderstanding associated with faulty interpretations of the new paradigms of emergence or self-organization now produced by the "sciences of complexity," we are in fact dealing with a new mechanistic approach that is to some extent even more radical than the old one, because reductionism is no longer needed. It is true that these mechanisms are sometimes described using probabilistic methods that make it possible to incorporate elements of uncertainty and randomness, which are due to our incomplete and perhaps irreducible ignorance of the sum total of all determining

We find it difficult to integrate the reality of these causes, which make the discoveries of the natural sciences (including the nature of human beings) possible, into our experience of social and moral life. The discovery of the biological factors that underlie our behavior—which is not, however, reduced to genetic factors, as in the caricature propagated by the current vogue—has spawned a crisis for the traditional foundations of law and morality. We are the heirs of a culture that began as theological and then became philosophical—idealist or dualist—postulating an immaterial soul or abstract universal reason as the ground of our freedom. In the West, this culture has reigned for more than two thousand years as the foundation of both moral and social life.

Today, after three hundred years of mechanistic science, we have learned more and more about how the body does what it does. We have increasing knowledge of how events take place in us and through us, including those we think we trigger freely, the behaviors we think we are the agents of. In other words, although we are far from being able to answer all of the questions that can be asked about our body and our mind,

factors and their details. But these methods also make it possible to limit these uncertainties and, as Nietzsche put it, to "tame chance." Even if the details of the emergent behaviors cannot be predicted quantitatively and precisely, they can be known qualitatively and with their generic properties. These properties are thus produced by a global determinism that makes it possible to predict their occurrence, if not yet their detailed parameters. The role of indeterminacy and chance is taken into account operationally, with no need for a statement about its irreducible, ontological character or that it is due only to a provisional deficiency of knowledge. Thus the novelty can itself be foreseen—as something new—within the contours defined by the known constraints. This allows us to understand Ecclesiastes' "there is nothing new under the sun" in a relativized and perhaps new fashion (see below, n. 112, the parable of the caper bush).

As long as the vitalist theories held sway, Life could serve as the medium or substrate of the mind and of the animate (and animating) soul in the body. This is precisely how it functioned for the proponents of *Lebensphilosophie*, from Leibniz and Kant through Bergson and Whitehead, by way of the Romantics, Nietzsche, and the German idealists. Even when the soul ceased to be an object of scientific investigation, and later in the modern mechanistic and materialist context of the life sciences, vitalist doctrines and the sciences of the mind did not stop playing on the ancient ambiguity inherited from the traditional oneness of the animate and living, on the one hand, and the inanimate and lifeless, on the other. Even Descartes' concept of the animal-machine or La Mettrie's human machine does not eliminate this ambiguity, because machines have always been an extension of the living human being. The concept of the machine—and, today, of the computer program—raises the question of the designer and the origin of whatever it is that informs the machine and gives it sense and purpose. This is where the mystery of "life," which is supposed to distinguish these "machines" from inanimate physical beings, might find a new refuge. But it is precisely the quest for natural physical mechanisms that differ from those of artificial machines, aimlessly generating structures that are preserved even as they reproduce and modify themselves—"self-organizing" mechanisms—that eliminate Life as an explanatory concept.

Today we can return, with even greater force, to Albert Szent-Györgi's seeming paradox: "Life as such does not exist." This does not mean our subjective and linguistic experience of life, which is opposed to death, of course, but life as an object of inquiry or as a concept with some explanatory value in modern biology. Even though they are still called the "life sciences," these fields of research no longer seek to understand the nature of some entity or essence known as "life," but the mechanisms by which certain compound physical bodies assemble themselves, grow, reproduce, and die. In fact, as I have suggested elsewhere ("Morts ou Vifs?" in Atlan, *L'Organisation biologique*, pp. 283–284), these disciplines could equally well be called the "death sciences"—the study of the death of organized physical objects. This is even truer today, now that we have learned about other mechanisms of physiological cellular disorganization, called "apoptosis" or programmed cell death, which are intrinsic elements of many normal processes in the development and growth of organisms. (For an introduction to this new field see, for example, one of the earliest articles on this topic: G. Williams, "Programmed Cell Death: Apoptosis and Oncogenesis," *Cell* 65 [1991], pp. 1097–1098; and the reviews by J.-C. Ameisen, "Les sociétés cellulaires: de l'interdépendance à la complexité" and "Le suicide cellulaire," in the "Sociétés cellulaires" section of *Pour la Science* [April 1998], pp. 6–12 and 60–66.) I have proposed replacing Bichat's definition of life as "a set of functions that resist death" by "a set of functions that can make use of death" (see "La vie et la mort: biologie ou éthique," in Atlan, *Entre le cristal et la fumée*, p. 278). Finally, we should not omit the classic definition—which is more than a sick joke—of life as a sexually transmitted fatal disease. Even though not only the duration of life, but also the quality of life, remains a constant preoccupation of physicians, politicians (i.e., citizens), and each and every one of us, for the biologist life as such does not exist as an object of inquiry.

the more we know about what our body can do and understand how it can do it, the less can we hold onto the classic idealist or dualist image of the human being as a soul or mind steering the body,[69] epitomized in the ancient metaphor of the charioteer and his team.

On the contrary, here we seem to have returned to Homeric Greece and the characters of the *Iliad*, as well as the biblical personages in Genesis and Exodus, where the links between body and mind were very different from those we habitually represent to ourselves today. For the notion of body has a history. It was conceived along with the idea of the immaterial soul that was supposed to inhabit it and govern it, more or less effectively. It is in this context that Plato pictures the body as the prison of the soul, which resides in and animates it, and as the main obstacle to the contemplative philosophy of which the soul alone is capable. This tradition dominated philosophy and then theology at least until the Renaissance, and then with greater confidence until the mechanistic scientific revolution of the seventeenth century. But another, almost forgotten, tradition was preserved as well: the nondualist philosophies of nature of the Stoics and Epicurus (among others), for whom no knowledge was possible without physics. Theirs was a science of undivided bodies and souls, because the various parts of the ancient soul, pneumatic, fiery, and bilious, were no less material than the bodily organs. (They were, however, composed of more subtle matter: fire, the substance of the gods; and air, the breath of the spirit.) This forgotten physicalist tradition was rediscovered and rejuvenated in the seventeenth century by Spinoza and subsequently upheld by the naturalist scholars and philosophers of the last three centuries. Today a new physicalism reigns in biology, especially in the neurosciences and the "cognitive sciences" (one branch of what are still known as the "sciences of the mind").

We also need to know that consciousness, as a psychological phenomenon that we experience, has not always existed. It seems to have appeared rather recently in the history of the human psyche. As consciousness of consciousness, as the possibility of an interior dialogue of self with self, it may be the product of a process that dates back no more than three thousand years or so. In a volume that is extraordinary in more ways than one, Julian Jaynes, drawing on the mythological texts of Greek and Hebrew antiquity, postulates that the very idea of a voluntary consciousness, to which we attach our experience of a subject able to conduct a dialogue with itself and to decide how to behave, emerged at a relatively recent date in the evolution of Homo sapiens.[70] He dates its emergence to the time of the *Odyssey* and the later biblical texts (the Prophets). By contrast, the characters of

69. Spinoza has at long last won this posthumous victory over Cartesian dualism and the idealism of Leibniz and then of Kant. In the *Ethics* he waxed ironic about this dualism, in which he saw the perfectly incomprehensible image of man in nature as "a kingdom within a kingdom." Categorically denying any explanatory value to the anthropomorphic invocation of final causes in nature, he endeavored to base the *Ethics* on a physics of the human body. This physics, which he himself described as weak, has given rise to much misunderstanding, because it has been taken for a mechanism or dynamic that should be compared with the physics of Descartes, Leibniz, and Newton. It is in fact a physical theory of the organism (and thus a first attempt at what we would call today the biophysics of organized systems) and has lost nothing of its evocative significance.

70. Julian Jaynes, *The Origin of Consciousness in the Breakdown of the Bicameral Mind* (Boston: Houghton Mifflin, 1976). Jaynes conceives of the transformation as the outcome of an internalization of voices and visions. His theory is that these were originally perceived as having their own external existence and attributed to natural entities—sacred mountains, plants, animals—or to human beings who thereby acquired the status of god-kings. We can have some idea of this by the way in which we relate to the modified states of consciousness that are associated, for example, with problems of communication between the right brain and left brain, or with chemically induced states. In some traditions, the ritual use of hallucinogenic plants constituted a transitional period during which the memory of the ancient bicameralism was

the *Iliad* and the earlier books of the Bible, composed before the notion of a real separation between body and soul emerged, lack psychological depth, in the modern sense of subjects who dialogue with themselves and weigh the pros and cons of alternatives before deciding on one course of action or another. The behavior of these more primitive characters is determined by the gods who possess them and who may exploit them to realize their own stratagems, their own intrigues. The heroes may suffer or rejoice, but they observe what is done by their instrumentality and have no sense of determining it.

When we come to contemporary biology and our knowledge of the mechanisms that determine what we are and what we do, the circle is closed: we have returned to the idea that we decide and act freely, when we are in fact determined and actuated rather than acting. All the same, there are several differences between this picture and the ancient myths.

periodically reactivated, even while reflexive consciousness seized an ever-more-dominant role, thanks to its greater adaptive efficacy in an increasingly populated and urbanized world, where individuals found themselves further removed from proximity to the god-king. The importance of the Australian aborigines' dreamland as a source of revelation to be applied to social, territorial, and ritual organization is probably a survival from that period, among a people who were long isolated from all contact with ancient and modern urban civilizations. The most ancient origins of the universal anthropological phenomenon of the "sacred" and "profane" lie in those modified states of consciousness, although this ritual usage itself has disappeared from most of the major religions. (For a discussion of this hypothesis, see Atlan, *Enlightenment*, pp. 337–350.) In the history of Judaism, the culmination of this development is marked by the concept of the "end of prophecy" (see Chapter 4).

The Hellenist James Redfield, analyzing Homer's language, observes (following others) that the mental reality of the Homeric hero is fundamentally different from our own. He identifies Socrates as the first occurrence of a mental self presented as distinct from the bodily self, but does not systematically track the differences between the *Iliad* and the *Odyssey*, which might be significant from an anthropological perspective. (See James Redfield, "Le sentiment homérique du Moi," *Le genre humain* 12 [1985], pp. 93–111. In addition to the reference to his *Nature and Culture in the Iliad: The Tragedy of Hector* [Chicago: University of Chicago Press, 1975; 2nd ed., Durham: Duke University Press, 1994], the article includes an extensive bibliography on the subject.) Redfield notes that it is always the Homeric hero's body that speaks to itself, through organs such as the *stethos* (rib cage), *thymos* (breath), and heart. Homeric language does not have a word for "mind" in the modern sense; the *noös*, located within the *stethos*, is employed to grasp the meaning of things and words, but in the manner of a sense organ that sees and perceives them. This *noös* is not, however, consciousness of what we understand. "When we are conscious of having noted some detail, the operation has already taken place; we are not conscious of having registered something until after the fact. We can talk about the *noös*, but not perceive it. That is why, even though it is a mental faculty, the *noös* cannot be identified with the mind" (Redfield, "Le sentiment homérique du Moi"). Similarly, the Homeric *psyche* is not a "soul" or a self capable of choice. The Homeric *psyche* is associated with existence after death, "an empty shell, an apparition endowed with speech," that exists only as a phantom in the Underworld. For Homer, what we designate as "mental" is a bodily function. Thus the Homeric epics are "prephilosophical," notably in their failure to distinguish between soul and body. Socrates was the first person for whom *psyche* designated something like a soul or spirit in the sense we attach to these words:

> Socrates holds that the "self" is not constituted by one's friends, possessions, clothes, body, or appetites, but by the *psyche*, an elusive entity, capable of choice and thus of virtue—a doctrine that would later have, as a corollary, the divorce from all of these external carnal objects that are incarnated in the world. From this revolutionary idea there emerged, after multiple avatars, our Western idea of the person, which often implies a sort of Manichean drama of a soul with desires (base and material) arrayed against our conscience or principles (lofty and spiritual) for control of a third term, the will. (ibid.)

The lack of any substantial distinction between soul and body was revived, in a philosophical context, by the Stoics and by Epicurus. It nevertheless remains true that this strictly philosophical opposition to Plato and Aristotle could not have been expressed before the post-Homeric experience of consciousness-of-consciousness made awareness of such a distinction possible. In Homer, according to Redfield, as long as "the interior self is nothing other than the organic self" (ibid.); as long as its *ego*, "an open arena of forces," is not limited; as long as the self is forced to redefine itself every day on the basis of what the gods make it do, the *noös* of human beings is like "the day that the father of men and gods brings them" (*Odyssey* 18, 136–137). The question of a possible distinction between soul and body cannot even be asked. In this sense, we are dealing with *prephilosophical* human *nature* rather than with a debate among philosophical schools.

This also makes it possible to understand that the transition from myth to philosophy did not take place suddenly, as a simple dichotomous choice between the "religious" and the "rational." For a number of centuries, and not only in ancient Greek culture, multiple forms of rationality coexisted and were interwoven, sometimes in the same authors, before the separation of *mythos* and *logos* appeared (see Chapter 4, §1).

First of all, it is clear that in the past, much more than today, bodies, with their pains, diseases, infirmities, limitations, and difficulties in accomplishing the tasks necessary for survival—and especially as compared to the lightness, rapidity, and apparent nonlocalization of thought—could readily be perceived as prisons of the soul and obstacles to knowledge, at least to the knowledge attained by pure contemplation of ideas. By contrast, the wager made by the empirical and logical method of modern science was to consider the sensory experience of bodies by bodies as a sine qua non of knowledge of the real world, under the control of reason. One of the technical consequences of the success of this method is our substantial success in releasing our bodies from many of their former constraints and miseries. This physical liberation, for all that it remains incomplete and relative, makes it easier to conceive of a mind whose existence is fused with the body, although the walls of that prison do not circumscribe its activity. At the same time, however, this eminently successful science teaches us that our body lives quite independently. Just as for the ancient Stoics, the role of the living soul is now played by something material: the matter that makes up our cells, molecules and macromolecules, DNA, proteins, and so on, whose chemical activity underlies the self-organizing properties of living bodies. As formerly, we learn from this that we are actuated and determined when we think that we decide for ourselves.

But this is where a second difference emerges between the Homeric conception and our own. If our behavior is the result of "possession" and outside activation, the puppeteers are no longer the gods of a mythical pantheon but the molecular reactions of our genes and our cellular metabolism. This is a difference of scale, because the gods were anthropomorphic and personified, whereas the molecular interactions produce our cognitive abilities, our language, our capacity for social communication, and, ultimately, our consciousness of ourselves—that is, our subjective experience of personal existence—mechanically and impersonally, as in a self-organizing machine.[71] There are both advantages and disadvantages to each of these two modes of possession—or rather determination—by mythical personal gods,

71. While avoiding both the reductive simplifications of classic materialism and the disembodied idealism of the philosophies of Mind, Spinoza's theory of three kinds of knowledge supports a formal concept of the reflexive nature of the mind, which is one with the body whose idea it is (in the sense of what permits it to be thought) and at the same time the idea of itself and of its oneness with the body. It is this reflexive determination of thought, simultaneously autonomous and one with the body, in that it is one and the same thing, that Spinoza, anticipating modern cognitive science, expresses by the formula of the "spiritual automaton" (*Treatise on the Emendation of the Intellect*, §85, trans. Shirley, p. 24 [Elwes' title: *On the Improvement of the Understanding*]). (The expression was taken over by Leibniz but given quite a different meaning in his idealist philosophy.) "Men judge of things according to their mental disposition, and rather imagine than understand" (*Ethics* I, Appendix). But the progression from this imaginative knowledge of the first type to the intuitive knowledge (knowledge of the third type) by which one has an adequate knowledge of "the singular essence of things"—and thus of beings and of one's own self—passes through knowledge of the second type, that is, rational comprehension that deals with "common notions." These are adequate ideas that are arranged *according to* the state of the brain (although not caused by that state, because it is always the same; see *Ethics* II 7, III 2). But they are "common" in the sense that they are produced by every human being (a unity of a mind and a body) in the same fashion as other human beings produce them, and thus in the same way as they are produced by nature as thought, in the infinite understanding that is one with infinite corporeal nature. The rational sequence of the proofs in the *Ethics* makes it possible to trace the logic of this progression, running from the appendix to Part I through Propositions 4 and 14 of Part V (and what follows them), by way of Proposition 38 of Part II. In this progression, the identity of mind and body retains the same character, even though the status of the ideas produced changes from inadequate, confused, and mutilated to adequate, inasmuch as they are produced in the infinite understanding. In all these cases, the human mind's capacity to form adequate rather than confused ideas is correlated with the complexity of the body to which it is joined, that is, to the fact that the brain can have a vast number of states, and, more generally, that "human bodies are capable of the greatest number of activities" (*Ethics* V 39, Note).

projections of ourselves in which we can recognize ourselves, or by impersonal mechanisms over which technology gives us some control.

Whatever the case, we are no readier today (or at least not yet) to cheerfully accept the return to unconscious determinism that the human and natural sciences have nevertheless uncovered for us. The problem is that in the meantime, thanks to the emergence of consciousness in the evolution of the human psyche, morality has changed its character for most of us; the morality of free will, of the will exercised by an immaterial soul and the responsibility borne by a subject, an autonomous causal agent, has replaced the morality of the sublime soul, of heroism, generosity, and submission to one's fate. One of the major tasks of the theology of the monotheistic religions has been reconciling this freedom of moral choice, which has become the condition of human salvation, with the divine omniscience and omnipotence that are the alpha and omega of their doctrines.

10. NATURE'S ULTIMATE TRICK: THE PARABLE OF THE DIVINE INTRIGUES (*'ALILOT*)

Given what we have seen about certain currents of kabbalistic thought (to be strictly distinguished from spiritualist theology), it is not astonishing that it features—as among the Stoics, but also as survives to the present day in the context of Hebrew monotheism—more than one mode of kinship with this notion of the nature *of* human beings and of nature *in* human beings, simultaneously actor and acted on by the human form of the "image of god(s)" (*ẓelem 'elohim*).[72] Here, as in every monism in which the distinctions between body and mind are not essential but depend on the perspective from which we observe their activities, we encounter the games that determinism plays with itself. In general, instead of an essential duality of matter (passive, created) and mind (active, creative), these distinctions are part of an "infinite chain of being"; what is taken as body and opacity, when contrasted to the light or soul that fills it, is itself mind and light when contrasted to the vessel that it fills and animates in its own turn. One consequence is that the soul is not thought of as an immaterial entity. In the Hebrew tradition the notion of soul has its own history; the soul was not always a "spirit," in the sense of incorporeal. In Genesis, the human body molded from the earth is animated in different ways by different souls, created from other elements; the *nefesh ḥayyah* (living soul), common to animals and human beings, is identified with the blood ("for the *nefesh* of the flesh is in the blood" [Lev. 17:11]). The *ruaḥ* (spirit) is breath and wind. The *neshamah* or intellectual soul (related to *neshimah* 'respiration'), perhaps immortal, is literally an in-spiration of life (*nishmat ḥayyim*), blown in through the nostrils.

72. See Chapter 3.

In this tradition we find the forgotten vision of an absolute natural determinism affecting human beings, expressed not in Greek or in Latin, of course, but through the images and characters of the Hebrew Bible, amplified by the parables of the Talmud and kabbalah. Among these images we have already met the *ẓelem 'elohim*, the form of the human body or the divine form in human beings, which makes them simultaneously actors and acted on; we will also encounter the image of the game or toy, the *sha'ashua'*, the plaything of the Creator and source of his delight, the game that Creator and creation play together and with Wisdom; and especially the terrible image of a crafty deity (or nature) that employs *'alilot*—pretexts—to make human beings responsible for the sins that he or it made them commit. This radical thesis is developed by Eliaschow in his *Book of Knowledge*. That author is heir to the long line of rabbis (including the Gaon of Vilna) for whom the doctrines of the Talmud and Midrash, as well as Lurianic kabbalah, refer to a world governed by necessity, in which everything that happens must happen and is foreseen from all eternity. A passage in Tractate *Avot*, classically invoked to affirm free will, is inverted into an assertion of absolute determinism: *hakol ẓafui ve-ha-reshut netunah*, "everything is foreseen but permission [*or* the possibility (to choose)] is granted."[73] The classical reading of this dictum by idealist theologians, following Maimonides, begins with the second half of this statement: "We are granted the possibility of choosing"; that is, we have free will to decide the sequence of events, for which we are thereby responsible. As for the first part, "everything is foreseen," it is taken as referring to the mystery of God's omniscience.[74] Eliaschow, by contrast, begins with an affirmation of uncompromising determinism: "Everything is foreseen, including what is done by our choices."[75] Although there is no denying that we have the possibility of choosing each time we make a choice, this is merely a game and in no way modifies the "everything is foreseen" character of rigorous determinism. Free will, which is perfectly real as an internal and subjective experience, is an illusion if we believe that it determines the sequence of events. In Eliaschow's universe, human responsibility, and especially guilt, are the result of one last trick played by creation, a plot, a terrible *'alilah* that toys with human beings, beginning with the first couple, Adam and Eve, and thereafter with all of their descendents, past, present, and future.

To develop this idea,[76] Eliaschow cites a midrashic interpretation (Midrash *Tanḥuma* on Gen. 39:1) of Psalms 66:5: "Come and see the works of God, terrible in his deeds (*nora' 'alilot*) among men." This phrase, "terrible [*or* terrifying] in his deeds" (*nora' 'alilah*), quoted in one version of the Yom Kippur liturgy,[77] anchors the midrash, which is itself terrifying.

73. M *Avot* 3:15.

74. Maimonides developed this thesis in the introduction to his commentary on *Avot*, known as the "Eight Chapters," as well as in his *Sefer ha-Madda'* (Book of Knowledge; see above, n. 27). The mystery of the omniscience of God, who is the *ex nihilo* creator of nature and its laws, derives from the fact that, for Maimonides, God's essence is identical with his knowledge, which is creative just as he is, and unlike human knowledge, which is distinct from the knower. As a consequence, "the knowledge attributed to [God's] essence has nothing in common with our knowledge, just as that essence is in no way like our essence" (*Guide* 3.20, p. 293). "If someone asks us about the nature of his knowledge, we answer that we cannot grasp it, any more than we can fully grasp his essence" (*Eight Chapters*, chap. 8).

This position has been criticized, notably (as we saw above) by the Maharal of Prague in his introduction to *Gevurot Hashem* and in *Derekh ḥayyim* (his commentary on *Avot* [5:6]), where he insists that "God's knowledge is not his essence" (*Tractate Avot with . . . Derekh Ḥayyim* [London, 1961], p. 233). Spinoza's criticism of the thesis of creative intellect is a key point in his attack on Maimonides' ontology, too.

Note that some of Maimonides' formulations, especially in *Sefer ha-Madda'*, may sow confusion and reinforce the thesis of those, like Leo Strauss and Shlomo Pines, who read the *Guide* as an "esoteric text" whose surface meaning conceals the author's true positions, as he alerts readers in the introduction to the *Guide*. (See, for example, Ira Robinson, Lawrence Kaplan, and Julien Bauer, eds., *The Thought of Moses Maimonides: Philosophical and Legal Studies = La Pensée de Maimonide: Études philosophiques et halakhiques* [Lewiston, NY: Mellen, 1990]). But for almost all of Maimonides' disciples and, over the centuries, mainstream Jewish theology, it is the surface level that carries the day.

75. Eliaschow, *Sefer ha-De'ah*, vol. 2, pp. 295–296 ("Explanation of the Tree of Knowledge," §8).

76. Ibid.

77. See above, n. 61; Chapter 1, n. 77. <The "orthodox" understanding of the phrase, of course, is "awesome in his deeds." But that is the entire point here.>

Depending on the context, the same word *'alilah* 'deed, maneuver' may mean "feat," "brilliant deed," or "pretext" (in the sense of perverse exploitation or bad faith). When associated with God, it is usually understood in the sense of "feat" or "great deed," that is, a remarkable action that attests to its author's intelligence and power. *'Alilah* is related to *'alul* 'expected consequence' and to *'illah* 'cause'. (In Hebrew, the scholastic notion of God as First Cause is *'illat ha-'illot* 'cause of causes'.) But the same word *'alilah*, when applied to human actions, may mean "deceit" or "trick," that is, the pretext by which a cunning and evil person spins a web of lies to trap his adversaries and justify the harm he does them. There are many examples of this. In Deuteronomy (22:14 and 17), *'alilot devarim* 'false charges' (*lit.* "pretexts of words") designates the wicked act of a husband who, wanting to divorce his wife because "he dislikes her," levels false charges against her.[78]

In modern Hebrew, the word *'alilah* means "plot," both the story line of a novel or play and the devious scheme of a conspirator. In most cases, of course, the praiseworthy sense of a brilliant action is intended when the term is applied to God, and that of perverse behavior when applied to human actions. In some places, though, the biblical text is ambiguous.[79]

There is at least one notable exception to this. When God speaks of his own treatment of Egypt and Pharaoh, he uses a verbal form of this same stem, *le-hit'olel*, whose regular connotation is inflicting harm or doing evil. In Exodus 10:1–2, God says that he has hardened the hearts of Pharaoh and his servants so that they will refuse to allow the Hebrews to depart and so that the full scenario of the ten plagues, with all the signs and prodigies accompanying them, can play out to the end. "You will recount for your children," he tells Moses, "how I made a mockery of the Egyptians"—or, in Rashi's gloss, how I "played" (*śiḥaḳti*) with them, as Balaam (who uses the same verb) accuses his ass of doing to him. This divine con-

78. The eleventh-century commentator Abraham Ibn Ezra (on Deut. 22:14) glosses *'alilot* as "causes" and *devarim* as "either true or false [*sc.* words]," i.e., good or bad arguments.

79. It would be impossible to analyze the many biblical occurrences of the word *'alilah* here, even if we limited ourselves to those that may be ambiguous. For example, David prays to be spared from doing evil: "Incline not my heart to an evil thing, to practice deeds of wickedness (*le-hit'olel 'alilot be-resha'*)" (Ps. 141:4). By asking God not to steer his heart toward evil, he is implicitly holding the deity responsible for any evil he himself might do. In fact, the expression *le-hit'olel 'alilot* is much stronger than simply "do evil." It reinforces the notion of *'alilot* as deceitful words or actions by making them the cognate accusative of a verb with the same stem, which already includes the connotation of a bad action (to "trick, deceive" or to "trifle with, mock"). It is in this sense that the biblical text employs it with God as the subject and the Egyptians as the object ("I made a mockery [*hit'allalti*] of the Egyptians" (Exod. 10:2). (See the main text immediately after this note.)

Another example, remarkable because of the interpretive technique on which it is based, occurs in Hannah's prayer of thanksgiving: "YHWH is a god of knowledge (*de'ot*), and by him actions (*'alilot*) are weighed" (1 Sam. 2:3). (On the plural form *de'ot*, see Chapter 1, n. 58.) The ambiguity of the meaning of these *'alilot* associated with YHWH is underlined by an exegetical tradition that employs the technique of *ḳere-ketib* (the text as read versus the text as written). In this verse the written text is *lo'* with an *alef* (לא), which expresses negation; thus "*'alilot* [plots and pretexts] are not appropriate to YHWH, the god of knowledge." But we are invited to read instead *lo* with a *waw* (לו), "to him," which transforms the meaning into "*'alilot* [brilliant feats] are appropriate to YHWH, the god of knowledge." Here the exegetical technique is used to attach two senses to *'alilot* and apply both of them to God, negating his association with the pejorative meaning and affirming his association with the positive one. But the very recourse to this technique, which leaves the written text unemended while proposing a different way of reading it, shows clearly that the ambiguity persists and in fact invites reader-interpreters to delve more deeply into the text. This is precisely what the midrash on the expression *nora' 'alilah* in Psalm 141, cited by Eliaschow, does explicitly, unabashedly attributing the pejorative sense to God-Elohim.

It is remarkable that this play on words is applied to a text that itself evokes, in self-referential fashion, the ambivalence of language. As at Babel, perverse use of language is associated with perversion of "spermatic knowledge," which is then found to have been turned into the "seed of falsehood" (see above, n. 52).

fession spawned classical commentaries that seek, grosso modo, to salvage the morality of free will by observing that Pharaoh hardened his heart on his own after each of the first five plagues and was deprived of his free will starting only with the sixth plague, as if, with the escalation of his wickedness, he had decided for himself that he would longer exercise choice. This type of exegesis, aimed at saving the morality of the plain sense (the *peshaṭ*) of the narrative, vanishes in the midrashic reading of Psalm 141, *nora' 'alilah 'al benei 'adam*: God is "terrible in his deeds among men." As this verse is read by Midrash *Tanḥuma* and developed by Eliaschow, Pharaoh's case is not all that exceptional. Human beings are the playthings of the events they think they determine. This can be extended to all human situations—not only to Pharaoh's ostensibly free exercise of his will during the earlier plagues but also to the entire cast of biblical characters, and, more generally, to all human beings and the consequences of their choices. Nevertheless, in what seems to be a manipulative if not indeed perverse fashion, they are held responsible for them and accounted guilty for their consequences.

In this reading, *nora' 'alilah*, despite referring to God's actions in nature, is understood not as "awe-inspiring in deed" but as "terrifying in plots"—terrifying in pretexts and bad faith. The verse in Psalms is taken as "Come and see the works of God, terrible in his *plots against* men." To avoid any doubt about its intentions, the midrash continues with a parable that gives us a clear picture of this terrifying behavior of God, who attempts to slough off his own responsibility onto the victim, using bad-faith arguments, pretexts and false charges, as in Deuteronomy. The tale is (by coincidence?) that of a faithless husband who has decided to divorce his wife (following the ancient judicial procedure, of course) and is looking for a pretext. He has already prepared the bill of divorce and needs only a reason to give it to her. He returns home and asks his wife to pour him a cup of wine; when she serves him he rejects it as tepid and invokes this lapse as an excuse for divorcing her. She can only retort sardonically that the document had been drawn up even before he entered the house, evidence that in his bad faith he was looking for the first available pretext.

This, according to the midrash, is the meaning of the verse in Psalms. The far-from-edifying parable of the false-hearted husband is invoked as a model of God's normal treatment of human beings, alluded to in this verse: God is "terrible in his plots—his pretexts and bad faith—against men." This thesis, heretical and scandalous on the face of it, is supported by other passages from the Midrash and Talmud that deal with situations comparable to Pharaoh's, in which God plays with and abuses human beings. The victims are none other than Adam himself, Moses, Joseph's brothers, and finally, by extension, all human beings who are punished, ostensibly for their actions, even though the sequence and consequences of these deeds were foreseen from all eternity.

The first and clearest example is Adam, whose transgression, according to Genesis, brought death into the world. In the midrash, Adam contests this charge, noting that death had been foreseen from all eternity by the Torah (which existed before the creation of the world), as proven by its institution of rites of death-related purity and impurity. If so, he cannot be held responsible for it. What is more, even his fall was inevitable, inbuilt in creation; it defines and shapes human nature just as much as his mortal character does. To hold Adam morally responsible for his fall and for the death of his descendents is thus an unfair charge, an *'alilat devarim*, an accusatory pretext, which Adam himself, speaking through the midrash composed by his descendents, protests vigorously. The same applies to the other examples. Moses' sin, which prevented him from entering the Land of Israel, was a mere pretext; it was foreseen from all eternity that he would not cross the Jordan. So too the sin of Joseph's brothers in selling him into slavery; it was a link in the chain of causes by which the earlier predictions of exile and slavery in Egypt, followed by the exodus and release from bondage—events foreseen from all eternity—were realized.

This notion rests strongly on a talmudic passage[80] that is one of the most explicit in assigning culpability for the evil committed by human beings to the author of Creation; for it was he, as Rashi notes, who created the *yeẓer ha-ra'*, the evil impulse, or, more precisely, the creative nature and dynamic of evil. The Talmud quotes three verses that exonerate Israel of responsibility for their sins and permit them to escape a judgment that would otherwise certainly condemn them. The first of these verses is Micah 4:6, where God confesses to having abused (*hare'oti*, from the stem *ra'* 'bad, evil') the exiled Israelites. Next comes a celebrated verse from Jeremiah 18:6, in which God admits that he manipulates the house of Israel like "clay in the hands of the potter." (Spinoza would take up this image to illustrate his thesis of absolute determinism and the illusory nature of free will.) The third verse is Ezekiel 36:26: "I will remove the heart of stone from your body and give you a heart of flesh." Even more explicit is the next verse, because it bears not only on inclinations—the heart of stone or flesh—but on the details of behavior: "I will put my spirit into you and cause you to follow my laws and observe my rules faithfully."[81]

This is certainly an appalling conception, scandalous to those who would hold fast to the classic moral theology of a personal God who punishes the evil and rewards the good that human beings freely choose to do. This, it seems, is the perspective of most of the book of Job and of a straightforward reading of the surface meaning of many verses of the Bible. But there is another way to read these texts, suggested by the first and last chapters of that same book of Job. Their lesson is that the questions raised by Job and his friends, however poignant, stem from a childish morality and magical thinking that expect God/nature to respond to human hopes and behavior as if he/it too were a person, a moral subject in dialogue with another moral subject. In fact, the determining causes of the experiences of a human being, in this case Job, are to be looked for in a natural history that complies with its own necessities and that, for better or worse, makes sport of Job's choice of virtue rather than of vice.

Finally, we must insist that this radical and absolutely determinist reading of the passage in *Avot* ("everything is foreseen but the possibility to choose is granted") is not theological. It must not be confused, for example, with a doctrine of predestination and grace, like those, Protestant

80. B *Berakhot* 31b–32a.

81. In this talmudic passage, R. Eleazar has the prophet Elijah hurl a serious indictment against YHWH: "Answer me, YHWH, answer me, that this people may know that you, YHWH, are *the* Elohim; for you are the one who turned (*hasibbota*) their hearts backward" (1 Kings 18:37). The word *hasibbota* expresses simultaneously the idea of *sibbah* 'cause' and *mesovev* 'turn, reverse' (but also 'effect', i.e., cause). The three verses quoted in the body of our text are then invoked to demonstrate that God admits the validity of the charge that he himself is the true cause of the Israelites' sins. But R. Eleazar goes on to assert that Moses uttered a similar accusation against heaven, assigning ultimate responsibility for the sin of the golden calf to the god of Israel, who had lavished a surfeit of silver and gold on the people. Another sage, R. Ḥiyya b. Abba, likens the case of the people to that of a man who spoiled his son: "He bathed him and anointed him and gave him plenty to eat and drink and hung a purse round his neck and set him down at the door of a bawdy house. How could the boy help sinning?" (B *Berakhot* 32a).

or Jansenist, that derive from the teaching of St. Augustine. Those are theological doctrines of salvation, whereas this kind of Jewish determinism is a theory of the sequence of cause and effect in the nature of things and events and in the knowledge we can have of them; in other words, it is a naturalist and epistemological theory of immanent causality. We should remember that the talmudic dictum that "everything is in the hands of heaven" is qualified by "except for the fear of heaven," a fear that is itself "the beginning of wisdom." As we shall see, this implies a doctrine of salvation—a doctrine of what is just and what is wicked—that is realized neither by grace nor by works, but by knowledge of one's works and self-awareness; that is, by acquiescence (in the Stoic sense) and the perspective on oneself that this knowledge makes possible.

II. "BEFORE CREATION": THE WORLD OF *TOHU* AND THE 974 LOST GENERATIONS

The human setting of this midrash, with its parable of the faithless husband, should not mislead us. It suggests an anthropomorphism carried to the nth degree, in which absolute determinism, hidden behind the mask of a god-judge, is the arbitrary product of the will of God, just as unintelligible as his decision to create the world. The radical character of this image, which casts God as a perverse and scheming husband, suggests that we should see it as a metaphorical anthropomorphic projection rather than as an introduction to the psychology of God or a reflection on the morality of his motives.[82] This is why Eliaschow does not linger on this image. Following the tradition of Lurianic kabbalah and the Gaon of Vilna, his *Book of Knowledge* relates to a historical process determined by the law of the "management of the world," a law that emerges from a primordial thought or wisdom that "precedes" creation—a wisdom that is produced by this determinism but that also makes it possible to think it. The evil that human beings do and of which they are the playthings does not originate in the perverse will of some superhuman being, but in a necessary primary process that precedes and governs the existence of the world and could not be other than it is. The absolute determinism in which "everything is foreseen" from all eternity is intelligible, at least in principle, because it is not the arbitrary and incomprehensible product of a Creator in the image of man. The role of the "esoteric meaning" of the Torah and the systematic extraction and exegesis of its myths of origin is to provide the pieces of this intelligibility. Primordial wisdom is certainly hidden and "unattainable" in that it does not allow itself to be mastered; but it is expressed and partially revealed in the paths of our knowledge, "straight and crooked" like the serpent.[83]

82. This is, incidentally, the import of the kabbalistic method in general, and one way in which it can be contrasted to the theology of a personal god. It describes the "direction of the world" (*hanhagat ha-'olam*) using models of cosmic mechanics and not as the exercise of more or less arbitrary power by an omnipotent personal deity, omniscient and infinitely good. In these models, the divine persons, who in fact reproduce human figures—the king, the father, the mother, the grandparents, and the other "faces" of the *Zohar* and the Lurianic kabbalah—are anthropomorphisms that are themselves subject to the primordial wisdom, which in turn proceeds from the infinite and "simple" (in the sense of undifferentiated) One, which has neither image nor form and is neither personal nor impersonal. It is through this "hidden wisdom" that the worlds are emanated, created, formed, and governed; these figures give individual forms to the simple infinite, which also fills the world of impersonal objects (see below, n. 98).

83. The repeated exhortation to "make known his deeds (*'alilotav*) among the nations" (Isa. 12:4, Ps. 9:12 and 105:1, 1 Chron. 16:8) shows the marks of the ambiguity we have underscored, in which divine works (*'alilot*) are sometimes understood as exploits or prodigies and sometimes as complex and manipulative machinations through which human beings act, but thanks to which they can also

Like the long exegetical tradition that it develops,[84] Eliaschow's *Book of Knowledge* makes the origin of sin antedate the sin by Adam and Eve and the sin by the serpent, and even the earlier "sin of the earth" in producing trees that bear fruit instead of trees that are entirely edible fruit (as prescribed by the creative word of Genesis [1:11–12]), in the *ẓimẓum* and the "shattering of the vessels." This is, of course, the myth of origin by which the Lurianic kabbalah describes the process that brings into existence all of the "worlds," in their fullness and diversity, as well as the singular objects and beings that we experience in our own existence. In this scheme, the shattering of the vessels produces a fragmented and scattered world in which the *sefirot* are isolated from one another—the world of "points" (*neḳuddot*) or world of chaos (*tohu*)—and is the origin of the existence of individual and separate objects and beings as well as of the ineluctable evil that ensued. This evil is to be understood as simultaneously evil that is committed—a sin or error whose origin lies in the deficient and incomplete character of every particular object, separated from everything that is not it—and evil that is suffered, or pain, whose origin lies in the former, that is, in the evil actions of nature.

For Eliaschow, the "cosmic" evil of the shattering of the vessels is realized in the history of human beings in a series of successive stages. The series begins, incidentally, long before this shattering of the vessels, with the formal duality of *ḥesed* and *gevurah*, "expansion" and "contraction," where evil does not exist (other than "in its source"), a limitation of no-limits, definition in the infinite, negation in affirmation, the unity of the one and the multiplicity of a thousand (the homographs *'alef* and *'elef* [אלף]). "After" the shattering of the vessels, and conditioned by it, the stages continues with the little-known history of the "974 generations of the world of *tohu*." The *Book of Knowledge* develops this theme, found in the Talmud and Midrash and based on a passage in Job:[85] "The old way trod by wicked men, who were snatched away before their time [*or* without time], their foundation like a flowing river."[86] The Talmud reads this as an allusion to the "974 generations who were to have been created before the world was created, but were not created: the Holy One, blessed be He, arose and planted them in every

plead their innocence. This is particularly evident in Psalm 103, the context of the verse "He made known his ways to Moses, his deeds (*'alilotav*) to the children of Israel" (Ps. 103:7). Knowledge of God's deeds is more than simply observation of the prodigies of nature. In this context it is also, and above all, comprehension of the twisted and complicated ways in which the plots that ensnare human beings unfold, even while they see themselves as the agents of history and as responsible for what is actually done *through* rather than *by* them.

The adjective "terrible" (*nora'*), associated with the divine plot and pretext, must also be understood as referring to this knowledge that we can and must have of it. The first sense of this word applies to what inspires fear (*yir'ah*), which is derived from the same root. But when "the fear of YHWH is the beginning of wisdom" (Ps. 111:10) or "the fear of YHWH is the beginning of knowledge" (Prov. 1:7), the reference is not fear of what tomorrow will bring, or of punishment in this world or the next. As the *Zohar* explains (Introduction, 11b), it designates instead a horrified fascination with the power of nature and the power in nature, an infinite power that nothing and no one can withstand. Thus the word *nora'* 'terrible, awesome', in the sense of that which inspires this sort of "fear" or "awe"—or *wonder*—applies especially to what arouses this frightened admiration by offering us sight of perfection. "Wherever completeness appears is called *awesome*, . . . [and] there is no awe except where completeness appears" (*Zohar*, Exodus 79a [trans. Matt, vol. 4, p. 426]). When YHWH's deeds are referred to as "terrible" (Exod. 34:10) the word has the sense of "awesome," the quality of the initial feeling of being overwhelmed by perfection, a sentiment that can serve as the driving force behind human apprehension(!) and knowledge of what takes place in the world. When Elohim is described as *nora' 'alilah*, we realize that the perfection involved applies to his plots as well.

84. See, for example, Scholem, *Kabbalah and Its Symbolism*.

85. See Eliaschow, *Sefer ha-De'ah*, "Explanation of the World of *Tohu*," Part 1, Introduction, §3, p. 3; Part 2, 2nd Explanation, 1st Branch, §§5–7; 2nd Branch, §2–3, pp. 20–29.

86. Job 22:15–16.

generation, and it is they who are the insolent [*or* violent] persons in each generation."[87] Rashi explains that the number 974 is derived from the thousand generations that should have been born before the giving of the Torah to prepare for its reception, as alluded to by the Psalmist: "The word that he commanded, for a thousand generations" (Ps. 105:8). But time pressed and it was impossible to wait beyond the twenty-six generations from Adam to Moses. The other 974 were thus "snatched away," as the book of Job puts it, and carried off in "a flowing river." Consequently they were left "without time" and the Talmud deems them not to have been created. For the Midrash,[88] by contrast, they had

87. B *Ḥagigah* 13b–14a; see also B *Shabbat* 88b.

88. *Gen. Rabbah* 28:4; Midrash *Shoḥer Ṭov* 90. In Job, the "wicked men" snatched away before their time evidently refer to the generation of the Flood (as noted, for example, by Ibn Ezra in his commentary). But the talmudic midrash understands the phrase as an allusion to the "thousand generations" or "two thousand years" that preceded the giving of the Torah. This theme is suggested by the image of "the word [i.e., the Torah] that He commanded for a thousand generations" (Ps. 105:8). This is in turn linked to the midrash about the two thousand years that were initially ordained for human beings to live "without Torah" and prepare themselves to receive it. (That this seems to imply only two years per generation should not particularly astonish us; in origin myths about "before" and "after" the creation of the universe the number of years has a value that is symbolic rather than temporal. Recall that the antediluvian generations from Adam to Noah lasted for centuries, in an original "heroic" age of humanity clearly different from our own.)

According to the biblical narrative, however, only 26 generations separate Adam from the revelation at Sinai. The missing 974 generations are transferred to that epoch of "two thousand years before the creation," when Wisdom was the Creator's plaything and took delight in the as-yet unborn "sons of man" (see *Gen. Rabbah* 1:1 [Prov. 8:30–31]; see also Atlan, *Enlightenment*, chap. 7, "Man-as-Game," and n. 8.134, p. 393).

Now, during this period "before the creation of the world," the only possible existence is that of the Torah itself, as the arena of the possibilities with which the Creator plays, before they are launched into their adventure of begetting one another. Consequently the missing 974 generations have no temporal existence and are scattered among the generations to come, as the arrogant, violent, and "wicked men," who did not become extinct even after the giving of the Torah.

According to a talmudic legend, this change of plan was not immediately understood by the angels of the heavenly host who attend on the Creator: "When Moses ascended on high" to receive the Torah, after those 26 generations had passed, the ministering angels protested God's intention to make humanity a gift of "that secret treasure, which you hid for 974 generations before the world was created." Moses, told by God to answer the angels, observed that they have nothing to do with the Torah, whose commandments and prohibitions, notably the Ten Commandments, relate to human actions such as idolatry, perjury, honoring one's parents, murder, theft, etc. This answer satisfied the angels, who then praised God and showered gifts on Moses (B *Shabbat* 88b). We may infer that what they had had in mind was another Torah, the Torah of "the world on high" or *'aẓilut*, of which the kabbalists speak, which is exclusively wisdom and tree of life, and which can be realized without being mediated by statutes that relate to permission and prohibition, to good and evil, because evil "on high" is not evil (see Chapter 1).

Commenting on this passage, Rashi cites the theme of the thousand generations in the anthropomorphic version of the midrash, with its image of the Creator's "initial intention," modified after he "saw" that it would not work: "Those generations would have been created during the two thousand years by which the Torah preceded the world, as we read, 'the word that He commanded for a thousand generations' (Ps. 105:8). The Holy One blessed be He saw that the world could not survive without Torah for such a long period, so he skipped over them and did not create them, and gave it after 26 generations. Thus 974 generations are missing from the thousand" (Rashi on B *Shabbat* 88b).

This anthropomorphic image of a personal Creator-God who modifies his initial plan in midstream recalls a better-known midrash about the two accounts of creation in Genesis (the Elohist and Jahwist, in the jargon of higher criticism): "If," considered the Creator, "I create the world on the basis of mercy alone [*ḥesed* and *raḥamim*, associated with the divine name YHWH], its sins will be great; on the basis of judgment alone [*din*, associated with the divine name Elohim], the world cannot exist. Hence I will create it on the basis of judgment and of mercy, and may it then stand!" (*Gen. Rabbah* 12:15). Clearly we are dealing not with a change of mind by a humanized Creator but with a rhetorical device that describes the *current* state of things as the result of a compromise between opposing logical and ontological constraints. The mechanistic cosmogony of the kabbalah, for example, allowed Moses Cordovero (*Shi'ur ḳomah*, chap. 60, pp. 130–135) to address this notion of the modification of the plan for creation outside of time, thereby eliminating theological discussions about God's will and the possibility of changes in it. In practice, the contrasts expressed by these images are located *inside* the divine essence, conceived as a divine *world*, that is, inside the system of the *sefirot*. These dramatizations, in which the Creator of the universe changes his mind and direction during the process of creation, culminate in the midrashic statement that "before the creation of the world he created worlds and destroyed worlds" (e.g., *Gen. Rabbah* 3:7). But what these narratives express is only the development, pictured in different contexts, of the ontological opposition between

some form of existence: a thousand generations "rose into thought" to be created, but 974 of them were "deleted." We will soon return to the classic midrashic idiom of "rising into thought to create [*or* to be created]" and its implications for the very notion of creation. Eliaschow, who borrows this expression, holds these generations to be both created and uncreated. The stream that carries them off is compared to waking up from a dream. Their potential and unrealized existence is distributed among the generations that do exist, in which history is produced by means of violence. In the *Book of Knowledge*, this created and uncreated existence is that of the "world of points" or "world of *tohu*," which "precedes" the emanation of the worlds from the infinite One.[89] This existence outside

the constraints of division implied by autonomous existence and the constraints of union implied by the reception of being, which is itself a precondition for coming into existence. We experience this contrast in temporal alternations of union and division and understand the changes of direction described by the Midrash as projections on the timeless of the temporal experiences of our life. It is as if the temporal destiny of being, with the contradictions produced by the bare fact of human existence, transforms a formal opposition in the worlds on high into a succession of changes in passing time. What is more, not only do existing creatures experience these changes; they are held accountable for them, as if they were their autonomous and true cause: "As for the world and its governance, everything has been handed over to this man. If he performs good actions, the governance [of the world] is made straight and correct; if he does evil, the governance is changed into rigor (*din*) and harshness. This is because his creator made him the ruler of everything that exists in this world, so that its governance, as well as all the categories of the worlds that are above him, as far as Binah, all [of this] is modified according to his action" (Cordovero, *Shi'ur ḳomah*, p. 134). These changes in the governance or direction (*hanhagah*) of the world, effected sometimes through the generosity (*ḥesed*) of the union in which being exists, sometimes through the rigor (*din*) of division, in which existing beings can generate themselves, determine the structure of the *sefirot* as described by Cordovero. In the same generation, also in Safed, Isaac Luria formalized this arrangement in the better-known mythical sequence of contraction (*ẓimẓum*), shattering of the vessels, and restoration and reintegration (*tiḳḳun*).

Finally, the fact that these changes in the management of the world were produced by human behavior, even before the creation of man—for the midrash assigns them to the epoch "before creation"—can be readily understood if we recall that "autonomous" human actions are themselves determined and foreseen from all eternity.

It is this type of compromise between opposing logical and ontological constraints that informs our midrash that Torah was not to have been given until after a thousand generations had passed. As infinite wisdom, the Torah is relevant in principle only to human beings who are already generated and unified in their multitude. This is the import of the number of one thousand (*'elef*) generations, in which the multitude is united in the one that is also designated by the letter *'alef* (see Eliaschow, *Sefer ha-De'ah*, vol. 1, p. 4.) Note that the number 26 is also symbolic of an organized whole, because it is the numerological value of the Tetragrammaton, YHWH, the divine name that combines the three tenses of the verb *to be*. Here too we encounter, in another fashion, the dialectic of rigorous law and benevolent generosity found in the midrash on the two divine names in the two accounts of creation. Humanity's self-generation would have had to be uncompromising and rigid, without the benevolent gift of a specific doctrine and especially without the assistance of a heteronomous law that lays on them, as on a child, protective duties and prohibitions that give them time to develop their autonomy later. Thus the gift of the Torah is a crowning achievement, the revelation of wisdom and tree of life opening on eternity, which human beings could have enjoyed at leisure, imitating the Creator. But this is clearly impossible; only dissolution can be born from the total division implied by autonomy. It is a utopian ideal, an adjunct of hope and nostalgia. Nevertheless, the very fact that this ideal can be conceived carries with it real efficacy for a self-generation that is only partially autonomous, because the nostalgic hope that results from it nourishes an impulse to life, to union, to good, and that has some chance of overmatching the evil impulse of death, division, and destruction. This is why these 974 generations, unavoidably bad because hurled "before their time" into separation and division, must be scattered and diluted as "wicked men" among all the created generations that are produced and amended by the yoke of the Law. Here, in the language of myth, we find Maimonides' idea, later taken up by Spinoza, that the moral law of the permitted and forbidden is a stopgap, a provisional morality, the consequence of an original fall, and not the ideal condition of a humanity liberated from the outset, knowing only the experience of the true and false and never having to deal with knowledge of good and evil.

89. Eliaschow, *Sefer ha-De'ah*, Introduction, vol. 1, p. 4. See also ibid., "Explanation of the World of Tohu," Part 2, 4th Explanation, 18th Branch, §1, 2:146:

> This is the location of the root of the roots of the thousand generations about whom it is said, "the word that he commanded for a thousand generations" (Ps. 105:8). For the Torah, which is the inwardness derived from the light of the Infinite that spreads in Knowledge, was given to the thousand generations that derive from Knowledge. It was given moreover through our teacher Moses,

time is related to the "hidden thought" of the "primordial man" (*'adam ḳadmon*) from which it proceeds.[90] This leads Eliaschow to the famous question about the reason for creation: "Why did God create the world?" Or, as it sometimes put, "Why is there something rather than nothing?" For the Talmud, this question is one of those that, from a certain point of view, ought not to be posed: "Whosoever speculates on four things, it were better for him had he not come into the world: What is above? What is beneath? What was before? And what will be hereafter?"[91] By contrast, kabbalistic works such as Ḥayyim Vital's *'Eẓ ḥayyim* and Eliaschow's *Sefer ha-De'ah* focus precisely on this question, relying on ancient midrashic texts that describe what existed "before creation." Thus we read about Wisdom playing with the world and humankind even before they were created (Prov. 8:31–32) and of the Creator who constructs and destroys a succession of worlds, one after another.

It seems appropriate to look at the strange idiom employed in passages that deal with the reason for creation—"it rose into thought [*or* into his thought] to create the world"—which brings us back to the question of the personal or impersonal character of all these images. The midrashic passages are interpreted, using kabbalistic typology, as allusions to the world of emanation (*'aẓilut*), the world of points, and the

> who contains all of them, because he is Knowledge itself, that is, Knowledge that is the root of the souls from where all the souls were drawn. The light of the Torah resides in this inwardness, as mentioned. This is why it was given only through him and also why it is through him that all of these thousand generations will be purified, because . . . even the 974 generations will be repaired in the end and the Torah will be given to them as well. . . . What we said, that our teacher Moses was the root of all these thousand generations, is the esoteric meaning of what our rabbis said [in *Sifre*], as Rashi cites [in his commentary on Deut. 1:11].

This verse speaks of the thousandfold blessing that Israel will receive. The midrash (*Sifre* Deuteronomy 11) expresses its amazement that the number is limited, when Abraham had already received the promise of an infinite number of blessings, like the dust of the earth (Gen. 13:16), the stars of heaven (Gen 22:17) and the sand of sea (Gen. 32:13). The answer is that these thousand blessings come from Moses and are unconditional, unlike the others, which come from the god of Abraham and Israel and depend on the Covenant, that is, on observance of the Torah (see Rashi and *Siftei ḥakhamim* on Deut. 1:11). This notion of Moses as "Knowledge itself, that is, Knowledge that is the root of the souls from where all the souls were drawn," goes far beyond the historical conception of the man Moses, even with his biblical designation "divine man" or "the man of [the] god(s)" (*'ish ha-'elohim*). In fact, it evokes what Spinoza makes of the figure of the "spirit of Christ," fully assimilated to the "idea of God," eternal and infinite, which, he says, guided the Patriarchs (long before Jesus came to the world) and made it possible for them to recover the freedom lost by Adam (*Ethics* IV 68, Note). This "idea of god" is nothing other than "a necessarily existing infinite mode" of thought (*Ethics* I 23; Letter 66 [64]); or again, "the eternal and infinite intellect" constituted by the sum of the individual eternal minds (*Ethics* V 40, Note), which he also calls the "Eternal Son of God" (not the same as "Christ in the flesh"), that is, "the Eternal Wisdom of God, which has manifested itself in all things and especially in the human mind, and above all in Christ Jesus" (Letter 21 [73]).

The irony of history is that Spinozism, so long anathematized, could, like kabbalah, serve as a philosophical anchor for a doctrinal rapprochement between Judaism and Christianity on the issue of the man-god and the Son of God! For this, however, one would have to conceive of God as the infinitude of what exists and not as a person. We shall return to this in Chapter 4, when we consider the question of Spinoza's "Christianity."

90. Eliaschow, *Sefer ha-De'ah*, "Explanation of the World of Tohu," Part 2, 4th Explanation, 18th Branch, §1, vol. 2, p. 147.

91. M *Ḥagigah* 2:1. This remark is generally interpreted as referring to things that the human mind can never grasp, because they lie beyond the boundaries of the universe. This does not mean, though, that it is impossible or forbidden to reflect on these subjects, as the endless discourses on them in the midrashic and kabbalistic literature demonstrate at great length. Rather, this mishnah warns us against the difficulty and danger of such reflection, which carries us outside the universe. It actually sympathizes with those who engage in it, who are then accounted, as it were, to have "not come into the world" (see Rashi on B *Ḥagigah* 11b). The *gemara* on this mishnah, however, distinguishes reflection "on what was before" from the other categories, because it concerns what has already happened and rests on more solid ground than reflection about an unknowable future "after" the end of time, or some elsewhere beyond the limits of the universe, "above" or "beneath." Nevertheless, reflection about the pre-creation past is also deemed to be dangerous—like mentioning the dunghill on which a palace has been erected (B *Ḥagigah* 16a).

world of *tohu*, which were produced by the primordial *Adam Ḳadmon* before the worlds of creation (*beri'ah*), of formation (*yeẓirah*), and of action (*'aśiyyah*). The "reason" for the creation of the world, and thus of the evil that appears in it (a subject that some would rather not talk about), is described by the expression "it rose into thought" (*'alah ba-maḥshavah*) to emanate, to create, to form, and so on; or "thus it rose into thought before me."[92]

Eliaschow and the tradition on which he draws do not understand this phrase as referring to a mere whim or God's arbitrary or inscrutable will, but rather to an internal and reflexive causality that evokes (though in a different way) the *causa sui*, the self-caused of the Neoplatonists and Spinoza. To the obvious question, "*Where* did it rise from?" when there was only the infinite and undifferentiated One with no above or below and no movement, the answer is, "From what was to be emanated/created, etc." As for "Whose thought?" it was that of the primordial man (the *Adam Ḳadmon*), an archetypal figure intermediate between the formless infinite and created objects with a form, through which the creative force is expressed. This "man" plays

92. See Eliaschow, *Sefer ha-De'ah*, vol. 1, p. 2 ("Explanation of the World of Tohu," Part 1, Introduction, §2). The idiom of "rising into thought" is sometimes impersonal (like the former phrase) and sometimes personal (like the latter). As stated above, the Midrash itself employs the impersonal form for the creation and destruction of the 974 generations, but a personal form in other instances. Various instances of these expressions in the kabbalistic literature exemplify Scholem's thesis that theistic language is sometimes used to express an immanentist pantheism. Authors who borrow passages from the Talmud and the Midrash that employ the personal form ("his thought" or "before me") often transform it into the impersonal form ("[the] thought"), as if the first were merely allegorical and the two forms interchangeable. For example, Eliaschow cites a midrash on the transition in Genesis from the name Elohim to YHWH (see above, n. 88), but in an impersonal wording that deviates from the text of *Genesis Rabbah* (12:15): "At the beginning it rose into thought to create with the category of rigor; but when he saw that the world could not endure, he associated the category of mercy with it" (*Sefer ha-De'ah*, "Explanation of the World of Tohu," Part 2, Explanation 4, Branch 14, §2, 2:120). Similarly, Ashlag (*Sullam* on *Zohar*, Genesis 15a) writes "it rose into thought," despite the personal form of the text of the *Zohar*. Long before, *Sefer Habahir* (§88) has "it rose into thought" in the context of the idea of an infinite thought toward which one ascends, in contrast to prophetic vision, which involves a descent. The phrase is also found in the Talmud (B *Berakhot* 61a), again with reference to an apparent change in the created world, from Adam "male and female" to Adam "in the image of Elohim": "At the beginning it rose into thought to create two; but then only one was created" (obviously before the doubling into Adam and Eve, who appear in the light of this midrash as unified in the image of Elohim). The second expression, in a personalized form, sometimes with "his will" instead of "his thought," can also be found; for example, at the very beginning of Ḥayyim Vital's *'Eẓ ha-ḥayyim* (*Sha'ar ha-kelalim*, chap. 1): "When it rose into his will, blessed be his name, to create the world in order to be benevolent to his creatures. . . ." Naphtali Bacharach, in his *'Emeḳ ha-melekh*, also used this form in the context of the initial game of the Infinite (see below, §14). We find it, in a variant wording, in *Genesis Rabbah* (44:23): "Thus it rose into the knowledge (*da'at*) of the 'Place' (*ha-maḳom* [a postbiblical name for God]). . . ." In the typology of the kabbalah, thought, will, and knowledge all refer to the first *sefirot*; but their use in these expressions is paradoxical, because they seem to exist before the emanation of the sefirotic worlds. As we have seen, Eliaschow solves the paradox (as does Ashlag in the *Sullam*) by referring these terms to the hidden thought mentioned at the beginning of the *Zohar*, which, related to *Adam Ḳadmon*, precedes all the worlds and whose first product is the world of *tohu*. He does so at the very beginning of his book (*Sefer ha-De'ah*, "Explanation of the World of Tohu," Part 1, General Introduction, §2, vol. 1, p. 2) by way of a preface to his exposition of the world of *tohu* and the 974 generations as the esoteric meaning (the *sod* or secret) of this idiom and of others, such as "he created worlds and destroyed them," that seem to speak of what was "before the creation of the world." Eliaschow cites the talmudic legend in which Moses, discovering that Rabbi Akiva would be able to deduce many teachings from the Torah that he himself did not know, is astonished that the Torah was given through him instead of that latter. Moses is even more astonished and shocked when he sees the excruciating martyrdom of the later sage: "Such Torah, and such a reward!" God answers both queries in the same way: "Be silent! Thus it rose into thought before me" (B *Menaḥot* 29b). (See Atlan, *SR* II, chap. 12.) Eliaschow explains what seems to be presented in this legend as the arbitrary decision by a personal God that produces such terrible evil in the world as the effect of the unavoidable and impersonal root of evil in the world of *tohu* and the 974 generations, coming from "the secret meaning of the thought above, the hidden brain of the primordial man" (*Sefer ha-De'ah*, "Explanation of the World of Tohu," Part 1, General Introduction, §2, vol. 1, p. 2).

two roles simultaneously, as the creative source and as the ultimate effect of creation in the perfection of its fulfillment outside time. It is into his infinite thought—"hidden wisdom" or "impenetrable darkness," according to the very first sentence of the *Zohar*[93]—that creation "rises," like a blueprint sketched from the building it describes. This process is explicitly and vividly described as the passage of light through the sensory orifices of this "man," notably through his "eyes," which receive what they see "rising before them" but which also express this creative force in "the look" (*le regard*) that emerges from them.[94]

In this system, where creation (*beri'ah*) is only one mode of the infinite process by which worlds come into existence, the notion of ex nihilo creation is to be seen in a perspective different from its classical theological sense of an absolute beginning that was the effect of an unfathomable decision, of the will of a Creator. The moment of *beri'ah* is precisely that of the negation of light, a negation that creates the nothingness of darkness. The *Bahir*[95] states this very clearly with regard to the creation of darkness and evil in Isaiah 45:7, where light is "formed" (*yoẓer*) and peace is "made" (*'ośeh*). Only darkness and the *tohu* from which evil precedes are "created" (*bore'*); this is why they do not have their own existence, inasmuch as they are only the absence of light and limitation of being. In this vision, it is the nothingness of finitude that is created, rather than being, which is eternal and infinite. Rather than creation *from* nothing it is creation *of* nothing, in an eternal process of differentiation and individualization. It is thus perfectly natural that evil is "created"; like chaos, it is nothing, a limitation of the luminous order and peace that permeate forms and actions, respectively.

Here we cannot go into the details of the reflexive causality that the *Book of Knowledge* applies to describe the coming into existence of individual beings and the origin of evil that inevitably follows from it. Suffice it to say that, at the end of this process, the *causa sui* immanent in these separate existences will ultimately be able to produce itself by eliminating evil, with the same inexorability with which evil is produced, and by two different paths. The first and more common path is a process in which evil annihilates itself, exhausting itself in a series of crimes and punishments; as Hillel put it when he saw a skull floating on the water, "because you drowned [others] they drowned you, and the end of those who drowned you [is that] they will be drowned."[96] On the other path, which is followed by the "righteous," it is weakened and mitigated by observance of the Torah and the precepts, by "knowledge and duties."

All of these developments illustrate Gershom Scholem's remark about the language of the kabbalists, which oscillates between a mythical representation of a cosmic process whose mechanical unfolding is intelligible (at least in its principles if not in the details of its realization) and a theistic notion of a personal God, acting intentionally through his will in pursuit of a goal that by definition is good. "The pantheistic tendencies in this line of thought are cloaked in theistic figures of speech, a device characteristic of a number of kabbalists."[97] As Scholem also underscores, the mythical component of the content of these developments is much more important than seems to be indicated by the

93. *Zohar*, Genesis 15a (trans. Matt, vol. 1, p. 107).
94. *Sefer ha-De'ah*, "Explanation of the World of Tohu," Part 1, General Introduction, §2, vol. 1, p. 2; see also ibid., vol. 2, p. 147 (Part 2, 4th Explanation, 18th Branch, §2).
95. *Sefer Habahir*, §§11–13. See also Chapter 1, §10.
96. M *Avot* 2:6 (in some texts 2:7).
97. Gershom Scholem, "Kabbalah and Pantheism," in *Kabbalah* (New York: Quadrangle, 1974), p. 147.

use of the hallowed words of the theistic lexicon, such as "the intention of his will" and "the cause of the end of the creation of the world."[98]

98. In practice, these two representations already coexist in the Talmud and Midrash, which sometimes interpret biblical texts as accounts of a cosmic process, an origin myth of the universe, humanity, and the people of Israel, and sometimes as dialogues between human characters and their god(s), subsumed under the name YHWH-Elohim, whose unity is then personified by the very fact of these dialogues. This personification is clearest in the sacrificial rites, religious duties, and prayers. Elsewhere (*SR* II, chap. 5, §6), building on the recurrent expression *'ani YHWH*, which can be rendered "'I' [is] YHWH," I have tried to show that the classical theistic notion camouflages the projection onto nature of our experience of individual existence in its pronominal modes of "I," "you," and "he." This projection is not so much an animist *belief* in the personal character of natural forces as it is an *active* enterprise to humanize nature, on the basis of the social and familial experiences of interpersonal relations. These experiences discover forces that, despite being personified as king, judge, father, mother, and other social and familial figures, are just as natural as the forces ("gods") of the earth, wind, rain, fire, and those that control animals and plants, <just as Igni is the Hindu god of fire and Bacchus the Roman god of wine>. The god of Israel is then experienced less as a personified and personal God and more as the God of (human) persons, "greater than all gods" (Exod. 18:11), whose power can be felt in the experience of a social and family life that is no less a natural phenomenon than is the nonhuman world. The efficacy of science and technology bears witness (were it needed) of the influence of human nature on human beings and nature. This humanization of nature by technology has taken the baton from animism and magic. Its success remains quite relative, however, because it has had to pay a steep price: the impersonalization of technology. The pejorative term "technoscience" is used to condemn this phenomenon, which makes people fear that the effective and operational humanization of nature will lead only to a depersonalizing naturalization of the human. This is what is still at stake in the adventure of knowledge into which the human race was thrown in its infancy, and of which we dream in myths such as those of Prometheus and of the Tree of Knowledge.

An entire tractate of the Talmud, *Ta'anit*, whose main topic is the rituals, prayers, and fasts instituted in times of drought, is riven by this tension between the personal and impersonal character of natural phenomena. It is clear, on the one hand, that rain is a natural phenomenon independent of human beings. But its timing and quantity affect the harvest and humans' ability to survive on the land they farm. In addition, this experience of nature that sends them the rain on which they depend is the occasion for introspection and a moral experience, as if the course of nature, which they know does not depend on human beings, could be diverted into a different, "supernatural" world by their moral behavior. In an earlier work I analyzed another example of this double mode of thought, the attribution of eclipses to severe moral failings, even in an age when their causes were known and their date could be calculated in advance (Atlan, *Enlightenment*, pp. 265–266 and n. 6.43, on the Maharal of Prague and B *Pesaḥim* 94b). According to the Talmud, "three keys remain in the hands of the Holy One blessed be He and were not entrusted to the hand of any messenger—the key of rain, the key of childbirth, and the key of the resurrection of the dead" (B *Ta'anit* 2a). In other words, these remain in the domain of a divine providence that is aware of human beings and their needs. As in the case of eclipses, the rabbis who composed this text did not necessarily believe that rainfall was causally determined by the moral level of human societies. On the contrary, like procreation (another factor in survival), rainfall is essential to their destiny. This is why it is linked, as a providential act affecting the fate of human beings, with the resurrection of the dead, which depends directly on who they are and what they make of themselves, in a system of human activity oriented toward the supernatural of a future world. But the operative difference between nature and this type of supernatural that is produced by artifice and morality is restored several pages later: "The day when rain fails is greater than the Resurrection of the Dead, because the Resurrection of the Dead is for the righteous only, whereas rain is both for the righteous and for the wicked" (ibid. 7a). In other words, it is the amoral nature of rain that forces us to accord it greater importance if we want our relationship to nature to be the occasion for moral progress.

Finally, note that the personified formulas of ritual should mislead us no more than the midrash about the machinations of the faithless husband. According to the sages, intention (*kavvanah*) in prayer is not the same as that in the study of Torah; although the former should not be neglected, the latter ultimately has a greater effect on the path of freedom (on this point, see the exposition by Rabbi Ḥayyim Volozhiner, *Nefesh ha-ḥayyim*, parts 3 and 4). In the words of the ritual and through the personal pronouns (*I*, *you*, and *he*), our relationship with nature is mediated, as it were, by those persons who incorporate us into human society and civilization: the mother who nourishes and the father who teaches, then the grandparents, then other persons of the same sex and the opposite sex, and finally figures of social authority and political sovereignty. We want to see ourselves, in relation to controlling nature, as children (of the "father") or as subjects (of the "king"); that is, as those who are unfree and subordinate to other persons with whom dialogue is possible, rather than in a relationship with the impersonal.

Of course, this dialogue is really a monologue we hold with ourselves (the Hebrew *hitpallel* 'pray' is a reflexive form) and does not correspond to any reality that is given and already there. Collectively, however, we may have enough power—given us by impersonal nature—to transform this subjective discourse into an intersubjective *social* reality. (Note too that, on the operational level, the science and technology of producing artifacts, taken to

This is the context in which Eliaschow takes up the metaphor of the "random drops" and applies it to the original and creative knowledge, not yet sexual, possessed by the *Adam Ḳadmon*, the primordial human being, through whom the infinite light passed and created the worlds as it emerged from his eyes, mouth, ears, and nostrils. Thus the circle is closed; the emission of "random drops" in the story of Adam, Eve, and the serpent no longer derives from the sin of the Tree of Knowledge but from an archetypal emission of "random drops" that is the source of a first version of the world, provisional and imperfect.[99]

12. WHAT FREEDOM?

Thus these kabbalistic myths, which may originate in the traditions of the "ancient Hebrews" to whom Spinoza refers,[100] describe a world that is totally determined, but whose determining causes are at least in part knowable and intelligible for the human beings on whom they act. This picture of the world, difficult and forgotten, was covered over by the simple idea of responsibility, inconceivable in the absence of free will. But it is incumbent on us to restore this picture, because it allows us to deal with the reality of nature, including human nature, as revealed (or constructed) by the modern life sciences.

have been produced by human beings, are extending this reality further and further today.)

All the same, the *given* reality on which we have the possibility of acting is the study of the Torah—in the broad sense of the "garment of the infinite" (see below, §14), intellection, wisdom, and science; and this study is supposed to make us know that reality. The personal and personifying intention of prayer as an exercise and practice through which we mold ourselves serves only as an antechamber or corridor leading to the freedom of the self-caused. At the end of this path, if there is one, we can conceive, along with Spinoza, that "the endeavor of the better part of ourselves is in harmony with the order of nature as a whole" (*Ethics* IV, Appendix, §32). Then we are supposed to turn our gaze beyond the figure of the Father and King, perceiving ourselves no longer as children but as fathers or mothers and emancipated slaves, because we no longer serve a king, but the Infinite.

99. See, for example, Eliaschow, *Sefer ha-De'ah*, vol. 1, p. 28 (Part 1, 2nd Explanation, 3rd Branch, §3). In addition,

> All the unions [in the world of "points" and of the "shattering of vessels"] were only of the male-and-female within the female only and of the male-and-female within the male only, and this is in truth only of the random drops themselves. The same holds for the random drops that come out of each face (*parẓuf*) individually, that they are only of the male-and-female within that face. . . . All of the unions mentioned by the Rabbi [Ḥayyim Vital in *'Eẓ ha-ḥayyim*] with regard to the world of points . . . and the shattering of the vessels . . . are unions of the male-and-female within the female by themselves, and of the male-and-female within the male by themselves, that is, actual random drops (*ṭippot ḳeri*), as was said. But the principal male-and-female in [the faces, which are called] the Crown of the right side and the Crown of the [left side], and also the Wisdom and Intelligence of the right side, which is the global father of these points, with the Wisdom and Intelligence of the [left side], which is the global mother of these points, stood in truth back to back. This is why the verse says of them, "who were snatched away before their time" (Job 22:16), for "time" alludes to sexual union, in keeping with the esoteric meaning of "it gives its fruit in its time" (Ps. 1:3). (ibid., 4th Branch, §2)

In this passage, the theme of the 974 lost generations is reintroduced through the verse from Job cited above (p. 102) and is associated with that of the random drops, now taken as the origin of the very first processes of the emanation of worlds through the *Adam Ḳadmon*.

100. In his letter to Oldenburg cited above (Letter 21 [73]), Spinoza explains that his idea of God "differs widely from that which is ordinarily defended by modern Christians. For I hold that God is of all things the cause immanent, as the phrase is, not transient. I say that all things are in God and move in God, thus agreeing with Paul, and, perhaps, with all the ancient philosophers, though the phraseology may be different; I will even venture to affirm that I agree with all the ancient Hebrews, insofar as one may judge from their traditions, though these are in many ways corrupted." It is in this context that he uses (Letters 23 [75] and 25 [78]) Jeremiah's metaphor of clay in the hand of the potter. We may assume, in our turn, that Spinoza knew both the biblical text and its application by the Talmud (B *Berakhot* 32a) and Midrash, inasmuch as he employs the same image to assert that human beings are "inexcusable" before God for what their nature makes them do (Letter 25 [78]). Talmud study often begins with tractate *Berakhot*, which comes first in the Talmud and is one of those most studied in yeshivas. So it is plausible that Spinoza had studied at least parts of it in his youth and adolescence and that it is the origin of his belief about the traditions of "all the ancient Hebrews." (In Chapter 4 we will consider other, not necessarily exclusive, hypotheses.) Later in their correspondence, Spinoza responds to Oldenburg's

objection that this idea would make all things and all human actions subject to a "fatalistic necessity." The result, according to Oldenburg, would be that "the sinews of all laws, of virtue, and of religion, are severed, and that all rewards and punishment are vain" (Letter 22 [74]). This is the classic objection, which cannot conceive the possibility of ethics without free will and accordingly rejects any pantheistic doctrine of absolute determinism, which Spinoza, in his letter, traces back to Paul, "perhaps all the ancient philosophers" and "all the ancient Hebrews." Spinoza's reply to this objection, in the next letter, is neither apodictic nor theoretical. He begins by explaining that in no sense would he make God subject to fate and then introduces the notion of free necessity, developed in the *Ethics*, and applies it to the knowledge by which God knows Himself: "All things follow with inevitable necessity from the nature of God, in the same way as everyone conceives that it follows from God's nature that God understands Himself. . . . No one conceives that God is under the compulsion of any fate, but that He understands Himself quite freely, though necessarily" (Letter 23 [75]).

His response to Oldenburg's objection invokes in particular the experience of "many persons":

> Further, this inevitable necessity in things does away neither with divine nor human laws. The principles of morality, whether they receive from God Himself the form of laws or institutions, or whether they do not, are still divine and salutary; whether we receive the good, which flows from virtue and the divine love, as from God in the capacity of a judge, or as from the necessity of the divine nature, it will in either case be equally desirable; on the other hand, the evils following from wicked actions and passions are not less to be feared because they are necessary consequences. Lastly, in our actions, whether they be necessary or contingent, we are led by hope and fear.
>
> Men are . . . without excuse before God [only] because they are in God's power, as clay is in the hands of the potter, who from the same lump makes vessels, some to honour, some to dishonour. If you will reflect a little on this, you will, I doubt not, easily be able to reply to any objections which may be urged against my opinion, as many persons have already discovered along with me.

(See also Letter 25 [78], where Spinoza explains the sense in which human beings *might* be able to justify themselves from the perspective of an anthropomorphic divine morality but in fact cannot do so.)

Rather than a demonstration, this is a review of the experience of moral life that can be had by a person like Spinoza himself, who understands the reality of this determinism as a consequence of "true philosophy": "I do not presume that I have found the best philosophy, [but] I know that I understand the true philosophy," as he wrote around the same time (Letter 74 [76]).

But this does not rule out recognizing the power of the emotions and the reality of proximate, partial, or constrained (unfree) causes, which make no less a contribution to determining human conduct, even if they are known inadequately. Rational knowledge of absolute determinism is not enough to liberate human beings from the power of their emotions; it is on this point that Spinoza explicitly parts company with the Stoics in the preface to *Ethics* V. This is why knowledge of the third kind, the subject of this last part of his work, is necessary. We must take account of the fact that we are always guided by our emotions and that we can be governed by reason only if it inspires emotions that are stronger than the passive emotions that enslave us. ("A true knowledge of good and evil cannot check any emotion by virtue of being true, but only insofar as it is considered as an emotion" [*Ethics* IV 14].) Whence the difficult path that leads from slavery to the freedom of the intellectual love of God, the logical progression followed by *Ethics* IV and V, as well as the necessity of a sort of provisional existential morality along the way, which Spinoza describes in this letter, independent of the level of knowledge and philosophical truth that one has achieved. He explicitly commends the recourse to this provisional morality: "The best we can do, therefore, so long as we do not possess a perfect knowledge of our emotions, is to frame a system of right conduct, or fixed practical precepts, to commit it to memory, and to apply it forthwith to the particular circumstances which now and again meet us in life, so that our imagination may become fully imbued therewith, and that it may be always ready to our hand" (*Ethics* V 10, Note). Like a philosopher who "understand[s] the true philosophy," he understands the truth of absolute determinism and knows that in these conditions the path of freedom, like "all things excellent, [is] as difficult as [it is] rare" (the last sentence of the *Ethics*). But this does not eliminate the moral and religious experience of many persons (described in his letter to Oldenburg), whether or not they are philosophers and embarked on this path. We shall return to this question of provisional morality and the salvation of the ignorant in Chapter 4.

In fact, it is precisely to such an image of the world that the physiological sciences lead us (or lead us back) today. We are controlled and determined by the chemistry of our metabolism, just as the ancients were controlled by gods, spirits, and demons. The difference is that our world, instead of being enchanted by the members of a pantheon or by angels and demons, is ruled by impersonal laws whose efficacy we increasingly experience. The idea of personified nature has been cast aside as magical thought and is raised today only in dreams, deliria, and hallucinations. The personal God fills the void for some; but this is only inside "religion," in Georges Bataille's sense, in some preternatural or supernatural domain that no longer has anything in common with the natural laws we are learning to discover. Although we still humanize nature, we do so by means of technology and the collective and anonymous control that it makes possible. No one can predict whether the depersonalization of "technoscience" leads ineluctably to the general violence of Babel or is only an unavoidable stage.

What is hard to understand in this picture, as in the myth of the sparks of randomness and the 974 lost generations, is not the possible awareness of some absolute and timeless determinism. On the contrary, every discovery of another physical or biological mechanism, especially when cast as a mathematical law or some timeless formal structure, gives us another experience of a determining cause, even if its field of application is local and limited. The problem with this mechanical representation of the things that control us has to do with our subjective and social experience of our freedom.

To what does this permission or possibility, this possibility of freedom that is a like a toy relegated to children, refer, if (for all serious purposes) everything is controlled? What remains of our freedom when our free will, as described by the commentaries on these divine *'alilot*, is reduced to a fraud, to a pretext employed by the sly and scheming human nature in human beings, playing a game with itself, to make human beings responsible for what it made them do?[101]

From the theological perspective of a moral subject challenging God, himself a moral subject and the first cause of everything that happens, this game takes the form of a set of false reasons, pretexts, and poor excuses aimed at justifying what cannot be justified. In fact, this perspective itself is an illusion, because there is nothing that needs to be justified, according to God's final answer to Job, which convicts him of speaking too much, without knowledge and without intelligence: "Where were you when I laid the earth's foundations?" (Job 38:4).[102] The "game" is not that of a morally perverse being, but rather, in the manner of a playful *causa sui*, the game of the infinitude of what exists, which plays "in itself, for itself, and by itself."[103]

101. Imagine that a child, who thinks that she freely determines her behavior by means of her will, possesses a sense of responsibility as a byproduct of her knowledge of acting voluntarily. The subjective reality of this awareness does not vanish, even if we have reasons to believe that her awareness that she decides and acts is in fact her own memory and perception of what her own body (brain and muscles) is doing, accompanying the action without determining it. The neurophysiologist B. Libet has obtained stunning results on the chronological sequence of the decision to act and the start of the action from experiments with voluntary subjects during brain surgery (which had to be conducted while they were conscious). Libet observed repeatedly that the conscious decision to move some part of the body corresponded to an electrical event in the brain that took place 200 to 300 milliseconds *after* the start of the movement (B. Libet, "Unconscious Cerebral Initiative and the Role of Conscious Will in Voluntary Action," *Behavioral and Brain Sciences* 8 [1985], pp. 529–566). This result gave rise to vigorous controversy, not so much because of the experimental method, which does not seem to have been criticized, but because of its incomprehensibility in the context of current theories of action, in which the conscious decision to act is held to be an efficient cause of the movement. This result has been employed, notably by Eccles, to support a dualist theory of mind and body. But it fits perfectly into a monism that identifies mind with body and implies, as Spinoza strongly affirmed, that modification of one cannot cause modification of the other, precisely because they are one and the same thing (*Ethics* III 2).

It follows that voluntary awareness or conscious volition, and the experience of the free determination of will that accompanies it, are merely a consciousness of past events, turned around and imaginatively projected onto the future. This reversal, although imagined, is constitutive of a temporal unit of the conscious and unconscious self, creating a specific human world in which individual consciousness can coincide with individual existence. (See Atlan, *Entre le cristal et la fumée*, chap. 5.)

102. See Job 38, especially vv. 2–5. In these verses, Job's counsel (Heb. *'ezah*, which can be read as the feminine form of *'ez* 'tree') is "darkened without knowledge" (v. 2), because he lacks "understanding" (*binah*) and speaks "words without knowledge" (*millin beli da'at*).

103. This expression is used by several kabbalists who develop this idea of a game that is both creative and created. (See David Cohen, *Kol ha-nevu'ah*, pp. 258–267, for a list of some of these authors, most of them direct or indirect disciples of Moses Cordovero's. We shall see that for Cordovero this theme took on the dimensions of an extraordinary history, simultaneously a myth of origin and a philosophical tale.)

But in that case, is there any room left (and if so, where?), in this vision in which free will is denied by the sequence of causes, for an ethics of freedom and responsibility?[104] Most preachers, moralists, and philosophers (religious or otherwise, theists or humanists) are unwilling to accept a determinism like this, which seems to deny any possibility of human freedom. An abyss of incomprehension increasingly separates our impersonal science from our ethics of a moral subject endowed with free will, which dissolves a bit more with every discovery of new unconscious determinations, biological, psychological, or social, of what we had always considered to be freely decided choices.

The opposition stiffens between science, which keeps discovering new mechanisms and reinforces our experience of a determinist nature that extends to living beings and to human existence, on the one side, and what we refer to as "values," which are supposed to be the arena in which we exercise our freedom and responsibility, on the other. The result is a crisis in which more science seems to connote less humanity.

It was possible to avoid this crisis or shunt it aside as long as biology was vitalist and spread its wings over philosophies of life that could serve as the ground of morality and law, like the system of the classical idealists, Kantian and post-Kantian.[105] Today, however, such a stance is increasingly difficult to maintain, because the mechanistic sciences of the living keep discovering the sway of nonliving physical and chemical factors, even in what was traditionally considered to be the arena of individual human existence that was immune to them.

Whence the nagging question about what room can still be allowed—de facto and de jure—to the free determination of our actions by our will, when we know more and more what our body can do and how, mechanically, it does it.

This question, although (as we shall see) poorly phrased, is the starting point of the growing distrust with the very notion of scientific progress. This distrust is exacerbated in proportion as the operational efficacy of science and technology constricts the resistance mechanisms and escape hatches that so-called humanist philosophies once afforded. This leads to the crisis of meaning and of values that is held to characterize Western civilization today. It also spawns antiscience movements and a return to various irrational mysticisms or ignorance, where paradox and the invocation of mystery keep traditional moral systems alive. This mysticism may take a brutal form as religious obscurantism, in which "Western" science

104. The Kantian answer to this question, which distinguishes reflective reason from determining reason—the former allows room for the exercise of human freedom in its supersensory domain—is having ever-diminished success in its resistance to the invasion of that domain by determining reason. Among the antinomies of pure reason, the existence of God and the immortality of the soul remain propositions that determining reason cannot declare to be true or false. But more and more it seems that this is no longer the case of human freedom as the source of free choice superimposed on natural causes. In fact, contemporary biological science, contrary to Kant's predictions, is increasingly discovering the nature of the mechanisms that determine, in a purely causal and not teleological fashion, the behavior of organisms, including the human organism. This is yet another confirmation of the radical stance taken by Spinoza, who not only denied that final causes have any explanatory value but also affirmed that human beings believe themselves to be free because they have "consciousness of their own actions, and ignorance of the causes by which they are conditioned" (*Ethics* II 35, Note).

105. For a brief history of these philosophies of Life, from Aristotle through Hegel and Bergson, see Georges Canguilhem, "Le concept et la vie," *Revue philosophique de Louvain* 64 (1966), repr. in *Etudes d'histoire et de philosophie des sciences*, 7th ed. (Paris: Vrin, 1994), pp. 335–364; translated (in part) as "The Concept of Life," in *A Vital Rationalist: Selected Writings from Georges Canguilhem*, ed. François Delaporte, trans. Arthur Goldhammer (New York: Zone Books, 1994), pp. 303–319.

and reason are utterly demonized. But it may also assume the subtler disguise of idealistic and romantic philosophies of mind, in which condemnation of technology and science is an indispensable preliminary to the "return" of the "subject" or "unveiling of being." Sometimes, because the undeniable successes of science and technology and of the rationality they express are difficult to ignore, they are paradoxically salvaged in the form of pseudosciences taken out of the critical context in which they developed and transformed in their turn, in a grand synthesis of science and mysticism, into dogmatic theories. Here God is confounded with the Big Bang, quantum mechanics "explains" parapsychology, and psychoanalysis permits the subject and its mental nature to make their long-awaited return. A swaggering and reductive vulgarization reinforces the ambivalence produced by the fascination and fear with the new fetish of science. Science is taken as the arena in which a new magic, mysterious and formidable, is produced, and which has the added advantage over ancient magic of being effective. The circle is closed when, to salvage some part of the reality of the wholly mental subject-spirit, the biology of living bodies is itself spiritualized; genes and the genome, invested with mysterious power over individual destiny, now replace the effects of the transmigration of souls. They are no longer just molecules, but the alchemical crucible of life and of the final causes of nature, where human freedom, which transcends sensory perception, is supposed to have its "vital" foundation. Philosophical materialism and idealism then appear as what they have always been since their origin in Aristotelian hylomorphism: the two faces of a single vitalist metaphysics, based on an improper generalization of the psychological and linguistic experiences of human beings endowed with intentionality.

Instead of giving in to these mirages, we can update the ancient philosophies of fate and eternal return[106]—which do not necessarily imply the passivity of a resigned fatalism—in order to produce a better formulation of the question about the conditions of our freedom.

This will allow us to understand how the absolute determinism of *'illot* and plots in no way suppresses freedom, anymore than the psychobiological mechanisms of our behavior do. To achieve this, however, we must first follow Spinoza and totally invert the picture of freedom we have inherited from philosophies of free will. We must no longer locate freedom in the interstices of determinism and our ignorance of causes. We must renounce the facile idea that chance and indeterminacy are prerequisites for the existence of human freedom. In other words, we must stop confusing freedom and free will—and not merely because, as in classical philosophy, free will is a naïve and low form of freedom when not informed by universal reason and when judgment is obscured by passion and opinion. On the contrary, we

106. These philosophies can be traced not only to Hesiod but also to Ecclesiastes' "what has been is what will be, and what has been done is what will be done; and there is nothing new under the sun" (Eccles. 1:9). The talmudic discussions of this verse, as with the dictum of tractate *Avot* that all is foreseen, focus on the meaning of the human actions that are still possible when all is vanity. As Ecclesiastes himself asks at the outset, "What does man gain by all the toil at which he toils under the sun?" (1:3). See, for example, B *Shabbat* 30b, as well as Rashi and Ibn Ezra on these verses. These discussions show clearly that the ancient rabbis shared this determinist and potentially absurd vision of the given of the human condition, which Ecclesiastes characterizes as "an unhappy business that Elohim has given to men to be busy with" (1:13). Thus it is not astonishing that man is denounced as "wicked from his youth" (Gen. 8:21). All the same, the commentators wanted to believe in the possibility of transforming this situation and did not despair of the future, of the world to come and to be built, looking for the means of this transformation in the Law and the precepts (the Torah and the *mizvot*), which are wisdom and practice. (See below, n. 112, on the role of observation in our appreciation of novelty.)

must begin by accepting what we encounter every day in our science of determinism, namely, that our subjective awareness of free choice is increasingly belied by our objective knowledge of impersonal laws and causes that determine these choices and show that they are not as free as we believe them to be.

Then we must ask ourselves about the possible role in our lives of these experiences of free choice, which have not been eliminated, even though our theoretical and general knowledge about the causes that determine our actions leads us to reject them as an illusion. Although they are illusory as a source of determination, our experiences of choice are real as subjective, intersubjective, and social experiences.

One classic answer to this question is given by Eliaschow in his *Book of Knowledge*,[107] in the distinction between the root cause (*'iḳḳar*) and secondary causes in a series of events whose outcome is absolutely determined from all eternity. Only the details of the sequence are shaped by the free choices of human agents. Thus human actions are only proximate and occasional causes from the perspective of the essential and change nothing in the final result. The final state is absolutely determined, but the ways of achieving it are undetermined. The contradiction between "everything is foreseen" and "choice is possible" is thus resolved by invoking a certain degree of indeterminacy.[108] As human agents, our knowledge of the causal sequence, in which we intervene at each instant, is limited to a few states that preceded the present state and several possible states from which we choose the next state. It is the indifferent character of the possible paths with regard to the "root" cause—the cause of the final result—that leaves room for human choice to be exercised in the "details" without modifying this determinism. Furthermore, it is with regard to these details that an action can be judged to be "good" or "bad." Because we are ignorant of this determinism that leads ineluctably to the same final result, "foreseen" from all eternity, we experience the indeterminacy of the multiple possible paths as an object on which we exercise our "free choice."

Still, this indeterminacy itself may not be real; our choice may be determined elsewhere, even in the details, by other factors that act on our decision-making, even though it is not determined in the context of the causal series in which the action is embedded. In all cases, whether the particulars of the choice are indeterminate or not, it is embedded in this causal series and *seems* to have the capacity (even if spurious) to alter its outcome. This is why the specific choice is the responsibility of the person who makes it. It is on the basis of this "detail" that he or she may be judged, according to the criteria of moral law, as good or bad, as just or wicked. Then we understand that it is only an *'alilah* or spurious reason, a pretext to burden us with a responsibility that is not really ours, because the result will be

107. Eliaschow develops a complicated dialectic of the general and the particular, of the potential and the actual, and especially of the temporal and the atemporal. This interpretive framework gives added meaning to the debates of Ecclesiastes. The appendix to this chapter presents several especially significant excerpts from his book.

108. It is easy to conceive such a process by imagining a physical system governed by a deterministic dynamic, that is, a mathematical law that describes its evolution from a given initial state. Such a system can sometimes be characterized by a unique final stable state (an attractor), toward which it inevitably moves and in which it stabilizes, whatever the initial state and the path (that is, the succession of states) taken to get there. In nonhuman physical systems, the role of indeterminate choice is played by some random fluctuation (produced, for example, by temperature) that changes the initial state, or a state on one path to a state on another path. In such a system, even though the sequence of states in time may be altered in this way, the final stable state, defined timelessly by its dynamic, is always the same.

the same no matter what. Agents and their choices are only the triggers and cannot be held to be truly responsible for the evil to which their actions may lead at the end of the day. Even if a choice limited to the details is not itself determined and is therefore "free," the subject bears responsibility only for the commission of one or several single actions at the moment of the impulse. That responsibility is always incommensurate with the stakes and consequences that follow them in any case. Of course, the pretext is all the more obvious if the details of these acts are themselves governed and determined by other causal series, of which we are quite unaware, even before they take place.[109]

Another answer to our question, which perhaps gets closer to the heart of things, invokes the timeless character of deterministic laws. Everything is foreseen "from all eternity" and not in the psychological experience of the mind of a being that exists in time and makes plans for the future. Contrary to appearances, there is nothing mysterious about the timelessness of determinism. It is, furthermore, the charge we level against this timelessness from the perspective of our experience of duration and the irreversibility of time. We experience it each time we learn something by deduction, as in mathematics. We see then, quite clearly, that "it is in the nature of reason to perceive things under a certain form of eternity (*sub species aeternitatis*)."[110] As Robert Misrahi notes, in his commentary on Spinoza, "the internal necessity of things is then grasped as 'eternity,' that is, as existence known in its essence without reference to time or to duration. This 'eternity' concerns all things, all beings, in what they have in common and in essence. Thus here, in *Ethics* II, eternity is a sort of gnoseological apperception, a sort of cognitive relationship between human beings and the world."[111] In practice, every mathematical physicist experiences this when contemplating a natural law that makes it possible to predict a temporal phenomenon. When a future event is predicted in this way, time no longer exists in our knowledge of that event; rather, the event exists from all eternity in the law that describes it. Clearly this does not mean we no longer exist in time and no longer perceive events differently as a function of whether they are past or future. In particular, the mathematical prediction of a future event does not eliminate our surprise at seeing this event take place if it is something we are not used to. More generally, scientific research offers us special occasions of this experience of novelty despite the predictive knowledge we can have of it. It is only too common that the theoretical prediction of the result of an experiment is not borne out, whether because the theory is wrong or irrelevant or because the experiment was

109. Thus the question arises of whether the pretext itself is determined; for example, whether the wife of the faithless husband could have behaved otherwise and whether the husband would then have been compelled to "choose" another pretext. We cannot discuss this possibility in detail here. Suffice it to say that this question concerns the character, determinate or otherwise, of random fluctuations in a dynamic system. The fluctuations are not determined by the dynamic but may be determined at a different level of organization and on a different timescale. For example, genetic mutations are random with regard to the functional organization of the organism in which they take place, but they are determined on another scale—that of DNA molecules—by factors such as radiation and chemical agents.

110. *Ethics* II 44, Corollary 2.

111. Benedictus de Spinoza, *Ethique*, trans. Robert Misrahi (Paris: Presses universitaires de France, 1990), p. 393 and n. 74. The logician and mathematician Jean Cavaillès focused on this relationship between reason and timelessness. "Science moves outside of time—if time means a reference to the lived experience of consciousness," he wrote in his Spinozist vein (Jean Cavaillès, *Sur la logique et la théorie de la science* [Paris: Vrin, 1997], p. 37; see below, Chapter 4, n. 55). Although it is difficult to generalize this statement to all the sciences, it holds true for the mathematicized predictive sciences, of which mathematical physics is the archetype.

mismanaged. It is the contrary, paradoxically, that astonishes us. We are surprised—and thrilled—when a complex and nontrivial theoretical prediction proves true. Our theoretical certainty that the prediction is valid does not eliminate our curious anticipation of the result. If the prediction is confirmed we are clearly satisfied—but also, for all that, a bit astonished that "it worked!" A fortiori, the broad generalization that "everything is foreseen" and "there is nothing new under the sun" does not abolish our experiences of the passage of time and of the newness that accompanies it.[112]

112. We cannot expect more than what is granted us, on the basis of our "toil under the sun." The rabbis countered this dictum of Ecclesiastes (1:3) with the observation that "under the sun man has no profit, but he does have [profit] before the sun" (B *Shabbat* 30b; see above, n. 106). According to Rashi, this means human toil on the path of the Torah, which "preceded" the creation of the world and thus was "before the sun." A midrash on this same text (*Eccles. Rabbah* 1:3) employs the expression "above the sun" to designate the "place" where human beings can expect to receive the fruits of their efforts, even if they must expect nothing "under the sun." The succession of lunar months is superimposed on the cyclical return of the seasons in the solar year "under the sun." The former belong to the world of the night and are thus partially immune to the unvarying return of the light of day. Consequently the succession of months is not imagined as a repeating cycle but as a succession of births and deaths, followed by rebirths and then again by deaths—as if the moon disappeared each month and was then renewed, as explicitly reflected in the Hebrew word *ḥodesh* 'month', from *ḥadash* 'new' (i.e., *ḥiddush ha-levanah* 'the renewal of the moon'). "Under the moon" we can have an experience of the new, along with corruption and death and the new life that emerges in its turn. The experience of eternal return and of the vanity of human efforts is overlaid by the prophetic hope of a new world. This is the "new song" (Isa. 42:10), "the new heaven and the new earth" (Isa. 66:22) that the prophet announces for the future age. This hope is illusory—Solomon teaches us in his wisdom—if we believe that it proceeds exclusively from our plans and our projections on the future that motivate our efforts, in ignorance of the causes of events and their full past history. But it is not an illusion if we can perceive the new in a different way, as a renewal, a *ḥiddush*, of what exists from all eternity. This possibility is suggested to us by our observation of the succession of the generations and the coming into existence of new beings, even though they are just the reproduction of what they resemble. But it is especially through the study of wisdom as applied to the world that we can have an experience of a discovery, a renewal, a *ḥiddush*, for all that this wisdom, in the infinitude of what it has expressed and will express one day, was already given, outside of time, to Moses from Sinai. (See below, n. 134. Also relevant is the story of Moses' inability to understand Rabbi Akiva's lecture to his pupils about laws given to "Moses from Sinai" [B *Menaḥot* 29]; see Atlan, *SR* II, chap. 12.) For the Midrash, every new interpretation by the sages, every new insight or new ways of expounding the law, all the novellae (*ḥiddushim*) of Torah and halakhah, were given to Moses from Sinai. Although they are thus covered by the dictum of "nothing new under the sun," these illuminations nevertheless constitute the "new song" announced by the prophet (*Eccles. Rabbah* 1:9–10).

In the talmudic passage about these contradictions between the text of Ecclesiastes and the teachings of the talmudic sages, as well as on the internal contradictions in that book, cited at the start of this note, Rabban Gamaliel gives several examples of the novelties of the future age; for example, "trees will bear fruit every day." A foolish student objects that this is impossible, because "there is nothing new under the sun." The rabbi responds with an explanation that is in keeping with his foolishness (see Chapter 4, n. 161), observing that there is already a tree, the caper bush, that is always in flower, although this is not the case with other trees. The same student poses the same objection to Rabban Gamaliel's other illustrations of new phenomena in the future; each time the sage shows him an existing (though approximate and uncommon) example of the future wonder in question. In other words, the response to the objection ratifies the principle that there can be nothing new but displaces it by giving it a different content, less foolish or more intelligent. In absolute terms, there will be nothing new in the future because the phenomenon already exists, or because something similar (although rare and uncommon) makes it possible for us to imagine it. The novelty predicted for the future is not so much that trees will always be in flower and bear fruit (that is already the case with the caper bush) but that we will observe something similar in all trees. The novelty is relative to the conditions of our experiences in the duration of our history. Although still an experience of something new, it does not make it impossible for us to know the phenomenon in the abstract, as an analogical generalization based on current isolated cases, independent of the conditions in which they are realized. Our a priori knowledge, theoretical and intuitive, *sub species aeternitatis*, "under the aspect of eternity," does not replace our existence in time.

Thus we understand in diverse ways—physics, for example, or phenomenology—that our perceptions of time and eternity are relative to the conditions of our experiences. The analysis of logical paradoxes, such as Newcomb's paradox, may make it possible to distinguish what Jean-Pierre Dupuy calls "occurring time and projected time" (J.-P. Dupuy, "Philosophical Foundations of the New Concept of Equilibrium in Social Sciences: Projected Equilibrium," *Philosophical Studies* 100 [2000], pp. 323–345).

The result is that our appreciation of the sense—both "direction" and "meaning"—or nonsense of our actions and our passions depends on our perspective on eternity and on our experiences, which themselves depend on the time in which

Finally, our knowledge (even if incomplete) of determining causes makes it possible, by extrapolation to infinity, to conceive full knowledge of the absolute determinism of nature. It is no longer imagined as some incomprehensible arbitrariness but as a timeless intelligible necessity that we understand in part. Rather than vainly attempting to push the limits of this determinism into the interstices of randomness and indeterminacy, which keep getting smaller and smaller, we can then postulate a priori that "everything is foreseen." Instead of passive resignation, our experience of the operation of rational knowledge can provide us, by extrapolation, with the perspective of a possible infinite knowledge of this absolute determinism—knowledge of self by self, in which what knows is what is known and also knowledge itself. Note that the assumption that knowledge of the absolute determinism of nature is possible, even if its realization is indefinitely and infinitely deferred, is an implicit methodological postulate that necessarily accompanies any project of rational knowledge. We cannot begin trying to understand the laws of nature unless we postulate that they exist and establish some a priori limit to their field of application.

Of course, as with any asymptote, we can have an idea of this infinite knowledge but can never attain it.[113] Still, this is enough to totally invert the relationship between acceptance of determinism and freedom. Far from being antithetical to freedom, it is in fact only the perspective of this infinite knowledge that makes it possible for us to conceive of the nature and reality of human freedom. Only this asymptotic approach to a potentially infinite knowledge of the absolute determinism of everything that exists can make it possible for us to entertain the possibility of a total or absolute freedom in which our being is fused with our knowledge. Unlike the classic opposition between free will and deterministic nature, where more knowledge seems to imply less humanity, the very experience of absolute and timeless determinism permits us to imagine, by extrapolation to infinity of full knowledge of determinism, the possibility of a freedom of another sort, something different from free will, in which more knowledge always means more freedom.

Here we cannot analyze in detail how this reversal, from opposition

we exist. Elsewhere (*Enlightenment*, pp. 265–266) I have analyzed how the Maharal of Prague expresses this dependence in the notion of a "separate world" and how he applies it in his theory of miracles. The separate world is "divine" in that it is the world of meaning, above the sun and stars, but it is human because the meaning is not given. Constructed by human existence, the meaning of this existence is "artificial" with regard to the natural world, and supernatural (in the sense of surreal) because it is nourished by the world of possibilities. Although the meanings are multiple, because there is more than one way to assign significance to things, not everything is possible in this separate world. Like the natural world, it obeys laws; in this regard it too is intelligible. Although it calls to mind Kant's suprasensible, the Maharal's separate world is very different, because its relationship to the natural world is not mediated by the living (*le vivant*), which is taken as the arena of the final cause and intentionality of nature from which human intentionality, and the free will supposed to accompany it, derive their source.

The relationship between the world of meaning and the world of nature is that of the possible to the real. We can know and understand possibilities, but they are not given to us in our necessarily limited experience of the real; this is why we can perceive them only as possibilities. The sole points of contact between the two worlds are the rare events that are located simultaneously in the impersonal and nonhuman chain of natural causes and in a personalized history or myth—individual or collective—that creates meaning. These points of contact can be called "miracles"—because rare—or signs (Hebrew *'otot*, which can also mean "miracles")—because they are meaningful, although borne by a reality that is not meaningful per se. This may lead us to think of Leibniz's pre-established harmony between matter, governed exclusively by mechanical causality, and mind, which is directed by final causes. But here it is a matter of an ad hoc and post-factum harmony, always under construction, rather than of a pre-established universal harmony. "If you conduct yourself randomly with me, I will conduct myself randomly with you" (Lev. 26:23–24) is the dialogue that God holds with human beings, in this vision of things where human beings determine whether or not things have meaning (see Chapter 6).

113. See above, p. 87.

to cooperation, might take place. Let us say only that the experience of free will still does not vanish. It is merely limited to the domain of the moment of decision, that is, to our temporal experience *in the present.* It has the same status as the sense of novelty conveyed by certain events in our existence in time, even though we can have timeless predictive knowledge of them and picture them as existing from all eternity in the law that describes them.[114] Our experiences of free choice retain their full reality and importance in our lives, on condition that we always look at them in their present and do not believe they determine the future. Although time is not an absolute, it does not disappear from our existence. Rather, what changes is how we look at things when our vantage point is this infinite knowledge toward which we proceed, though we never attain it.[115]

This also makes it possible to understand the possible ontological role of the daily ethical experiences that guide our existence. Ignorance that knows itself to be such, even confusedly, is different from ignorance that is ignorant of itself. Our capacity to conceive of causality is accompanied by recognition of a deficiency when we do not know the causes of certain phenomena, including the causes of some of our own actions. The origin of this capacity is of little moment here; whether, with Kant, it is one of the categories of our understanding or, as Hume would have it, it is no more than a quasi-animal faculty of memorizing repeated associations of similar pairs of consecutive events. In any case, this notion of absolute determinism of which we are partially unaware leads us, by another path, to the Maimonidean interpretation of the fall of the first man, which we mentioned at the start of this chapter.

In practice, when a subject says "I" but does not know the causes that make it act, it may believe, on ac-

114. See above, n. 112.

115. This alteration of how we look at things played an important role for the Stoics, because it was none other than the "greatness of soul," as underscored by Pierre Hadot in his presentation of "physics as a spiritual exercise for Marcus Aurelius" (Pierre Hadot, *Exercices spirituels et Philosophie antique*, 3rd ed. [Paris: Institut d'études augustiniennes, 1993]). Traditionally, "greatness of soul is the virtue or knowledge that raises us above what may happen, both the good and the bad" (ibid., p. 128). For the Stoic emperor, it was the practice of the "physical" knowledge of things, that is, of what they are "in relationship to the entire universe and to human beings," that produces this transformation of how we look at things, so that we view the present as the only true temporal reality that actually belongs to us (ibid., pp. 123–133). Just as, qua individual in infinite space, each of us occupies an infinitesimally small place compared to the entire universe, so too in the infinite duration of past and future we possess only the infinitesimal instant of the present of our sensation. The rest "is but opinion," if we think it pertains to us as past history or as a future object of desire and will. As Marcus Aurelius puts it, "the present time is of equal duration for all, while that which we lose is not ours; and consequently what is lost is obviously a mere moment. No man can lose either the past or the future. For how can a man be deprived of what he does not possess? . . . For it is but of the present that a man can be deprived, if, as is the fact, it is this alone that he has, and what he has not a man cannot lose" (Marcus Aurelius, *Meditations* II 14, in *The Communings with Himself of Marcus Aurelius Antoninus*, trans. C. R. Haines, LCL [London: Heinemann, 1916], p. 39; modified slightly). Again,

> In taking umbrage at anything, you forget this, that everything happens in accordance with the Universal Nature; and [forget] this, that the wrong-doing is another's; and this furthermore [you have forgotten] that all that happens, always did happen [so] and will happen so, and is at this moment happening everywhere. And you forget how strong is the kinship between [a] man and mankind, for it is a community not of corpuscles, of seed or blood, but of intelligence. And you forget this too, that each man's intelligence is God and has emanated from Him; and forget this, that nothing is a man's own, but that his baby, his body, his very soul came forth from Him; and this, that everything is but opinion; and this, that it is only the present moment that a man lives and the present moment only that he loses. (ibid., XII 26, p. 335–337)

Still, because he thinks about the true nature of things and the "greatness of soul" he achieves, "it is in your power to rid yourself of many unnecessary troubles, for they exist wholly in the imagination. You will at once set your feet in a large room by embracing the whole Universe in your mind and including in your purview time everlasting, and by observing the rapid change in every part of everything . . ." (ibid., IX 32, p. 251 [modified]).

count of this ignorance, that it is the cause of its own actions. What is more, it may judge that these acts are good or bad, on the basis of some criterion or other. The entire moral problem derives from this curious association between the idea of self and the knowledge of good and evil that results from ignorance of the true and false. Ignorance of causes is associated with self-awareness, that is, with the memories and the imagined future that constitute the identity of an *I*. The idea of a self, an *ego*, as a free agent, as the absolute origin of a causal series, takes the place—psychologically and socially—of the causes that act in and on it and of which it is unaware. Animals' ignorance of their ignorance of causes allows them to act even though the idea of a free and responsible subject never penetrates the mind of the body on which they produce their effects. But the large human brain and its much greater capacity for memory change the situation completely. Our memory allows us to be aware that we do not always know the causes of our actions. Our awareness of this ignorance then becomes knowledge of the good and evil that we do as if we were free agents, the cause of our own behavior. This is not necessarily an illusion, but rather an expansion of self-consciousness. The moral problem of the responsibility of the subject, supposed to be free to act and to do good or evil, is no less real because it results from this conjunction of ignorance of true causes with self-awareness/memory. This conjunction is real. It is, indeed, what makes human existence so rich and special. We cannot call it an illusion, except from the perspective of an infinite and omniscient reality that is itself illusory when projected onto the finitude of human existence.

13. THE CALL OF ETHICS

Contrary to the received idea that confounds freedom with free will, we see that increased knowledge of the mechanisms and factors that determine our actions can propel us toward greater freedom, in an oscillating movement between the experience of our finitude and our intuition of the infinite, between our existence in time and our knowledge of the timeless. It is there, between the infinite we conceive of and the finite we achieve, that the demands or call of ethics is located.

Rather than disappearing, as we might expect it to, because of the ostensible incompatibility of determinism and freedom, ethics actually grows and matures. It becomes the code of a local and subjective responsibility, limited in reality but potentially infinite in principle, because it is associated with the infinite search for knowledge and for the laws, human and nonhuman, that govern the world. In existence as given, this limited responsibility increases in proportion as this knowledge advances.

As in the philosophy of Emmanuel Levinas, but in a different way and perhaps more radically, it is not free will that is the ground of responsibility; rather, responsibility comes first—though without necessary guilt—in the world given to us, which is both determined and intelligible. This is an ethics with clear Stoic and Spinozist accents; as we shall see, however, the metaphor of the game helps us avoid confusing it (if such help be necessary) with fatalism; that is, with a moral code of passive resignation to ineluctable fate.

The relationship between ethics and the experience of alterity seems to be very different from what we have learned from Levinas, although the ethical injunction retains its supernatural or surreal character and is always at risk of disappearing in an excessively deep and overly definitive concern for natural "enrootedness."

In a 1957 paper on the idea of infinity, Levinas denounced an entire current of Western philosophy that he believed had been brilliantly encapsulated by Heidegger.[116] Here Levinas effects a double break. There is a theoretical discontinuity with regard to the ontology of the Self (*le même*), which makes Levinas postulate the ethical injunction as beyond being; responsibility, the challenge to respond to the other, precedes knowledge of self and freedom. But there is also a discontinuity in his philosophical descent from Heidegger, whose key ideas are reviewed in this paper.[117]

The opening to infinity produced by the face of the Other makes it possible to save a human nature that is not totally enrooted and is only a reflection or guardian of the being that is totally shut up in its ipseity. This enrootedness in nature, this familiarity with the soil, is a burden, a hidden millstone of this

116. E. Levinas, "La philosophie et l'idée de l'infini," *Revue de métaphysique et de morale* (1957); English version: "Philosophy and the Idea of Infinity," in Emmanuel Levinas, *Collected Philosophical Papers*, trans. Alphonso Lingis (Dordrecht: Martinus Nijhoff, 1987), pp. 47–59.

117. The passage that follows gives an idea of the subtlety of Levinas's analyses. But these analyses themselves open the way to going further still (*un nouveau dépassement*), so that (without returning to Heidegger) for the moral subject that perceives himself or herself *in front* of the Neuter, the Neuter is even more other than the Other.

> To be sure, for Heidegger man's freedom depends on the light [(*sic*) French: *liberté*] of Being, and thus does not seem to be a principle [note: Cf. *Being and Time* §62, on *Entschlossenheit* (resoluteness)]. But that was also the case in classical idealism, where free will was considered the lowest form of freedom and true freedom obeyed universal reason. The Heideggerian freedom is obedient, but obedience makes it arise and does not put it into question, does not reveal its injustice. . . . *Being and Time*, Heidegger's first and principal work, perhaps always maintained but one thesis: Being is inseparable from the comprehension of Being; Being already invokes subjectivity. But Being *is not* a being. It is a neuter which orders thought and beings, but which hardens the will instead of making it ashamed. The consciousness of his finitude does not come to man from the idea of infinity, that is, is not revealed as an imperfection, does not refer to the Good, does not know itself to be wicked. Heideggerian philosophy precisely marks the apogee of a thought in which the finite does not refer to the infinite (prolonging certain tendencies of Kantian philosophy: the separation between the understanding and reason, diverse themes of transcendental dialectics), in which every deficiency is but weakness and every fault committed against oneself—the outcome of a long tradition of pride, heroism, domination, and cruelty.
>
> Heideggerian ontology subordinates the relation with the other to the relation with the neuter, Being, and it thus continues to exalt the will to power, whose legitimacy the other alone can unsettle, troubling good conscience. When Heidegger calls attention to the forgetting of Being, veiled by the diverse realities it illuminates, a forgetting for which the philosophy developed from Socrates on would be guilty, when he deplores the orientation of the intellect toward technology, he maintains a regime of power more inhuman than mechanism (and which perhaps does not have the same source as it; it is not sure that National Socialism arises from the mechanist reification of men, and that it does not rest on peasant enrootedness and a feudal adoration of subjugated men for the masters and lords who command them). This is an existence which takes itself to be natural, for whom its place in the sun, its ground, its *site*, orient all significance—a pagan *existing*. Being directs its building and cultivating, in the midst of a familiar landscape, on a maternal earth. Anonymous, neuter, it directs it, ethically indifferent, as a heroic freedom, foreign to all guilt with regard to the other.
>
> Indeed this earth-maternity determines the whole Western civilization of property, exploitation, political tyranny, and war. Heidegger does not discuss the pretechnological power of possession effected in the enrootedness of perception (which no one has described so brilliantly as he), in which the most abstract geometrical space is in the last analysis embedded, but which cannot find any place in the whole infinity of mathematical extension. The Heideggerian analyses of the world which in *Being and Time* were based on . . . fabricated things [*l'attirail des choses fabriquées*] are in this philosophy borne by the vision of the lofty landscapes of nature, an impersonal fecundity, matrix of particular beings, inexhaustible matter of things.
>
> Heidegger does not only sum up a whole evolution of Western philosophy. He exalts it by showing in the most pathetic way its anti-religious essence become a religion in reverse. The lucid sobriety of those who call themselves friends of truth and enemies of opinion would then have a mysterious prolongation! In Heidegger atheism is a paganism, the pre-Socratic texts anti-Scriptures. Heidegger shows in what intoxication the lucid sobriety of philosophers is steeped.
>
> To conclude, the well-known theses of Heideggerian philosophy—the preeminence of Being over beings, of ontology over metaphysics—end up affirming a tradition in which the Self dominates the other, in which freedom, even the freedom that is identical with reason, precedes justice. Does not justice consist in putting the obligation with regard to the other before obligations to oneself, in putting the other before the Self (ibid., pp. 51–53). <I have replaced Lingis's rendering of *le Même* as "the same" with "the Self," which strikes me as a better translation and is in any case better suited to Atlan's reading of the text.>

ontology of the Self, which Levinas wants to be rid of—and help us be rid of—by perceiving, in the look of the Other, the call of ethics.

But there seems to be a certain ambiguity here. Surely it is the contrivances of technology that permit us today to personify nature and feel at home in it, to be much more comfortable than we are with the brute nature of the mineral, vegetable, and animal. Where is the Self and where is the Other? Can we truly come out of the Self when we are still "among ourselves," human subjects with human subjects, shielded from the even greater otherness of nonhuman nature?

The Other that is imposed on us before any reflection and for which we are supposed to be responsible is not only the other person. It is also the impersonal and mechanical that acts in the other person, as well as in animals, plants, rocks, and storms, and even in ourselves—in brief, in all of the nonhuman phenomena with which Job is confronted in the "answer" he finally accepts at the end of his book. The alterity of the amoral and impersonal is even more radical for me, as a personal subject, than is that of another person, because the latter is at least in my own image and my semblance. The neuter, the "indifferent" in the Stoic sense—that is, what nature deals me and is neither virtue nor vice, because it does not depend on me—is even more Other than the face of another person, on which I can always project virtues and vices. Yes, I must already be a moral subject. But this is the case for every human individual who learned in infancy to say "I" and "you" and "he/she," whatever content an individual attaches, incidentally, to good or evil, permitted or forbidden, for himself or herself and for others. The encounter with the neuter of that amoral id or it, which can work harm without being morally bad, is an encounter with an alterity that is more radical than the face of another person is. My encounter with nature is not an encounter with the Self unless I perceive myself as a neuter. To avoid the pain of these potentially harmful encounters, the Stoics recommended an ascesis that aimed at making them indifferent to indifferent (i.e., irrelevant) things, neutral with regard to what is neutral, accepting with equal tranquility of the soul everything that nature brought them, identifying their vision with that of nature so that they would no longer be strangers in the universe.[118] This ascesis, supposed to rest on physical knowledge of particular events, resembles Spinoza's third kind of knowledge, the only sort that includes adequate knowledge of particulars. We know, however, that Spinoza was careful to keep his distance from this attitude, noting that it supposed a spurious power of the human will to eliminate the force of the emotions merely by so decreeing.[119]

118. Hadot, *Exercices spirituels*, pp. 128–130.

119. *Ethics* V, Preface.

Every naturalist tradition, whether philosophical or mystical, conceiving of human beings *exclusively* as intimates of nonhuman nature, exposes them to the risk either of being annihilated in the latter or of perceiving themselves as its omnipotent origin, perhaps as the continuation of a childhood fantasy. It is Levinas's merit that he saw these dangers clearly and tried to warn us about them. But his initial intuition of an opposition between the closed totality of the verb *to be* and the opening to an infinity for which being is not enough is probably more fundamental. It is not between the Self and the Other, but rather between the neutral (or neuter) and the personal, the id and the ego, perhaps within the Self, that we encounter this opposition. This opposition pits what *is* against the *I* that *is not*, but is overlaid like a void or vain breath that adds *nothing* to its substance. Echoing Ecclesiastes, the midrashic work *Tanna de-vei Eliyahu* recommends that we begin each day by acknowledging our nullity: "Man's superiority over beasts is nothing (*'ayin*), because all is vanity (*hevel*)" (Eccles. 3:19).[120] The *I* (*'ani*, spelled *alef-nun-yod*) that distinguishes human beings from the animals is ultimately a permutation of a "nothing" (*'ayin*, spelled *alef-yod-nun*), a void where there is nothing and that is nothing. And for this I, "all"—everything it takes to be real from the perspective of its ego—is, just like that ego, mere breath and vanity. But it is precisely as such that it nevertheless says "I." After having said "I," it comes by nature to say "you" and "he" or "she" and finally "we." From here one can go on to characterize the experience of this *'ani*, which is "nothing," as the exception that escapes the whole. In the typology of the *sefirot*, it is either the first or the last *sefirah*, the initial zero or the zero of ten. As the first and infinite unity, it is designated in Hebrew by *alef*, a letter with no phonetic value, merely a placeholder for the vowels, which are indicated by "points." When it is the tenth, it is designated by *yod*, a letter/numeral whose shape is close to a point.[121] It escapes this "all" that is a nothing for it, just as it is itself a nothing as compared to the all. Like the zero that adds nothing, the zero of ten, the sign of the whole, remains outside the ten; it is only a sign and adds nothing to it. Thus with regard to this superiority of man over beasts, which is nothing, "all [that exists] is vanity, except for the pure soul that must render account before Your throne of glory."[122]

To put it another way, in this headlong flight from one nothing to another nothing, from one breath to another breath (*havel havalim*, the "breath of breaths" or "vanity of vanities" of Ecclesiastes), the *I* does exist—but as a moral subject that must account for its actions. Compared to this radical alterity of the impersonal, that of the "you" or the "he/she" is still very relative. There we are still "among ourselves" in a certain fashion within the Self.

But this does not necessarily mean that the perceiving consciousness of the other person (and not only her speech and logos) calls to us with less force, or that we are any less duty-bound to take account of her than of impersonal, blind and mute nature. Our responsibility vis-à-vis others, extended to that extreme other of amoral and impersonal nature, does not imply a leveling that reduces everything to some abstract Nature or in which everything—human and nonhuman, personal and impersonal—is fused. (The excesses of "ecologism" or "deep ecology"—ecology-as-ideology—in its nonhuman form are a terrible caricature of a naturalist ethics of responsibility.) Although human reality, with its experience of subjects—linguistic, psychological, judicial—is produced by nature, it institutes a different order of relationships, one that is both more and less familiar than that of nonhuman realities. Such experiences of persons may make us conceive the subject as "a god of persons," an *'ani* in our own image, whom Moses and the prophets proclaimed—evidently with some justification, if we judge by the timeless efficacy of this god's persistence—to be "the greatest of all gods."[123] We should recall that, for the prophets of Israel, the image and the semblance operate in both directions: from God toward human beings, in Genesis, where male and female are

120. *Tanna de-vei Eliyahu*, chap. 21. <Note that Hebrew *hevel* also means "breath.">

121. <In the "Hindi" or "Eastern Arabic" system of numerals employed in the Middle East, the zero is itself a point: • . >

122. *Tanna de-vei Eliyahu*, chap. 21, in the version found, inter alia, in *Siddur Rinnat Yisrael, nusaḥ Sefard*, ed. and annot. Shlomo Tal, 4th ed. (Jerusalem: Moreshet, 1995), p. 14. R. Jacob Meshullam Ornstein, in his *Yeshu'ot Ya'aḳov* commentary on *Tanna de-vei Eliyahu*, notes that *'ayin* can be read as an acronym of *'aval yesh neshamah*, "but there is a soul," and that this soul constitutes man's superiority to the beasts, contrary to the literal meaning of the text.

123. See Atlan, *SR* II, chap. 5.

created in the image and semblance of Elohim; and also from the human toward God, in Ezekiel's vision of the divine chariot. At the zenith of his step-by-step ascent through the divine worlds, whose structure will later serve as the model for the esoteric cosmogonies of the Talmud and kabbalah, the prophet Ezekiel encounters "the semblance of a throne" and, above it, "the semblance of a human form" (Ezek. 1:27). Man is created in the semblance of Elohim, while the god of the prophet's revelation is in the semblance of man. The creation of man completes and recapitulates the action of nature. (Elohim is explicitly understood as *natura naturans* by Joseph Gikatilla in his *Ginnat 'egoz*, cited above.) But prophetic morality tends to personify this same nature. Can we say, though, that in this personification the other human being is still an Other that is different from the Self?

The "look" of the other person (who looks at me), who has a human form and face, is like my own "look" seen from the outside, as in a mirror. This "exteriority" is clearly not nothing, as regards distance and alterity, and Levinas's fine analysis of the other person's perceiving consciousness or "look" (*le regard d'autrui*) remains relevant. In reality, however, as long as we remain within the personal and the human, the distance is both greater and smaller. For me the other person is not only an other but bad, even if it is not his fault. In this, what is more, I am like him—and consequently I can love him, or at least make an effort to do so. The injunction to love my fellow as myself is neither absurd (even though love cannot be commanded) nor tautological, because it concerns someone who is my fellow and mate (*réa'* רֵעַ),[124] in that he, like me, is bad (*ra'* רַע) from his youth.[125] Because this other is also one's fellow, the amoral and impersonal is more other for the moral subject than the other is.

The responsibility that I bear, as a person, toward that impersonal for which I am accountable is imposed first of all on the child and its imagination, in its fantasies of omnipotence. Our sense of free will and voluntary decisions is merely a weakened and socialized form thereof. With maturity and adulthood, the nature of this responsibility changes. It is no longer the exercise of a will to power that is only an illusory power of the will, but rather restraint, limitation, obedience to the law. The entire issue of achieving what is allegedly true freedom turns then on the transformation of the heteronomy of the law into autonomy, by an increase of knowledge and internalization of the required action that has become necessary—a transformation, in Spinoza's terms, of inadequate and alienating causality into an adequate causality of the self-caused, the only place where freedom can be exercised as "free necessity."

This transformation necessarily involves a lesson whose manifold character—simultaneously real and unreal, central and marginal, indispensable and superfluous—seems to be expressed by the metaphor of the game, as we shall see presently.[126]

Like the ancient philosophers' practice of "spiritual" physical exercises[127] and perhaps also—although in a different way—like the "rules of life" recommended by Spinoza as

124. In Leviticus 19:18, "love your fellow as yourself," "fellow" renders the Hebrew word *réa'*.

125. See Genesis 8:21.

126. See also Atlan, *Enlightenment*, chap. 7, "Man-as-Game."

127. See Hadot, *Exercices spirituels*. In that important work he demonstrates the central role of these "spiritual" exercises for the Stoics and Epicurus, and later for Plotinus and his disciples. We learn just how incomprehensible these doctrines are if seen as no more than abstract speculations that do not aim at salvation and how their preemption by the speculative theology of the Middle Ages caused the central role of these exercises to be forgotten, leaving what remains of their writings incomprehensible to us.

a way to develop good habits,[128] the practice of rituals and piety is a tutorial of sorts[129] in which we play the role of a free person, both wise and just, while perhaps waiting to actually become a bit more so.[130]

14. THE GAMES OF THE INFINITE—IN ITSELF, BY ITSELF, AND FOR ITSELF

This theme of the game that nature plays with itself, in which the role of each human being depends on his or her relationship to the Torah as theoretical and practical wisdom, was developed by some kabbalists as a myth of origins that is at least as important as the Lurianic *ẓimẓum* or contraction (though less well

128. *Ethics* V 10, Note (see also above, n. 100).

129. In the rabbinic literature we find many expressions of this idea that the acceptance of the Torah by human beings, often described as an amorous relationship (see for example, Scholem, *Kabbalah and its Symbolism*, pp. 55–56), is the path of true freedom, completing the initial liberation from slavery. One well-known passage of this sort is the talmudic homily that expounds Exodus 32:16 ("The tablets were the work of Elohim, and the writing was the writing of Elohim, engraved on the tablets") by a play on words: "Read not *ḥarut* 'engraved' but *ḥerut* 'freedom', for the only free man is he who occupies himself with the study of the Torah" (M *Avot* 6:2). The indelible engraving of the letters and words of the written Torah is the model for this freedom that lies outside time and outside existence, when our existence is modulated in time by the novellae (*ḥiddushim*) of the Oral Law, that is, by the new meanings that can be based on these letters. The law, having become the object of constantly renewed study and a means toward eternal and infinite wisdom, is transformed into a path of liberation. Obedience to the law becomes joyful because it is not compelled. Obedience inspired by fear is replaced by loving acceptance of its true nature; the "yoke of the commandments" (*'ol miẓvot*) is replaced by the "joy of the obligation" (*śimḥah shel miẓvah*). This idea is taken up by several authors in the metaphor of a game that the Infinite plays with himself, with the Torah, and with the human beings who study and practice it joyfully. This image of an amorous game that human beings play with the Torah and with the Infinite suggests a link to amorous knowledge, which clearly recalls, in certain aspects, the "intellectual love of God," the only true freedom for Spinoza. "The love of God towards men and the intellectual love of the mind towards God are identical" (*Ethics* V 36, Corollary), because "the intellectual love of the mind towards God is that very love of God whereby God loves himself, not insofar as he is infinite, but insofar as he can be explained through the essence of the human mind regarded under the form of eternity; in other words, the intellectual love of the mind towards God is part of the infinite love wherewith God loves himself" (*Ethics* V 36; see also the note ad loc., where Spinoza, associating this love with "our salvation, or blessedness, or freedom," considers that "this love or blessedness is, in the Bible, called Glory—and not undeservedly"). <See Atlan, *SR* II, chap. 4.> The "intellectual love of God"—both the idea and the phrase—can be found in authors whom Spinoza is known to have read, such as Ḥasdai Crescas, whom he mentions in his "Letter on the Infinite," and Judah Abrabanel (Leone Ebreo), whose *Dialoghi di Amore* was in his library. (See also H. A. Wolfson, *The Philosophy of Spinoza* [Cambridge, MA: Harvard University Press, 1934/1962], vol. 2, pp. 303–325.) As so often, Spinoza transforms this idea when he incorporates it into his own theory of the three kinds of knowledge, as he mentions in this note; what is more, the notion of "Glory" is explicitly linked to it in his own theory of the emotions in *Ethics* III. Finally, note that Maimonides, in a totally different philosophical context, describes the study of the Torah as the union of the agent intellect with the human soul, borrowing (as did Aquinas) an ancient Aristotelian image. Abraham Abulafia later transformed the same image by incorporating it into his kabbalah and eroticizing it, in the mode of other Gnostic, Neoplatonic, and mystical philosophers, both Muslim and Christian (including Meister Eckhart). As Idel underscores, this is a profound divergence from Maimonides' thought, which Abulafia claims to continue (see Moshe Idel, *The Mystical Experience in Abraham Abulafia*, trans. Jonathan Chipman [Albany: State University of New York Press, 1988], chap. 4, "Erotic Images for the Ecstatic Experience"). This suggests that for these authors (as in all erotic sacred literature) this love, though "intellectual," is inconceivable without the experience of physical love.

130. But unlike what might be, for Spinoza, a provisional morality that is relatively arbitrary and whose efficacy is merely subjective, this lesson would already include at least something of the goal toward which it leads. This is why we should not be surprised that the so-called theurgic streams of kabbalah, whose links with ancient natural magic are justly highlighted, emphasize the study and practice of the Torah precepts as means of acting on the structures of the divine and activating their forces. (See: Moshe Idel, *Kabbalah: New Perspectives* [New Haven: Yale University Press, 1988]; Charles Mopsik, *Les grands textes de la Cabale: Les rites qui font Dieu: Pratiques religieuses et efficacité théurgique dans la Cabale, des origines au milieu du XVIII*^{e} *siècle* [Lagrasse: Verdier, 1993].) Major philosophers of antiquity, especially those affiliated with Neoplatonic schools, including Plotinus and Proclus, recognized this theurgic side of pagan cults, for which they had great respect (see, for example, Emile Bréhier, *The History of Philosophy*, trans. Joseph Thomas [Chicago: University of Chicago Press, 1963], vol. 1, chap. 7).

known).[131] It is worked out at some length in an unpublished manuscript by Moses Cordovero, introduced to the public by Bracha Sack.[132] She includes a large excerpt from his *Sefer 'Elimah*,[133] in which Cordovero expounds the well-known eighth chapter of Proverbs on the role of wisdom in the divine and human order of creation. Wisdom (*ḥokhmah*) in this book is generally interpreted as referring to the infinite and eternal primordial wisdom that is the origin of all things and the ground of their intelligibility. Elsewhere it is called "Torah," it too infinite and universal, of which the Mosaic Torah we know is only a particular instance, adapted to the specific temporal circumstances of its reception by Moses and the Hebrews who had just been freed from bondage in Egypt.[134] In this chapter (Prov. 8:30–31), Wisdom's relationship both with the Creator and with creation is pictured as a game, designated by the Hebrew words *sha'ashua'* and *śeḥok*, two terms for the delight of playing, with the respective connotations of amusement and laughter. The Torah as game and pleasure (*sha'ashua'*) can also be found in several verses in Psalms;[135] the sexual overtones of this link between the human mind and divine wisdom have inspired many commentators. But Cordovero expanded this notion into a myth of origins that describes, in the manner of a Neoplatonic procession, how the Infinite, one and undifferentiated, produced, stage by stage, the possibility of another, whose progressive diversification is the origin of the hierarchical multiplicity of the ladder of beings, worlds, and things. As a myth of origins, this myth of play and Torah in some ways complements the myth of the *ẓimẓum*. It tells of a cosmic diversion that begins as the game played

131. The *ẓimẓum* is the initial withdrawal of the "simple infinite light" into itself so as to leave room for a dark void in which the world could be created. See, for example, Scholem, *Kabbalah and Its Symbolism*.

132. Bracha Sack, *The Kabbalah of Rabbi Moshe Cordovero* (Hebrew) (Beer Sheva: Ben-Gurion University of the Negev Press, 1995).

133. Cordovero interprets the game with Wisdom as an allusion to the game of the Infinite that is at the origin of all existents, but also with them:

> The aspect of the Ein Sof vis-à-vis the existents is beneficent in two respects. The first is that the Ein Sof exists for them wherever they are and permits them to exist and survive; this is the secret of the expansion (המשכה) of the infinite light on them. But this is not a game (*or* plaything, source of delight; Heb. *sha'ashua'*), because [the *Ein Sof*] is beneficent to them on account of its own good and draws near to them on its own account; whereas the existents, from their own perspective, are remote. . . . The true game is that the lower effects [i.e., objects in the lower world] repair themselves, rise up from the depths of their condition, and are refined until they draw near their cause and all together receive its good in a perfect equality, so that the *Ein Sof* is in *Malkhut* [Sovereignty, the lowest or tenth *sefirah*] just as it is in *Keter* [Crown, the first *sefirah*] and all the *sefirot*. Then it is the game and absolute good that it emanates (מטיב) to all else. It is with regard to this secret that we read, "I was his delight" (Prov. 8:30.). There are many sorts of delights, because whatever was distant and drew closer is a greater source of delights and its appearance to the face of the emanated good is a greater novelty. This will be when the existents are refined and all together constitute a single body and single *sefirah*. They are refined until they return to the existence that was theirs at the time when existence was that of the primordial Torah. (Moses Cordovero, *Sefer 'Elimah*, chap. 17, p. 176a; quoted by Sack, *Rabbi Moshe Cordovero*, p. 172)

134. See, especially, how the Maharal of Prague explains the curious expression, "Moses received the Torah *from* Sinai," in his commentary on M *Avot* 1:1. For him, Mt. Sinai, which is referred to as the "mountain of Elohim," just as Moses is the "man of Elohim," was a necessary intermediary, in the process of revelation, between the potential infinitude of the Torah as given and the finitude of the Torah as received, which is the "Torah of Moses" (see Chapter 1, n. 87). See also Scholem, *The Kabbalah and Its Symbolism*, chap. 2.

135. In Psalm 119 we encounter the main elements of this story, namely the psalmist's delight in the Torah of YHWH, overlaying an alphabetical acrostic. This very long poem praises all facets of the Torah, evoked in many different guises (law, witness, ordinances, commandments, paths, judgments, justice, etc.) in its 176 verses. The pleasures or delights (*sha'ashu'im*) of the game with the Torah are celebrated several times: "I take delight in your laws" (v. 16); "Your teaching [Torah] is my delight" (v. 70); "I will delight in your commandments, which I love" (v. 47); "I have longed for your deliverance, O YHWH; your teaching is my delight" (v. 174). Superimposed on this explicit content, the acrostic refers to the same experience, but in its abstract and timeless dimension, as suggested by the associations of

by the Infinite in itself, by itself, and for itself. But in the absence of a partner the game is unsatisfactory; whence the need for the Infinite to call into existence something other than itself, "outside" itself, although still "in itself," as indicated by the designation of the first world, *'azilut*, traditionally rendered "emanation" and related to *'ezel* 'alongside, with'. Thus the game consists of bestowing good, of pouring out a stream of light to make this other exist in all possible forms, however diversified and distant from their infinite source. But this is not the end of this game, which can conclude only through the reverse movement—the Neoplatonic "conversion"?—by which the other, now separate and self-subsistent, returns to be reintegrated with itself and with its infinite source.

Yehuda Liebes has summarized this myth as follows:

> The cause of creation is inherent in the Holy One blessed be He's need for laughter and play with a sexual quality. To fill this need he—the Holy One blessed be He—needed entities other than himself. At first this game necessarily takes place in the depths of the unity of the Emanator-deity, but by means of this game entities are created that increasingly take on independent being, rung after rung, and thus assume a more active role in the game and amusement, which becomes a mutual and dialectic erotic game. The object with which the game is played is the Torah, which exists in every rung of existence at a different spiritual level, and is also responsible for the existence of that rung of existence. This game reaches the acme of its perfection in the level of Israel, and especially in the level of the mystics who know how to raise the Torah, along with the existents linked to it, level by level, and to return them to their divine source; for God's erotic game has a dialectic character; it is built like heartbeats and encompasses rungs in which the entities are separated from the divine unity and rungs in which they are reincorporated into it.[136]

The account of the different stages of the game, first of the Infinite with itself, then between the Infinite and the other that the game brings into existence, is another way of expressing the tension between the theistic creationist aspects and the pantheist immanent aspects of many kabbalistic texts. Various scholars have noted that this tension is especially prominent in Cordovero.[137] Viewed by many as one of the most precise and rigorous kabbalists, Cordovero devotes much of his magnum opus, *Pardes rimmonim* (Garden of Pomegranates), to this problem and to apparently formal questions about the nature of the *sefirot* vis-à-vis the *Ein Sof* (the Infinite) and the production of *Keter* ("crown"), the first *sefirah*, which is also sometimes referred to as the *Ein Sof*.

The essential role of the Torah in each stage of the process is one of the most striking features of this myth. In particular, during the initial stage—which is also the ultimate stage—when the game is played by the *Ein Sof* in itself, the primordial Torah serves as the first "other"

the letters of the alphabet. The psalm consists of twenty-two strophes, each with eight verses that begin with the same letter of the alphabet. Thus each letter is repeated eight times at the start of each verse of that strophe. As we shall see, the primordial Torah initially produced by the game, the "garment of the Infinite" through which the Infinite can be attained, is also woven of the letters of the alphabet and the infinite meanings they produce in their multiple combinations.

136. Yehuda Liebes, quoted by Sack, *Rabbi Moshe Cordovero*, p. 141, n. 7.

137. In addition to the works by Scholem and Idel already cited, see, for example, David Cohen (known as "the Nazirite" and the leading disciple and editor of the works of Abraham Isaac Hacohen Kook) in his *Kol ha-nevu'ah*; see also J. Ben Shlomo, "Rabbi Moses Cordovero's Theory of Divinity," Ph.D. dissertation, Hebrew University of Jerusalem, 1965 (Hebrew).

that the game produces "outside" the *Ein Sof*. But this primordial Torah is clearly different, at least in its appearance, from the text of the Mosaic Torah that we know as the Pentateuch. An especially vivid description can be found in a work by the seventeenth-century German kabbalist Naphtali Bacharach, *'Emeḳ ha-melekh* (Valley of the King), inspired by the doctrines of both Cordovero and Isaac Luria.[138] In its first part, *Sha'ashu'ei ha-melekh* (The King's Games [*or* Delights]), the author describes the process of *ẓimẓum* by which the undifferentiated Infinite that "fills all" contracts and withdraws into itself to produce a void that the worlds can occupy. Preceding and setting off this process, however, is a *sha'ashua'*, a game "from parts of itself to itself" in which "a power of estimation [i.e., of measurement and limitation *(shi'ur)*] is born from the engraved letters, which is the potential Torah, latent in the alphabets."[139] He then refers to the beginning of the *Zohar* on Genesis, where the very first movement is referred to as the "engraving" of the letters of the royal seal. As for what triggers the *sha'ashua'*, the game itself, Bacharach repeats the classic phrase (mentioned above) used to answer questions about "pre-Creation," that is, why everything began: "In the beginning the game rose into his undifferentiated will." This is a paradoxical expression, as we have seen, because it presupposes something "below" from which it can "rise." Bacharach himself explains this, referring to the post-Creation world and in particular to the righteous who will inhabit it and who will arouse desire above by their own arousal below, in the manner of the woman's climax (*mayin nuqbin*) that precedes and triggers the man's climax. He offers an even more vivid description of the details of this initial differentiation as a trigger that generates "points" in the undifferentiated light:

> This contraction is produced by a movement of joy, as in a man who, in his great joy, laughs and quivers; that is, it sparkles and makes the points shine like lightning, and then connects the points with one another until they form letters in their image and semblance, as his wisdom (may He be blessed) decreed. There is no place to ask, with regard to this semblance, why the Infinite created the form of the letters in this semblance rather than another, because the first world is very much hidden and their meaning will be uncovered only in the future, by Rabbi Akiva son of Joseph, who disclosed the meaning of the crowns on the letters, which are the secret of the light of the garment of the Infinite. They are called crowns in relation to the letters that are beneath them.[140] As for the light of these points, how they expanded and flowed so that form of the letters was constituted from them, this was revealed by the prophet Elijah to our master, the light of the seven days, Rabbi Isaac Luria, whose words can neither be added to nor subtracted from.[141]

Here again we encounter the theme of the Torah as the garment of the Infinite, that is, the very first combinations of letters mentioned by *Sefer Yeẓirah*, the different ways of forming "231 gates" by pairing each of the twenty-two letters of the alphabet with all the others. It is from these letters and letter-pairs that the Torah is produced "above" as the source of everything that exists "below," where the Torah takes on other forms, including the one we know, in which the letters

138. Naphtali ben Jacob Elḥanan Bacharach, *'Emeḳ ha-melekh* (Amsterdam, 1653; repr. Jerusalem, 1994).

139. Ibid., chap. 2 (fol. 2ba).

140. This alludes to the talmudic legend that Rabbi Akiva expounded new teachings from the serifs ("crowns") that God attached to the letters of the Torah, teachings that Moses himself did not understand, even though they were attributed to him (see above, n. 112).

141. *'Emeḳ ha-melekh*, chap. 3 (fol. 2bb).

are no longer "arranged according to the alphabet" but according to the combination of words that are linked together. On the contrary,

above there is nothing—only letters of light that exist in the secret of the appearance of Adam . . . in the form of male and female. . . . And this image and semblance descend through all the worlds, even the world of the angels, . . . some of whom have the form of men, and some the form of women. But they are all only letters, which are nourished by and take delight in the light of that paper, that is, in the light of the essence that His regard and influence [i.e., His providence] project on them, may He be blessed, through many partitions and divisions.[142]

In *Pardes rimmonim*, Cordovero alludes to this primordial Torah in a slightly less metaphorical fashion and analyzes the several levels of studying and understanding it. He draws on a passage from the *Zohar*[143] about the inner meaning of the Torah, which lies beyond the narratives and the laws. Four levels permit access to this meaning, beginning with study of the biblical text, then of the Mishnah, then of the kabbalah. These are the inner levels of the Torah, which the *Zohar* compares to garments layered one over another and that one discards, one after the other, as one advances in knowledge. Under the garment of the written text is its "body" (the laws), which is a garment for its "soul" (the hidden meaning of which the kabbalah speaks), and which is in turn a garment for the "soul of its soul," which is "the spiritual essence of the letters, their existence, their combinations with one another and their affinities, so that someone who descended into the depths of the matter would be able to create worlds.[144]

This structure of successive layers is the same as that of the human body and of the system of *sefirot*, such that "the Torah itself is a garment for the essence of the Divine, in the same way that the *sefirot* are the garment for the essence [of things]."[145] The fourth level can truly be attained only in the world to come and brings "the joy hoped for after the resurrection."[146] This is why the *Zohar* does not even give it a name. At this level, even more universal and more inward than *Sod*, the hidden meaning of the kabbalah, the Torah is reduced to the twenty-two letters of the alphabet and to all of the permutations and combinations that set them in motion. As such, this inner level of the Torah is the closest to the infinite light of the *Ein Sof* for which it is the innermost "garment."

Drawing on the passage from *'Emeḳ ha-melekh* quoted above, we can portray the passage from the one and undifferentiated Infinite to the first entity that exists both in itself and outside itself as a movement from a static state of rest, where nothing happens, to an incipient motion that may be compared to the quivering laughter that produces the initial source of all heterogeneity, pictured first as points and then as lines that form letters. But this image is permanently linked to that of the *sha'ashua'*, the erotic game and laughter of the primordial man, both male and female, whose form seems to direct these movements, even before Adam and Eve and all other creatures existed as independent entities.

142. Ibid., chap. 4 (fol. 3aa).
143. *Zohar*, Numbers 152.
144. Moses Cordovero, *Pardes rimmonim*, §27, Letters, §1. Here Cordovero is alluding, inter alia, to the creation of golems and other living beings by means of combinations of letters and the use of *Sefer Yeẓirah* (see Chapter 1).
145. Ibid.
146. Ibid.

In another work, *Shi'ur ḳomah*, introducing a thirty-four-page explanation of the word *Torah*, Cordovero transfers this story to a distinctly human context (he calls it a *mashal*, a metaphor or parable).[147] This is a narrative full of fantastic details, set on a remote island that is home to an extremely wise man who plays with and derives pleasure from his wisdom, constructing many buildings and chambers of precious materials, one below another, and creating many species of birds whose song ravishes him and who accompany him as he moves through these rooms. This leads into the author's long, literal, and detailed exegesis of his own story, transforming it into a philosophical tale. The same theme of play, *sha'ashua'*, is repeated several times in this book, notably in chapter 54, with regard to the word *hashgaḥah* 'providence' (in the sense of both "observing" and "controlling"), as a sequel to the "arousal" (*hit'orerut*) of desire; and then in chapter 62, on the question of the *shi'ur* 'measure', as a sequel to the "material form."

This myth, like all myths of origin, does not just provide a naïve popular image, vividly anthropomorphic, of the origins of the world. It also attaches meaning to our present experiences by locating them in a timeless structure that establishes relationships among them.

The Torah, cast in these accounts as primordial wisdom, *also* refers, in one way or another, to the Torah of Moses and Israel that we know, with its stories and laws. In its universal and abstract root it refers to the infinite game of the permutations of the letters before it is constrained by the limits of speech. As such, it is the "garment of infinite divinity" and the "Torah of *'aẓilut*," to be distinguished from the one we know, the "Torah of creation (*beri'ah*)," which concerns the things of our own world. But the two are not really separate, because the latter serves as the entryway to the former, as explained, for example, by Naḥmanides, who wrote about the hidden meaning of the "Torah of *'aẓilut*" in the introduction to his commentary on the Pentateuch. As Bracha Sack notes, Cordovero insists on the

> essential internal link between the Torah in the possession of Israel and the Torah at the level of divinity, [because] the Torah is one and unchanging. . . . The emphasis that the Torah of Israel is the Torah of *'aẓilut* may imply a cautious attempt to attenuate the conclusions that might be drawn from the idea of the Torah of *'aẓilut* and the Torah of *beri'ah* [creation] held by the author of the *Ra'aya' meheimna'* and *Tiḳḳunei ha-Zohar* [works usually appended to the *Zohar*]—conclusions that might lead to neglect of the actual Torah given to Israel, with all its stringencies.[148]

However, the absence of an imposed and mandatory univocal meaning in a text that, at its source, is composed only of possible combinations of letters allows (at least in principle) almost unlimited creative and original exegesis based on reconstruction and decomposition of the text. For the scholar-sages who devote themselves to the study and renewal of this text, interpreting it means selecting one sense out of all the possible senses, produced by some combination of letters, which can always be linked to the familiar canonical written text by means of permutations and numerological equivalences.

To put this another way, human participation in the infinite divine game, in the form of new interpretations of the Torah and its elevation to its source, implies the initial descent by the Torah into the concrete and divided world of human action. This is why angels may not take part in

147. Moses Cordovero, *Shi'ur ḳomah*, §13, "Torah," chaps. 1–50.

148. Sack, *Rabbi Moshe Cordovero*, p. 144.

the infinite *sha'ashua'*, as Cordovero underscores; they are not bound by the Law. Their status is fixed once and for all; although initially it may be loftier than that of human beings, they cannot alter it and rise higher to reach the common source of the Torah, the *sefirot*, and Israel, where the infinite *sha'ashua'* attains its fullness.[149]

Now we can understand the metaphor of the Torah of letters as the "garment" of the *Ein Sof*. These infinite possible readings and meanings available for interpretation are in fact a ritual of dressing up, with the double function of concealing and revealing. The one and undifferentiated Infinite is hidden and unattainable under its garment of letters but allows itself to be "read" and even understood through them.

15. INCORPORATIONS

In this context we see how the concrete practice of Torah and the precepts can serve as a lesson in freedom, both individual and collective, whose degree of success clearly depends on the "elevation" attained.

One aspect of this lesson, both mental and physical, makes the expression "spiritual exercises" just as inadequate as it was in ancient materialist or physicalist philosophies such as Stoicism and Epicureanism. The role of physics as the science of nature—that is, of bodies—is unmistakable in the content of these exercises by philosophers whose activities were not limited to teaching in the Academy, the Lyceum, the Stoa, or the Garden, inasmuch as they could be those, for example, of Marcus Aurelius, an emperor who was the disciple of a slave, Epictetus. Here we reach the limit where ostensibly "spiritual" exercises are in fact exercises of the body to strengthen the mind, while, vice versa, exercises of thought—of memory, reading, repetition—fortify the body through the habits they instill. The monotheistic religions derived from biblical Hebraism reinvented, each in its own way, these "spiritual" exercises that always involve words, thoughts, and deeds.

The Christian dogma of the incarnation, which seems to pick up this theme, is probably one of the most obvious roots of Christian kabbalah in Jewish kabbalah. The difference is that in the former it is the incarnation of divinity in the person of God (the body of Christ), whereas in the latter it is an incarnation of the divine world in human bodies and persons. This produces contrasting attitudes toward our own body. In one case the movement is from below to above, of return, facilitated by the incorporation—the ingestion—of the person of Christ; the life of the human body partakes of that of the divine body, with the risk that the former may be disincarnated by the latter. This attitude finds its extreme expression in the human in-corporation par excellence, when a new body takes shape and develops in the body of a pregnant woman. This "secret of gestation" (*sod ha-'ibbur*), transposed from the human sphere to the divine worlds, is addressed by a kabbalistic literature that is at least as extensive as that about spermatic knowledge.[150] For Christianity, it is appropriate that the incarnation of the person of God in a child conceived and carried in the body of a woman must itself be conceived as effected without the participation of a male

149. Ibid., pp. 144–146.

150. An idea of its magnitude can be had from the works of Gershom Scholem and from Idel's *Mystical Experience in Abraham Abulafia*, pp. 203–225. On the "pregnant year" (the Hebrew term for a leap year) and application of this "secret" to the cycle of the seasons, see Atlan, *SR* II, chap. 9.

body. It remains, at least in part, disincarnated, as the act of the Holy Spirit, the only male partner of that conception.

In the rabbinic Judaism created by the talmudic sages, by contrast, the movement is from above to below, with the projection of divine life onto the various movements of our bodies, but also with a risk of obsession, that is, confusion of the relative and the absolute. Thus in both cases there is a danger of idolatry—of an abstract divine person (the personal God of religion) for one, of a multiplicity of deities incarnated in human activity for the other. It is perhaps this double danger that another approach, obedience to the one God as carried to the extreme by the religion of submission, Islam, seeks to avoid; but it may run up against the new peril of combining both idolatries.

Put another way, the monotheistic religions have never been totally freed from their fixation with idolatry. As a result, perhaps paradoxically, the demonstrative polytheism of Far Eastern and animist traditions, as well of the ancient Greeks, has much to teach us about oneness. Note, however, that there is no Jewish *Kama Sutra* or Tantra yoga, and that Dionysiac cults and sacred prostitutes have no place in Jewish rites, with the possible exception of the post-Sabbatean heresies (the Donmeh in Turkey and Frankists in Germany).[151] As Rabbi Naḥman of Bratslav, comparing rabbinic mysticism to other mysticisms of illumination, all of which rise toward the sublime peaks of ecstasy, is supposed to have said, "we have specialized in descent."[152] Perhaps this is linked in some fashion to the character of the ancient Israelite sanctuary, limited to a single temple, to a unique location on a hill in Jerusalem, and to the confinement of the erotic image of the embracing angels—the cherubim—to the innermost chamber of this temple.[153]

In any case, the emphasis that some authors place on the multiform image of the erotic game that is the origin of all things—the game of the Infinite with itself, of the Infinite with Wisdom, of Wisdom with human beings, a game played by the nonhuman with the human, by the personal with the impersonal, by the finite with the infinite—indicates that, with regard to freedom, individuals who are aware of the absolute natural determinism that governs their body and mind need not be resigned to their fate. The "possibility given," the "power"[154] to choose, is tantamount to permission to play at being free. This permission, though purely ludic and thus unreal, nevertheless institutes an arena of real play,[155] a place

151. See Gershom Scholem, *Sabbatai Sevi, the Mystical Messiah (1626–1676)* (Princeton, NJ: Princeton University Press, 1973).

152. Atlan, *Enlightenment*, pp. 347–348.

153. In the Holy of Holies—the innermost chamber of the unique sanctuary, the Temple of Jerusalem—which was entered only by the high priest and only once a year, on the Day of Atonement, two golden cherubim embraced above the Ark that contained the Tablets of the Law. According to R. Kattina, at the three pilgrimage festivals (Passover in the spring, Shavuot at the beginning of the harvest season, and Sukkot in the autumn) the curtain was drawn aside so that the assembled pilgrims could view the embracing cherubim. "Look!" they were told, "you are beloved before God as in the love between man and woman" (B *Yoma* 54a). Among the various human activities that were ritualized in the sacred precincts—assembly, singing and dancing, reading and studying the Torah—the table of the shewbread, in an inner courtyard, represented food; even further inside, as in the most sacred place of any house, the embracing cherubim represented Eros. (See: Roland Goetschel, "Les métaphores du chérubin dans la pensée juive," in *L'Interdit de la représentation* [Paris: Le Seuil, 1984], pp. 89–104; Bernard [Dov] Hercenberg, "La nudité et le caché," in *Le Corps*, ed. Jean Halpérin and Nicolas Weill [Paris: Albin-Michel, 1996], pp. 141–162.) Similarly, in the Song of Songs, which, according to Rabbi Akiva, is the holiest text of all, Eros represents the love between the people of Israel and its god.

154. The Hebrew word *reshut* can mean "authorization," "permission," "authority," "power" (in the sense of sovereignty), and even "domain" (see next note).

155. This arena is duplicated in time and space by the laws that regulate the carrying of objects from one place to another on the

Sabbath. The talmudic notion of the "rest" that is appropriate to this day includes a ban on moving objects from one domain to another. These domains are defined spatially as those areas under the control of the individual and those under the control of the collective—the private "domain of the one" (*reshut ha-yaḥid*) and the public "domain of the many" (*reshut ha-rabbim*). Along with all of the rituals that define and institute the "rest" of the seventh day, application of these rules makes the Sabbath an arena for a sort of timeless game, an image of the world to come, in which we can perceive the eternity of a "future" world that is already here. Slightly more than twenty-four hours a week are "stolen" from passing time, from the time of the hourglass, in which the sand (Hebrew *ḥol*, which also means the "profane," as opposed to the sacred, outside time) flows imperturbably. As such the game is doubled; the spatial relationship between the inside of the individual subject (the *yaḥid* authorized to exercise its power) and the social and natural outside of the power of the other, in the plural (the *rabbim*, the multiplicity of other persons and of the forces of nature) institutes a temporal alternation between the "domain/power of the one" and the "domain/power of the many."

for exercising a sovereignty and authority that strongly resemble a child making believe that it is the master of an empire, the famous "kingdom within a kingdom" that Spinoza mocked (start of *Ethics* III), which we imagine to be real and absolute even though we really do not yet know enough about it.

Rather than resignation, the metaphor of play reflects the responsibility that is determined for us, a responsibility we have not chosen, as in a game we are forced to play. The prize is the possibility of humanizing what is initially given as impersonal. The game is passive, because imposed on us; but the better we know its rules—the more aware we are of "what the body can do"—the more active is our part.

APPENDIX: THE DIALECTIC OF ABSOLUTE DETERMINISM AND CHOICE IN THE *BOOK OF KNOWLEDGE* AND IN ECCLESIASTES

The answers to the question of whether choice remains possible in the context of absolute determinism, suggested by Rabbi Shlomo Eliaschow in *Sefer ha-De'ah*, are clearly informed, like all his work, by the rich mythology of the Midrash, reinterpreted and rationalized through the interpretive framework of the Lurianic kabbalah. Pursuing as demythologized a reading as possible, we can recognize several elements or perspectives, different in form, that address different levels of comprehension and experience, both of our exercise of free choice and of the absolute determinism of nature.

Beyond the distinction between the essential (that is, the universal), which is subject to absolute determinism, and the details (the particular), which may remain objects of choice, Eliaschow also refers to the classic distinction between the potential, which is predetermined, and the actual realization, which results from a "choice" by a creature. What is more, we see that both the "details" and the "actual realization" that is "chosen" are produced by the sequence of determining causes in which they are embedded and which is realized by these choices. The excerpts presented below give some idea of this complex dialectic of determinism and the free choice that translates it into reality. A major difficulty is that Eliaschow, like most kabbalists, employs language in which a mechanical description of the impersonal processes "of emanation, creation, formation, and implementation" from the undifferentiated and impersonal infinite One are mixed up with an anthropomorphic theistic description of creation by the will of a personified "emanator." We have seen, however, that the use of theistic personifications, qualified by expressions such as *kivyakhol* ("as it were" or "if one may say so"), need not mislead us and that, as Scholem writes, in fact it conceals a notion of the infinite God as immanent in the nature of things and the nature of beings.

Here, then, are several extracts from this book, which has not yet been translated from Hebrew.

It follows from all this that even [with regard to] those things that happen and are done by choice, nevertheless the essence of those things themselves is [that they happen] by destiny, including the entire affair of the serpent and the extreme pollution that emerged and was made in the world, from which derive all the impurity and evil and travails and the great and terrible evils that are done in the world since the day that God created man in the world and until the coming of the Messiah. For even though all of them were caused and came about as a consequence of the sin, which was by the man's choice and schemes, nevertheless at the core of all things are the decrees of the Emanator on high, may His name be blessed, and from his actions.[1]

But it is not only a matter of the "essence of things." The details of events are also determined:

We learn from all this that all of the terrible events that have happened to us since we became a nation and until the present—all is caused by the world of *tohu* and all that is done [*kol ha-ma'aśeh*] on the earth was arranged and fixed from that time. There is nothing new under the sun because all is arranged from the eternity that was before us and everything is done in its season and time, in that time and season set for it.[2]

Ultimately, all of this nevertheless takes place through the "free" choice (*beḥirah*) of created beings, including the serpent! Eliaschow cites many commentaries about the prelapsarian serpent, "the king of the animals" who was not bad—quite the contrary—and was predisposed to evil only by the rigor (*din*) from which he proceeds and whose lofty root is in "the holy serpent, above, above, in the Holy of Holies—that is, in superior knowledge." This is also the common root of Pharaoh and Moses, to the second of whom the serpent corresponds in the animal kingdom.

From the aspect of knowledge, the serpent was the Tree of Knowledge, good and bad, in the animal part of creation; this is why it is said of it that "the serpent was the most cunning of all the animals of the field," because it originated from the secret of knowledge. And this is also what they said in the Midrash (*Gen. Rabbah* 20:5), and in the Gemara (B *Soṭah* 9b), namely, that before the sin the serpent was the king over all the animals and beasts and walked erect like a human being, because its root is of the aspect of knowledge and knowledge, as is known, extends from one edge to the other edge and rises upward over everything.[3] And because it is derived from knowledge, which has two parts, *ḥesed* and *gevurah*, consequently it also had the power of choice to go to one side or the other. But it was still not totally bad. Inasmuch as it was produced by virtue of the harsh rigors (*dinnim ḳashim*) and was a vessel ready and prepared for this, Sammael came down and clothed himself in it and went to Eve and seduced her. As it says in *Pirḳe de-Rabbi Eliezer* (chapter 13), all the serpent's wickedness was from Sammael, who clothed himself in it. It offers a parable, like a person who is possessed by a spirit of madness, and everything he does is not of his own accord but on account of the spirit of madness that is in him.[4]

1. Ibid., *Sefer ha-De'ah*, "Explanation of the Tree of Knowledge," Section 8, vol. 2, p. 296a.
2. Ibid., "Explanation of the World of *Tohu*," Part 2, Explanation 4, Branch 18, §3, vol. 2, p. 147ab.
3. An allusion to the constellation Draco.
4. Ibid., Branch 4, vol. 2, p. 80.

Thus the serpent's "power of choice to go to one side or the other," like Adam and Eve's ability to "choose" to disobey, does not mean that they were not manipulated, because their disobedience existed from all eternity in "the decrees of the emanator on high." Indeed, this world of *tohu* is one way of referring to the process of creation as if it were both free and determined, in that the coming-into-being of autonomous creatures necessarily implies their separation from their source. It is this separation that constitutes the reality of their "wickedness," which accompanies them even before they exist.

Eliaschow says this explicitly when he refers to the wickedness of the 974 generations who were both created and not created.[5] These primordial generations epitomize an evil whose possibility preexists the existence of created beings, because it is intrinsic to the process of separation (the *dinnim ḳashim* or "harsh rigors") by which autonomous creatures come into being; all the impurity and evil in this world was produced by the "wickedness of the 974 generations, which did evil of their own accord and volition."[6] These generations, which in fact represent only possibilities of existence, are themselves endowed with "choice." This is why they are "bad," because they "could have" chosen good. Eliaschow goes on to describe what would have happened in this eventuality:

> The 974 generations could have taken hold of their root and been strengthened with power and might to annihilate themselves before the divinity and unity of the one lord, the *Ein Sof*, may his name be blessed; . . . then his glory would have spread in his light through the entire space of the *ẓimẓum* and the *tohu* would not have been made at all, Heaven forbid, but the light of the reintegration (*tiḳḳun*) would have begun to spread at once. . . .[7]

To put it another way, although these potential separate existences were ipso facto seeds of evil as well, in theory they might have refrained from bringing this evil into existence by being reunited with their source, "annihilating themselves" so that they were no longer separate. But this is only a theoretical possibility and can be contemplated only in retrospect, as a rhetorical device; it implies the immediate suppression ("Heaven forbid") of the entire lengthy process by which separate beings had to come into existence. This is why the evocation of this possibility does not contradict the many affirmations that all the suffering in the world of *tohu*, as well as Adam and Eve's transgression, were determined by "decrees" made "on high," ineluctably contained in the "decision" to bring objects and beings into existence as autonomous realities. The result is a hierarchy of evil, whose levels correspond to different stages of the process. The possibility of apparently free choice is given each time, but the choice that is actually made is necessarily predisposed toward evil by the preceding stage. The first stage is the possibility of separate existence, represented by the 974 created and uncreated generations. The second stage is the impersonal evil that results from it in the world produced by the creation recounted in Genesis. This evil is that of all the approximations and imperfections of the created world: first of all the "disobedience of the earth," which, according to the Midrash, produced trees that bore individual fruits instead of trees that are entirely edible fruit, as initially ordained; and then the

5. See ibid., Explanation 2, Branch 1, §5, vol. 2, p. 20.
6. Ibid., Branch 17, §4, vol. 2, p. 141b.
7. Ibid.

cunning of the serpent that led Adam and Eve to disobey. The last stage is the specifically human evil produced by the sin of the Tree of Knowledge. In each stage, choosing good and life remains a theoretical possibility; but the contrary choice is determined by the effects of all the previous choices, which are themselves stages in the process of the separation and autonomization of the beings called into existence. The narrative, with its apparently temporal sequence, presents a hierarchy of the levels of evil. At each stage we can imagine that "it could have been otherwise" ("Heaven forbid") even though it could not in fact have been otherwise.[8]

It is in this sense that creatures "choose." Their status as autonomous beings condemns them to bear responsibility for what they are and what they do—that is, the tragedy of their separation and of the divisions that ensue—as if they determined themselves and their actions. In other words, free choice is a tautology when applied to beings that are postulated as necessarily autonomous and separate even before they exist, who are "created and uncreated." And the same status is transferred to the created beings that do exist; Adam's "choice" of sin was determined from all eternity by the "previous" choices of those created/uncreated generations, acting in and through him. The subjective experience of the choice to implement the action and actualize the potential of sin remains present, but it cannot modify in any way the eternal "decrees of the emanator on high." Between the 974 generations that exist only in potential and the human creatures that are "possessed" and determined by them, the same pattern is found in the intermediate hierarchy of determining causes and determined effects in the world of *tohu*, where we meet the serpent and Satan—and they, too, choose evil because it has been determined that they will to do so. The following passages, tracing existing things to the causes of their existence, which determine them in their being and in their actions, suggest as much.

> All events whatsoever, both in their whole and in their particulars, have their cause in the world of *tohu*. For every sin and every transgression are produced and caused by virtue of the pollution that adheres to the root of each human being when he descends into the world of *tohu*. This is the fermenting passion, the impurity and filth of every soul, that draws him ever on to do evil. Of this it was said, "the devisings of [*or* effects produced by; *yeẓer*, from *y.ẓ.r* 'make, produce'] man's mind are evil from his youth" (Gen. 8:21), meaning from his birth and from his root. This is the force of the natural evil that adheres to human beings from their birth and from their root.[9]

Eliaschow goes on to observe that the impulse toward good that also exists would have made it relatively easy to avoid this evil, had the effects of the sin of the Tree of Knowledge (which also has its roots in the world of *tohu*) not been added. He describes these effects as the creation of "a man who is depraved through and through (*'adam beliya'al kullo*)," a sort of "bad angel" who seizes possession of all creation. This angel is composed of many parts, each extremely powerful, which correspond to all the elements of good that can be found in all the worlds, in their totality, in their details, and in the details of their details, as Ecclesiastes says: "God has set the one [the

8. On this subject see also ibid., Explanation 1, Branch 5, pp. 33–34, where Eliaschow explains why this apparently free choice must be able "to go to one side or the other," according to different categories that we can conceive in the process by which the undifferentiated infinite One reveals itself in multiplicity, bringing into existence the four worlds of *'aẓilut, beri'ah, yeẓirah*, and *'aśiyyah* (Emanation, Creation, Formation, and Making), and the multiple beings that proceed from them.

9. Ibid., Explanation 4, Branch 18, §4, vol. 2, p. 148b.

day of prosperity] over against the other [the day of adversity]" (Eccles. 7:14). Ultimately, this depraved man

enters and spreads through [the world and] man too and unites with the evil force that adheres to him from his birth, as mentioned. Because he is a bad and most terrible angel it is very difficult for a man to overcome him. It is on account of him that it is written, "there is no righteous man on earth who does good and never sins" (Eccles. 7:20), for (see Rashi on B *Yoma* 20a) he makes him sin despite himself, in some sin or other.[10]

These texts are difficult to translate because several key words have multiple meanings, as we have already indicated. In particular, *ḥeṭ* ' 'sin' can also mean "mistake" or "slip-up," as in the case of an archer who misses the bull's-eye. Similarly, *ṭov* and *ra'* designate good and evil as moral categories, but also prosperity and adversity, happiness and unhappiness, good and bad, pleasure and pain, both felt and inflicted. In this last excerpt, Eliaschow supports his exposition by citing a comment by Rashi that must seem particularly harsh for those who are not familiar with this world in which good and bad actions are predetermined and where, consequently, salvation is not achieved directly and simply through the effects of these actions on their authors and their environment. Modern history offers abundant examples of the evil that human beings do without wanting to, believing that they are acting for the good of humanity and promoting justice and happiness. Just as the road to hell is paved with good intentions, the opposite may also be the case: the bad intentions that accompany transgressions—themselves perhaps determined by impulses—may produce salutary effects, for the agent or for others. In this case, inserting these effects into the chain of causes brings to light a striking "injustice," if it is only the intention that counts. This sort of uncoupling of the consequences of acts and the salvation of their agents is illustrated by Rashi's comment on a particularly edifying passage in the Talmud. From the histories of Saul and David—the former who lost his throne on account of a single transgression, and the latter who kept his despite two grave sins—R. Huna draws the following conclusion: "How little does he whom his master supports need to grieve or trouble himself!"[11] Rashi explains that one who is helped by God can be confident and need not worry that he will be struck by some punishment. This doctrine, like others we have cited, has provoked many questions among the theologians of a personal God who punishes and rewards, because of the implication that justice is not blind and that the verdict and sentence are influenced by whether or not a person is already and a priori "supported by God," whatever the rectitude of his or her actions.

10. Ibid.
11. B *Yoma* 22b.

But this involves attaching equal reality to absolute determinism and to the choices by which "creation" or "creatures" determine themselves to be "just" or "wicked," thereby adding the final element to their existence. Awareness of absolute determinism can lead to the erroneous belief that whether one is "just" (obeying the law and its obligations) or "wicked" (sinning, thereby producing evil and ultimately destroying oneself) makes no difference, because the final result is the same, including the details:

From the start of the creation of the world the Holy One blessed be He has been surveying the deeds of the righteous and the deeds of the wicked. . . . One might erroneously say that observing the Torah and the precepts and sinning and receiving punishment are equivalent, and that righteousness and wickedness are the same, because through both one and the other his will is accomplished.[12]

That is, by one path or another the hard rigors are weakened and the evil inherent in the creation of separate beings is finally destroyed, no matter what happens, when the latter complete their coming-into-existence. But Eliaschow, following the Midrash, insists that this is a misapprehension; the path of the just is nevertheless preferable to that of the wicked and only the former is really "good." He develops the idea that even though both paths lead to the repair of the shattered vessels and are thus equivalent from the perspective of the final result, they are not the same and are not equivalent with regard to the process of repair and reintegration itself.

Two kinds of repairs (*tikkunim*) are always at work, in every season and every moment. One is by all the tribulations and punishments in this world and in the world of the souls. . . . The second *tikkun* is through the Torah and precepts that are performed constantly by the community of Israel, in any case.[13]

In both cases, there is a process of "repair" by which the rigors of division come into existence, despite the separation of beings who have been divided from their source. It makes no difference what "choices" bring them into existence; their emergence from potential to actual is effected "by the creatures themselves, through the combination and refinement done by them and caused by them." Put another way, the individual's participation in the sequence of causes is what endows it with a separate existence, to the extent that this separation can still be "reattached" so that the dividing rigor can be "weakened" by the chain that links them. In both cases, the rigors that are the source of evil and unhappiness are weakened by the very act that brings them into existence. In the first case, evil is indeed produced, but only to destroy itself qua evil; what remains at the end is "refined" and "purified," like metal that has been smelted. In the second case, the separate act also takes place but is immediately constrained by the Torah and precepts that link it to everything else. Note that the path of the Torah and its precepts is not that of isolated individuals but of a community, whose members are mutually responsible for the success or failure of their journey. This community is called "Israel"; but this is a tautology, inasmuch as Israel is defined as a collective by the Covenant that binds together its individuals and clans around the Law and its precepts. This is an approximate historical definition (is Israel Israel?), but also ahistorical and archetypical, open in principle to every human being. Although the process of emanation and creation—that is, the coming-into-existence of separate beings—is ineluctable in both cases, the modes are quite different. Eliaschow clarifies this difference in several ways, appealing notably to Moses Cordovero's theory of "channels" (*zinnorot*). Schematically and less technically, one can say that the path followed by the "wicked" permits them to come into existence as a link in the chain, while the path followed by the "just" permits them to find their place in the infinite totality of being, which is structured like a tree. This "tree" will not be completed until the "end" of this infinite process; but in the meantime the just produce fruits that they too can enjoy, at the same time as they fertilize everything else that exists. The path of sin creates only the capital, but the path of the just creates both the capital and profits—the fruits of their endeavor.[14]

12. *Sefer ha-De'ah*, "Explanation of the World of *Tohu*," Part 2, Explanation 4, Branch 18, §4, vol. 2, pp. 149b–150a.
13. Ibid., Branch 19, §1, vol. 2, p. 154a.
14. Ibid., pp. 154–155. <This passage depends on the polysemy of the Hebrew *perot*, which means both "fruits" and "profits.">

These two ways of structuring the "tree" are represented by the metaphor of the "repairs made to the trees of life and knowledge."[15] The Tree of Life, entirely "good," is that of the "Torah and good deeds." The Tree of Knowledge, too, can be repaired and configured by the Torah and good deeds, but it can also be cultivated by the path of evil and the suffering it causes. This is why the Tree of Knowledge of "good and evil" is also the Tree, both "good and bad," of Knowledge.

At their root, however, these two trees are really only one, because

they are the two crowns of knowledge, the crown of generosity (*ḥasadim*) and the crown of rigor (*gevurot*), which in the root are one, both of them in the midline, that is, in the knowledge above, which is the inner knowledge. But when they emerge from knowledge and spread out with their branches, generosity goes to the right and rigor to the left, and both of them are sacred. But when they extend downward in the worlds of *beri'ah, yeẓirah*, and *'aśiyyah*, they become the Tree of Life and the Tree of Knowledge.[16]

These passages give us some idea of the dialectic of determinism and choice developed by the *Book of Knowledge*. In the language of personalized theism, we might say that both the just and the wicked perform God's will, but the former do so by linking into and merging with the organized infinite whole that proceeds from the undifferentiated infinite One, whereas the latter merely hold their place in the infinite sequence of causes. God's will—the bringing into existence of autonomous creatures, separate beings with the possibility of evil implied by division—is accomplished in either case. This is possible only if they derive their existence from themselves; and this is precisely the object of the possibility of choice granted them, so that they can themselves determine the nature of their existence. Finally, returning to the question of language—theistic and theological, or immanent and impersonal—the next excerpt from the *Book of Knowledge* clearly illustrates this association of personified language with an account of mutually determining impersonal mechanisms in the hierarchy of processes that endow objects and beings with their limited and separate existence:

We learn from all these midrashim that all the events whose cause is stated in the Torah—their sole first cause is the decree [*or* destiny]. Both the events and the causes are among the secrets of his direction of the world, his name be blessed. They are hidden with him from all eternity, that is, before the world was created. And the cause of all is in the world of the points, in their death and shattering, and in everything done to them in the world of *tohu*.[17] For it is in the world of points that there emerged the roots of everything that would emerge in the future in all four worlds, in their totality and in their details and in their most minute details, for all eternity. Their emergence is effected by the secret of the thought on high of *Adam Ḳadmon*, the primordial man, that is, by the hidden wisdom that is in him. And there the emanator was as it were thinking thoughts about all creation, how to found it and how to make it exist, as well as all the processions of its direction, in their totality and in their details, from the start of its root through the end of its final purpose. There he observed and watched everything and from there everything emerged by will, thought, and decree from him, may his name be blessed.[18]

But all this still does not explain how the choice of one path or another, the righteous path or the wicked,

15. Ibid., §2, vol. 2, pp. 155–156.
16. Ibid., Branch 19, §2, vol. 2, p. 155b.
17. An allusion to the Lurianic "shattering of the vessels."
18. "Explanation of the World of *Tohu*," Part 2, Explanation 4, Branch 18, §3, vol. 2, p. 148a.

each of which produces different local effects, is nevertheless determined at its source, even though its realization seems to be indeterminate. This difficulty is overcome by the nondeterministic role assigned to the choices in the mechanism of determinism. This role is restricted to the domain of "souls," to the exclusion of the domains of angels, demons, and worlds—that is, to the exclusion of everything that is concretely determined by the forces of nature. Only souls have the possibility of choosing "to go to one side or the other." Their choices do not determine their actual movements, however, but only accompany them in some fashion. Nothing essential is modified in the sequence of events. Only details of the structure of worlds will be damaged or repaired, depending on what the choice has impressed on the *mode* of realization—and not on its content—depending on whether it is effected intentionally and in a context of separation or reunion.[19] This reminds us of Spinoza,[20] for whom the action per se, which is a link in the chain of causes, is ethically neither good nor evil, even though it may be good or bad through its consequences. It is the intention that counts, as we say; that is, the image of self and others associated with the action. Here too there is a difference of principle between the "righteous" and the "wicked," although they are compelled to carry out the same action. The good intention of the righteous is equivalent to knowledge that there is a determinism by which the intention is realized, because their path is that of wisdom as well as of the good actions that flow from it and that they can predict. The bad intention of the wicked is merely the result of their enslavement to the chain of causes, of which they know little more than the passions that govern them. This is also why there are degrees in the freedom of the righteous and the enslavement of the wicked, depending on the level of knowledge and wisdom they have attained. Only full and infinite knowledge of absolute determinism would allow the righteous to be totally free. But this is impossible for finite and separate beings, because it would require them to be fully identical with the infinite from which they, like all other creatures, proceed. As we have seen, the result of this separation, inherent in creation itself, is that there is no righteous person who is immune to error and sin, despite himself. But as we have also seen, this does not mean that the choice of one path or another—even if, ultimately, it is only a matter of intention—makes no difference in the mode of being of the person who makes the choice. Despite appearances, this is not a fatalist outlook or a doctrine of predestination. In fact, the logical difficulties of this attitude are not truly overcome (no more so than for Spinoza) unless we understand determinism not as a temporal process that shapes the future as a function of the past but as timeless, intelligible only *sub species aeternitatis*, like a mathematical law. Our experience of choice is an illusion, from the perspective of the infinite and of eternity, but this does not mean it is not real from the point of view of our existence as finite beings in time. Eliaschow suggests this several times. For example, returning to the choices of "souls," who alone have the ability to choose "to go to one side or the other," and to the 974 created/uncreated generations that preceded Genesis, he quotes the *Zohar*'s observation[21] that the malefactors in this world, although they have the capacity to choose, are pushed to evil by their supernal soul "even before they come into the world."[22] Insofar as choice is a causal factor, its source is in the primordial eternity "before the creation of the

19. Ibid., Branch 17, §3, vol. 2, p. 140.
20. *Ethics* IV 59, Proof.
21. *Zohar*, Leviticus 61b.
22. *Sefer ha-De'ah*, "Explanation of the World of *Tohu*," Part 2, Explanation 4, Branch 17, §3, vol. 2, pp. 140–141.

world"; but insofar as choice is part of lived existence, it is not a causal factor. These two levels of knowledge and intelligibility, temporal and timeless, are equally real, although different in the experience we can have of them.

This vision of things, in which our knowledge and understanding of the world transcends our experience of the temporal succession of events, calls our attention to the statement by Ecclesiastes:

> He has made everything beautiful in its time; he has also put "world-time" (*'olam*[23]) into the minds of men, so that they cannot find out what Elohim has done from the beginning to the end. (Eccles. 3:11)

It is the sense of time, of the world as temporality, that Elohim has placed in our minds and that prevents us from knowing his works *sub species aeternitatis*, "under the aspect of eternity," "from the beginning to the end." It is this sense of time, too, that makes us believe our own efforts can determine the future. This is what Ecclesiastes—the wise King Solomon—understood in the evening of his life, when he reflected on the nature of "the business that Elohim has given the sons of man to engage in" (3:10). This "business" is precisely to lead our lives in time while acting in view of a future we think we determine. This is why, implanting the sense of time in us, "Elohim has made it so, in order that men should fear him" (3:14). The human condition, compounded of the sense of time, fear of the unknown future, and the belief that our actions determine this future, is a "bad business that Elohim has given the sons of man to engage in" (1:13). The word used in both 1:13 and 3:10 for "business," *'inyan* (and the related verb *la'anot be-* 'to engage in'), can be translated even more radically. *'Inyan* can also be understand as an "interest" (in the sense of attention or curiosity) by which God snares human beings (*la'anot be-* 'to show interest in'), as well as in the sense of "to test" them and even "to torture" them (which would be the homograph *le'annot* [לענות]). This is the message of a later passage (which employs a different Hebrew word for "test"):

> Elohim God will judge the righteous and the wicked, for he has appointed a time for every matter, and for every work. I said in my heart with regard to the sons of men that Elohim is testing them to show them that they are but beasts. For the fate of the sons of men and the fate of beasts is the same (*mikreh 'eḥad*). (3:17–19)

In other words, we are tested by this bad business in which we are engaged, even though, because of the time-based perception that is our lot, we can never discover what this business is, absolutely and from all eternity; that is, "what Elohim has done from the beginning to the end." It is still the same old *'alilah*, the same fraud that victimizes us as moral subjects who can choose in time and are held accountable for the effects of our choices. But the wise person can at least

23. <*Temps-monde* in Prof. Atlan's French rendering of the verse. When it was pointed out to him that although *'olam* means "world" in postbiblical Hebrew, this sense is scarcely attested in the Bible, where the meaning is generally closer to "eternity," he added the following note.> I understand *'olam* here as "world-time," meaning the whole of the universe as viewed in time rather than in space. The usual rendering, "eternity," is misleading, especially in this context, inasmuch as time comes into being with the world, whereas eternity is timeless, given "from before the creation of the world" or, as this verse puts it (in contrast to the *'olam* in the minds of men) "from the beginning to the end" (and not known to men). Spinoza emphasizes this point in several passages of the *Ethics*, notably in his references to *sub species aeternitatis*. This difference between time and eternity supports the interpretation presented here.

discover this truth, although it does not modify in the slightest his ultimate lot, which remains death and, like the beasts, the return to dust. Then the wise individual who, like Ecclesiastes himself, has set his mind "to seek and to search out by wisdom all that is done under heaven" (1:13) and has thereby become aware of this bad business that Elohim gave to human beings, ultimately understands that "there is nothing better for them than to be happy and enjoy themselves as long as they live; [and] . . . that whatever Elohim does will endure forever [i.e., in the future, in *time*]; nothing can be added to it, nor anything taken from it. Elohim has made it so, in order that men should fear him" (3:12, 14).

The flight of time is, in fact, eternity: "That which is, already has been; that which is to be, already has been; and Elohim will seek what has flown away" (3:15). Finally, after drawing various other conclusions, such as "wisdom is superior to folly" (2:13), we reach the peroration of the book: "The end of the matter, when all has been heard: Fear God, and keep his commandments; for this is the whole of man. For Elohim will bring every deed into judgment, with every secret thing, whether good or evil" (12:13–14).

Even here, in the Hebrew text of this verse (*'et kol ma'aśeh ha-'elohim yavi' ve-mishpaṭ*), we encounter the same ambiguity about what will be judged, the deeds of human beings or those of God, as noted by commentators who propose construing the sentence differently: "Every work of God will be brought into judgment for every secret thing, whether good or evil."[24]

Thus it is in the same movement, as it were, that God endowed human beings with a sense of time, a sense of free choice, and the sense of responsibility that stems from it. But rational knowledge and wisdom also allow us to arrive at the form of eternity that logical and mathematical determination suggests to us. This knowledge puts our experiences of free choice into the proper perspective, which in turn transforms the meaning of our responsibility.

24. See, among others, the interpretation by Rabbi Ḥayyim Volozhiner, *Nefesh ha-ḥayyim* 1.12.

CHAPTER 3

SPIRITS AND DEMONS: SUBJECT AND SHADOW

1. HIDDEN CAUSES

The search for the hidden causes of things and events is always ambiguous. The cause, being hidden, is not evident and unmistakable. In addition, there is always the question of whether the cause discovered by science or lore—a necessarily esoteric lore, because it concerns what is hidden—is in fact an efficient cause and not just a verbal explanation that supports a reason that only looks like a cause. This is why explanation by hidden causes is typical of both magical thinking and scientific knowledge.

In the science of antiquity, the Middle Ages, and the Renaissance, demons and spirits played the role of hidden causes; they were the elements of interpretive systems that could make sense of events and perceived objects only in the context of a set of processes that could generate them, although the spirits and demons themselves could not be perceived by the senses. Just as today, causes could be framed only in the abstract, as explanatory schemes or ideas, based on the observation of other events, perhaps correlated with the first set, that were interpreted as their effects. Thus the search for and identification of hidden causes seems to be a perennial part of the human enterprise of explaining, with the goal of acquiring control at some level and some degree of reality over objects and events in space and time.

Elsewhere I have tried to analyze the differences between magical explanations and scientific explanations.[1] These differences involve the nature of what are considered to be valid and legitimate causal explanations rather than the postulate of intelligibility itself. In all cases, we explain objects that we see, hear, and so on by means of other things, some of which can be thought of and produced only in our mind and can never be directly perceived. Genes in biology and elementary particles and fields in physics are abstractions of this

1. See Atlan, *Enlightenment*, chap. 4, with regard to "electrogenic demons," and chap. 5. See also Atlan, preface to *Le Golem*, on the similarity and differences between the spurious efficacy of magical formulas and the real efficacy of the equations of mathematical physics.

kind, although the ways in which these two sciences employ them in their explanatory schemes are not the same.

The simplest causal relationship between two events, of which one is a cause and the other the effect, has had a troubled and controversial existence since David Hume's well-known and devastating critique: we never observe a causal relationship, but merely infer it from the habit of repeated associations and from the routine (because pragmatic and effective) use of these regularities and repetitions, which is almost instinctive to human behavior.

For scholasticism and precritical philosophy, here following Aristotle, the only form of knowledge was through causes. A treatise like the *Liber de Causis* (Book of Causes) could exert a decisive influence on philosophy and theology in the Middle Ages.[2] But the critical philosophers, and later the positivists, from Hume through Kant and Wittgenstein, inquired about the nature and legitimacy of causal relationships as a mode of knowledge or category of reason. Among the positivists, Wittgenstein posed the question categorically: "There is no possible way of making an inference from the existence of one situation to the existence of another, entirely different situation. There is no causal nexus to justify such an inference. We cannot infer the events of the future from those of the present. Belief in the causal nexus is superstition."[3]

Under the influence of physics, the "queen of the exact sciences," which is often cited as the paradigmatic science of our modern age, the notion of cause has lost its explanatory and predictive value, to be supplanted by physical or mathematical law. Science has finally begun to separate itself from magic. In physics, explanation by causes, more or less occult, has been replaced by general laws that explain the progress of phenomena and the occurrence of their particular instances, starting from the conditions in which they take place and are observed. But this remains an ideal of formal science, which the other empirical sciences, notably the life sciences, have not been able to achieve so quickly. At the beginning of the twentieth century Pierre Duhem, relying largely on Claude Bernard, noted the vast difference between the status of fact and theory in mathematical physics and their status in physiology, where the search for causes remained the order of the day.[4] This remains the case today. Cer-

2. See: Etienne Gilson, *La philosophie au moyen âge* (Paris: Payot, 1976); Alain de Libera, *Albert le grand et la philosophie* (Paris: Vrin, 1990); W. Beierwaltes, "Maître Eckhart et le *Liber de causis*," in *Voici Maître Eckhart*, ed. Emilie Zum Brunn (Grenoble: Jérôme Millon, 1994), pp. 284–300; and Cristina d'Ancona Costa, *Recherches sur le* Liber de Causis (Paris: Vrin, 1995).

For medieval philosophy, knowledge by causes was scarcely distinct from theology, because it had to lead back, in one way or another, to the First Cause or cause of all causes. Through the *Liber de causis*, a compilation based on Proclus' pagan *Elements of Theology*, Muslim, Jewish, and Christian theology recovered that work and integrated it into the ambient Aristotelianism. As a result, one of the core problems of Neoplatonism, the production of multiplicity from the One, and its solution by means of a hierarchy of emanating causes continued to permeate philosophical and scientific thought until the seventeenth century. It was only following the mechanical revolution and the successive proclamations of the principle of sufficient reason (by Descartes, Spinoza, and Leibniz) that causality was severed from the theological context and "naturalized." The multiplicity of causes was reduced to one, the efficient cause. But this was possible only thanks to the mathematization of mechanics, which introduced a series of new problems, or rather new ways of posing the old problems. These, including dualism and the relationship between body and mind, still haunt philosophy today. While the natural sciences were developing in their modern, empirical and logical form, culminating in the effective operationalism that characterizes them today, theology, expelled from the academy, had to sneak back in through the window of idealism and vitalism.

3. Wittgenstein, *Tractatus logico-philosophicus*, trans. D. F. Pears and B. F. McGuinness (London: Routledge and Kegan Paul, 1961), 5.135–136.

4. Pierre Duhem, *The Aim and Structure of Physical Theory*, trans. Philip P. Wiener (Princeton, NJ: Princeton University

tain factors—once upon a time they were "humors"; today they are molecules—are regularly invoked as the determining causes of most biological phenomena. As such, they are not immune to Hume's skepticism about the reality of causal relationships, which are never perceived directly in nature; and their deduction by Kant as a category of understanding is not enough to establish their empirical relevance. When determining factors are identified, in the form, for example, of specific molecules, their causal role (which in any case is only partial) in the occurrence of a particular structure in or particular behavior by an organism is usually inferred only by means of correlations. Such induction may be reinforced when it leads to effective action through preventive or therapeutic experimental treatments whose logic is based on the hypothesis of the causal relationship in question. But an abusive generalization to which biological theory is not always immune leads to an attempt to explain all the phenomena associated with living beings by invocation of factors of the same type; the causal role of these factors or, worse still, of similar factors that are conjectured to exist but have not yet been identified is extended to all biological phenomena, while observations that cannot easily be fit into the dominant explanatory framework may be neglected. In that case, the accepted theory can hardly be defended against the criticisms of Hume or Wittgenstein. No one knows whether the sciences of the living, and even more so the sciences of human beings, will ever be as formalized as physics is. Should they attain this condition, it might then be possible to announce the final disappearance (tomorrow or the day after) of the spirits and demons that continue to populate our mental universe and to satisfy our (perhaps animal) need to explain everything at any cost by specifying the cause. The techniques of mathematical modeling, which rely in part on the theory of dynamic systems, give some hope that this may happen and these sciences too may be liberated from the pursuit of hidden causes, a quest that we have inherited, albeit transformed, from the magical science of yesteryear. Conversely, it may well be that mathematical and even cybernetic formalization will reach its limits and prove unable to model relatively complex and unique systems, notably because of the underdetermination of theories by facts.[5]

There is no doubt that modern science has broken with magical explanations, where causes are animate and personal and where, as for Aristotle, the final cause is the most important of all because it explains things by demonstrating their meaning. Nevertheless, the debate continues about the legitimacy of the vestiges of causality in contemporary science, even reduced to the efficient and formal causes, and about the difficulty (if not indeed the impossibility) of eliminating all causalism from scientific discourse.[6] Whatever its form, explicit and acknowledged or implicit and heuristic, the quest for the hidden causes of phenomena continues to inform and motivate

Press, 1991), pp. 180–183; see also pp. 26ff. on the secondary, parasitical effects of explanation in physics, compared to the representative portion of the theory, to which the latter owes its power and fruitfulness. Here I should mention the analyses that led me to conclude, in an earlier work, that "total explanation is mystical" (*Enlightenment*, p. 264) and that "explanation is a bonus in the sciences" (ibid., pp. 191–193).

5. Atlan, *Tout, non, peut-être*, pp. 130–164.

6. In this regard see Mario Bunge, *Causality and Modern Science*, 3rd edition (New York: Dover, 1979), which remains an excellent summary of the subject. <The long-needed update to take into account the new questions raised by recent discoveries in biology and molecular genetics, whose lack was lamented by Prof. Atlan in the French text, appeared in 2009.>

Positivists and postpositivists such as Arthur W. Burks and Herbert Simon have endeavored to formalize causal explanations themselves in order to save the intuition of causality while ridding it of prescientific elements. See: A. W. Burks, "The Logic of Causal Propositions," *Mind* 60 (1951), pp. 363–382; H. Simon, "On the Definition of the Causal Relation," *Journal of Philosophy* 49 (1952), pp. 517–528.

scientific activity and (perhaps even more so) technological research. In the absence of a definitive principle of explanation that can provide a *reason* for things, the *cause* is the only handle we have for acting on a phenomenon in order to make it occur or to avert it, depending on whether we consider it to be desirable or not. The scientific practices for transforming nature, such as contemporary biology and especially medicine, are always looking for such causes; so are the human sciences, which are supposed to cast light on social and political actions. In these domains the problems we have noted are overcome, to a greater or lesser extent, by a pragmatism that prefers operational efficacy to the completeness or sometimes even coherence of a causal explanation: a factor on which we can act and whose alteration modifies the phenomenon in question is considered to be the cause (generally partial) of a phenomenon, a physiological state, or an illness. Of course this must transpire with some regularity. The observed modification must be "reproducible," as we say, in similar situations that are as identical as we can make them. We can often evaluate this reproducibility only by statistical methods, with all their limitations and the interpretive problems they spawn. In other words, we admit the reality of "true" causes, which are both efficient and explanatory, while recognizing the difficulty of conceiving them precisely and adequately.[7]

Thus the quest for causes that still motivates the nonmathematical sciences, despite its claim to be modest and pragmatic, is marked by such a strong desire for action and meaning that it contains the same errors of judgment philosophers have been denouncing for centuries—the confusion between partial cause and complete cause, a disproportion of cause and effect,[8] and, more recently, a confounding of statistical correlation with causality.[9] The last of these has turned into a veritable disaster in biology and medicine, ever since statistical tests became the technique of choice for uncovering *the* cause of every illness and behavior—in short, of every event with an emotional content, positive or negative, that cannot be left to fate and the inexplicable, or even to the complexity of multiple causes and their interactions. From this point of view, even if the known or supposed mechanisms by which causes produce their effects are different in our contemporary scientific and technical discourse than they were in the natural magic of the past, the stakes of blind belief or critical distance remain the same when it comes to the hidden order and determining causes we discover in our knowledge of nature.

Phenomena that have been brought under only partial control are still explained by means of generalizations that tend to go beyond the limited scope of the effective empirical knowledge we can have of them. These global explanations then take on an air of the ancient invocation of spirits and demons (beneficent or otherwise) that are supposed to lie at the origin of events (fortunate or otherwise), or a whiff of the soporific virtues of opium, or the tone of the elusive speeches of Molière's physicians as they sagely explain "why your daughter is mute." Depending on the

7. This fits in perfectly with Spinoza's theory of knowledge, which borrows the scholastic adage that we can have knowledge only through causes but which distinguishes adequate from inadequate causes. An adequate cause is one whose existence is both necessary and sufficient for the effect to be produced. It is thus a total cause; the supposed effect can neither be observed nor explained in its absence. Inadequate causes are partial causes that by themselves neither produce nor explain an observed phenomenon.

8. Hume offers a brilliant analysis of this in his criticism of the "argument from design" (*An Inquiry Concerning Human Understanding*, Section XI).

9. Daniel Schwartz, *Le Jeu de la science et du hasard* (Paris: Flammarion, 1994). See also: the meticulous and still relevant criticisms by R. C. Lewontin, "The Analysis of Variants and the Analysis of Causes," *American Journal of Human Genetics* 26 (1974), pp. 400–411; Atlan, *SR* II, chap. 8.

fashion in medicine and physiology, it may be hormones that explain everything, as in the 1950s, or, as today, genes, as part of the ubiquitous notion of the "genetic program."

The notorious inadequacy of these global explanations, despite the undeniable but limited successes of the sciences and technology on which they are based, is regularly underscored;[10] but the temptations of superstition seem to be too strong. Fashionable effects are associated with it and scientific models or theories supply the food, generally poorly digested, for the ramblings of magical thought, as formerly, despite the modern vocabulary they employ.

2. SPINOZA AND THE SPIRITS

The learned notion of demons as the hidden causes of certain phenomena is very old. The Pythagoreans already asserted that "the whole air is full of souls, and that these are those which are accounted daemones and heroes. Also, that it is by them that dreams are sent among men, and also the tokens of disease and health; these last too, being sent not only to men, but to sheep also and other cattle. Also, that it is they who are concerned with purifications, and expiations, and all kinds of divination, and oracular predictions, and things of that kind."[11]

Until the seventeenth century the existence of spirits, animal and other, was commonly accepted. It was part (forgive the pun) of the spirit of the age. In this context, Spinoza's critique of the role of animal spirits in the union of soul and body is particularly instructive, because it directly concerns the status of spirits, ghosts, and demons, ancient and modern, in explanations of the phenomena they are supposed to cause.

Spinoza, despite his high regard for Descartes as scientist and philosopher, was nevertheless astonished by the latter's notion that soul and body interact through the pineal gland and animal spirits. He could hardly believe that "so great a man," "a philosopher who had stoutly asserted that he would draw no conclusions which do not follow from self-evident premises and would affirm nothing which he did not clearly and distinctly perceive, and who had so often taken to task the scholastics for wishing to explain obscurities through occult qualities, could maintain a hypothesis beside which occult qualities are commonplace."[12]

We also know his replies to his correspondent Hugo Boxel, who had solicited his opinion about ghosts and spirits.[13] Spinoza begins by mentioning that those who accept the existence of ghosts seem to hold that "something exists of nature unknown."[14] He continues: "If philosophers choose to call things which we do not know 'ghosts,' I shall not deny the existence of such, for there are an infinity of things, which I cannot make out." In other words, ghosts

10. The criticism of these notions can be traced back to the first systematic expositions of molecular biology as a general theory of the structures and functions of living organisms, such as: François Jacob, *The Logic of Life: A History of Heredity*, trans. Betty E. Spillmann (New York: Pantheon, 1973); Atlan, *L'Organisation biologique*; idem, "Du code génétique aux codes culturels," in *Encyclopédie philosophique*, ed. A. Jacob (Paris: PUF, 1989), pp. 419–430. It has recently been revived, in more acute fashion, in the wake of the media circus associated with the human genome project; see Henri Atlan, "Cartographie génétique fonctionnelle et séquençage de nucléotides ou programme 'Génome humain'?" *Médecine/Sciences* 8 (1992), pp. 262–263. See above, Chapter 1, n. 6.

11. Diogenes Laertius, *The Lives and Opinions of Eminent Philosophers*, "Life of Pythagoras" (trans. C. D. Yonge).

12. *Ethics* V, Preface.

13. Letters 55–60 (51–56).

14. Letter 56 (52). In another letter (58 [54]) Spinoza speaks ironically of stories in which "there are ghosts of every kind, but perhaps none of the female sex," repeating Boxel's own phrase in his previous letter (57 [53]).

and spirits are a pseudo-explanation of obscure things, of things one cannot fathom, by means of a hypothesis that is even more unintelligible. This is why his letters to Boxel endeavor to demonstrate the obscurity and unintelligibility of the qualities the latter attributes to ghosts and spirits and the absurd and superstitious character of the reasons given for accepting their reality.

But what is also interesting in this correspondence is one of the main arguments advanced by Boxel, who was asking Spinoza not so much about the existence of ghosts, spirits, and bodiless souls, which he doubted not for a moment, as about the properties that could legitimately and reasonably be ascribed to them. He concedes that "as for evil spirits, who torture wretched men in this life and the next, and as for magic, I believe the stories of them to be fables."[15]

15. Letter 57 [53] <corrected on the basis of the Latin>. In addition to the "learned Lavater," whom Boxel called to witness, we should mention a mid–seventeenth-century controversy that was simultaneously scientific, philosophical, and theological, associated with a 1664 trial for witchcraft in England and revived for contemporary readers by Thomas Harmon Jobe ("The Devil in Restoration Science: the Glanvill-Webster Debate," *Isis* 72 [1981], pp. 343–356). The protagonists on one side were Joseph Glanvill and Henry More, members of the newly founded Royal Society; their opponent was John Webster, a physician and Paracelsian. The dispute did not center on the reality of spirits—all accepted their existence—but on which scientific discipline should study them. For Glanvill and More it was a case of diabolical phenomena, which could occur only as a result of impure and illicit (sexual) commerce between spirits and matter, typical of demonic possession and sorcery, and to which women were particularly disposed. We must remember that in that age not all sciences enjoyed equal legitimacy from a theological perspective. Robert Fludd and Johannes Kepler clashed about whether use of computation in astronomy was legitimate or diabolical (see Atlan, *Enlightenment*, pp. 172–176 and 223–224). Although both Fludd and Kepler were raised on hermetic, alchemical, and kabbalistic science, Kepler had achieved notable success by employing mathematics to describe the motion of celestial bodies. For Fludd, however, only a "qualitative" geometrical mathematics was compatible with divine science. The division of numbers, as of matter, was unworthy of it. Similarly, Henry More, like the other Cambridge Neoplatonists, rejected Descartes' materialist mechanics and saw Spinozism as a materialist "mysticism" and perversion of the Hebrew kabbalah (see Chapter 4, n. 239). On the other hand, he defended the new mechanical science, which he harmonized with orthodox Anglican theology by means of a Platonic notion of the separation of the soul and body, with which he associated the concept of a "spiritual substance" supposed to explain the phenomena of all the various kinds of apparitions attributed to ghosts and spirits. By contrast, Paracelsian "chemists" such as Webster remained faithful to the alchemical tradition for which matter was no less pure than spirit. The "feminine" nature of its secrets did not entail an impure and diabolical science of a different sort than that with which the soul, in its divine nature, contemplated the work of the Creator. Like all other causes, spirits—the unseen causes of apparitions and mysterious phenomena—had to be the objects of normal science; there was no need to call in witches and demons. This was the background of the controversy that erupted after Glanvill, a philosopher and clergyman with a scientific bent, published a book about witches and witchcraft, *Some Philosophical Considerations Touching Witchcraft and Witches*, a bestseller when it appeared in 1666 and revised and expanded several times thereafter. Paradoxically, it would seem, the main attack on this book, which strikes us as sensible and as prefiguring the Enlightenment, at least socially and politically, was launched by Webster, a proponent of the old science and natural magic of the Renaissance. This was not the first time the old science and natural magic, which had sent Giordano Bruno to the stake, were accused of heresy and the smell of hellfire. It was precisely such charges against Webster that Henry More employed to defend Glanvill's book, adding his warrant as a member of the Royal Society, one of the crucibles of the New Science. The latter, to be sure, investigated material manifestations of spirits, now thought to be immaterial, but it did so in order to combat them, as a specific branch of science, that of the interaction with matter by the devil and witches, philosophically and theologically dangerous, and not to be confounded with the pure and disembodied science of God's manifestations in nature. In the preface of his *Antidote Against Atheism* (1653), More counseled that students of the spirit world take on the "plain shape of a mere naturalist" and combine observation with case studies. Following him, in 1668 Glanvill called on the Royal Society to conduct a systematic study of the spirit world so as to draw a map of the Country of Spirits.

> After considerable scrutiny, Glanvill satisfied himself that the existence of persons who made contracts with the Devil, had familiar spirits, transformed themselves into animals, and performed many "praeternatural things," was proved by the "Attestation of Thousands of Eye and Ear witnesses, and those not of the easily deceivable Vulgar only, but of wise and grave Discerners; and that, when no Interest could oblidge them to agree together in a common Lye." (Jobe, "The Devil in Restoration Science," p. 347, quoting Glanvill)

For Webster, by contrast, who was pursuing a natural philosophy, apparitions and extraordinary phenomena had to have causes that, although hidden and thus the objects of occult science, were nevertheless natural. Traditional science, meaning natural magic, had to make it possible to explain them without having to invoke either the devil or his subordinate

But he is upset with Spinoza's reply, which rejects the very existence of spirits, because this contradicts what has been accepted by the philosophical tradition since classical antiquity, as well as the opinion of "modern authors"; and "to contradict so many trustworthy historians, Fathers, and other persons placed in authority would argue a remarkable shamelessness." In other words, it would be mad to deny what the whole world accepts on the basis of established facts that have always been considered to be the effects of these bodiless souls. The only question, for Boxel, involves the enlargement of the domain of these effects; it is 1674, less than fifty years after a series of witch hunts had victimized peasant women almost everywhere in civilized Europe. Spinoza does not contest the observed "facts," but he is opposed to their false causal interpretation by a hypothesis that is more obscure than what it seeks to explain.

The question was sufficiently important for Spinoza to have taken it up in the lectures for his circle, published as the *Short Treatise on God, Man, and his Well-Being*. In chapter 22 of part 2 he seems to hold "[animal] spirits" to be self-evident realities.[16] But in chapter 25, "Of Devils," he takes up the question of false causality. After attacking the supposed properties of the devil and demons in order to demonstrate their incompatibility with the ontology and theology of a loving God who is one and perfect, he concludes by rejecting the supposed need to "posit Devils in order to find [in them the] causes of hate, envy, anger, and such passions. We have come to know them sufficiently without the aid of such fictions."[17]

demons: "Where there is no natural agent, there the Devil can operate nothing at all, and if there be a natural agent, his concurrence is not necessary" (ibid., p. 350, quoting Webster). Note that in this context, contrary to what we saw in Chapter 1 about the existence of "good" demons and their king Ashmedai, who are not bad "at their source," here demons, like the witches who are possessed by them, are thought of as creatures of the devil. As a physician, Webster diagnosed those who believed in material manifestations of the devil through the agency of demons and sorcery as suffering the disease of the mind and "black bile," known since Aristotle, that was called "melancholy." More and Glanvill countered that Webster himself was possessed by the devil, like his master Paracelsus and all adepts of his occult science.

Thus spirits and demons were still a very real presence in seventeenth-century science, as indicated by several titles of the "scholarly" literature on the subject, in which theological preoccupations with the interpretation of Scripture were clearly evident. In 1668 Glanvill published an expanded version of *An Antidote*, under the title *A Blow at Modern Sadducism*. After his death in 1680, More supplemented it with an unfinished work by Glanvill, "A Proof of the Existence of Apparitions, Spirits and Witches, out of Holy Scripture," and published the whole as *Sadducismus Triumphatus; or, Full and Plain Evidence Concerning Witches and Apparitions. In Two Parts, The First Treating of their Possibility, the Second of their Real Existence.* This book, reprinted several times until 1726, was intended to silence the Paracelsians and in particular to refute John Webster's 1677 work, *The Displaying of Supposed Witchcraft. Wherein is Infirmed that there are many sorts of Deceivers and Imposters, and Divers Persons under a passive Delusion of Melancholy and Fancy*. (On all of this literature, see Jobe, "The Devil in Restoration Science.") Finally, it might be useful to recall that spirits and demons remained associated with the stereotypes of seventeenth-century Puritan thought, where, notably, "all witchcraft comes from carnal lust, which is in women insatiable" (Heinrich Kramer [Institoris] and Jacob Sprenger, *Malleus Maleficarum* [1486], cited by Evelyn Fox-Keller, *Reflections on Gender and Science* [New Haven, CT: Yale University Press, 1985], p. 60).

16. Curley, p. 140. Note that for Descartes, as previously for the alchemists, these spirits are material, although composed of a subtle matter, like the gases of alchemy. ("Spirit of salt," an old term for hydrochloric acid and its vapors, as well as "sublimation," the direct passage from the solid to the gaseous state, are linguistic holdovers of these ideas—not to mention alcoholic "spirits.") Spinoza abandoned the reliance on "animal spirits" when he wrote the *Ethics*, but Guéroult thought it necessary to retrieve them in order to interpret Spinoza's physics and what he called his "neurocerebral model" so as to differentiate it from ostensibly parallel ideas in Descartes. For Guéroult, the fluid parts of the human body (*Ethics* II, postulates before Proposition 14) are "the equivalent of the animal spirits." Although Spinoza does not employ this term, Guéroult does, "in order to place the doctrine in the context of the theories of the epoch and for simplicity's sake" (Martial Guéroult, *Spinoza* [Paris: Aubier, 1968], vol. 2, p. 202). Nevertheless, the fact that Spinoza abandoned this terminology, except in the text where he criticized Descartes, seems to indicate the discontinuity he perceived between his own "clear and distinct" ideas about the union of body and mind and the confused ideas conveyed by the classical notion of animal spirits.

17. Curley, p. 145.

3. CAUSES IN NATURE AND ANIMATION OF SUBJECTS

A century earlier, in another context, Rabbi Judah Loew ben Bezalel, the Maharal of Prague, confronted the problem of reconciling the mathematical laws of celestial mechanics with the causes traditionally employed to explain natural phenomena such as eclipses, which the Talmud ascribes to egregiously immoral conduct by human beings.

His solution was to separate the domains of impersonal nature, which is determined by these laws, from the domain of human experience, with its moral and social dimension. The only possible relationship between these two domains was one of analogy, which might allow the projection onto natural phenomena of a humanized and personalized meaning.[18] As we know, even this analogy disappeared from classical science after it defined itself as the science of the objective phenomena of physical nature in which human subjectivity is merely an epiphenomenon. As an effect of nature in human beings, not an adequate cause, subjectivity came to be perceived by most people in an incomplete and confused manner, through the veils of illusion and the passions.

18. See R. Judah Loew ben Bezalel, *Be'er ha-golah*; cited in Atlan, *Enlightenment*, nn. 6.43–44 (pp. 282–283).

Today, more than three hundred years after Spinoza, we find ourselves on the threshold of a *new* New Science, which is trying to extend its scope to the human reality, in its all of its biological, social, and moral dimensions, without, however, returning to the total theological syntheses of the Middle Ages and Renaissance. In this context it may be of interest to reexamine some of the terminology of the various demonological "sciences" of bygone times.

Although the technical nature of causality has changed, its existential stakes remain the same. This is why ancient thinkers' critical reflections about the attitude—liberating or enslaving, clear-sighted or blind—that should be adopted or avoided about the effect of gods, astral influence, and other angels or demons supposed to determine our destiny, in whole or in part, in reality or in illusion, remain relevant. All of the many schools that followed Plato and Aristotle—Stoics and Epicureans, Cynics and Skeptics—asked these questions and proposed answers that have nourished Western philosophical thought for more than two thousand years. The discussions conducted in the rabbinical academies of the Land of Israel and Babylonia, whose edited records compose the two Talmuds, could not avoid echoing these reflections. They are found in the biblical and postbiblical literature in Hebrew and Aramaic, adapted to the ethical, legal, and political objectives pursued by the Sages and their disciples in the Jewish communities that were subject to Babylonian, then Greek, and then Roman and Parthian rule.

4. "NATURAL MAGIC" AND THE SCIENCE OF BYGONE TIMES

To begin with we should recall that, like astrology and alchemy, the lore of demons and angels, of genii and spirits, was an integral part of ancient science. Gnostic texts are full of it.[19] This involved elements of a composite lore that the Gnostic and Neoplatonic traditions traced back to the primordial knowledge, Gnosis,

19. Both demonology and angelology were elements of the syncretic Gnostic cosmogony that drew on Greek, Jewish, and Chaldean magical and astrological sources. Following the lead of the pagan Stoic and Neoplatonist philosophers, the Gnostics transformed the gods of Greek and Egyptian mythology (whom they assimilated to biblical figures) into allegorical representations of the forces that control the universe. This produced, inter alia, "the Gnostic habit of attributing

to the planetary entities of paganism the status and role of the angels in the Jewish demonology of the apocalypses. Already reduced to an astral function by Mathematicians, the Greek god as reused by the Gnostics gains a renewal of personality as the archon of intermediate space" (M. Tardieu, "The Gnostics and the Mythologies of Paganism," in *Mythologies*, ed. Yves Bonnefoy, Eng. ed. Wendy Doniger, trans. Gerald Honigsblum et al. [Chicago: University of Chicago Press, 1991], vol. 2, pp. 677–680, on p. 677). The human soul, wandering in this space, searches for the repose of returning to the One, which the initiatory knowledge, Gnosis, allows it to achieve. "The gods of paganism thus served the Gnostics merely as a means to illustrate this dialectic. In the same way that they are used by their contemporaries who are magicians, philosophers, and astrologers, so the Gnostics use these henceforth supernumerary entities of a drama, whether classifying them in catalogs of demons or dissolving them in allegorism" (ibid., p. 680).

20. See, for example, the treatise on the gods, demons, and spirits attributed to Iamblichus, *On the Mysteries*, trans. Emma C. Clarke, John M. Dillon, and Jackson P. Hershbell (Atlanta: Society of Biblical Literature, 2003).

21. See Elaine Pagels, *The Gnostic Gospels* (New York: Random House, 1979, 1989).

22. See F. A. Yates, *Giordano Bruno and the Hermetic Tradition* (Chicago: University of Chicago Press, 1964).

23. Gershom Scholem, *Kabbalah* (New York: Quadrangle, 1974), part 2, chap. 7, "Demonology in Kabbalah"; idem, *Abraham Cohen Herrera* (Jerusalem: Mossad Bialik, 1978) (Hebrew).

24. Alexandre Koyré, *Mystiques, spirituels, alchimistes du XVIe siècle allemande* (Paris: Gallimard, 1933, 1977); Walter Pagel, "Religious Motives in the Medical Biology of the Seventeenth Century," *Bulletin of the Institute of the History of Medicine* 3 (1935), pp. 97–128, 213–231, 265–312; idem, "William Harvey and the Purpose of Circulation," *Isis* 42 (1951), pp. 22–38; idem, "Paracelsus and the Neoplatonic and Gnostic Tradition," *Ambix* 8 (1960), pp. 125–166; A. G. Debus, *The English Paracelsians* (New York: Franklin Watts, 1966).

25. L. J. Rather, "Alchemistry, the Kabbala, the Analogy of the 'Creative Word' and the Origins of Molecular Biology," *Episteme* (Milan) (6) (1972), pp. 83–103.

26. Walter Pagel, "The Prime Matter of Paracelsus," *Ambix* 9 (1961), pp. 117–135.

of the ancient Egyptians and Chaldeans.[20] This tradition, later taken up and transformed by Christian theology,[21] was one of the sources of the ancient Hebrew science of the Talmud and kabbalah. This same myth-impregnated lore was rediscovered by the European Renaissance, with the help of kabbalist-alchemist-hermeticists such as Pico della Mirandola, Giordano Bruno, Paracelsus, and many others.[22]

Well into the seventeenth century, we should note, kabbalists who depicted themselves as philosophers and who wanted to distance themselves from the irrational tendencies of other streams of kabbalah included angels and spirits as essential elements of their cosmogony. Some, who saw themselves as engaged in philosophical inquiry, actually wrote full-length treatises on demonology.[23] This should not astonish us, of course. It is perfectly consistent with the rational and "modern" character of these preoccupations in Renaissance science and "chemical" medicine, as evidenced by the theories of Paracelsus and his followers.[24] Similarly, as we have seen, the widespread invocation of animal spirits as the cause of the motion of living bodies fit perfectly into the naturalist rationalism that characterized the magical and animist proto-science that we today dismiss as "mystical."

Note too that alchemical imagery continued to permeate some branches of science, to some extent unconsciously, well into the nineteenth and even twentieth centuries. In a thought-provoking and well-documented article, L. J. Rather shows that the famous dream in which Kekulé discovered the structure of the benzene ring was in fact alchemical. The mythical dragon or serpent Ouroboros, which swallows its own tail, and the symbol of the circle were almost second-nature for Kekulé, who, for all that he was a nineteenth-century chemist, had grown up on alchemical texts and traditions.[25] The search for the chemical formula was identified with the search for matter itself, in the pure tradition of the Paracelsian theory of prime matter, the *tria prima* or three principles (salt, sulfur, and mercury).[26] This prime matter united abstract configurations of forms and principles, "creative words," images, and their concrete realization in material objects. The idea that knowledge of matter can be obtained from knowledge of creative formulas can be traced back to the ancient notions of the creative logos and seminal reason, which Paracelsus turned into the germs of intelligent life, containing—like the serpent that swallows its own tail—their ultimate material production in their original state. For Rather, this idea of formulas or words, elements of a creative natural language, an avatar of the Stoic *logos spermatikos*, is also the origin, albeit unconscious, of the formal processes of twentieth-century molecular biology and its preliminaries in the molecular physiology of the nineteenth century. It is clear that the power of linguistic metaphors in biology[27]—sometimes with explicit reference to the Chinese book of divination, the *I Ching*, and invocation of an analogy between its sixty-four hexagrams and the sixty-four codons of the genetic code—reinforces the interpretation. Of course, neither Kekulé nor any contemporary biologist would ascribe any scientific truth value to these images in themselves before waking reason and experimentation confirmed their validity. This is what distinguishes the magical use of formulas that are supposed to have an inherent creative power from their scientific application by means of the experiments they inspire and whose logic we interpret through them. This distinction is all the more striking, perhaps, when we consider the extreme abstraction of the equations of mathematical physics, whose concrete material efficacy is no more obvious than that of magical spells.[28] Nevertheless the idea that matter is form and language, a priori strange for a modern mind that still holds more or less to dualism, runs through more than two thousand years of science and natural philosophy, despite the breaks represented by successive scientific revolutions.

So it is not surprising that several kabbalists[29] present their work not as expressions of irrational mystical currents within a religious tradition but as science and natural philosophy inherited from a primordial Adamic wisdom and transmitted to Moses, the prophets, and the sages of the Talmud and Midrash. Their classical opposition to philosophy must not delude us. They objected only to Aristotle, because of his dominant role in all the philosophy and theology of the Middle Ages, and especially in Jewish scholastic theology through Maimonides. On the contrary, important elements of Neoplatonism and Stoicism were integrated into their doctrines, both physical and metaphysical.

Thus for Israel Sarug, a student of Isaac Luria's in Safed and one of the most important disseminators of the Lurianic kabbalah in Europe, "there was no difference between kabbalah and philosophy."[30] His disciple, Abraham Cohen Herrera of Amsterdam, wrote two treatises in Spanish that expounded and explained ideas of the Safed kabbalah using concepts explicitly borrowed from Plotinus and Proclus. One of these, the *Casa del Divinidad* (House of Divinity), was a systematic exposition of angels and spirits.[31] Just like the motions and configuration of the stars and like the principles and elements of Paracelsian alchemy, angels and demons were thought to be hidden

27. See the contributions by Roman Jakobson and François Jacob, *Critique* 322 (1974).

28. It is interesting to analyze, for example, the similarities and differences between a magical *fiat lux* that engenders light by the utterance of a formula imitating the initial *fiat lux* of Genesis and operational use of Maxwell's equations in the physics of electromagnetic radiation (see Atlan, preface to *Le Golem*).

29. Including Ḥasdai Crescas in the fourteenth century, Moses Cordovero and his disciples, the Maharal of Prague in the sixteenth century, Moses Ḥayyim Luzzatto in the eighteenth century, and others.

30. Scholem, *Abraham Cohen Herrera*, p. 34.

31. Ibid., p. 26.

causes.[32] Gershom Scholem dates the last original text in kabbalistic demonology to 1709.[33] But he concludes his chapter on the topic with the remark that "belief in demons remained a folk superstition among some Jews in some countries down to the present. The rich demonology in I. Bashevis Singer stories reflects the syncretism of Slavic and Jewish elements in Polish Jewish folklore."[34]

5. "THE SPIRITS TELL"

Late in his life Yehudah Petaya (1859–1942), a Baghdad kabbalist and author of a voluminous commentary on Ḥayyim Vital's *'Ez̧ ḥayyim*, published a small book with the double title *Minḥat yehudah–Ha-ruḥot mesapperot* (Judah's Offering–The Spirits Tell).[35] There he recounts, inter alia, his experiences as a healer and presents case histories from his practice. His treatment methods relied on traditional medicine strongly influenced by practical kabbalah, employing plants, verbal communication, and dream interpretation. He tells how he confronted demons and spirits that appeared to patients in their dreams or that possessed them and spoke through their mouths. One of the main problems was, accordingly, identifying and naming the spirit, generally of a dead person, which had thus been reincarnated as it were. Only in this way could the healer learn the spirit's previous history and understand the purpose and mechanism of its reincarnation. Similarly, when a spirit appears in a dream the healer has to determine whether it is a demon or an angel. The words of an angel figure in prophetic dreams are supposed to convey true messages, on condition, of course, that one knows how to interpret them. By contrast, the words of a demon must always be suspected as false and cunning; and this is where the healer must apply all his skill. For cases in which his colleagues, although leading rabbis, had failed, he reveals the tricks he used to force demons and spirits to reveal their past identity and current need. A major part of his technique involved dream interpretation, based on the theories of the Talmud, which devotes many pages of tractate *Berakhot* to this topic. He also considers one of the inherent problems of dream interpretation: why dreams are discreet and allusive and do not convey their messages clearly. He offers three reasons for this: dreams often have to do with events of the distant past; they may call to mind some sin (which one would rather not acknowledge); and they may involve sexual relations between husband and wife.[36]

Despite appearances, this book probably owes nothing to Freud, considering when and where its author practiced. Instead, it is evidence of the internalization and psychologization of kabbalistic motifs that can also be found among the Hasidim of Eastern Europe, starting in the eighteenth century. Demons and spirits came to be identified with the souls and minds of the men and women through whom they spoke. For example, one of Petaya's greatest successes was his identification and rescue of the soul of the false messiah Shabbetai Z̧evi, reincarnated in a young man (who was not aware of being possessed); the poor fellow was tormented by heretical thoughts whose origin he could not fathom.[37]

32. In their "chemical" theory of disease, Paracelsus and some of his disciples, such as Duchesne and Fludd (the latter was a contemporary and friend of Harvey, the discoverer of the circulatory system), identified the four elements with demons or bad spirits, which were the causes and essences of disease (see Debus, *The English Paracelsians*, pp. 26–32, 90, 112, 120).

33. Scholem, *Kabbalah*, p. 325.

34. Ibid, p. 326.

35. Yehudah Petaya, *Minḥat yehudah–Haruḥot mesapperot* (Baghdad, 1933; Jerusalem, 1956).

36. Ibid., p. 34.

37. Ibid., pp. 86–91.

Here, as with astrology, what was formerly a scientific explanation is transformed into superstition and magical thinking, to which some people remain attached, with varying degrees of tenacity, because of their need to explain the unknown at any cost—here by means of the hidden causes suggested by these traditions. Note, however, that when these prescientific explanatory schemes are integrated into the culture of some societies they may continue to play a not-insignificant role in therapeutic practices, especially by means of what is called, with no good idea of how it works, the placebo effect. These schemes are all the more effective because they feed the beliefs and worldviews held by members of these societies, both patients and therapists.[38]

6. THE DETERMINING CAUSES OF NATURE AND THEIR REPRESENTATIONS IN CLASSICAL ANTIQUITY

There are two types of lessons to be drawn from these stories.

On the one hand, we learn the extent to which the desire for knowledge of causes, which still motivates most scientific inquiry, leads us down a path full of pitfalls and that has never split off from the hallucinatory rationalizations of myth, dreams, or imagination.[39]

On the other hand, we also learn to take these ancient explanatory systems, which today have become inexhaustible reservoirs of superstition, and place them in the perspective of the movement of creative knowledge that spawned them before they were diverted into the arrogance of "revealed wisdom." Just like evil, which, at its source, is not evil, superstition, at its source, is only one aspect of an indefinitely repeated pursuit of truth, which necessarily goes astray when formulated as an established lore, as some kind of off-the-rack explanation, a manual, a catechism, today's scientific truth, tomorrow's naïve error or superstition.

Surviving various transformations, this identification of spirits and demons as the hidden causes of natural phenomena remained the basis of ostensibly scientific doctrines well into the seventeenth century. It did not fully forfeit its legitimacy until the next century, notably after Lavoisier's chemistry showed up alchemy for parascientific occultism. Today, on the threshold of the *new* New Science, when some see a need to restore to the world the enchantment of which it was deprived by the mechanical science of the eighteenth and nineteenth centuries, it is appropriate to cast an inquisitive and detached look at some elements of this prescientific, magical, and animate world. Human beings could feel at home there, thanks to the meaning they discovered in it, despite the

38. As we have just seen with regard to Petaya, it is in psychotherapy, and therapy in general, that magical practices, both prescientific and nonscientific, may continue to have a certain efficacy, all the more so when they are set in a traditional social and cultural context that invokes explanatory schemes in which everyone believes—including potential patients, of course. Elsewhere I have shown that the relative success of a therapeutic technique, probably due to the undoubted though limited power of the placebo effect and transfer phenomena, offers absolutely no evidence for the scientific truth of the theoretical corpus on which it is based. This explains the many forms of alternative medicine that have proliferated even in our own societies, supplanting or in association with scientific medicine (see Henri Atlan, "L'argument d'efficacité: médecine scientifique contre nostalgie scientifico-mystique," in *La pensée scientifique et les parasciences* [Paris: Albin Michel, 1993], pp. 183–193). Conversely, traditional doctrines may retain their value when transposed and internalized in a new context where they are still accepted, even if the therapeutic techniques they ground are no longer deemed effective. This is how, since the eleventh century, following Maimonides, it has been "forbidden to heal using the medical lore of the Talmud" (see Atlan, *Enlightenment*, p. 268 and n. 6.52, p. 284).

39. See Manuel de Diéguez, *Le Mythe rationnel de l'Occident* (Paris: PUF, 1980); Atlan, *Enlightenment*, passim.

limited (to say the least) means at their disposal for exerting real control over hidden forces. Of course we are not contemplating a return to magical formulas and amulets in an attempt to act on objects. But perhaps we would be well advised to re-enchant the world with stories of fairies, elves, imps, and sympathetic demons—like the Ashmedai of the legend—whom we know to be fictions, rather than with genes, molecules, quarks, and the Big Bang, to which we ascribe the virtues of hidden causes that are even more miraculous than those they really possess in the limited theoretical and operational domains of their scientific relevance, theoretical and operational.

It is with this in mind that I have collected several texts of talmudic and kabbalistic demonology. I suggest we see them not as a manual to be studied so that it can be put into practice but as a mode of inquiring into the desirable relations between human beings (as a space in which their moral and legal meaning can be constructed), on one side; and demons, angels, and other "natural" forces, human and nonhuman, which are supposed to act in and around human beings, on the other side.

Before we look at several of these texts we should review a classic talmudic discussion of the power of natural determinants—here, the motion of the planets and constellations—and the possibility of modifying their influence. For the authors of the Talmud and *Zohar* it was not a question, as one would say today, of "believing" in astrology, but of knowing how to integrate their doctrines with the scientific knowledge of their age. They could no more dismiss astrology than we can "not believe" in physics and chemistry today. The question is not one of accepting the content of scientific knowledge but of determining its relevant domain of application, of knowing how to deal with it and how to integrate it into our life and thought. Astrology is no longer part of the body of scientific knowledge. Consequently we have no reason to "believe" in it, no more today than yesterday, as a lore that has the status of revealed truth. But the talmudic sages' discussions related to Babylonian astrology, widely accepted in their time as the science of natural determinism.

In the Babylonian Talmud, tractate *Shabbat*, we read that when the mother of the future R. Naḥman bar Yiẓḥaḳ was pregnant, astrologers predicted that her son would be a thief. She consequently decided to raise him in strict piety, and in particular to make sure that he always kept his head covered.[40] This narrative exemplifies the celebrated aphorism that launches the discussion: *'ein mazzal le-Yisrael*, "Israel has no [governing] constellation," meaning that it is not subject to astral determinism. This contrasts with the supposed dependence of all earthly life on the influence of the stars: "There is not a single stalk of grass that does not have a *mazzal* [planet or constellation] in the firmament that strikes it and says, 'Grow.'"[41]

From this story we learn that although the Sages accepted natural determinism, they nevertheless believed that it could be modified, chiefly through education and by study and practice of the Torah, which constitutes the behavior that defines a human type called "Israel." Note, however, that one sage, R. Ḥanina, seems to have rejected this idea: *yesh mazzal le-Yisrael*, he maintained—"Israel is subject to astral deter-

40. B *Shabbat* 156. Note in passing that this is the only place in the Talmud that refers explicitly to covering the head. Clearly this custom, which has become the most visible sign of Jewish orthodoxy, was only a pious custom in antiquity, like the lengthy prayers of the ancient pietists (B *Berakhot* 32b).

41. *Gen. Rabbah* 10:6.

minism." He is the very same sage who taught that "everything is in the hands of heaven except the fear of heaven."[42] In other words, R. Ḥanina, too, entertains (though in a different fashion) a possibility at least of circumventing, if not escaping, natural determinism. As long as the fear of heaven remains in our power, we possess a key that allows us to act on everything else that is controlled by heaven. In practice, awareness of the overwhelming reality of this determinism and the awed and reverential fear inspired by this awareness are what constitute, in the words of Psalms and Proverbs, "the beginning of wisdom" or "beginning of knowledge."[43] As this wisdom or knowledge expands we gain the ability to know more and more of these determining causes and, consequently, to have some control over them. In principle, an apparently unlimited potential to modify destiny is allotted to the sage who is also a righteous person, that is, to one who combines great wisdom with moral perfection. This is expressed by the saying that the Talmud attributes to God: "I rule man; who rules Me? The righteous man: for I make a decree and he [can] annul it."[44]

7. THE "GENETIC" CAUSES OF SCHOLASTICISM

Talmudic demonology, with its fantastic stories (like "the sparks of randomness" and the tale of King Solomon and Ashmedai, the king of the demons, which we reviewed in Chapter 1), fits into this context of the knowledge and perhaps control of the hidden causes of things. But its determining causes are not those associated with the stars. Whereas the latter must themselves obey eternal and immutable laws,[45] demons (and angels) are partial and sometimes haphazard causes, themselves the effects of other causes, some unknown, some known or at least suspected, as in the case of certain human actions. What is more, one of the most striking characteristics of these demons or spirits (animal, vegetable, or other) is the type of causality attributed to them. They are not only hidden causes invoked as reasons that *explain* certain phenomena, but also causes that *produce* or *generate* their effects. This is why sages and magi could have knowledge of them that was not only speculative but also active and practical. As for the ancient Stoics, this knowledge encompassed active and productive causes, seminal reasons or *logoi spermatikoi*, both spells and beings, spells that create reality, in a literally spermatic knowledge.[46]

42. B *Berakhot* 33b (also *Megillah* 25a, *Niddah* 16b).

43. "The beginning of wisdom is the fear of YHWH" (Ps. 111:10); "fear of YHWH is the beginning of knowledge" (Prov. 1:7). See Chapter 2, n. 83, on the nature of this "fear" as the experience of perfection in its relations with knowledge. For Plato, too, philosophy springs from the sense of wonder. Brought down from heaven by Iris, the messenger of the gods, it is the "daughter of Thaumas" (*Thaeatetus* 155–156). "The sense of wonder is the mark of the philosopher. Philosophy indeed has no other origin, and he was a good genealogist who made Iris the daughter of Thaumas" (155d [trans. Francis Cornford]).

44. B *Mo'ed katan* 16b; see also Rashi on 2 Sam. 23:3; *Zohar, Leviticus*, p. 15a.

45. This difference has become more evident since astronomy matured into a mathematical science in which the motion of stars can be computed and their positions predicted long in advance. This development significantly altered how we think of the link between human actions on earth and astral motions in the heavens. As we have seen, the Maharal of Prague confronted this problem at the end of the sixteenth century, raised not by astrology, which he rejected out of hand, but by an inverted causality in which human misconduct was supposed to trigger certain heavenly events (such as eclipses). After the Copernican revolution and the consolidation of Keplerian and Newtonian celestial mechanics, it was much easier to reject any form of direct causal linkage between the behavior of the stars and the behavior of human beings than to stop looking for hidden causes of a magical nature behind human and nonhuman events on earth.

46. See above (n. 25) for Rather's thesis about the significance of this notion, which was explicit until the time of Paracelsus and his alchemist disciples and then continued to implicitly direct modern science in some processes, conscious or unconscious, of discovery.

We are no longer accustomed to this type of causality, because, since Descartes, we have been tugged about by a dualism of body and mind, of causes and reasons, on the battlefield of the symmetrically opposed metaphysics of the French materialism of the eighteenth-century *philosophes* and of post-Leibniz German idealism. Nevertheless, the contemporary biological monism seems to re-orient us toward the type of monism, neither idealist nor materialist, that Spinoza, building on but also in opposition to Descartes, attempted to reconstitute but which, in the ensuing three-plus centuries, has never managed to escape its marginal status—glorious or damned—vis-à-vis the dominant currents of philosophy. This species of creative determinist causes, "genetic" in the traditional Scholastic sense of "producing a genesis," resembles what modern biology teaches us by means of sometimes indiscriminate metaphors, like that of the program of the same name. Molecular biology has accustomed us to think of physical entities as simultaneously causes of and reasons for our normal and pathological states. So-called information-carrying molecules or "genes" have an abstract structure described by a sequence of signs that produce concrete effects on the physical structure and function of organisms. If these metaphors are taken literally, they may lead us back to a sort of pre-formationism in which the homunculus thought to be contained in the seed has been replaced by the genome, to which we now attribute autonomous activity that it does not really have.[47]

Geoffroy Saint-Hilaire saw the inadequate nature of the causal relationship used by preformationism, in which the cause contains the effect and there is no distinction between complete and partial causes.[48] In his critique of causality, Hume, too, demonstrated the illusory character of invoking causes that are not proportional to their effects, employed notably in the fallacious argument from design to prove the existence of God as cause, based on observation of nature as effect.[49]

Renewing our acquaintance with the spirits and demons of ancient science could make it easier for us to identify these inadequacies and avoid falling back into the precritical confusion to which we are constantly exposed by the desire, both scientific and magical, to identify *the* cause of everything, whatever the cost.

8. ELEMENTS OF TALMUDIC DEMONOLOGY

One of the privileged domains in which the effects of demons and spirits of all sorts were supposed to be observed was medicine, as can still be seen in the works of Paracelsus and his disciples, who are nevertheless considered to be the forerunners of modern experimental medicine. The Talmud sometimes designates illnesses by the name of the demon that produces them. The story of Solomon and Ashmedai is inserted into an analysis of the effects of a demon named Kordiakos,[50]

47. R. C. Lewontin notes the inert character of DNA molecules, which can do nothing by themselves—not even replicate—without the activity of regulatory proteins that have enzymatic activity. Although the structure of the proteins is encoded in the DNA, only proteins can do something, such as produce other proteins or DNA molecules. (See: R. C. Lewontin, "The Dream of the Human Genome," *New York Review of Books*, May 28, 1992, pp. 31–40; idem, *Biology as Ideology: the Doctrine of DNA* [New York: HarperCollins, 1991, 1995], chap. 3.) See also n. 10, and Chapter 1, nn. 5 and 6.

48. Geoffroy Saint-Hilaire, *Vie, travaux et doctrine scientifique d'Etienne Geoffroy Saint-Hilaire* (1847), quoted in Georges Lapassade, Jacques Piquemal, Georges Canguilhem, and Jacques Ulmann, *Du développement à l'évolution au XIXe siècle* (Paris: PUF, 2003 [1962]), pp. 14–16.

49. Hume, *An Inquiry Concerning Human Understanding*, Section XI.

50. See below, n. 72.

supposed to produce behavioral problems that may render their victims legally incompetent. But medicine was not the only area where demons were supposed to be afoot. The talmudic and kabbalistic literature deals with them and, as always, is at pains to distinguish not only the real from the illusory but especially what may be invoked in the service of the "good" (unification, the repair of the shattered vessels) from what is only a source of damage and destruction. With regard to the reality of created beings, demons have the special status of unfinished potential beings, created (according to some) at twilight on the eve of the first Sabbath,[51] that is, when the sixth and last "day" of Genesis was over but the Sabbath, as the conclusion of this activity, had not yet begun. They are described as souls or spirits associated with bodies whose creation was interrupted and which consequently remain in a state of potentiality and pure desire. We find a generic definition of the talmudic sages' conception of the properties of what they called "demons" (*shedim*) in tractate *Ḥagigah*:

> Six things are said concerning demons: in regard to three, they are like the ministering angels; and in regard to three, like human beings. In regard to three they are like the ministering angels: they have wings like the ministering angels; and they fly from one end of the world to the other like the ministering angels; and they know what will happen like the ministering angels. . . . And in regard to three, they are like human beings: they eat and drink like human beings; they propagate like human beings; and they die like human beings.[52]

Not all of these properties necessarily coexist. And one of the six immediately gives rise to an objection: How could they possibly know the future? The answer given is that "they hear from behind the veil like the ministering angels"; that is, they too have an indirect knowledge of hidden things, the type of knowledge that is the object of various practices of divination. This is associated with or competes with the knowledge of the sage, who also attempts to know the future through application of his intellect, which is both deductive and intuitive, and which enables him to "see" the future consequences inherent in the past and present. "Who is wise?" asks the Talmud, and answers, "He who sees what is [about to be] born."[53] And again, "Who is wise? He who learns from every person."[54]

This specific and ambiguous relationship between knowledge through demons—which implies knowledge *of* demons—and the knowledge that is pursued by the sage and that he is supposed to learn from the Torah, culminates in King Solomon, and especially in his adventure with the king of the demons. The ambiguity of this relationship derives purely and simply from the fact that the primary nature of demons is destructive, the source of diverse kinds of damage, including disease and other catastrophes. Their very name, *shed*, evokes the verb *shod* or *shadod* 'overpower, destroy violently, ravage'. Their company is necessarily perilous, both for physical and mental health and with regard to the knowledge that can be acquired through them, which is a potential source of error because always a medium of cunning and deceit. Nevertheless, they are not necessarily bad in every case and all circumstances. Indeed this knowledge, carefully controlled, can be a path—sometimes the only one—for the assimilation of divine wisdom into human society.

51. M *Avot* 5:8.
52. B *Ḥagigah* 16a.
53. B *Tamid* 32a (cf. M *Avot* 2:9).
54. M *Avot* 4:1.

9. DIVINATION AND PROPHECY

The same ambivalence accompanies the divinatory practices against which the Mosaic Law specifically warns the Israelites.[55] Deuteronomy associates the use of oracular knowledge derived from omens and necromancy, or any other form of divination and witchcraft, with the idolatrous practices of the peoples of Canaan, considered to be deadly. These practices are presented as the reason these nations were condemned and supplanted by a new society, Israel, organized around this law, the Torah.

Nevertheless, as Naḥmanides emphasizes, the Torah does not absolutely condemn this form of serpentine, divinatory knowledge, assembled from conjectures and interpretations, because it is an integral part of all knowledge. He maintains that the various magical practices forbidden in the Bible are not of equal impurity.[56] In his commentary on the verses in question—which denounce anyone "who consigns his son or daughter to the fire, or who is an augur, a soothsayer (*me'onen; lit.* cloudwatcher), a diviner (*menaḥesh*), a sorcerer, one who casts spells, or one who consults ghosts or familiar spirits, or one who inquires

55. Deut. 18:9–12.

56. According to the *Zohar*, the practice of magic and use of techniques of divination and sorcery are a consequence of Adam and Eve's eating the fruit of the Tree of Knowledge (see below, n. 107) and as such an element of human nature. It is forbidden to Israel only because it is often expressed in a context of idolatry, of which it may be a manifestation. This emerges quite clearly from a passage in the Talmud that denounces it as an idolatrous practice, which "weakens the heavenly retinue" by using the powers of that retinue in its own way (B *Sanhedrin* 67b).

But this very same passage draws a number of distinctions among different practices, not all of which are condemned equally. We must remember, first of all, that only Israel is forbidden the worship of pagan gods, and consequently these magical practices to the extent that they express it; but they are considered to be the normal lot of human beings in their relations with the gods, including "the sun and the moon and the stars, all the host of heaven, . . . which YHWH your god has allotted to all the peoples under the whole heaven" (Deut. 4:19). What is more, sorcery and magic are distinguished from recourse to demons. The latter is only a particular and therefore limited means of activating a specific hidden cause to produce an effect. On the contrary, magicians achieve their ends without demonic assistance, thanks to the power and scope of their lore. What magicians do is extraordinary only because this lore is generally unavailable to the common run of mortals.

Having established this, the Talmud goes on to distinguish three types of magic, all of them designated by the same word but judged differently by the law. The first—absolutely forbidden to Israel—is punishable by death, as stipulated by the Bible, because it is assimilated to idolatrous practices such as sexual relations with animals, condemned in the same passage (Exodus 22). Even those who engage in magic are not culpable, according to Rabbi Akiva, except to the extent that their practices are effective (M *Sanhedrin* 7:11 [67a]). The sorcerer who manipulates natural objects and exerts real control over them must be distinguished from the illusionist who merely hoodwinks the audience. Hebrew uses the same word for both, *kashaf* or *mekhashef*, just as English refers to both as "magicians"; but only the former incurs the death penalty. The same word is employed in Exodus (7:11) for the wise men and magicians who are privy to the arcane lore of Egypt and are summoned by Pharaoh to counter Moses and Aaron. Returning to the talmudic passage in question, we read that this idolatrous magic, reprehensible because it weakens the heavenly retinue, can be combated and overcome in its own domain of effectiveness; a righteous and especially pious sage, Rabbi Ḥanina, defeats it easily, thanks to his bond to the unity from which the forces of this heavenly retinue proceed.

A second type of magic is also forbidden a priori, although its practitioners have no liability under the law: illusionists who try to make people believe they have powers they do not really possess.

Finally—what at first sight is surprising but which can be readily understood in the context of the present chapter—the text mentions a third type of magic, which is totally permissible, "such as was performed by R. Ḥanina and R. Oshaia, who spent every Sabbath eve studying the *Book of Creation*, by means of which they created a fat calf and ate it" for their Sabbath meal. This permissible magic is defined not by its content but by its practitioners: eminent sages who, thanks to their lore—the Torah—were able to create living beings and even a man (as the Talmud itself informs us two pages earlier; see Chapter 1 on Rava's creation of an artificial man). It is also interesting to see that one of the sages in this story is the same Rabbi Ḥanina who overcame magic directed against him and who, parting company with the other rabbis, asserted that astral determinism applies even to Israel, while allowing a possible limited escape through knowledge of this determinism that begins in "awe" or a sense of wonder, the "fear of heaven" that is the "beginning of wisdom." (See above, §6, and nn. 42 and 43.) For Rabbi Ḥanina, the affirmation of determinism, like that which proceeds from scientific knowledge today, is not fatalism. On the contrary, as we saw in Chapter 2, it goes hand in hand with the quest for knowledge of this determinism as a condition of human freedom.

of the dead" (Deut. 18:9–11)—he notes that observing the shapes of clouds and studying birds' wings and calls, with the goal of achieving foreknowledge of future events, are normal human activities. Unlike child sacrifice, for example, and the other practices that the Bible denounces as abominations, these customs were not what condemned the Canaanites to be dispossessed of their land. In fact, they derive from the ancient lore of stars and angels and of the different ways in which their powers determine creatures and objects on earth.[57] Naḥmanides explains that every nation has its own *śar* or tutelary angel, a prince of the heavenly retinue, through whom the forces that govern and protect it are channeled. It is knowledge of these forces, which are closely related to astrology, that underlies divinatory practices. These are forbidden to Israel only because it does not have a *śar* that links it to the world above. Israel's particular god, the Name that Moses teaches along with the laws that accompany it, rules the entirety of "heaven and earth." Instead of divinatory practices, Israel can employ the various forms of prophecy that accompanied its history until the early years of the Second Temple and can inquire of the oracle in the Temple in Jerusalem (the *Urim ve-Tummim*),[58] strictly limited to the High Priest and to certain questions of collective interest.[59] Solomon's wisdom, too, drew on the science of other nations. Hence it is not astonishing that the sages of Israel who came after Solomon—the rabbis of the Talmud and those who followed them, including Naḥmanides himself—also learned from it. This is why they are so ambivalent about these traditions and practices. In the same passage of his commentary, Naḥmanides refers the word *menaḥesh* 'diviner' not to *naḥash* or 'serpent' but to the verb *ḥash* 'hurry, hasten', as in Psalm 119:60, "I hasten and do not delay [to keep your commandments]." He holds that divination can be justified, in some fashion, by our impatience to know the future. The several categories of oracle include the *yidde'oni* 'one who consults ghosts', a word derived from the same root as *yeda'* 'knowledge'. Finally, even the invocation of ghosts and spirits is not forbidden in all circumstances. According to one talmudic interpretation, the ban applies only to the practice of opening oneself to possession by an "impure spirit" (a "graveyard demon," as Rashi puts it) by fasting and spending the night in a cemetery.[60] It is remarkable that Rabbi Akiva contrasts this to one "who fasts that a pure spirit [Rashi: "a prophecy of the Shekhinah (divine presence)"] may rest upon him." The former, although forbidden, is effective. The second no longer is, explains the Talmud, because "your iniquities have made a separation between you and your god" (Isa. 59:2). But it would be a prophetic practice much more effective than the impure version had it remained linked to

Finally, an important talmudic principle is learned from the admonition that the Israelites "not learn to do the abominable practices of those nations" (Deut. 18:9): "It is written, 'you shall not learn to do'; but you may learn in order to understand and instruct" (B *Shabbat* 75a). On the one hand, there is a duty to learn whatever is needed to understand what these practices are all about, notably with regard to the astronomical calculations that are employed by the sorcerers, magicians, and astrologers of pagan nations as well; on the other hand, there is also an obligation to teach and legislate with full knowledge of what details of these practices are permitted or forbidden.

57. See also Naḥmanides' commentary on Leviticus 16:8, with regard to the Yom Kippur ritual of the scapegoat, chosen at random from two *śe'irim* or goats, where he refers also to Leviticus 17:7, in which the same word designates a type of demon (see below, n. 73).

58. Exod. 28:30. In Chapter 5 we will see how the *Urim ve-Tummim* were employed to divide the land among the tribes in the time of Joshua. Three techniques of permitted oracular knowledge appropriate to the God of Israel are mentioned in 1 Samuel 28:6. King Saul, anxious before the impending battle with the Philistines, "inquires of God." But "YHWH did not answer him, either by dreams or by Urim or by prophets."

59. See also Rashi and Naḥmanides on Numbers 23:23; B *Yoma* 71a and 73a.

60. B *Sanhedrin* 65b.

the study and observance of the Torah. To illustrate the power of the Torah as a potential source of efficacious knowledge, the Talmud adduces here the famous episodes (which we mentioned in Chapter 1) of Rava, who created an artificial man, and of the two sages who, by combining the letters of *Sefer Yeẓirah*, created a fat calf for their Sabbath meal every week.

Finally, divinatory practices that involve relations with spirits and demons are reported in the Talmud as having been conducted by some sages in certain circumstances[61] and are introduced in a few kabbalistic texts by Isaac Luria and Ḥayyim Vital.[62]

Thus the ancient biblical context of the Torah's prohibition of these practices prescribes a different form of divination for Israel, restricted to special situations and strictly controlled by a form of critical wisdom derived from the judicial tradition instituted by the Mosaic Law itself. Prophecy,[63] which continued until the first years of the Second Temple period, is permitted; so is inquiry of the *Urim ve-Tummim*, the oracular device worn by the high priest,[64] though only within the Temple compound and limited in its application. An intermediate link between serpentine sorcery and prophecy is Balaam, the prophet of the Gentiles. Hired by the king of Moab to curse Israel and bring about its destruction, the force and truth of his own prophecy compel him to bless the people instead.[65] It is this same Balaam whose degree of prophecy is compared to that of Moses (to his disadvantage in some points but his advantage in others) and who is also described as a magician and expert in the practice of serpents (*neḥashim*), meaning divination.[66]

The Mosaic Law institutes, in some way, a process whereby this mode of knowledge,

61. B *Berakhot* 18b.

62. Cited, for example, by Petaya, *Haruḥot mesapperot*, pp. 65–66.

63. This is a "blameless" or "innocent" practice, according to Deuteronomy 18:13, in contrast to the soothsaying and divination that the next verse outlaws for Israel. According to Naḥmanides, here Israel is being enjoined to remain "perfect" or "whole" in its "fear of YHWH" and not to base its conduct on some divinatory knowledge of the future. The "fear"—the sense of wonder—aroused by the majesty of creation and its determining causes must remain whole and complete, just as the very perfection (*shelemut*) of these causes, of "everything that is in the hands of heaven," is complete (*shelemah*) (see above, n. 43; and Chapter 2, n. 83). We should recall, nevertheless, that this is merely the "beginning of wisdom." The sage, "who sees what is [about to be] born" (above, n. 53), conceives of the future that is potential in the present not through prophetic vision, which is fluid and ambiguous like dreams and hallucinations, but through his experiences and reflexive knowledge of the determining causes, which give him clear insight into the sequence of causes and effects in the fabric of the universe.

64. See Exodus 28:30. The Talmud (B *Yoma* 71a and 73a) gives details of how the *Urim ve-Tummim* were interrogated and deciphered. Words were formed by combinations of the letters in the names of the tribes of Israel, which were engraved on the twelve precious stones inlaid in the breastplate of the High Priest (see Chapter 5). The Talmud (ibid. 73b) also explains the derivation of the term: "*Urim* [from *'or* 'light'] because their words are enlightening; *tummim* [from *tam* 'complete, whole'] because their words are fulfilled."

65. Numbers 22:5–24:25.

66. See Numbers 23:23 and the commentaries by Rashi and Naḥmanides.

The story of Balaam, idolator and magician, but also a true prophet, has inspired many commentaries. In the thirteenth century Moses de Leon (the presumed author of the *Zohar)* summarized the elements that are among the most suggestive in his *Sheḳel ha-ḳodesh*. He begins by noting that Balaam's prophetic level was as high as Moses'. Like him, he saw through a "transparent glass," clearly and distinctly, and not, like the other prophets, through a "clouded glass." But this prophecy was not personalized for him, unlike God's interchange with Moses, "face to face" (Deut. 34:10), or call to him, *va-yiḳra' 'elav*, "He called to him" (Lev. 1:1). On the contrary, Balaam's eyes are opened as if in a revelation that falls on him "at random"—*va-yiḳḳar 'Elohim 'el Bile'am*, "God happened on [*or* met unexpectedly] Balaam" (Num. 23:4). The summons to Moses, *va-yiḳra'* (ויקרא), with its terminal *aleph*, the sign of unity, is contrasted to *va-yiḳḳar* (ויקר), which expresses, according to Rashi, chance or coincidence (*'ar'ai*), contempt or disdain (*genai*), and the impurity of seminal emission (*ḳeri).* This theme is developed by de Leon, who explains that Balaam achieved his elevated degree of prophecy by means of "impure" magical practices that drew down the impure spirit on him. Here he is referring to the talmudic passage (B *Sanhedrin* 105a) in which some of the Sages hold that Balaam "practiced enchantment by means of his penis," while others maintain that "he committed bestiality with his ass," that famous ass who, according to the biblical account, spoke prophetically

although still assigned the highest value in principle, was increasingly limited and controlled with regard to its content, until it vanished after the destruction of the First Temple and was replaced by the combination of textual hermeneutics, reason, and empiricism that constitutes the talmudic tradition. The Babylonian exile, the centuries of the Second Temple, and the long exile after its destruction saw the creation of the new institution of "rabbis"—judges and scholars—who displaced and replaced most of the traditional authority of the priests after the prophets lost their capacity for prophecy.

The age of prophecy was not an era marked by the simple and luminous discovery of a transparent truth. Rather, it was a period of conflict between true prophets and false prophets, conducted according to the rules for verification laid down by Deuteronomy.[67] The ambiguous nature of prophecy, with regard to the truth of its content, is expressed in famous talmudic dicta such as "the sage is superior to the prophet" and "since the Temple was destroyed, prophecy has been taken from prophets and given to fools and children."[68]

After the end of prophecy, the judicial status of serpentine knowledge obtained through omens and divination seems to be summed up in the rule that one may interpret omens if one does not trust them enough to rely on them when deciding how to act. In other words, it is acceptable to employ this type of knowledge on condition that you not allow yourself to be governed by what you think it tells you.[69]

This is why the legends of the Talmud, Midrash, and kabbalah

(Num. 22:23–36; it saw the angel before Balaam did and, after "YHWH opened its mouth," spoke to him). De Leon's formulation is very precise:

> This is the rule for magicians and sorcerers: they do not achieve what they seek before they make themselves impure. Whereas when the holy prophets, who serve the Highest, wish to attain the degree of the prophetic spirit and achieve what they seek, they hallow themselves and engage in sacred matters so as to find and attain their goal. But those magicians, soothsayers, and diviners make themselves impure with a certain form of impurity so that the impure spirit will descend on them. For it is as a function of their conduct and actions that the spirit descends on them, from the side of the sacred or from the side of impurity. Thus the wicked Balaam was a magician and diviner and could not achieve what he sought until he defiled himself with his ass, and then he attracted the impure spirit to himself. The secret of the matter is that the wicked Balaam's degree of prophecy proceeded only from the side of impurity. It was with that spirit that he attained knowledge on the impure side, like our teacher Moses in the degree of the sacred: Moses at the highest level of the sacred, Balaam at the lowest level of impurity. (Moses de Leon, *Sheḳel-Haḳ-ḳōdesh*, ed. A. W. Greenup [London, 1911; repr. Jerusalem, 1969], "Gate of Foundation, Second Part," p. 18 [see also *Zohar*, Numbers 193b])

Right before this (p. 17) de Leon quotes the verse that "Elohim happened on Balaam" and, like Rashi, glosses the verb as an allusion to nocturnal emissions and impurity (*ḳeri*). It was by means of this practice, perhaps continued with the ass whose animal nature responded by breaking into speech, that Balaam triggered his serpentine and random prophecy. He excelled in this, just as Moses excelled in intentional prophecy that was personalized by the call and face-to-face dialoguer. Note that Balaam's sexual abuse of his ass is mentioned in the Talmud (B *Sanhedrin* 105) with reference to Balaam's description of himself (Num. 24:16) as one who "knows the knowledge of the upper regions (*or* knowledge of the highest [*yodea' da'at 'elyon*])." This is not astonishing if we recall that sexual connotations are never far away in the context of knowing and knowledge. We know too that Moses, like the members of the wilderness generation and, more broadly, the "sparks of randomness," proceeded from this same *da'at 'elyon* (see Chapter 1, §10). This is another indication of Balaam's status as Moses' mirror image, of the sorcerer as the inversion of the sage.

What emerges from all this is that access to supernal knowledge is not a matter of intellectual effort. It depends on practices and exercises that are both spiritual and physical, which can be "pure"—oriented toward life and unity—or "impure"—marked by death and decomposition. The truth of this knowledge is not necessarily incompatible with the impurity of such practices and of the spirit that accompanies them, as shown by the case of Balaam, whose prophecies are true, perhaps even despite himself. Solomon's wisdom, attached to the side of good and the side of evil and nourished in part by Ashmedai and his demon cohorts, is another example (see Chapter 1, §7, n. 55; below, n. 115). Moses de Leon justifies this situation by the nature of "fire"—the words of the Torah—which cannot be affected by impurity. This is another application of the principle we have already encountered, namely, that evil at its source is not evil.

67. Deut. 18:15–22.

68. B *Bava batra* 12a, 12b.

69. See B *Ḥullin* 95b and Rashi ad loc., where Abraham's servant Eliezer (Gen. 24:14) is offered as a counter-example. Sent to find a wife for

his master's son, he determines his behavior by a sign that he reads as an omen: among all the young women he will soon encounter at the well, he will speak only to the one who offers him water to drink.

70. B *Bava meẓi'a* 59b.

71. B *Yevamot* 122a.

continue to refer to these processes of illumination and inspiration as the last vestiges of prophecy in a world that no longer has "true" prophets and as always mixed with uncertainty, "impurity" (that is, death), and dross. Several levels of generality and universality are assigned to the various revelatory phenomena of this type, ranging from dreams, at the lowest and most personal level, "one sixtieth of prophecy," to the *bat ḳol* or "daughter of a voice," a heavenly declaration of collective import heard by an entire group, at the highest, and passing through the utterances of *maggidim*, angels (or demons) that convey knowledge to scholars. We encounter this distrust in the famous talmudic episode in which Rabbi Eliezer ben Horkanos, finding himself in the minority on a question of law, invokes the authority of a *bat ḳol* to prove that his view is the correct halakhah. All the sages present hear the voice but nevertheless reject his opinion, on the grounds that "the Torah is not in heaven."[70] This phrase, borrowed from a verse in Deuteronomy (30:12), is understood as an invitation to the sages to find the Torah in their own interpretations, "very close . . . , in [their] mouth and in [their] heart" (ibid., v. 14), rather than in some new revelation sent down from heaven.

10. THE SHADOW OF A SHADOW

Another talmudic passage in which this ambivalent combination of respect and distrust appears clearly opens unsuspected horizons to the possible relations between shadows and demons. The text in question follows a discussion where, in the absence of the normally required testimony of two eyewitnesses, the announcement of a man's death by a *bat ḳol* is accepted as sufficient to permit his widow to remarry.[71] In this context (legal and not mystical), Rashi explains that the *bat ḳol* in question is an utterance by "a voice that they heard without being able to distinguish the human form from which it came." On the one hand, in this particular case such extremely shaky "testimony" is accepted willingly, because otherwise the woman would be an *agunah* or "chained woman" and forbidden to remarry—a status that talmudic law seeks to avoid at all cost. A woman whose husband has disappeared is neither widow nor divorcee and may not remarry until her husband has been located and given her a divorce (perhaps under constraint) or until his death has been proven. On the other hand, an objection is raised to so ready an acceptance of this testimony that could release her by establishing the death of her husband: perhaps the voice is that of a demon?

It is in this utterly real context that the Talmud takes up the question of how we can distinguish a human being from a demon in human form. An amazing method is suggested for determining whether a fugitive vision in human shape is a demon: look for its shadow! Although it is true that demons can adopt human shape, they do not cast a shadow. After the objection is raised that demons do have shadows, an expert demonologist is summoned to determine the question. His name is *Yonatan Shida*, "Jonathan the demon"; Rashi, perhaps with a smile, enlightens us that "he was a demon, or [perhaps] an expert on them." The possibility that this informant was himself a demon is particularly interesting, because it means that were are learning to recognize a demon "self-referentially," as it were, through knowledge acquired from a demon himself. In any case, we now learn from this expert that although a demon may have a shadow, his shadow cannot have a shadow.[72]

This clearly suggests that if the ability to cast a shadow is a sign of reality, depth, and opacity, in contrast to the phantomlike nature of a spirit or demon, it is a sign of even greater depth and reality when the shadow itself casts a shadow. But we must try to understand what this procedure and expertise mean for the sages of the Talmud and Midrash, who considered demons to be real beings, invisible, imperceptible "breaths," disembodied spirits wandering in search of a body, created at the very end of the sixth day of Creation, just before the start of the first Sabbath, when there was no time left to form their bodies.[73]

But the *Zohar*[74] delves even deeper into this ambiguity, playing on the similarity between *shed* 'demon', *shad* 'breast',[75] and *shaddai* (which is a homograph of *shadai* 'my breasts'), a divine name that reflects the nourishing sufficiency of the earthly effects of the deity (as if from *dai* 'sufficient'). Elsewhere, the *Zohar* suggests that "demon" may also designate a real human being, or rather part of his or her soul. The same applies to *ruaḥ* 'spirit', which may mean part of the human soul (the "affective soul" of Aristotelian terminology)[76] or an incorporeal being whose effects are frequently associated with those of demons and angels. Thus a *shed* or demon, associated with and integrated into the human soul of a righteous person, may be transformed into an angel that guides the human being it occupies to live the life of a sage. More precisely, the *Zohar* sees this as the origin of the mundane knowledge that could be attained by the talmudic sages, whom it calls here "masters of philosophy (*filosofiya*), astrologers of Israel, who know what was and what will be, from the signs of the sun and moon and their eclipses, and every planet and constellation, and what is seen in the world."[77] On the other hand, if the moral

72. This same colloquy is found in B *Giṭṭin* 66a, several pages before the story of Solomon and Ashmedai, with regard to the similar issue of a man who calls out from a pit that whoever hears him should write a bill of divorce for his wife: perhaps the voice was that of a demon? On the next page, the Talmud takes up the question of diseases that may leave a man deranged, to the point of rendering him legally incompetent to issue a divorce. These pathological states are ascribed to demons, like the one named Kordiakos, who seizes hold of those who have drunk too much new wine straight from the vat (see Rashi on M *Giṭṭin* 7:1 [67b]). The Gemara explains that knowing the name of the demon responsible for an illness is indispensable for writing amulets to cure it. The story of Solomon and Ashmedai follows soon after. It gives way to several pages that take up the "medical" thread and expound techniques for treating diverse pathologies. These pages are part of talmudic medicine, whose application was banned by the rabbis of the Middle Ages (see Atlan, *Enlightenment*, p. 268 and note 6.52, p. 284).

73. See above, p. 157, the passage from B *Ḥagigah* 16a. Naḥmanides (on Lev. 17:7) quotes this talmudic passage and then explains that demons are composed of only two elements, fire and air, whereas all other creatures are composed of all four elements (earth and water as well). This verse speaks of the "goats" or "fawns" (*śe'irim*) to whom the Israelites were henceforth forbidden to offer sacrifice. For Rashi, the word means "demons." Abraham Ibn Ezra explains the metonymy: "madmen see them in the shape of goats" and also, perhaps, because of the way in which one's skin bristles (like goat hair) if one sees them.

74. *Zohar*, Genesis 27a.

75. This comparison is found in a talmudic discussion of the manna. Fallen from heaven, the manna was a sort of prophetic equivalent of food. It was intended for the generation of the wilderness, the "generation of knowledge," in which the sin of the Tree of Knowledge, which also involved an element of nourishment (see Chapter 2), was being repaired. The Bible says that manna tasted like *leshad ha-shamen* 'cream of oil' (Num. 11:8). Playing on the two senses of the homograph *shed/shad* and construing the first word as *le-shad* 'like a *shed/shad*', the Talmud compares the flavor of manna successively to that of the breast for a nursing baby and to that of a demon: "Rabbi Abbahu said, . . . 'Just as the infant tastes several flavors in the breast, so did Israel find many flavors in the manna as long as they were eating it.' Some say: [*Le-shad* means] 'a real demon; just as a demon changes into many colors, so did the manna change into many flavors'" (B *Yoma* 75a).

76. See Henri Atlan, "Souls and Body in the Genesis," *Korot* 8(5–6) (1982), pp. 115–121.

77. *Zohar*, Deuteronomy 277a.

qualities, body, and soul of the individual with which the demon has merged do not allow him to rise above his own self-interest, the demon guides his life into harmful activity that ultimately leads to his own destruction.

For the *Zohar*, the possible transformation of demonic power into divine power is explicitly compared to the reversible metamorphosis of serpent and Moses' staff, called *maṭṭeh ha-'elohim* 'the staff of [the] god(s)'.[78] We are still dealing with the theme of the *naḥash*, of cunning knowledge, which must sometimes be twisted and "serpentine" in order to discover, from what is actually seen, that which is hidden in nature. This knowledge may be union with the nature thus disclosed; but it may also be a source of control and exploitation that are uninhibited in proportion as the knowledge is effective. What is more, because this knowledge is always unreliable, like prophecy itself, it is important to know how to distinguish it from the direct and "normal" knowledge of things and of the causes that produce them.

This explains the role of shadows and images in diagnosing the real or illusory character of what is known in this way. That demons do not cast a shadow is clear evidence of their unreal, phantomlike character, belonging to the domain of dreams or hallucinations, so far as we can see them. But there remains the possibility of an imaginative vision of the shadow itself. And this is why one should hunt out and test the image, or the "shadow of the shadow," as explained in slightly greater detail by a passage in tractate *'Avodah zarah* that discusses the shadow of a tree and the shadow of its shadow. Rashi explains that this expression refers to the elongated shadow produced when the sun is low in the sky, longer but not so dense as the shadow produced by the sun at the zenith. For Rashi, the shadow itself cannot be longer than the height of the tree. The "shadow of the shadow" is the penumbra that is longer, less dense, and less dark than the true shadow.[79]

Thus looking for the shadow of the shadow of a demon implies that one is not satisfied with noting the presence of a shadow and goes on to study the variations in the shadow produced by the motion of the sun. Following the advice of the expert demonologist of the Talmud and recognizing a demon by the absence of the shadow of the shadow—even though said demon may take a human shape and may even cast a shadow in the sun (or rather in the moon, in the passage cited above from tractate *Yevamot*)—is equivalent to verifying indirectly, by noting that the shadow does not vary in size and intensity, that it has different causes, perhaps imaginary or hallucinatory, than those that habitually produce the shadows of objects and corporeal beings. This is an application of the "experimental method" to verify not only that the shadow exists but also that it is produced according to the rules, conforming, as it were, to the standard laws of umbral mechanics.

11. SHADOW AND SHADE, INSIDE AND OUTSIDE

On the same page of tractate *Yevamot* that is uncertain about the heavenly or demonic origin of the voice, we encounter (still in a legal context) another example of possible confusion between a human being and a demon in human form. This is the actual case of a man bitten by a serpent who called out, from a distance, that he was dying, after first identifying himself as X son of Y. But when the judges or witnesses reached the corpse they could not verify his identity. Here too

78. Exod. 17:9; *Zohar*, Genesis 27a (see Chapter 1, n. 78).

79. B *'Avodah zarah* 48b.

there is doubt, as in the previous case of the widow, as to the identity of the person who spoke. The law is that X the son of Y is considered to have died; his wife is recognized as a widow and may remarry. In passing, we learn that this extraordinary event occurred in a place called Ẓalmon, whose name recalls both *ẓel* 'shadow, shade'[80] and *ẓelem* 'image, form', a wink toward the close meanings that the *Zohar* carries further.[81]

This linguistic proximity is already suggested by the phrase "their shadow (*ẓillam*) has departed from them" (Num. 14:9), which the Bible applies to human beings whose days are numbered and who are accounted as little better than phantoms or walking dead, despite their physical presence. *Ẓillam* is a homograph of *ẓelem* (צלם), hinting perhaps that the divine form of their bodies has left them, in some unseen fashion, even though they are still seen moving about as if nothing had happened. The *Zohar* develops this idea, affirming that the approach of death involves the progressive departure of two aspects of the *ẓelem*, to which the Song of Songs (2:17) alludes when it evokes the day when "the shadows flee."

Here we need to realize that for a culture in a hot country, where the sun beats down fiercely, shade and shadows are associated with a protective power. This power is then transposed to the "clouds of glory" or "wings" of the divine presence, in whose shade human beings are nourished and sheltered, as in the *sukkot*—booths or tabernacles—that housed the Israelites in the wilderness. Hence the disappearance of the shadow is a presage of death, as if one's guardian angel has departed. Accordingly the shadow has a dual attribute, both cognitive and practical. On the one hand, observing and studying it allow us to discover the hidden causes of certain things (indicating, for example, the activity of demons). On the other hand, its very existence is paradigmatic of the efficacy and security of our bond to the fire of our life and the light of our inspirations.[82]

As we shall see, it is the essential reality of our person, simultaneously active and passive, subject and object, that is illuminated as it were by this metaphor of the shadow of the shadow. The metaphor of God who shields human beings in his shade, his protective shadow, is read literally and inverted; the shape and motion of the shadow are determined by the person who produces it. To put it another way, if man is a shadow, an image of the deity, the latter is in turn a shadow of man, the shadow of a shadow.

This idea is employed by a midrash quoted by Ḥayyim Volozhiner to explain the celebrated "I am who I am" (Exod. 3:14), of which a more literal rendering is "I shall be what I shall be." This phrase, far from stating the motionless identity of being, as is commonly thought, expresses rather a dynamic reciprocity of god and man. According to this midrash, when God answers Moses, who has inquired as to his name, "Go tell Israel that my name is 'I shall be what I

80. <Note that the distinction between "shade" and "shadow," obvious in English, does not exist in either Hebrew or French.>

81. *Zohar*, Leviticus 13b and 43a.

82. Consider the phrases "the shadow of Shaddai" (Ps. 91:1) and "YHWH is your guardian, YHWH is your shade at your right hand" (Ps. 121:5). On the cognitive plane, shadows can be thought of either as darkness or as a means of making what is illuminated visible. Shadows play this dual role in Plato's famous cave, too (*Republic* VII, 514a–517a); it would be first of all "the shadows, then the images of human beings and other objects reflected in the water, and then the objects themselves" that a prisoner in this cave would see most clearly if suddenly forced to raise his eyes toward the light, because the glare would keep him from seeing objects immediately (516a, trans. Paul Shorey; in *Collected Dialogues*, ed. Edith Hamilton and Huntington Cairns [Princeton: Princeton University Press, 1961]). All knowledge necessarily implies boundaries by which things are known, the shadow around what is lit up. Depending on the orientation of the light, however, the shadow may conceal in order to better reveal, making it possible to see the light without being blinded; or, on the contrary, it may merely obscure things and keep them from being seen.

shall be,'" the meaning is "just as you are with me, I am with you." The protective divine shadow is then invoked to support this reading: "YHWH is your shade on your right hand" (Ps. 121:5). What is meant by "YHWH is your shade"? asks the midrash, and replies that the sense is "like your shadow": "just as your shadow smiles at you when you smile at it, and cries back at you when you cry, and if you look at it angrily or cordially it does the same to you, so too . . . 'YHWH the Lord is your shade/shadow.' Just as you are with him, he is with you."[83] Moshe Idel cites this midrash as quoted by Meir Ibn Gabbai[84] and compares it with the metaphor of the shadow in an Assyrian proverb and in a Gnostic treatise, where the king is God's shadow on earth and man is the king's shadow—thus the "shadow of a shadow." Idel cogently observes that here this metaphor is inverted to illustrate the idea, widespread in the kabbalah, that the activity of divinity "above" reflects human activity "below."

12. SUBJECT AND SHADOW

This midrash and its mirror-image interpretation of "I shall be what I shall be" fits into the overall theme of Volozhiner's work, that of a radical human responsibility in which the structure and activity of the "upper worlds" depend on human actions. In this framework it is easy to understand how angels and demons, which are parts of this structure, may be determined and sometimes even created by human actions, as the Midrash and kabbalistic writings frequently affirm.

But these demons and spirits, as effective as they may be, have no material reality beyond the shadows of the bodies that produce them. This leads to the delusion denounced by Deuteronomy (32:17) as the acme of idolatry: sacrificing to "demons that are no gods (*shedim lo' 'eloah*), to gods they had never known (*'elohim lo' yeda'um*)"; worshiping, that is, not just incomplete and limited divinities like the sun and moon, but their shadows, which lack even their efficacy.[85] This clearly supposes the possible existence of demons who do partake of divinity, who are "gods" (*'elohim*), as Ibn Ezra notes in his commentary on the verse. This is why it is important to know them and to know how to distinguish among them. For it is fear and ignorance of the hidden realities that make us take every "shadow" for a "real" demon, without seeing that this shadow does not cast its own shadow and is mere form without matter, or matter dissociated from form, whose shadow has no shadow.

Finally, all of this also implies that the divine image or form from which man was created (the *ẓelem 'elohim* of Genesis) is itself a reproduction of the human image, like a man's shadow. The *Zohar* invokes the phonetic similarity of *ẓel* 'shadow, shade' and *ẓelem* 'image'—the divine image of the human body—to describe the hidden causes that determine the specifically human aspect of our lived experience. Neither soul nor body, but the "garment of the soul" located at the junction between the two, the *ẓelem* as an organizing form is also the theater of the internal drama, the arena of the conflict between good and evil.[86]

This particular relationship between the *ẓelem*, the divine image of the human form, and the creative impulses for good and bad

83. Ḥayyim Volozhiner, *Nefesh ha-ḥayyim* 1.7 (p. 26).

84. Moshe Idel, *Kabbalah: New Perspectives* (New Haven: Yale University Press, 1988), pp. 176 and 180. This metaphor of God and man as reflecting each other can also be found in the mystical Islamic tradition of Sufism, for example in the *Fuṣūṣ al-ḥikam* of Ibn Arabi (1165–1240); see Ibn al-'Arabī, *The Bezels of Wisdom*, trans. R. W. J. Austin (New York: Paulist Press, 1980), pp. 45ff.

85. *Sifre*, quoted by Maimonides in *Guide* 3.46.

86. *Zohar*, Leviticus 13b.

(the *yezer tov* and *yezer ra'*) that direct human beings, like angels and demons that possess them, is expounded in detail in the foundational book of the Lurianic kabbalah, Ḥayyim Vital's *'Eẓ ḥayyim*:

In sum the form (*ẓurah*) is the root, which is called "man" (*'adam*) and is composed of *yeḥidah, ḥayyah, neshamah, ruaḥ*, and *nefesh*. The *ẓelem* itself is also divided into five categories corresponding to these five parts [of the soul, here called "form"] and it is the garment of the form. This form can never subsist without this *ẓelem*. This *ẓelem* is the *yezer tov* and *yezer ra'*, which are angels within man himself, but not the man himself. It is the breath of a skeleton [*or* fleeing vapor of self],[87] which is wrapped in the vessel of the body. This is why it is called *ẓelem* [image], because it is something intermediate between form and matter. It resembles both of them, and is an image and icon of both.

87. The Aramaic *garmei* reproduces the double meaning of the Hebrew *'eẓem*, both "bone" and "self" (in the sense of "essence" or "substance").

88. Ḥayyim Vital, *'Eẓ ḥayyim* (Jerusalem, 1988), Sha'ar ha-ẓelem, end of chap. 1 (2:49–50).

You also know, however, that the precepts (*mizvot*) exist only to refine and purify *ẓelem* and matter. But form has absolutely no need of repair (*tikkun*) and wraps itself in *ẓelem* and matter only to draw the light to them in order to repair them. Understand this well, because this is the reason for the soul's descent into this world, to repair and purify, just as the exile of the divine presence purifies the [fallen] sparks, as is known.[88]

The text adds that some angels are larger than the human soul and consequently are not wrapped in a body, but others are smaller than the soul, "because man (*'adam*) himself is the form, and these angels that accompany him are the 613 organs of the *ẓelem*, which are his servants and vehicle."

What emerges from this scheme is that man is the battlefield of the impulses, which clash in an arena, the *ẓelem*, that is located between the body and the soul and is neither one nor the other. Man himself is the prize fought over, the confrontation zone of the forces (or "angels") that originated outside him but now envelope his soul (his very self) and are wrapped in his body. "Servants and vehicle" of the self, they drag him now in one direction and now in the other, depending on the balance of power among the elements that compose him, the outcome of the previous history of the universe that has led to this individual's existence.

It is by practicing the *mizvot* or precepts of the Law (of which there are 613, corresponding to the 613 "organs" and "sinews" that make up the body and the *ẓelem* that inhabits it) that a man can hope to be able to steer his vehicle and order his servants in the direction of redressing and repairing the damage they have already wrought, instead of allowing himself to be maneuvered by them, the random issue of the events that brought them together in him. This text refers only to angels, and not to demons, because it is describing a universal archetype in which evil, still at its source, is not intrinsically evil. Like Satan in the book of Job, or Ashmedai, the demon king who is subdued by and subdues Solomon, the demons of the evil impulse (the *yezer ra'*), in their source "above," are just another species of angel. For this model, the elements of the moral problem, the good and evil aspects of human behavior, are determined outside the realm of individual initiative and are in some sense imposed on human beings by forces that penetrate them and act through them. It is clear that this is relevant to the absolute determinism of human behavior we

encountered in the previous chapter. As we saw there, the ethical code that follows from it holds an individual accountable for acts he did not determine and that are his only inasmuch as he is the seat and instrument (as well as spectator) of these acts, even if he sometimes has the experience of choosing them.

Note, incidentally, that in this context we can understand the divine character of the *ẓelem*, the "divine image" of Genesis by which the first man and woman were created, as expressing their power, which is divine insofar as it derives itself from the two "creative" impulses that epitomize the effects of all these angels. In this sense the *ẓelem* is indeed divine, just as all angels are divine (which is why they may be called *'elohim*), and also reflects their multiplicity: "Let *us* make man in *our* image," says Genesis, a plural that the Midrash interprets as that of the assembled angels of the heavenly host, with whom the Creator took counsel.

13. KNOWLEDGE WITHOUT CONSCIENCE?

This returns us to the key idea that evil at its source is not evil, that death and destruction, as principles, are integral parts of the processes of life and construction to which they are opposed. Pure and impure are ethical categories of human behavior, not part of the nature of things.

This is why there is nothing irrevocably bad about keeping company with demons, especially if the goal is to increase one's knowledge. As we have seen, "demonic" knowledge is not necessarily to be rejected, because it can be reworked and in some way purified when applied with the goal of reintegrating the worlds. An appropriate goal and moral context may make it possible to avoid the temptations of exercising unchecked control in the service of special interests.

We might say that behind the symbolism and ambiguity of the serpent, demons, and the Tree of Knowledge there stands Rabelais's classic warning, "knowledge without conscience [*science sans conscience*] is but the ruin of the soul."[89] This is indeed true, although in a different context and application, because the science in question is not the objective and impersonal science of the modern age. Its attempts to control nature are successful thanks to its method, which is based on an operational consensus that imposes a distance between the object of our knowledge and our subjective inner life. This distance is sometimes perceived as a form of depersonalization, in that the individual human being is no longer at the center of the universe. The most spectacular products of technology merely reinforce the notion that they are fabricated by some impersonal overseer of our lives, while no one seems to be able to guide or even predict the direction in which we are being carried. No one can be assured, in particular, that the benefits of technological advances will not be swept away and buried by major catastrophes, only some of which can be predicted, and even then with no certainty as to their magnitude and the risks of their occurrence. But this situation, as worrisome as it is, can be dealt with, at least in principle, by endeavoring to limit the hubris of technology and its fantasies of omnipotence. This is the objective of diverse institutions founded to promote an ethical approach to scientific research and applications. The separation of domains actually increases the effectiveness of such an ethical approach; it is just because the object of modern science is depersonalized, although produced by human beings, that we can readily admit the possibility of a personal ethics that has a different origin (in history, in tradi-

89. Rabelais, *Gargantua and Pantagruel*, chap. 2, VIII, trans. Urquhart and Motteux.

tions and religions that have produced laws, in myths, and in philosophies) and that functions as a necessary counterweight or barrier.

Unlike contemporary science, the science of bygone days pursued a goal that was simultaneously cognitive, practical, and axiological. In addition to their supposed efficacy, alchemy or astrology furnished not only knowledge of the hidden processes underlying nature but also salvation through the magus, whose person expanded to fill the universe. Jewish kabbalah was part of this movement, but with a fundamental difference that was also a source of tension: its constant concern to avoid identification with alien magic, considered to be idolatry and the enemy par excellence. In this it clearly resembles the dogma of the Scholastic theology of the monotheistic religions. But unlike that, and more closely related to the ancient Hebraism of the Bible, it strove to combat idolatry on its own ground, by showing that its own endeavor to integrate the personal and impersonal increased its effectiveness in realizing both cognitive and ethical objectives.[90]

The union of the cognitive and the ethical, sundered by contemporary science, is mandated by every attempt to produce a philosophy of the moral subject. Hence reflection on this ancient wisdom remains instructive, even if its operational value and efficacy as the science of nature vanished long since.

14. WISDOM OF THE LEFT SIDE AND WISDOM OF THE RIGHT SIDE

So it is inside itself that this science of bodies and minds, seeking to control nature and itself, encounters a rupture between the straightforward and cunning uses of the intellect. In his commentary on the *Zohar*, the *Sullam* (Ladder),[91] Rabbi Yehuda Ashlag presents it as a dialectic tension between the wisdom of the right side and the wisdom of the left side. It is also this rupture that the practices aiming at "union" or "reunification," exemplified in many places in the *Zohar* and subsequently taken up by all those inspired by it, seek to overcome; because the disclosure of concealed things entails both recognition of multiplicity and differentiation between causes and their effects, it must always be accompanied by a return to their single origin, worked by means of a project to unify contraries and weaken oppositions. Ashlag presents this project as a cooperative endeavor, based on compromise, by the two types of wisdom. The wisdom of the left side, which is divided and selfish by its very nature, renounces flowing downward from the head. This allows it to receive from the wisdom of the right side, which flows naturally downward. Conversely, the wisdom of the left side can then serve as a source of wisdom that is constructed upward, on the basis of the divided and scattered experiences of the objects of existence.

This rather complicated theme recurs throughout the work of Ashlag (1884–1954), one of the last major kabbalistic authors. We should pause to examine this distinction, which plays a key role in the exegetical system of the *Sullam*. Ashlag never gives an explicit source for the terms "wisdom of the left side/right side."[92] Perhaps we can

90. Ever since the theme of the golem first appeared in the Talmud, the capacity to manufacture an animal or artificial man has often been presented as the classic emblem of this lore, the touchstone both of the truth of its doctrine and of its operational efficacy and moral excellence (see Chapter 1; and Atlan, preface to *Le Golem*).

91. Yehuda Ashlag, *Zohar with the Sullam Commentary* (Jerusalem, 1945) (Hebrew).

92. We may identify an implicit origin outside the kabbalistic literature, in that the word *ḥokhmah* 'wisdom' can also mean "skill" or "craft," as in "come let us deal shrewdly (*nitḥakkemah*) with them" (Exod. 1:10). There is also the talmudic idiom of going to the right side or left side of the Torah

trace it to the very first sentence of the *Zohar* on Genesis,[93] where the primordial creative wisdom seems to be called a "spark of impenetrable darkness" or "a lamp of impenetrability," which combines the light of *ḥesed* ("generosity," "effusion"), characteristic of the right side, with the darkness of *din-gevurah* ("judgment" or "rigor"), characteristic of the left side. Many kabbalistic texts associate analytical intelligence (*binah*), which circumscribes, defines, and distinguishes, with the origin of *din* and, equally, of evil. For example, according to the *Zohar*: "Even though this river [which "flowed out of Eden to water the garden" (Gen. 2:10) and is compared to this "intelligence," considered also to be a nursing "mother"] is not *din*, the *dinnim* [rigors] come from its side."[94] "Wisdom" and "intelligence" (*ḥokhmah* and *binah*) are like the father and mother of the child "knowledge" (*da'at*). Their union is such that there is an aspect of wisdom in intelligence and vice versa. It is this *ḥokhmah* in *binah* (and thus on the left side) that is split into the wisdom of the left side (where it is located) and the wisdom of the right side (from which it proceeds). This is precisely how the *Sullam* expounds the passage from the *Zohar*.[95]

The same theme, in various forms, appears in many other parts of the *Zohar*, and in particular in the dramatic episode that opens the section known as the *'Idra Rabba* (Great Assembly).[96] Here we are told that Rabbi Simeon bar Yoḥai, preparing himself to reveal to his disciples the most esoteric aspects of his doctrine, is torn by the dilemma of not disclosing them (and thereby preventing wisdom from spreading through the world) or disclosing them (thereby accepting the risk that this wisdom will go astray when received by those unworthy of it). To avoid this peril he exhorts his disciples, already carefully selected, to prepare themselves by means of *yiḥudim* ("acts of unification"); they join together in pronouncing a curse on anyone who employs the divine Name to forge "a graven or molten image." These are the idols banned by the Ten Commandments; their fabrication is not forbidden per se and in all circumstances, but only insofar as they are considered to be representations of the deity. For Ashlag, here Rabbi Simeon is alluding to the effusion of the "wisdom of the left side from above to below," because this leads to the crystallization in images of what ought to remain hidden, even when it is revealed.[97]

More generally, Ashlag develops the theme of the two species of wisdom into a master plan and applies it in various contexts, in each of which it takes on a different meaning. To grasp its richness, we need to understand how its structure and dialectic are integrated into the more general framework of the system of the *sefirot* and the "faces" or "configurations" (*parẓufim*) that serve as the armature of the various kabbalistic treatises.

The system of the ten *sefirot* and their configurations can be seen as a dialectic arrangement of the two "sides," as moments of thought or aspects of reality, variously expressed according to the context: the one and the multiple, the same and the different, the infinite and the finite, effusion and retention, union and division, light and darkness, affirmation and negation, being and nonbeing, and so on. In the kabbalistic system, these two "sides" designate the anthropomorphic

(or perhaps of directing the Torah to the right or left side): "To those who go to the right side [of the Torah] it is a medicine of life; to those who go to the left side [of the Torah] it is a deadly poison" (B *Shabbat* 88b; see below, p. 175). See also Chapter 1, n. 68, for the relationship among wisdom, cunning, and nudity, according to B *Soṭah* 21b.

93. *Zohar*, Genesis 15a (trans. Matt, vol. 1, p. 107).

94. *Zohar*, Leviticus 98b.

95. Ashlag, *Sullam*, vol. 14, p. 69.

96. *Zohar*, Numbers 127b–128a.

97. Ashlag, *Sullam*, vol. 15, pp. 40–43.

notions of "right" and "left," or *ḥesed* ("generosity," "effusion") and *din-gevurah* ("judgment," "rigor," "limitation"), or male (on the right side, that of *ḥesed*) and female (on the left side, that of *gevurah*). This last pair expresses the dialectic interlinking of these notions, because *gevurah*, though feminine, is commonly considered to be a property of the *gever* 'man' and *gibbor* 'hero' (respectively גבורה, גבר, and גבור). More generally, the left side is the source of evil, because the division and multiplicity that characterize it can degenerate into opposition and violence; but the right side can also produce the evil of nondifferentiation. On the other hand, the left side, which is also the place of "fire," is the side of "life," as change and motion (and of death, which always accompanies its renewal).[98] The right side, that of "water," is the nourishing source that permits this motion to continue. In Ashlag's system, these two sides are represented by the "desire to give" and the "desire to receive," which, although opposed, clearly cannot exist without each other. His entire interpretation of kabbalistic cosmogony and psychology rests on this definition, which he develops throughout his work, from the sixteen volumes of his *Talmud 'eśer ha-sefirot* to the short treatise *Matan Torah* and the summary in the introduction to the *Sullam*. The desire to receive is on the left side, that of division, where the pleasure with what is received is exhausted in its satisfaction and can be resumed only with a new and different object of desire. Its symbol is the "closed door." But this desire to receive can be satisfied only as a response to a "desire to give," and that in turn can be satisfied only if the gift is received. Thus converted into satisfaction of the desire to give, the desire to receive is no longer shut behind what has been received. Now its symbol is the "open door," because it is not exhausted by a particular object that it desires and receives and is transformed into a desire to permit the desire to give to be satisfied indefinitely, by being received.

It is in this context that Ashlag introduces the notions of "wisdom of the right side" and "wisdom of the left side."[99] This dialectic applies also, and perhaps especially, to knowledge as the discovery and reception (*ḳabbalah*) of creative wisdom, and to thought as exploration and acceptance in speech. Next, borrowing the distinction between "direct light" and "reflected light," developed by Moses Cordovero[100] and later in the Lurianic kabbalah, he associates the direct reception of the wisdom of the left side ("downward") with the sin of the Tree of Knowledge.[101] The effusion (of the right side) that permits this reception is then harnessed in the service of division, which is apt to remain fixed. But the reception of wisdom cannot be limited to the wisdom of the right side, because in that case it would concern only the undifferentiated One and would in some sense be tautological and purposeless. It is also necessary for the latter to sway the former into its service, to prevent it from flowing downward and to uncover it as reflected in its objects, "upward." The aborted sacrifice of Isaac, "the father of the left side," bound tightly for slaughter by Abraham (the father of the right side who initiated the repair of the sin by shattering his father's idols and then "by going down to Egypt"), is classically interpreted as the binding of the left side by the right side. This theme is developed at length in the *Sullam*'s exposition of the *Zohar*'s explanation of the meaning of the "voices of the shofar," the rhythmical sounds played on the ram's

98. See Atlan, *Entre le cristal et la fumée*, chap. 13.

99. E.g., Ashlag, *Sullam*, vol. 1, pp. 55–60 and 121–125; vol. 2, pp. 60–71; vol. 3, pp. 42–49.

100. E.g., in his *Pardes rimmonim* (Korets, 1780), chap. 15.

101. Ashlag, *Sullam*, vol. 4, pp. 43–53 (on *Zohar*, Genesis 81b–82b); vol. 14, p. 86.

horn on Rosh Hashanah, which is also the "day of judgment."[102] The result is the reintegration of the left and right sides in the culmination represented by the third patriarch, Jacob-Israel. Wisdom of the left side is severed from its source above, and that source is thereafter hidden in a "modesty" that guarantees its survival. It is then taken up again and clothed in the wisdom of the right side, which allows it to flow, but only by reflection (in both senses of the term) and "upward."

This complicated scheme corresponds to fairly common experiences of our cognitive lives. One of the most obvious, for Ashlag, is that of thought (as an infinite and impersonal activity), which is expressed in words, by means of the intellect, and in discourse, uttered by the voice.[103] The words and letters are the discrete vehicles without which no thought can be expressed. Thought must pass through the analytic intelligence that verifies, distinguishes, and separates. But the content of thought is expressed in sentences, which connect and wrap the words and the letters together by the continuity of the voice that speaks them or the text in which they are written, and not in isolated words or individual letters. This set of gaps and links between words and letters, which is also known as "white space," is the source of meaning.[104] The voice that connects, modulates, and consolidates words by speaking them in a sentence does the work of the wisdom of the right side, which "binds up" the wisdom of the left side and prevents it from flowing just as it is, as a succession of individual utterances. Wrapped together in the stream of the voice (or against the background of the page), it is then reflected in the words and what they designate, so that the meaning of the content of thought can appear.

102. *Zohar*, Exodus 184b; Leviticus 18a–19a; Numbers 98a–100b; and Ashlag, *Sullam* ad locc.
103. Ashlag, *Sullam*, vol. 2, pp. 282–283.
104. See Atlan, *Enlightenment*, pp. 59 and 117; see also *SR* II, chap. 12.

In contrast to this normal revelation of thought by the intelligence of speech or text, some hallucinations produce the experience of an uninterrupted stream of images and words, each perceived with extreme intensity in moments that seem to be outside time. However, the succession of these moments is such that the images and words are linked only by fugitive associations, no sooner glimpsed than forgotten. Unlike the unit of meaning contained in a sentence or speech, this stream does not connect the words and images that compose it. They remain discrete, bright sparks whose separation is in proportion to the intensity with which they are perceived—the fireworks of the "wisdom of the left side falling downward."

15. KNOWLEDGE "BELOW" AND UNITY "ABOVE"

This scheme also applies to the a priori reasons that the intellect sometimes comes up with in its rational flashes that trigger purely theoretical discourses. What these discover is certainly structured and meaningful, but its relationship to concrete objects is far from obvious. Theoretical intelligence, which analyzes, defines, and links words in sequence, produces abstract entities that can be used in formal systems or, arbitrarily, in a posteriori rationalizations; it does not produce the reason of things. There, too, at a certain moment the wisdom of the left side must be severed from its theoretical source so that it can be controlled by the immediate intuition of things and serve as practical reflection that moves upward. This is what Ashlag seems to say when he locates the "lights" or the "secret" of numbers and calculation in the "head" (the first three *sefirot*) of this wisdom of the left side.[105]

To put it another way, logic must bow to and serve empiricism so that its fitness to things may be revealed. This is why it is important to measure and weigh things using an honest scale, in which the two sides align in the middle when the pans balance.[106] This metaphor of the balance is employed at the start of the section of the *Zohar* known as *Sifra' de-Ẓeni'uta'* (Book of Modesty), which develops the dialectic of a wisdom that must be clothed for the truth of things to appear through, and thanks to, these garments.[107] Thus logic and numbers, reflected upward, seem

105. Ashlag, *Sullam*, vol. 12, pp. 9–10 (on *Zohar*, Numbers 221b). In another context, that of sacrifices, he explains that "the head of the wisdom of the left side, of division, is the impure fat, while its body, associated with the right side by the median line, is the pure fat)" (ibid. on *Zohar ḥadash*, Genesis, vol. 17, p. 79).

106. Ibid., vol. 2, pp. 302–304.

107. See Atlan, *Enlightenment*, chap. 9. This dialectic is also suggested by the shadow in the story of the Tree of Knowledge, with its double role: obscuring, but also as the backdrop that makes it possible to perceive distinct objects (see above, n. 82).

When Adam and Eve "knew that they were naked" and covered themselves in fig leaves (Gen. 3:7) they attempted, according to the *Zohar*, "to be covered by [*or* cover themselves, hide behind] the shade of that tree from which they had eaten, [which (i.e., the shade) is] called 'leaves of the tree'" (*Zohar*, Genesis 36b [trans. Matt, vol. 1, p. 229])—leaves that, as the *Zohar* goes on to state, convey knowledge of magic; after clothing themselves in fig leaves "they discovered every kind of sorcery and magic" (ibid., 53b [p. 296]). Ashlag links these two statements (*Sullam*, vol. 3, p. 181), associating them in the image that views witchcraft as the shadow of knowledge. He contrasts this shadow, which obscures the light of knowledge, with the "shadow of the sacred" that hides in order to disclose. This is the opposition between the forbidden magic of idolatrous witchcraft, which "weakens the heavenly retinue," and the permitted magic of the righteous, associated with the unity from which this retinue proceeds (see above, n. 56).

We must realize that the difference between the two does not really apply to the content of the knowledge or even to how it is used, as shown by the dialogue between Rabbi Simeon bar Yoḥai and his son Rabbi Eleazar in *Tiḳḳunei ha-Zohar* (see above, p. 58). The difference lies in the context and intention of these practices: on the one hand, the shadow of submission to the *miẓvot*—the duties and prohibitions prescribed by the Torah as a path of "holiness"; on the other hand, the shadow of sin, which also produces knowledge, but over the long term engenders a different reality that is opposed to the former. We see clearly how the law, with its prohibitions, casts a shadow over the spontaneous creativity of the impulses. These impulses, whether "good" or "bad," impulses of life or of death, *yeẓer tov* or *yeẓer ra'*, are creative forces (see Atlan, *Entre le cristal et la fumée*, chap. 13), as indicated by their designation *yeẓer*, which derives from the verb *yaẓar* 'make, create', and also means "created being." Sin—at least the first time, before it becomes habit, a veritable addiction—permits discovery, an opening toward the unknown, knowledge of what had been hidden. But this knowledge "weakens the heavenly retinue" by diffusing "the shadow of death" (*ẓalmavvet*) over everything that is not it.

> Unlike the shadow of the sacred, meaning that produced by a *miẓvah*, as in the case of Moses [whose face radiated light because he *obscured* his face in the incident of the burning bush], the shadow that is produced by a transgression is the origin of the darkness that obscures the faces of human beings—what is called "shadow of death." (*Sullam* 3:182, on *Zohar*, Genesis 53b)

But when the transgression is repeated and does not return to its source, it reinforces the lethal impulse that produced it and created the dependence of the addiction, in which the desire is exhausted by repetition. Similarly, the knowledge possessed by the idolatrous magician is dangerous precisely to the extent that it is true and efficacious, constructing its own reality and setting it up as an absolute. This knowledge, on "the other side," replaces the knowledge of the righteous man, which is potentially open to the infinite, and prevents it from establishing a truly nourishing and life-bearing reality for the long run for whatever is not it.

In certain very specific circumstances, the Torah itself transforms transgressions into obligations. We have already mentioned several practices normally forbidden but prescribed inside the Sanctuary, such as inquiring of oracles (the *Urim ve-Tummim*), or burning the incense and tending the perpetual fire on the Sabbath. Similarly, incest with one's sister, extended to one's sister-in-law, becomes the obligation of levirate marriage (Deut. 25:5–9); the prohibition on wearing garments that combine linen and wool (Lev. 19:19)—which, according to the *Zohar*, is emblematic of Abel's offering of animals and Cain's of plants—becomes an obligation as part of the ancient practice of attaching ritual fringes, which must be of wool, to one's garment, even if that is made of linen (Num. 15:18; see *Zohar*, Leviticus 87a–b).

On a more personal level, the Mishnah tells us that repairing a transgression by its return (Heb. *teshuvah*, normally translated as "repentance") to its source is not possible for one who says, "I will transgress and repent, transgress and repent" (M *Yoma* 8:9 [85b]). The repeated violation of the prohibition, when recurrent, loses its quality of transgression and becomes, by force of habit, "like something permitted" (ibid., 87a). Through repetition, the transgression constructs a separate reality of the "other side," the mirror image of the sacred, of the "holiness" that is constructed by the routine of obligation.

to come from the bodies of things and not only from the intellectual ability to cut up and divide. Effecting the division, "severing the wisdom of the left side," like tricking the trickster, is a way to use these abilities appropriately and dynamically, without freezing them in delirious—because petrified—theoretical rationalizations. According to the *Sullam*, wisdom of the left side must submit to this trial in order to be received.[108] It is the acid test to which theory must submit. But it is also the moral test that consists of overcoming the temptation of remaining in the here and now of illumination (great or small), motivated by a desire to receive that is nourished only by itself and not by a desire to make it possible to satisfy the desire to give.

We see that this language, simultaneously complex and metaphorical, abstract and anthropomorphic, makes it possible to keep playing with connotations that are both cognitive and ethical. It serves as the vehicle for an ethics of knowledge in which the perspectives of the moral subject and the knowing subject occupy the same place in the operational model. But the model does not confound these two perspectives, and rightly so, because they face in opposite directions. The Psalmist juxtaposes "truth [that] springs up from the earth" with "justice (*ẓedeq*) [that] looks down from heaven" (Ps. 85:12 [11]), meaning generous justice or just generosity. Corresponding to the self-control of the pursuit of knowledge of the truth, which looks for its foundations in objects rather than in a priori principles, is the ethical summons or demand, the benevolence that falls from heaven, with no roots in the earth.

The same interpretive scheme is applied to the story of Babel, where it is necessary to confound the languages "on earth" in order to preserve the unity of language "in heaven." This is a dialectic that pits division, the necessary proliferation of languages, against the spurious transparency of a unique language, or the multiple languages of demons against the angels' sacred language—languages that switch roles depending on whether they are spoken "above" or "below."

The same model applies to the biblical and talmudic symbolism of the signs of the animals that may be eaten.[109] Leviticus 11 and Deuteronomy 14 prescribe that only quadrupeds that chew their cud and have cloven hooves may be consumed as food. According to the Talmud,[110] animals that satisfy the first of these requirements do not have incisors in their upper jaw. According to the biblical commentary by the kabbalist Menaḥem Recanati (late thirteenth and early fourteenth centuries), this anatomy corresponds to a structure with *gevurah* in the feet and *ḥesed* in the head.[111] This arrangement defines edible animals because we are meant to reproduce it in our humanity: division, by the power (*gevurah*) of measuring, distinguishing, and limiting, "in the feet," that is, in our conduct; and integration "in the head," meaning an unlimited effusion (*ḥesed*) of thought, free association, and the unification of contraries. This is contrasted with an inverse structure, with

108. Ashlag, *Sullam*, vol. 5, pp. 8–13. See above, p. 171 and below, n. 119, for the application of this pattern to the binding of Isaac. This interpretation can be compared to how the Sufi mystic Ibn al-Arabi interpreted the trial to which Abraham was subjected when enjoined to sacrifice his son. More than as a moral test of religious fidelity, intended to assess his unconditional obedience, Ibn al-Arabi views it as a test of the quality of his knowledge of God. Was Abraham capable of understanding that God did not demand that he sacrifice his son and that it was rather a question of the definitive abrogation of human sacrifice and its replacement by animal sacrifice? (Ibn al-ʿArabī, *The Bezels of Wisdom*, "The Wisdom of Reality in the Word of Isaac," pp. 98ff.).

109. <Standard translations render the adjectives *ṭahor* and *ṭame'* in this context as "clean" animals and "unclean" animals; but the more precise meanings of (ritually) "pure" and "impure," associated respectively with life and death, would suit Prof. Atlan's meaning better.>

110. B *Ḥullin* 59.

111. See Atlan, *Le cristal et la fumée*, p. 267 n. 1.

division—taboos and restriction of thought—above, and unity—a "mixed multitude" (*'erev rav*) or disordered jumble of values, behavior, and bodies—below. We are invited to counter "all behaviors are permitted, but not all thoughts are permitted" by "all thoughts are permitted, but not all actions and physical actions are permitted."[112] "Below," on earth, unity is an illusion if one begins with it, because it is a jumble that leads to violence, of which the generations of the Flood and the Tower of Babel are the symmetrical archetypes—the violence of all against all in the mixed multitude before the Flood, the violence of all against the individual in the totalitarian unity of the builders of the Tower.

Finally, the symbolism of the *shamir* used to hew the stones for Solomon's Temple seems to reproduce at least some elements of this structure. How is it possible to quarry and dress the stones—in short, to build—on the earth, "below," without violence and without destruction, so that the resulting edifice will be open to the infinite rather than closed and finite? This is one of the meanings of the biblical admonition, "if you make for me an altar of stones, do not build it of hewn stones; for by wielding your tool upon them you have profaned them" (Exod. 20:22 [25]). There too, the legend teaches us that demons—here through their king Ashmedai—can be useful in bringing this task, which requires the skill and cunning of the "wisdom of the right side," to a successful conclusion, while foiling the skill and cunning of the "wisdom of the left side."[113]

But beginning this task is a difficult undertaking because, even though the Torah, as unified wisdom, is its supreme value, the Torah can be both a "drug of life" and a "drug of death"—a drug of life for those who "go to its right side," a drug of death for those who "go to its left side";[114] or a drug of life for those who are deserving and of death for those who are not. Now it is precisely the acquisition of such merit or uprightness that is the prize of relating to the Torah as a guidebook or teaching that leads the student-sage along the path of knowledge.

Hence there is a circularity, a clear source of tension, which we can try to escape by establishing an unmistakable demarcation between the domains: the cognitive on one side, the axiological on the other; objective science on one side, morality and/or the personal deity on the other. This segregation of the camps is certainly the best solution today, when it is required by the constant advance of science and technology toward increased control and ever more impersonal objectivity. But it may also be the easiest solution, if we forget that it is employed *against* the enduring need for synthesis and for grand unified theories and ideologies. This, incidentally, is why the separation of domains and methods is always being challenged by the cross-purposes of unspoken or implied stakes in the minds of the individuals who are its agents. What is more, in this voluntary divorce from science, the historical roots of ethics are constantly being called into question—along with ethics itself, as practical morality and religion—by the exercise of the critical reason cultivated by the methods of the objective sciences. So we are always yearning for a wisdom that would be both objective rational knowledge and the emergence of the subject, of self and other, simultaneously the source and target of this knowledge.[115]

112. Ibid., chap. 12.

113. See above, Chapter 1.

114. B *Shabbat* 88b.

115. In this yearning we constantly encounter the ambiguity of the brazen serpent (see Chapter 2) and of the Tree of Knowledge. It is this multiple ambiguity of science and of intellectual wisdom, simultaneously demonic and divine, cunning and enlightening, twisted and straight, constructive and destructive, that is dramatized by the talmudic legend in which King Solomon and the king of demons become almost indistinguishable (see Chapter 1). The former,

16. THE EGO AND THE SUBJECT

Today, however, the demons have changed their nature, because they are no longer personified. Unlike the world of Paracelsus and the alchemists, our physical and chemical universe is no longer populated by angels or spirits. Only human beings still have individual existence, and then only in their social and legal dimensions and their subjectivity.

Now we find ourselves in a shared disenchanted and depersonalized nature of objects, but vested with a knowledge of its laws that gives us a technical control never known in the past. In this nature, however, we continue to perceive ourselves and to behave as animate persons and autonomous subjects, even if we know that this autonomy is dominated by the effects of these same natural laws and their determining causes. The only demons that survive are inside us, like the *daimones* of the ancient Greeks or the *shedim* of the *Zohar* and Lurianic kabbalah, shadow-images that live in us and animate us, at the junction between our body and our mind.

This means the nature of the subject, too, has changed from that entertained by the ideal of the free human being, who is subject and author of his or her own history. It is no longer an absolute agent, the originator of a causal sequence of actions. Instead, it is a particular instantiation of the totality of nature, the focal point of interacting force fields and dynamic processes, where a subjective and intersubjective individual existence can be built and experienced.

This is why a clear-cut dichotomy of the objectivity of science versus the subjectivity of ethics may be misleading. Such partitioning is convenient and even necessary to ward off the confusions and jumble of uncontrolled hallucinations. But it also tends to give credence to the idea that the only straightforward, transparent, and rigorous—in a word, "true"—wisdom is that of science, whereas ethics, not to mention politics and law, belongs to the wisdom of the left side, fluid, cunning, rhetorical, constructing truths that are only approximate and always debatable and controversial. It is Themis against Metis, or Apollo in one corner and Dionysus in the other.

But today we know that things are not so simple. Scientific objectivity is the result of the consensus of a society of scholars, which has its own specific forms of discourse and rhetoric; its operational truth sometimes proceeds by twisting paths. By contrast, despite their origins in myth and dubious foundations, law and ethics can be built and rebuilt, little by little and step by step, overcoming misconceptions and exploiting the possibilities of debate and intersubjectivity. The

whose kingdom is identified with the House of peace, decency, light, life, and fecundity, must employ trickery to acquire the knowledge possessed by the latter, so that he will be able to build this house and thereafter to rule it in peace and justice. This is an enterprise full of danger, the greatest of which is falling victim in turn to Ashmedai's cunning, to the point of resembling his twin and being mistaken for him. Ashmedai finds a way to purloin the "seal of Solomon," king of Jerusalem, and employs it to assume the appearance of its rightful owner and replace him on his throne.

Closing the circle of demons, shadows, and the divine image of man, the *Zohar*, after expounding on "man walks about as a mere shadow" (Ps. 39:7 [6]), quotes explicitly from the "sorcery-book of Ashmedai" (*Zohar*, Leviticus 43a). This homily develops the role of the divine image (*ẕelem*), split into two shadows/images (*ẕelalim/ẕelamim*) that are united in it in the life and death of every human being. (See also *Zohar*, Leviticus 13b, on the "two shadows" mentioned by Solomon in the Song of Songs [2:17].) The *Zohar* presents the "book of Ashmedai" as a source of information for those who would engage in magical practices of the left side, teaching them how to attach their dual *ẕelamim* to "two spirits" that can reveal to them, "impurely," the good and bad that will befall them. The *Zohar* seems to entertain no doubts about the value of this mythical book as a source of knowledge, even though implementing its instructions in magical practices can be done only "in impurity," that is, in close proximity to death (see above, n. 66).

rhetoric of debate may produce a form of truth that seems to spring up from the earth after having been cast out of heaven.[116] In both cases, the inherent misunderstandings of intersubjectivity and the imprecision of natural language, which always exists even if formalized languages can relegate it to the background, are transformed and in some sense rectified. They are harnessed in the service of the desire for peace or harmony, or at least for reciprocal understanding of the disharmony, when individuals are compelled to open up to whatever is not themselves.

116. See Psalm 85:12; Atlan, *Enlightenment*, chap. 9.

So the contest between Themis and Metis plays out on the inside, even if they operate in different ways, both objective knowledge of *is* and the subjective knowledge of *ought*. In both cases, the motivating force is the desire of individuals who know. It is at the core of "biblical" knowledge, the desire/knowledge imbued by Eros, that these antagonistic aspects of wisdom confront each other. Rather than an opposition between objective science and subjective knowledge, this is a very different opposition, one that cuts across both of them. The prize of the contest between knowledge of the right side and knowledge of the left side is this desire for knowledge, because it covers both aspects or is expressed in two ways, which we can designate the *desire of the ego* and the *desire of the subject*. Here *of* must be understood as the sign both of the true genitive, expressing ownership, and of the accusative: both desire that is experienced *by* the ego or subject and desire that is focused *on* the ego or subject. It is through the desire or desires that a subject or ego seeks to satisfy that it strives to bring itself into existence. Its own existence, whether as ego or as subject, is the cynosure of its desire. For the *ego* is not the *subject*, and they do not necessarily have the same desire, although one cannot exist without the other.

This mode of existence of the autonomous subject is also suggested by the notion of the subject-shadow, discussed above, recognized as a human subject only if its body casts a shadow. You would think that the sight of a body in full light, with its animating spirit or *daimon*, would suffice to indicate the presence of an autonomous subject. In fact, we must always suspect that the body has been enslaved, that it is possessed by its *daimon*; this is precisely what was recognized by those whom Ricœur accordingly labeled the "masters of suspicion": Nietzsche, Marx, and Freud. But this does not necessarily imply the "disappearance of the subject." Rather, the subject appears only in the shadow of this body, as an accompanying reflection, distinct from it but unable to exist without it. One can admit the existence of a subject, tied to the body and the forces that act on it, only after observing some effect worked by this body and marked by its individuality, just as its shadow is a specific effect worked on the undifferentiated light.

Thus a sort of balance of power is established in each individual, between the weight of the ego and the weight of the subject. Observations of animal behavior give us a first and tenuous image of this conflict and its possible resolution. Individuals seem to be determined simultaneously by their own needs and by those of the species, expressed in their behavior, which sometimes seems to run counter to their particular interests. This way of looking at the matter is clearly a teleological and anthropomorphic projection of what we think are the individual's interests or needs. In practice, this behavior, determined by physical and chemical mechanisms that are studied by the biological sciences, is produced both by the history of the individual and by the history of the species that engendered it, as well as its ancestors, siblings, and descendents.

Transposed to the human species, this relationship between individual and group is not only biological but also cultural. The same individual can be perceived—and perceive itself—as ego or as subject, depending on whether the accent is on its selfish (egoistical) behavior, which seems to respond to the interests of its own survival, or its altruistic actions, sometimes taken despite itself, which seem to be motivated by the interests of the species, as mediated by the culture of its social group. The subject differs from the ego in that its desire expresses its needs and interests both as an individual and as an instance of the humanity that informs it, all at the same time and sometimes in conflict with one other. Language and thought, as expressions of the desire/knowledge through which it relates to what is not itself, clearly play a decisive role in this expansion of the ego to the subject. The ego, by means of memory and imagination, projects itself and expands in time and space. This is what allows us to project our immediate sensations of pain and pleasure onto an imagined future and onto another who resembles us and speaks to us. The *moral subject* is an ego that has transformed its pleasure and pain into good and evil, by means of what I have referred to elsewhere as "the different levels of ethics."[117] This transformation takes place in childhood and adolescence, through education and through the internalization of collective values that activate imagination, reason, dreams, and hallucination in varying proportions, depending on the culture. But the *thinking subject*, which may achieve a form of wisdom and salvation through knowledge, is an ego whose perceptions and sensations are increasingly influenced, modeled, and shaped by its thoughts. "Man under the empire of reason," for Spinoza, is the ideal example of the thinking subject, because his ego is then totally identified with the dictates of the second kind of knowledge, knowledge common to his own thought and to that of all other thinking beings that can be conceived. We experience this form of knowledge, which transcends the limits that our physical sensations and emotions impose on our more immediate, image-based knowledge, by means of rational thought, which is logical and mathematical. The subject is first of all an ego that expands and goes beyond itself in some fashion, simply because it thinks and knows more, in space and time, than what it can perceive with its senses and emotions.

This is why philosophers have long held rational thought, logical and mathematical in nature, to be a privileged experience of the transcendental character of the subject as opposed to the ego. This insight, expressed in diverse ways and in various ontological systems, is found among the Platonists and Neoplatonists of antiquity, notably Proclus,[118] as well as among the Stoics; but also, unmistakably, in the geometrical method of the seventeenth century, based on proofs (which Spinoza calls "the eyes of this mind" [*Ethics* V 23, Note]), and among post-Kantian philosophers such as Hegel and Husserl.

But this is precisely where the (apparent?) optimism of Spinoza and especially of the Enlightenment philosophers about the liberating effect of reason must be tempered by Freud's pessimism concerning the pitfalls that even a rational man may encounter on his path. The ego does not allow itself to be guided by knowledge as easily as a linear and simplifying notion would have us imagine. Although the cognitive and practical aspects of human nature have been radically transformed by the unprecedented expansion of

117. See Atlan, *Tout, non, peut-être*; *SR* II, chap. 2.
118. See especially Proclus, *Commentary on the First Book of Euclid's Elements*, trans. Glenn R Morrow (Princeton, NJ: Princeton University Press, 1970).

knowledge in the natural sciences wrought by reason during the last 150 years, the evolution of the affective and moral dimensions has been much slower.

This is where the dialectic of Metis and Themis, of crooked (left-side) wisdom and straight (right-side) wisdom, reasserts its claims, as they exchange roles so that each prevents the other from reducing all that exists to its own dimensions. When our ego is expanded by the operation of our intellect, the latter tends to be trimmed back to the dimensions of the ego. The distance between openness to reason and the closure of rationalization is not great. Rather than making the ego come to where the id was, as Freud would have it (*Wo Es war soll Ich werden*), rational knowledge may enlist in the service of the ego and bind it even more tightly in its own enslavement. Liberation by knowledge requires the difficult exercise of binding the bound, of inverting the serpent's crooked cunning by means of an even greater cunning, so that both of them are made straight.[119]

This is because mere awareness of its thinking activity does not suffice to turn the ego into a subject, neither of its thought nor of its action. The subject of the Cartesian cogito is generally just a grammatical subject that can easily be transformed into a predicate. "I think" almost always means "thought thinks in me" or "there is thought in me," or, as we say, "an idea came to me." The true subject, the agent of thinking, is not me, but the encounter within me between the activity of my brain and the activities of everyone and everything that surrounds me and is not me.

From here, the self in which "it thinks," the ego that experiences, feels, and perceives itself as the place of this thinking, is taken in two opposing but mutually supportive directions. One of them turns this thinking into its own, by clothing thought in its own discourse—in language—and particularly in its own emotions, thereby instrumentalizing it. The ego is then perceived as its active author, when in truth it has the alienated status of the selfish master vis-à-vis the slave that defines him as such and whom he cannot do without. The contrary motion consists of expanding the ego's capacity as a "thinking place" to the maximum, accepting in some fashion that this thought is not its own, that the "slave" has been liberated and is becoming more powerful. Then this place of thinking, necessarily limited, is expanded by virtue of the content of thought until it perceives itself as much larger than itself. This experience seems to have mystical connotations, if expressed in terms of fusion or of identification with, for example, Husserl's transcendental subject or the infinite immanence of

119. This is also suggested by Abraham's binding of Isaac, which the *Sullam* interprets as an episode in which the wisdom of the right side shackles the wisdom of the left side to keep it from growing and developing a "head." On the first level, it is necessary to prevent "left wisdom" from expanding in the service of a separate "head"—that is, of an ego that perceives itself as a source of power, dangerous in proportion as it has access to power and can wield it as it pleases. Next, this potentially destructive wisdom must be transformed into nurturing wisdom, no longer serving the power of an ego but rather the infinite and universal creative force. To accomplish this, the wisdom of the right side (Abraham), generous but undifferentiated, begins by depriving the wisdom of the left side (Isaac) of its capacity for autonomous thinking (its "head"). Abraham is thus raised as far as his own head, which serves both of them as a common source and makes it possible for Isaac to convey his operational and effective wisdom, now wrapped in effusions of Abraham's generosity, in the middle way that integrates them, Jacob-Israel. The ego, having been "beheaded," becomes a subject. The trial (*nissayon*) of Abraham and Isaac is not meant to test their blind faith or absolute obedience. Rather, it is the experience <also *nissayon*> by which the ego (that of Abraham and Isaac together) elevates itself (*mitnaśśeh*). Ashlag (*Sullam* on *Zohar*, Leviticus 18a) associates the homophonous roots *n.s.h* 'try, test' and *n.ś.'* 'rise, elevate' and proposes that we read Genesis 22:1, "Elohim tested Abraham," as "Elohim elevated Abraham."

Spinoza's intellectual love of God. In fact, if there is fusion, it is only that of the ego with the subject that it gradually comes to perceive in itself, as the subject grows until it effects its release from the ego. The mystical fusion with the infinite can then be understood as a projection, as an asymptotic objective that is never reached, in this expansion that turns ego into subject. The mathematical experience, once again, is a special example of this. It demonstrates the two interwoven aspects of construction and discovery that characterize our two opposing but complementary movements: possessive use, where what is possessed is not what one thinks it is; and disinterested opening, identification with the other with whom—and with which—thinking is thought.

17. THE STAKES OF THE INFINITE

For kabbalistic thought—unlike the psychological theologies of the Creator as a personal God in the image of man—the infinite subject of thought is not an ego; neither is it the subject of the action that created the world or in the world. For example, in the first chapter of Vital's *'Ez ḥayyim*, a systematic exposition of the Lurianic system, the first "movement" of creation and differentiation in the infinite primordial One (the *Ein Sof*) is epitomized in the midrashic phrase "it rose into his thought to create the world."[120] Vital expounds the Sages' admonition, "Whosoever speculates on four things—What is above? What is beneath? What was before? And what will be hereafter?—it were better for him had he not come into the world."[121] For we can speak only of what is in the world. Any imagination that projects us outside its limits, in time or in space, has the effect of making us not of this world and simultaneously annihilates any possible meaning of what we can say. This is why the curious expression "it *rose* into his will" (or "into his thought") does not designate a psychological process of voluntary action preceded by a deliberate voluntary decision. If so, we would have to ask *where* it rose from, even though, "before" the world, by definition only the Creator himself existed. But the only place from which it could "rise" is the world itself. This is why the processes by which multiplicity is produced by the One are described as inside the being of the world, from which we cannot escape in any case. The description of these processes of emanation, creation, formation, and action, as well of those that "precede" emanation—the subject matter of most treatises of speculative kabbalah—appears to offer an answer to the ostensibly forbidden question of before and after, above and below. The paradox disappears when we realize that the question itself, along with its answer, can be formulated only within our experience of the world, as intimated by the very phrase "it rose into his thought" (where we must supply "from the world") "to create the world."

The allegory of play used by Moses Cordovero and some of Isaac Luria's disciples is even more radical.[122] To protect himself from falling into the trap of talking about the "outside" (what is before, after, above,

120. Vital, *'Ez ḥayyim, Sha'ar ha-kelalim*, vol. 1, p. 1.

121. M *Ḥagigah* 2:1 (11b); see above, p. 105.

122. See Chapter 2, §4. Compare, for example, Moses Cordovero, *Shi'ur ḳomah*, §§13 and 29. In the first part of Naphtali Bacharach's *'Emeḳ ha-melekh* (Amsterdam, 1653; repr. Jerusalem, 1994), entitled *Sha'ashu'ei ha-melekh* (The King's Games [*or* Delights]), a set of points and then letters of light, in the shape of "Adam, male and female," are compared to a garment (*malbush*) in which the Infinite simultaneously covers and reveals itself. In Chapter 2 we saw how these treatises propose the metaphor of a Torah, or primordial wisdom, whose letters, arranged in a different order than in the Mosaic Torah we know,

below), Cordovero describes how all things come into being as the result of a game of the Infinite, in itself and with itself, which causes something other than itself to emerge so that it can amuse itself with the encounter. The first elements of this original multiplicity are none other than the letters of the alphabet; the game is that of the infinite combinations they make possible, while the pleasure or amusement (*sha'ashua'*) that accompanies it is that of knowledge, both abstract and sexual, of the encounter of the one with another, "in itself and by itself."

These anthropomorphic images, all of them like myths of origin, make it possible to avoid sinking into the outer-world spawned by the theology of a God who is simultaneously transcendent and personal. These narratives remain resolutely metaphorical and in-the-world. As we saw in the previous chapter, it would be a mistake to understand them as referring to a voluntarist psychology of a Creator endowed with the attributes of an individual subject, which can never be more than an idealized projection of our own ego.

As for annihilation of the ego, the goal of certain Indian doctrines and some currents of Hasidism, it too is quite ambiguous. Such annihilation can be the fruit only of prolonged asceticism and of an insight that is inevitably someone's. And that is precisely what we are supposed to be looking for; it is when our consciousness is "awakened" to our true nature that the illusions and blindness of the ego vanish. But this awakening to the impersonal infinite is reserved exclusively for those who achieve it, that is, some particular person who remains engaged in her individual existence, even if she changes her name when she is transfigured.

We can leave behind neither the impersonal world in which we exist nor the individual existence in which we perceive ourselves. But if we must have anthropomorphism, then, rather than projecting the image of another thinking and acting subject outside ourselves, it is better to see ourselves as the partners of an infinite subject who plays and takes infinite pleasure in knowing itself, in itself, and by itself, as if it were another. As finite subjects, we serve as the arena in which this game is played, but are also the players on that field; or as the stage of the drama, as well as the actors who trod the boards. But we are the leading actors, of course, rather than inanimate props or spear carriers with no speaking role. For it is through speech and in speech that we experience ourselves as real persons and subjects. This is why the relations between Adam's children and the demons and angels that form them and play with them cannot avoid the pitfalls of the games of language.

weave the garment of the *Ein Sof* and are the instrument for creating worlds that are infinitely renewed.

The *Zohar* mentions the Creator's game and amusement with the letters briefly (Introduction, pp. 1a and 2b), but these authors developed the trope at length. According to Scholem:

> In a way this very naturalistic view of the original nature of the Torah seems reminiscent of Democritus' theory of the atoms. The Greek word *stoicheion* has the meanings: letter and element, or atom. According to Democritus, the diverse attributes of things are explained by the diverse movements of the same atoms. This concordance between letters as the elements of the world of language and atoms as the elements of reality was already noted by certain of the Greek philosophers. Aristotle's succinct formulation: "Tragedy and comedy come from the same letters" [*De Generatione et corruptione* 315b] not only amplified Democritus' idea but stated a principle which recurs in the Kabbalistic theory of the Torah; namely, that the same letters in different combinations reproduce the different aspects of the world. (Gershom Scholem, *On the Kabbalah and Its Symbolism*; trans. Ralph Manheim [New York: Schocken, 1965], p. 77)

Ḥayyim Volozhiner picks up this theme in *Nefesh ha-ḥayyim* (part 2, chap. 17; part 4, chap. 26–29). He explains that the particular thoughts that come to our minds derive from this infinite wellspring of letter combinations in the Torah that is the garment of the Infinite.

18. ANGELS AND DEMONS IN THE GENERATION OF BABEL

According to the Talmud, angels, unlike demons, do not understand Aramaic.[123] Other talmudic passages disagree; the angel Gabriel, for example, is said to have taught Joseph the seventy languages spoken at Babel, including Aramaic.[124] To harmonize this, the *Zohar* suggests that they account that language of little worth and pay no attention to what people say in it.[125] Only biblical Hebrew, the Sacred Tongue, the "language of interiority,"[126] can spur them into action. This is why the Talmud warns against asking angels for something you need (it being understood that the request is made in the vernacular of the day, which was Aramaic); they will ignore you.[127]

But this refers only to angels and not to human beings, for whom Aramaic has a high status, especially for translation.[128] What is more, we cannot avoid dealing with demons, whether to keep them from harming us or to make use of their services; and this, in principle, requires fluency in Aramaic,[129] which is the language closest to Hebrew, the "inverse" of the Sacred Tongue, a language of the outside and of the "other side" (*'aḥorayim*) of objects and creatures and of the one who does not show his face.[130]

Finally, the use of Hebrew was not always recommended; it was not always innocuous, precisely because it was the language of the angels.[131] This, at least, is what the *Zohar* suggests in one of its glosses on the tower of Babel.[132] The one language spoken by the people of Babel before they were dispersed was Hebrew. Consequently they could enlist the angels in their service and their enterprise had every chance of succeeding and enduring. According to the *Zohar* here, what distinguishes demons from angels is that their effects are transitory and self-destructing.[133] Had the builders of the tower spoken Aramaic (or any language other than Hebrew), they would not have been able to recruit the angels and would have had only a demonic workforce at their command. Their project would have had only a very relative importance, because it would not have been able to endure.

But the language of angels, the "Sacred Tongue" (or, better, the "language of the sacred"), is effective because it expresses things as they really are, because the relations among its words express the relations among things.[134] Hence its magical formulas

123. B *Shabbat* 12b.
124. B *Soṭah* 36b.
125. *Zohar*, Genesis 89a.
126. Ḥayyim Vital, *Sha'ar ha-kavvanot* (Jerusalem, n.d.), *Derush ha-ḳaddish*, p. 15b.
127. B *Shabbat* 12b.
128. *Zohar*, Genesis 89b.
129. The kabbalistic therapist Yehuda Petaya exploited this to exorcise the demons who appeared in dreams or visions. They tried to pass themselves off as angels or spirits through whom prophetic truths are revealed, relying on the talmudic evaluation of dreams as the sixtieth part of prophecy. By communicating with them in Aramaic, he writes, he unmasked them as demons. The visions they spawned could then be diagnosed as pathological hallucinations, if experienced during waking hours, and the dreams as nightmares to be interpreted, rather than as prophetic dreams (Petaya, *Haruḥot mesapperot*, pp. 31–34).
130. Vital, *Sha'ar ha-kavvanot*, p. 15b. <The "other side" (*'aḥorayim*) alludes to Exodus 33:23, where God tells Moses, "I will take away my hand, and you shall see my back (*'aḥorai*); but my face will not be seen.">
131. More specifically, in the talmudic era of the Oral Law, Hebrew was the language used for writing and by the intellectual aristocracy of the "companions" (*ḥaverim*), as the Sages referred to themselves among themselves, as opposed to the unlettered, the *'ammei ha-'areẓ* or "people of the land."
132. *Zohar*, Genesis 74b–75a (*Sitrei Torah*).
133. Ibid.
134. See Rashi on Genesis 2:23 and *Gen. Rabbah* 18:4 (repeated in 31:8): The Hebrew word *'ishah* 'woman' corresponds in sound to *'ish* 'man'; but this phonetic similarity does not exist in other ancient languages—from which we are to learn that Hebrew describes the world as it is and that "the world was created in the sacred tongue [i.e., Hebrew]."

allow one to manipulate things by manipulating words. Just as demons are fallen and scattered angels, cut off from the nurturing unity, the spells that bind them are in the lesser languages that have only local and momentary effect.[135] By the same token, the only knowledge that can be acquired through them is "knowledge of the left side," which is itself divided. The people of Babel, with their unique language, tried to profit from their access to unified, effective, and enduring knowledge and to construct their own unity by making it descend from heaven, from the infinite heights of its normal abode, to the earth of their own habitation. "Come, let us build ourselves a city, and a tower with its top in the heavens, to make a name for ourselves, lest we be scattered all over the earth" (Gen. 11:4)—that is, "Come, let us make a name for ourselves; to that end let us build a city and a tower that will carry us up to heaven, from which we can draw knowledge." In this, says the *Zohar*,[136] they reproduced Adam's fall by the Tree of Knowledge, formally defined as the deed of the "primordial serpent . . . joined below and . . . separated above."[137]

Here we encounter once again Rabbi Ashlag's notion, presented above: the cunning of the left side triumphs when it exploits and twists the wisdom of the right side and diverts it to its own benefit. The only possible method of repair, then, relies on the inverted cunning that employs division—of languages, of individuals, and of things—to prevent this corruption, even if it means postponing access to wisdom. Patiently rebuilding the crooked wisdom of the left side, this time starting with its *disjecta membra* and proceeding upward, from below to above, the ensuing generations prepared for the generation of the wilderness that received the Mosaic Torah at Sinai, in which the wisdom of both sides, right and left, could be rejoined and effuse freely.

This enterprise must be resumed in diverse contexts after the giving of the Torah, every time Adam's fall is replicated. The breaking of the first set of tablets followed the episode of the Golden Calf and was not truly repaired until the construction of Solomon's Temple in Jerusalem. The Babylonian exile followed the destruction of this temple, until the Second Temple was built. The current exile succeeded the destruction of the Second Temple. In this project of upward reconstruction, which begins with divided knowledge, demons can be employed because they are, as it were, in their own element. Their world is that of our local science, which divides and distinguishes in order to understand and transform things on the basis of their essences and the hidden causes of their existence, without excessive effort to attach them to some common origin. The association with them is even more valuable when the light of the Torah remains accessible, even if in weakened form, and can protect us if necessary against their harmful effects. This is the origin of narratives such as Solomon's recruitment of Ashmedai for building the First Temple, and, in general, the use of techniques associated with magic, astrology, chiromancy, physiognomy, dream interpretation, and other manipulations of science, letters, and names, which were part of the wisdom of the Torah for the talmudic sages and kabbalistic masters. As we have seen, the difference between forbidden practices and permitted ones is

135. The idea that by manipulating formulas one can act on objects is shared by science and the magic from which it emerged (see Yates, *Giordano Bruno*). The difference between the magus and the modern physicist is that the latter knows that the efficacy of formulas is indirect; writing Maxwell's equations or reciting them aloud will not produce light (see Atlan, preface to *Le Golem*).

136. *Zohar*, Genesis 75a (*Sitrei Torah*). <See also (and more explicitly) *Gen. Rabbah* 38:9.>

137. *Zohar*, Introduction, 12b (trans. Matt, vol. 1, p. 85).

mainly the aim of unification sought by the latter, which is why they are called "acts of unification" (*yiḥudim*).[138]

This system of fall and re-ascent also draws on the midrash about Truth's career among human beings. Truth was not given to Adam as soon as he was created. First it had to be hurled down to the earth, because of the human capacity to deface it by means of error, falsehood, and violence, so that it could later sprout from the earth along with justice and peace.[139] Only then could a repaired wisdom spread without limits, downward, and unity be constructed below but not at the expense of the unity above. This, according to the *Zohar*,[140] is the hidden meaning of the verse "on that day YHWH will be one and his name one" (Zech. 14:9), which the kabbalistic meditation recited at the start of the Sabbath (in the Sephardi rite) quotes as the prefiguration of this completed repair.

In a theological context, this verse is generally understood as the proclamation of a monotheism that is not yet fully fledged, as if to say, "God is one, but his name is not yet one." The reading suggested by the *Zohar* is just the opposite; it does not involve the unity of "God" as Infinite (*Ein Sof*), as transcending all being, because we can say nothing about that, not even that he has a name.[141] It is rather the unity of being in world-time and in action. What this passage (like others) means, in this reading,[142] can be summed up as follows. The letters that represent the verb *to be* in all three tenses, "above"—*YHWH*—were sundered, and humans "below" have constructed many names from that *being*. Some of them, like the "God" of the monotheistic religions, refer to an absolute unity, or at least to an aspiration to such unity. In truth, however, these are only names, in various languages and based on different experiences. Their proclaimed unity is only nominal, as shown, among other things, by the multiplicity and violence of the monotheistic creeds. But on that day, says Zechariah, not only will his name be one, but the being of which this name is the name will also be one.

19. "GOD" AND THE NAMES OF THE NAME

The texts we have just mentioned belong to a context in which the relationship between God and the Name is the opposite of the standard concept, which is based on the theological notion of one God whose name, like that of a person, designates and invokes. For the people of Israel, the name

138. For example, the seventeenth-century kabbalist Abraham Azulai (in *Or ha-ḥammah*, a commentary on the *Zohar*) explains a celebrated passage about the techniques of astrological physiognomy and chiromancy, and their origins in an Adamic primordial science collected in the "Book of Solomon" (*Raza' de-razin*, in *Zohar*, Exodus 70), as an exposition of techniques of divination (*niḥush*), which are part of the science of the Torah, which encompasses everything. Hence these techniques ought not to be left to the outside (centrifugal) forces that would relegate them to the serpent (*naḥash*).

139. *Gen. Rabbah* 8 (see Atlan, *Enlightenment*, chap. 9).

140. *Zohar*, Exodus 135a (see below, n. 142).

141. "The source that has no beginning has no specific name; and the Tetragrammaton, and a fortiori all other designations, are restricted to a created glory" (Hai Gaon, quoted by Cordovero, *Pardes rimmonim*, §11.1, p. 64b).

142. The text of the *Zohar* that suggests this interpretation is as follows (where *it* refers to the Sabbath and to the tenth *sefirah*, Malkhut):

> Just as they unite above, so She unites below in the mystery of *one*, to be with them above, one corresponding to one. The blessed Holy One, one above, does not sit upon His Throne of Glory until She becomes, by mystery of *one*, like Him, to be one with one. (This explains the esoteric meaning of "YHWH one and his name one" [Zech. 14:9].) (*Zohar*, Exodus 135a [trans. Matt, vol. 5, p. 251])

See also Rashi's commentary on "the throne of Yah" (Exod. 17:16). As long as evil can reign on earth ("as long as the name of Amalek is not blotted out"), the name is not complete; YaH is only half of the Tetragrammaton. Nor is the "throne" complete; the Hebrew word here is *kes* (כס), missing the final *aleph* of the usual word for throne, *kisse'* (כסא).

YHWH stands for god(s) (*'elohim*). It is not the name of the deity. In fact, as written, it is not a noun at all, but a superposition of all three tenses of the verb *to be* (*hyh* 'was', *hwh* 'is', *yhyh* 'will be'). It is a name only when pronounced *'adonai*, whose literal meaning is "my lord(s)." (This is why it is usually translated, by approximation only, as "the Eternal" or "the Lord.") As a name of power and control it designates an infinite whole that makes everything that exists in time exist eternally. Nor is it the name of the Infinite (*Ein Sof*), because that has no other name by which it could be grasped, shown, or designated. The word "Infinite" allows us to have an idea of it but not to experience or come to terms with it through some makeshift image.[143] But this is not the case with the Name, YHWH, whose letters and structure can be comprehended through their multiple instantiations in the hierarchy of the worlds. The entire system of *sefirot* and "configurations" or "faces" (*parẓufim*) reproduces its structure over and over again, in all existing beings, both personal and impersonal. Vocalizing its four consonants with each of the nine Hebrew vowels in turn (*yuhuwuhu*, *yihiwihi*, and so on) produces nine variants. These, plus the classical vocalization *Yahweh* (or the unvocalized consonants), designate the ten *sefirot* of the kabbalah. In this scheme,[144] the possible superiority of the personal over the impersonal is *in* the world, associated with the world of gods and human beings and not with the infinite power in the being that makes them exist. The difference between the personal and impersonal is found in *natura naturata*, "natured nature" or "imprinted nature" (*ha-ṭeva' ha-muṭba'*, in Gikatilla's terms[145]), and not in the *natura naturans* designated by the Tetragrammaton as a verb, the agent of this existence.[146]

Abraham Ibn Ezra—biblical commentator, kabbalist, and grammarian—expounds this relationship between the Name and divinity, where, ultimately, it is God—the noun "God"—that serves as a name for the Tetragrammaton. In his commentary, Ibn Ezra endeavors to explain the first of the Ten Commandments (in Hebrew, the "ten statements"), which does not take the form of an imperative at all: "I [am] YHWH your god(s), who brought you out of the land of Egypt, from the house of bondage" (Exod. 20:2).[147] For Ibn Ezra this is a precept that governs the "thoughts of the heart" (mental acts) and constitutes the "root of the next nine commandments." "This statement means

143. YHWH is also referred to as the *shem havayah* 'name of being'; or, as Joseph Gikatilla writes (*Ginnat 'egoz* [Jerusalem, 1989], p. 24), "the truth of being and the name of being are the very same thing." The infinite—above, or beyond, or the source of being—is single and undifferentiated and cannot be grasped in its singularity because it has no structure that can be represented or conceived. For Ibn Ezra (long commentary on Exod. 3:15), the Tetragrammaton YHWH, the name *Ehyeh* 'I will be' (Exod. 3:15), and *Yah* are proper nouns; but all the other designations, including *'Elohim*, are attributes. The three proper nouns consist of the letters of the verb *to be*; but Ibn Ezra, who was also a grammarian, notes that the letters of which they are composed (*aleph, heh, waw, yod*) can serve, as prefix or suffix, to express the definite article, pronominal forms of verbs, or the coordinate conjunction—that is, as signs of relations rather than of things.

144. See Atlan, *SR* II, chap. 5.

145. Gikatilla, *Ginnat 'egoz*, p. 34.

146. <The Hebrew word פעל stands for both "verb" (*po'al*) and "agent" (*po'el*).>

147. In this verse the copula must be supplied, as is usually the case in Hebrew. Hence it could just as well be rendered "'I' is YHWH" or even "I *YHWH*," taking into account that the Tetragrammaton is a contraction of the three tenses of the verb to be (see Atlan, *SR* II, chap. 5).

In the same spirit we can understand the *Zohar*'s commentary on "that you may . . . understand that I am He. Before me no god was formed, nor shall there be any after me" (Isa. 43:10). Beyond the banal and redundant affirmation of a jealous monotheism, the *Zohar* focuses on "I am he" (*'ani hu'*), literally "I [am/is] he." In the *Zohar*'s reading, what the verse is emphasizing, and what is so important but so difficult to fathom, is that "I" and "he," for all that they designate different categories or *sefirot*, are nevertheless joined: "I is really he" (*'ani hu' mamash*), according to R. Eleazar (*Zohar*, Leviticus 86a).

that everyone must believe, with the indubitable belief of his heart, that this glorious Name alone, which is written but not uttered, is his god."

Other authors often take the First Commandment as an injunction to "believe in God." As we see, however, it is not at all about that. For Ibn Ezra it is not a question of "believing that God exists" (whatever that may mean); but, much more specifically, of accepting the Name YHWH, which is written but not uttered, as one's only god. The reality of gods is not in doubt; that is fully experienced by the "men of knowledge" (*'anshei da'at*) to whom the Torah is addressed, as Ibn Ezra wrote several lines previously. There is no denying the existence of natural forces that direct the course of events. What is in question, rather, is how we conceive of these forces, how we direct our thoughts and feelings (our "heart") with regard to them. The commandments of the Torah are addressed only to men of knowledge, endowed with common sense and a capacity for discretion or judgment (*shikkul da'at*, literally "weighing of knowledge"). There is no sense in which these commandments can contradict this capacity for judgment. All of the Ten Commandments, except for the fourth, are compatible with this capacity; this is why all accept them as components of "natural morality." Only with the Fourth Commandment, to observe and sanctify the Sabbath day, do we have to look for a hidden meaning, symbolic or other, because it too may not conflict with our capacity for knowledge and judgment. It is in this context that Ibn Ezra understands the First Commandment as a summons to renounce all gods and replace them with this Name. And this Name, as we have seen, is not a noun that refers to someone or something. It is, instead, a formula whose structure expresses the activity of everything that is. If we insist on uttering this formula, on giving it a name, we must resort to a name of power, "my lord(s)," which designates the sovereignty habitually attributed to the forces, human and nonhuman, that govern us.

This does not reify the formula, as would happen were we to pronounce it as written and, instead of the verb it suggests, transform it into the name of an object or person. Inasmuch as it is a verb in an indefinite form, the infinitude of its activity is suggested. Inasmuch as it is not a noun, it can refer to what it is not, that is, to an infinite that does not exist but whose potential of existence is expressed in this activity. Finally, when it is turned into a noun, it is a denomination of power and control, that is, precisely the one of the many names that designates a relationship of comprehension and submission to what makes what is be.

From this perspective, Zechariah's "on that day YHWH will be one and his name one" takes on a new meaning, that propounded by the *Zohar*. "His name" is "God," which is not really one, because it can be said in various languages: theos, Zeus, God, Gott, Allah, . . . Elohim. The Hebrew *'elohim*, treated sometimes as a singular and sometimes as a plural,[148] is a noun that designates the deity (or deities) of each individual, including personal gods in the sense of the god of some person—the god of Abraham, the god of Isaac, the god of Jacob, the god of Simeon bar Yoḥai, the god of Rabbi Meir, the god of the congregation of Israel. This name functions as a place, a presence, or "throne" where, through the activity and structure of YHWH, being is manifested in its "glory" and where the infinite power of the verb YHWH is expressed in a finite manner. Of course we fervently hope that human peace and concord will make this name one and singular. But the issue goes beyond the fusion and unification of divine names. In

148. See Chapter 4, n. 18, for Rashi's explanation of this phenomenon.

fact, this unification could be effected through violence, as shown by the history of the Tower of Babel. As long as the worlds are divided and separated into human and nonhuman, personal and impersonal, as long as they are sundered within themselves and severed from one another, it is YHWH—the structure and activity of being itself—that is split. The letters YH are separated from WH, as in the midrashic reading of the phrase *kes yah* 'throne of God' (Exod. 17:16).[149]

Sometimes YHW is separated from H.[150] Unity "above" can exist only when the unity below, the unity of the "throne of glory" that serves to name the unity above, is realized. This is a necessary condition, but not a sufficient one. The unity below must also be effected as a means of the unity above and not become an end in itself, as was attempted by the people of Babel, who wanted to make themselves a unique "name" through their unique language. The violence of the closed and all-encompassing unit is the "other side" that preys on the unity below. This is why the aim of observing the precepts is not merely to institute a monotheistic creed that worships one god (we are only too aware of the violence to which such religions are prone). The goal of the commandments is to unify YH with WH, in the interiority of each individual, as stated by the kabbalistic formula that is recited before the performance of a precept. The unity below—the minimum of peace required for this internalization—can then be nourished in turn by this unity of being and endure thanks to it. This is what the *Zohar* on the verse from Zechariah is saying. On that day, not only will the name of the Name be one, but being itself—the noun-verb YHWH—will also be unified.

This connection between observance of the precepts and the unification of YHWH is stated explicitly in a frequent interpretation of the last clause of Exodus 3:15—"this is my name for ever, and this is my memorial [*or* appellation] throughout the generations"—which serves as a pendant to the more famous "I shall be what I shall be" of the previous verse. It too is understood in diverse ways. For Naḥmanides, emphasizing the start of the verse, "my name" refers to the god of Abraham and the god of Isaac, and "my memorial" to the god of Jacob. The Talmud, as well as the midrash cited by Rashi here, sees the two separate nouns as an allusion to the distinction between the Tetragrammaton as written and as pronounced—a distinction that will vanish in the world to come, "on that day" when he "will be one and his name one."[151] According to Gikatilla, "my name" alludes to the prohibitions, which are associated with YH, and "my memorial" to the prescriptive commandments, which are associated with WH.[152] To put it another way, what is identified here as "my name" is not YHWH, but the deictic *zeh* 'this/that' that precedes it (i.e., "my name is 'this'"). The content of this name is the positive and negative commandments, the prescriptions and prohibitions, with a certain asymmetry between the "name" per se, which is "concealed" in the negativity of the prohibitions—as suggested by the play on words of "this is my name *le-'olam*," which, in addition to "forever," can also be read as "to be concealed"[153]—and *zekher* 'memorial', which expresses openly, through the positive commandments, what is buried in memory.

149. See above, n. 142.

150. See *Zohar*, Exodus 135a and n. 142 above.

151. B *Pesaḥim* 50a.

152. Gikatilla, *Ginnat 'egoz*, pp. 297–298. This interpretation rests on the numerological value of שמי (*shemi*) 'my name', 350, which, added to that of YH, equals 365, the number of negative commandments; and of זכרי (*zikhri*) 'my memorial', 237, which, added to WH, equals 248, the number of positive commandments.

153. <Because of the defective spelling לעלם instead of the standard לעולם.> See Rashi on Exodus, 3:15; Atlan, *SR* II, chap. 9.

Finally, the nature and properties of YHWH are also indicated by how the angels and demons relate to this name. In the book of Joshua we are told that an angel in human shape appears to Joshua during the night between two battles. Joshua, believing the stranger to be a man, asks whether he is friend or foe. The angel replies by identifying himself as "commander of YHWH's army" (Josh. 5:13). Joshua falls on his face and (according to the Talmud) hears an indictment of his sins. The Talmud is astonished that he took the stranger at his word; it could have been a demon who was feigning a false identity! To which the text replies, "we have a tradition that even demons do not pronounce the heavenly name [i.e., YHWH] in vain."[154]

154. B *Sanhedrin* 44a–b.

CHAPTER 4

MYTH AND PHILOSOPHY, KABBALAH, "EXPANDED SPINOZISM"

1. PHILOSOPHICAL QUESTIONS, MYTHICAL ANSWERS

As promised in the Introduction, we have now reached the point where it may be helpful to step back and define the nature of the relationship we have established between philosophical reflection about the sciences of the living and traditional readings of the biblical myth, sometimes by means of legends and other myths. In particular, we must justify Spinoza's place at the center of this nexus, which at first sight may be all the more astonishing because his absolute rationalism is held to be incompatible with any attention to mythology.

Conventional wisdom has it that philosophy was born in Greece, in the sixth century BCE, and supplanted a different mode of thought—what we call mythological. Plato and Aristotle succeeded Homer and Hesiod. The old myths were discarded or interpreted allegorically. Later, in the Christian West, this dichotomy led to the familiar and increasingly sharp distinction between religious thought and rational thought. It would be wrong, however, to project onto the past our modern notion of two tendencies or two mental attitudes, one of which is supposed to be the domain of reason, of philosophy and science, and the other that of religion and myth, in the sense of fable and superstition. The birth of Greek philosophy was not the sudden appearance of universal and timeless Reason, falling from heaven in sixth-century Greece, after which fables and myths withdrew in favor of the truth of science and philosophy. Decades of Hellenic studies have profoundly modified this older vision of the "Greek miracle." Some no longer even raise the question of a transition from myth to reason, the issue having been diluted (if not indeed drowned) in all of the social and political changes that took place in the Greek cities in the sixth and fifth centuries BCE.[1] "Myth is not an essence but a catchall," says Paul Veyne, "and reason is dissipated into a thousand little arbitrary

1. Pierre Vidal-Naquet cites the eminent English classicist Francis Cornford to bolster this approach:

> In fact, for an entire historical school the problem we are used to setting up as "from myth to reason" does not arise. This is what F. M. Cornford tried to show throughout his fundamental life's work, including the posthumously published *Principium Sapien-*

rationalities."[2] "Myth and *logos* are not opposites, like truth and error. Myth was a subject of serious reflection, and the Greeks still had not tired of it six hundred years after the movement of the Sophists, which we have called their *Aufklärung*."[3] In this vision, taken to the extreme, we are dealing in both cases with constructs of the human mind and imagination, sufficiently complicated to be "interesting" and to attract various forms of belief, which are also diverse and of varying degrees of complexity, to cluster around themselves.

Reason is multiple; we know this even better today. In an earlier work, subtitled an "intercritique of science and myth,"[4] I endeavored to analyze the differences in how reason is employed in the natural sciences, in the human sciences, and in the rationality of certain formalized mystical traditions. But these historians teach us that reason already had multiple forms in ancient Greece. Along with the philosophers' progress and their

> first steps among the "physicists" of the school of Miletus, in the sixth century before our era, we must leave room for the inquiries of historians, for the skill of physicians, for the proofs of the geometers, for the theoretical constructions of astronomy and optics. Although all of these disciplines can be inscribed in a single field of *episteme*, with its lines of force and limits—what we have called the spiritual universe of the *polis*—they nevertheless do not obey exactly the same norms, do not employ identical procedures of reasoning, nor reserve equal room for the techniques of empirical research. Thus multiple reasons.[5]

Still, something happened. There was a "mental mutation,"[6] or rather an evolution, over several centuries, in the form of the spoken and written word, as the power of god-kings gave way to that of citizens. The birth of philosophy appeared "to be composed of two major mental transformations: positive thought, excluding any form of supernatural and rejecting the implicit comparison established by myth between physical phenomena and divine agents; and abstract thought, analyzing the reality of this power of change that myth gave it and challenging the ancient image of the union of contraries in favor of a categorical formulation of the principle of identity."[7] Philosophy rationalizes myth. This means that, addressing the same topics—notably the origin of an ordered cosmos of multiplicity from the One and chaos—philosophy inaugurated a new form of discourse. The rationalized myth

> took the form of an explicitly formulated problem. Myths were accounts, not solutions to problems. They told of the sequence of actions by which the king or the god imposed order, as these actions were mimed out

tiae. In the passage from myth to reason, myth was not left behind, and what we call "reason" for the Greeks is often myth.

Cornford intended to combat the theory of the "Greek miracle," and so he wished to restore the bond of historical continuity between philosophical reflection and mythic-religious thought. Between the Hittite and Babylonian myths and Anaximander's cosmogony, Hesiod's Theogony supplies the missing link. There is no basic difference among the following three stories:

1. The Babylonian foundation myth in the *Enuma Elish*, which describes the slaughter of the monster Tiamat by the god Marduk and the creation of the world by Marduk out of the corpse.
2. The *Theogony*'s myth of Zeus's killing Typhon and drawing out, if not the world, at least the winds.
3. Anaximander's cosmogony, in which the different qualities, such as hot and cold, dry and moist, and so on, emerge two-by-two from the *apeiron* (the infinite).

(Pierre Vidal-Naquet, *The Black Hunter*, trans. Andrew Szegedy-Maszak [Baltimore: Johns Hopkins University Press, 1986], p. 253)

2. Paul Veyne, *Did the Greeks Believe in Their Myths? An Essay on the Constitutive Imagination*, trans. Paula Wissing (Chicago: University of Chicago Press, 1988), p. 122.
3. Ibid., p. 1.
4. Atlan, *Enlightenment*.
5. Jean Pierre Vernant, in Jean Pierre Vernant and Pierre Vidal-Naquet, *La Grèce ancienne* (Paris: Le Seuil, 1990), pp. 9–10.
6. Jean-Pierre Vernant, *Myth and Thought Among the Greeks* (London: Routledge & Kegan Paul, 1983), p. 343 (first published as "La formation de la pensée positive dans la Grèce archaïque" [1965]).
7. Ibid., p. 351.

in ritual. The problem was resolved without ever having been posed. However, in Greece, where the new political forms had triumphed with the development of the city, only a few traces of the ancient royal rituals remained, and even their meaning had been forgotten. The memory of the king who was creator of order and maker of time had disappeared. The connection between the mythical exploit of the sovereign that is symbolized by his victory over the dragon, and the organization of the cosmic phenomena, is no longer apparent. When the natural order and atmospheric phenomena (rains, winds, storms, and thunderbolts) become independent from the functions of the king, they ceased to be intelligible in the language of myth in which they had been described hitherto. They are now seen as "questions" which are the subject of open discussion. It is these questions (the genesis of the cosmic order and the explanation for *meteora*) that, in their new form as problems, constitute the subject matter for the earliest philosophical thought. Thus the philosopher takes over from the old king-magician, the master of time: he constructs a theory to explain the very same phenomena that in times gone by the king had brought about.*

*And he too sometimes brings them about: Empedocles knows the art of stopping the winds and changing the rain into dry weather. Cf. L. Gernet, "Les origines de la philosophie," *Bulletin de l'enseignement public du Maroc* 183 (1945), 9 [Vernant's own note].[8]

All the same, what could function as an explanatory system in one age does so less and less today, and it no longer works in the same way. This is not just a change in theory, or in the spirit of the age or a paradigm, in the sense in which Michel Foucault and Thomas Kuhn analyzed scientific revolutions. The replacement of mythology by philosophy was a more profound alteration, which seemed to affect reality itself—or, at least, the structures and mechanisms of perception that make it possible for us to say "this is real" and that differentiate the objects we encounter when awake from those we dream about, the shared outside world from the individual consciousness, perception from hallucination. The Greek philosophers themselves wondered about the nature of this change. Starting from their own perception of reality—which they knew to be different from that of their predecessors—as historians, geometers, and physicists, they looked in the myths, trying to distinguish accounts of actual events, the stuff of their history, from what was merely plausible, and finally from what, totally implausible, was either an old wives' tale or an allegory about the hidden origins and nature of things. It is notable that they did *not* simply ignore them as insubstantial productions—what we would today dismiss as wild imagination. Our modern explanations, according to which myth plays a zoetic function—Bergson's "fabulation"—or a social function, to organize and preserve power, are inadequate, because endowed with the properties and thus the defects of functional explanations. Although they are interesting and useful structures for *describing* the plausible roles that myth played in ancient societies, they say nothing about the *mechanisms* of internalization by which individuals made them their own.

8. Ibid., pp. 347–348.

At least one ancient philosopher, Epicurus, had the merit of proposing an explanation that resembled a myth but fit naturally into his own naturalist, materialist, and mechanistic philosophy. As versified by Lucretius,

Even in those days would the race of man
Be seeing excelling visages of gods

With mind awake; and in his sleeps, yet more—
Bodies of wondrous growth. . . .
. . . .

. . . Besides, men marked
How in a fixed order rolled around
The systems of the sky, and changed times
On annual seasons, nor were able then
To know thereof the causes. Therefore 'twas
Men would take refuge in consigning all
Unto divinities, and in feigning all
Was guided by their nod.[9]

There is no doubt that this attitude was due to their ignorance of natural causes, those made known by Epicurus himself (for this very reason referred to as "divine" by Lucretius!):

Or how were known [to the gods]
Ever the energies of primal germs,
And what those germs, by interchange of place,
Could thus produce, if nature's self had not
Given example for creating all?[10]

But as for knowing "what cause hath spread divinities of gods abroad through mighty nations,"[11] ignorance of the causes of these phenomena is insufficient. The men and women of those days had to have a different sort of imagination, so that they saw, waking, what we normally see only in our dreams.

Thus at the dawn of philosophy we must recognize multiple historical dimensions, not only social and political but also psychological and perhaps even biological. On the one hand, the advent of Greek philosophy is "a historical fact with its roots in the past, growing out of the past as well as away from it. This mental change appears as an integral part of transformations which between the seventh and the sixth centuries occurred at every level in the Greek societies—in the political institutions of the city, in law, in economic life, in money."[12] On the other hand, we cannot ignore the "mental" aspect of this change, where we find Epicurus and his case for believing in gods in a transformation of human imagination. In modern parlance this is a "mutation" in the sense this term has acquired since the appearance of the biological theory of evolution: the gradual replacement of a majority trait in a population at a particular moment by another trait, which a change in ambient conditions favors and selects at the expense of the former. This is precisely the thesis of the neurologist and psychologist Julian Jaynes, whom we have already mentioned.[13] For Jaynes, as for the historians, these favorable conditions are associated with the disappearance of the god-kings that accompanied the establishment of the Greek cities and the remoteness of their pres-

9. Lucretius, *De rerum natura*, V 1169–1171, 1183–1187 (LCL, trans. William Ellery Leonard).
10. Ibid., V 181–186.
11. Ibid., V 1161.
12. Vernant, *Myth and Thought*, p. 365.
13. Julian Jaynes, *The Origin of Consciousness in the Breakdown of the Bicameral Mind* (Boston: Houghton Mifflin, 1976).

ence when they reigned, outside Greece, over great empires. These conditions made it possible for "mutants" who internalized the "voice of the right brain" to progressively replace those who, as in a dream or hallucination, as well in syndromes of traumatic disconnection of the cerebral hemispheres, assigned that voice an external origin, even more "divine" because they were penetrated and possessed by it.

In the discourses that accompany this transformation, the posing of philosophical questions accompanies the exclusion, as an acceptable answer, of supernatural agents that are active in nature—gods—who are the protagonists of mythical narratives that recount their adventures and their effects on the history of things and human beings. But this rejection of the supernatural is not just the abandonment of one explanatory system. It implies a change in the reality status that the human mind assigns to this divine world, a change that cannot be adequately explained by social and political transformations. What is more, the change took much longer than that which led to the birth of Greek philosophy, properly speaking, in the seventh and sixth centuries. It did not really begin until Plato and Aristotle in the fifth and fourth centuries and was not complete until the disappearance of Greco-Roman paganism in the early Christian centuries.

The severing of myth from reason was difficult and took place only at the conclusion of several centuries of evolution during which "rationality and traditional beliefs coexisted, at times in conflict but always in one way or another intertwined, in a contest that was various, to be sure, yet always integrated, because it was a product of the *polis*. In later centuries one part of this originally integrated whole would impose itself as a model of intelligibility, while another part would be portrayed as an absurd fable; but at this stage the two components still supported each other, by which I mean they were complementary and mutually reinforcing even in their divergence."[14]

The world remained divine for a long time after Plato and Aristotle. Mathematics, as it developed after Thales and Pythagoras, was still imbued with elements that we would deem mystical today. The search for causes in a world peopled by gods was still "theological" and continued to be so for many centuries. But when this evolution was completed, in the first centuries of the Christian era, the very notion of nature was structured differently; the causes and reasons that could be entertained were no longer the same and the gods—replaced by the one God—no longer spoke. I subscribe, along with others, to the notion that the transition from mythology to philosophy, from the world of gods to the world of nature and reason, was not limited to ancient Greece. The process goes far beyond Greco-Roman antiquity and can be observed, in diverse forms, in diverse civilizations (Semitic, Indian, Chinese, and others). Whatever the mechanism may have been, we see the traces of a major anthropological episode in human history. This episode, evidence of an evolution in the mental nature of the human species and in the relations between what we today call the "sacred" and the "profane," was expressed in various ways, as a function of the language and of the historical and geographical conditions of different societies.[15]

14. Jean-Pierre Vernant, "Foreword," in Maurice Olender, *The Languages of Paradise: Aryans and Semites, A Match Made in Heaven*, trans. Arthur Goldhammer, rev. and augm. ed. (New York: Other Press, 2002), p. ix.

15. Atlan, *Enlightenment*, pp. 337–348. Whether or not one accepts Jaynes's hypothesis, it is also to this evolution that we must attribute the appearance, as a historical phenomenon, of the self-conscious subject—a development traditionally assigned to some time in the period between Plato and Augustine (see Chapter 2, n. 70).

Only so-called primitive peoples, for whom the divinatory institution of the shaman remained at the center of the structure of the cosmos and society, seem to have escaped (until modern times) this desacralization of the universe. Perhaps they could do so thanks to their maintenance of a decentralized tribal organization, with no urban concentrations and in close contact with nature, which favors the concrete and personified proximity of confused sources of power and might that are simultaneously human and natural—what we deem "supernatural."

In ancient Greece, this phenomenon was expressed by the existence of persons whom the historians locate at the junction, still linked to the world of the divine, where the first philosophers appeared. These were the *sophoi*, the "sages" or "wise men," whose heirs, occupying an ostensibly lesser (but really much more ambitious) rung, were the *philosophoi* or "lovers of wisdom."

These sages are semi-legendary figures in whom the traits of the soothsayer and poet are associated with those of the mathematician and inspired thinker. Along with Solon, one of the "Seven Sages," we may mention Pythagoras, Thales of Miletus, Heraclitus of Ephesus, and even Parmenides, all of whom played a decisive role as forerunners of the "divine Plato," despite the rupture introduced by the Socratic method. Jean-Pierre Vernant has written of these pioneers of philosophy, who were often grouped in sects and confraternities of initiates, such as the Pythagoreans:

> The form of the poem in which even a doctrine as abstract as that of Parmenides is still presented reflects how much of the character of a religious revelation was retained in early philosophy. Like the diviner and the poet, and still to some extent confused with them, the sage was originally defined as the exceptional being who has the power to see and reveal the invisible. When the philosopher seeks to describe his own work, the nature of his spiritual activity, and the object of his inquiries, he employs the religious vocabulary of the sects and brotherhoods. He presents himself as one who is elect, a θείος ἀνέϱ [divine man] who enjoys divine grace. He undertakes a mystic journey into the beyond, following a path of inquiry which is reminiscent of the way of the mysteries and at the end of which he obtains, through a kind of [initiation rite], a vision such as that of the last stage of initiation.[16]

The Heraclitean *logos* is not just the reason of our logic; it is also the word that initiates are allowed to hear. This is why historians speak of a mystical current in Greek philosophy or reinsert it into a broader context of Greek shamanism.

Now it is noteworthy that, despite the undoubted new departure in the nature of written texts represented by Plato, Aristotle, and the schools that came after them, these crossroads figures did not disappear with the founding of the Academy, the Lyceum, the Stoa, and the Garden. It was merely that their status was transformed; it required a process of several centuries, during which the world of the divine faded as a concrete reality, for them to be rendered progressively irrelevant.

Thus the development of Greek philosophy, an integral part of the social and political transformations of the polis, opened up the sects and democratized knowledge. But "philosophy was [not] simply the reflection of these other developments."[17] The content of knowledge was also transformed. The "mental mutation" spread to the majority, until the primary facts of the nature of reality, visible and invisible, were profoundly modified. The disappearance of shamanism is as univer-

16. Vernant, *Myth and Thought*, p. 353.
17. Ibid., p. 365.

sal as shamanism itself and seems to have characterized the human mind in every latitude. For a longer or shorter period, as a function of the historical circumstances of each society and its culture, the divine seems to have remained a concrete presence, directly perceived in a somewhat confused and mysterious fashion by all, but quite distinct for those whose visual and auditory acuity was particularly developed and gave them access to it.

Within the ancient Hebrew culture, this break corresponds to "the end of prophecy."[18] Some centuries later, the Muslim world and then Christian Europe rediscovered Greek philosophy and rewrote it to suit their monotheistic cultures. Finally, in the seventeenth century, the scientific and philosophical revolution that underlies modern rationalism began to emerge.

It was then that Spinoza, in his *Tractatus Theologicus-Politicus*, seeking to divorce philosophy and theology, undertook a comparative analysis of what he called the two modes of revelation of the "word of God," by natural light and by prophetic revelation as it has come down to us in the "Holy Scriptures." Like the ancients with their myths, he too wondered about the

18. See, for example, B *Soṭah* 48b. *Pace* the conventional wisdom, the biblical universe is polytheistic. It is a world of *'elohim* 'god(s)', a plural form that is sometimes construed in the plural and sometimes in the singular. This was the background against which Hebrew monotheism was proclaimed; it is precisely this initial polytheism that provides its meaning. Note that, in the Hebrew Bible, the plural construction of *'elohim* is not restricted to the *'elohim 'aḥerim* 'other gods' that the Hebrews were forbidden to worship. The first such occurrence is in Genesis, when Abraham attributes his travails to his god: "When *'elohim* caused me to wander (*hit'u* [a plural form]) from my father's house" (Gen. 20:13). Rashi, in his commentary, tells readers, "do not be surprised [that the verb is in the plural], for in many passages words denoting divinity or denoting authority are inflected in the plural." He goes on to cite several examples in which the word takes a plural verb, as here, or is modified by a plural adjective, as in *'elohim ḥayyim* 'living god(s)'. The same applies to *'adon* 'lord', from which is derived the divine appellation *'adonai* 'my lord(s)' (and not *'adoni* 'my lord'), or the phrase, applied to the God of Israel, *'adonei ha-'adonim* 'lord(s) of the lords', a plural noun that is, however, conjugated in the singular.

The god of Israel, YHWH, the god of the patriarchs Abraham, Isaac, and Jacob, and especially of Moses, is proclaimed to be "greater than all *'elohim*" (Exod. 18:11; cf. 2 Chron. 2:4) before he becomes the amalgamation thereof (see *SR* II, chap. 5). "Monotheism" is a construct produced by some streams of postbiblical Judaism, a paradoxical theology of the one and personal God, both immanent and transcendent. By contrast, the divine system of the *sefirot*, of the "faces" and "worlds" of the kabbalah, and the divine substance of Judah Halevi's *Kuzari* are in direct continuity with this original biblical polytheism, even though the Mosaic exhortation "Hear O Israel" and philosophical intuition later unified this multiplicity into a monism of infinite being and substance. This is why we can compare the polytheistic underpinnings of Hebrew monotheism with Greek polytheism. (See Chapter 3, §19.)

What Vernant says of the beliefs of the ancient Greeks (Jean-Pierre Vernant, *Entre mythe et politique* [Paris: Le Seuil, 1996], pp. 238–243) applies equally to the ancient Hebrews. It was not adherence to dogmas that can be stated in explicit propositions that constitute a credo, as we understand the term today, but an all-encompassing association of three interrelated forms of expression: rituals, myths, and images (whence the Hebrews' distrust of images). In the Bible, the Israelites' rebellions against their god YHWH are not motivated by a rejection of dogma or by atheism in the modern sense. (On the contrary, it was their loyalty to YHWH and his law that caused the Romans to despise the Jews as atheists who did not believe in the gods, including their apotheosized emperors.) Their repeated insubordination expressed their fidelity to the way of life and to the various rituals, sacrificial and other, that the peoples around them considered to be normal. If there was a lack of faith here, it was not in a theological sense, but a lack of confidence in the saving virtues of the law of Moses and his invisible deity.

Finally, a comparison between the mental universe of the ancient Hebrews and the cosmos of the polytheistic nations of antiquity from whom they set themselves apart, or whom they imitated, is justified because in both cases we are dealing with a world in which the main sources of knowledge and action derive from practices of inspiration and revelation. As we saw in Chapter 3, Hebrew prophecy was part of these rites, alongside and in competition with the traditions of divination and dream interpretation, which were found in all climes, including Greek temples (see, for example: J.-P. Vernant et al., *Divination et rationalité* [Paris: Le Seuil, 1974]; Raymond Bloch, *La Divination dans l'antiquité* [Paris: PUF, 1984]). Elsewhere (Atlan, *Enlightenment*) I have referred to the probable role of modified states of consciousness, dreams, and hallucinations, produced by the ritual use of hallucinogenic plants, as the common origin of all of these practices, including in ancient Israel. On possible traces of this origin in the biblical narrative, see Benny Shanon, "Biblical Entheogens: A Speculative Hypothesis," *Time and Mind* 1(1) (2008), pp. 51–74.

truth of the events recounted by the Bible or about their barely plausible or totally implausible character.

Since then, philosophy has moved further and further from theology and seems to be even more remote from mythology. This is certainly the case if we view mythology as the sum total of the narratives and discourses organized around the foundation myths inherited from ancient traditions. But the separation is not as radical as it appears, provided that we take the mythological to be a mode or style of discourse. Discourses that are not necessarily built around ancient myths may nevertheless share their polysemic and metaphoric character. Although this style is distinguished from the philosophical and scientific style, whose model remains the *mos geometricus* or geometrical mode, philosophers have never stopped using it, openly or surreptitiously. As we shall see, in the *Tractatus* Spinoza, in pursuit of aims more political than theological, invented the mythical figure of Jesus Christ as the perfect Spinozist philosopher, through whose mouth God taught the universal philosophical religion to ignorant peoples, so that they would understand it and submit to it willingly. He took and reused an ancient myth, that of the start of the Gospel of John, about the Logos-son-of-God, an intermediary between God and human beings; later he transformed and geometrized it into the "infinite understanding of God" of the *Ethics*.

Philosophical treatises always include elements of myth, in the sense of discursive elements that are outside any empirical or logical knowledge of a scientific character and that come from some unreal other place, sometimes—though not always—from the ancient myths themselves, and that play the same role of giving meaning. How, then, can we differentiate philosophical discourse from myth, if their styles are always interwoven? Is it a question of quantity? Of the balance between rigorous geometrical development as against metaphorical narrative, analogies, and parables?

This question applies explicitly to kabbalistic authors, who, like the ancient Neoplatonists, are sometimes considered to be philosophers and sometimes to be the heirs of ancient mythological traditions, reworked to a greater or lesser extent. Modern historians and commentators, following Gershom Scholem, have delved into the mythical or philosophical character of their writings. Moses Cordovero, for example, is considered to be a philosophical kabbalist, with a lesser aliquot of myth, as opposed to his friend and contemporary in sixteenth-century Safed, Isaac Luria, in whom the mythical elements predominate. This philosophical tendency is explicitly affirmed by authors such as Israel Sarug and Abraham Cohen Herrera, disciples of Cordovero and Luria.[19] Nevertheless, original mythical elements occupy an important place in Cordovero's writings, notably the myth of the game and of the Infinite's delight "in itself, for itself, and by itself."[20] He offers this account as an allegory to describe the process by which the one and undifferentiated Infinite produces what becomes the origin of the

19. See Chapter 3, §4. Before them, this was the case with Solomon Ibn Gabirol, a twelfth-century poet and kabbalist, author of *Keter malkhut* (Crown of Kingship), known to Latin Europe as the Arab Neoplatonist Avicebron, whose *Fons vitae* (Well of Life) was translated into Latin. It was also the case with Ḥasdai Crescas, a fourteenth-century kabbalistic defender of Orthodoxy and fierce opponent of Maimonides, but also a philosopher and mathematician critical of Aristotle, whom Spinoza seems to appreciate in his "Letter on the Infinite." On these questions, see, in addition to the works cited above and those by Scholem, Moshe Idel, "Jewish Kabbalah and Platonism in the Middle Ages and Renaissance," in *Neoplatonism and Jewish Thought*, ed. Lenn E. Goodman (Albany: State University of New York Press, 1992), pp. 319–351.

20. See Chapter 2, §14.

worlds and their multiplicity, through the infinite combinations of the letters of a primordial Torah. It is with regard to the relative weight of myth and philosophical argumentation in his work that Bracha Sack[21] repeats Yehuda Liebes's remark that, for Cordovero, only the question was philosophical.[22]

More generally, it seems clear that the philosophical analysis inaugurated by Maimonides in his interpretation of the biblical text, through the lens of the Talmud and Midrash, gave these kabbalists a renewed interest in philosophical inquiry.[23] Some of them, like Abraham Abulafia, integrated it into in their own systems. Others saw it as a challenge to be taken up, to show that their tradition answered these questions better than Aristotle in the mirror of Maimonides could. The latter attitude seems to be more common;[24] Abulafia, both a disseminator of Maimonides' philosophy and a "prophetic" kabbalist, a practitioner of the exercises in which the mystic pursues identification with words and combinations of letters, is exceptional and constitutes his own "trend in Jewish mysticism."[25]

All these authors took up the challenge and engaged in philosophical inquiry, but offered their own answers, inherited from the "received tradition" (Hebrew *ḳabbalah*)—even if these answers agree or at least harmonize with other currents of Greek philosophy, such as Neoplatonism and Stoicism. As we have seen, according to Sack and Liebes the questions are philosophical, but the answers are derived from myth.

This observation, certainly justified in context, seems to have much broader scope and applies to every school of philosophy. All answers proposed by philosophical systems contain mythical elements, whether explicit, in the form of an allegory, or implicit, in the form of traditional cosmogonies or ontologies. Critical inquiry postulates only the principles of identity and noncontradiction, set in motion by doubt. It always has the advantage over affirmation of not bearing the burden of proof. On the contrary, any attempt at an answer, positive or negative, employs, in addition to these principles, empirical evidence of existence or an explanatory logic that invokes causes and reasons. This is where habit often supplies preconceived notions, which do not vanish even when critical judgment is applied to them. More generally, unanswered questions

21. Bracha Sack, *The Kabbalah of Rabbi Moshe Cordovero* (Hebrew) (Beer Sheva: Ben-Gurion University of the Negev Press, 1995).

22. Yehuda Liebes, "In His Image: On the Kabbalistic Writings by the Author of *'Emeḳ ha-melekh*," *Jerusalem Studies in Jewish Thought* 11 (1993), pp. 104 and 117–120 (Hebrew).

23. See Moshe Idel, *Maïmonide et la mystique juive*, trans. Charles Mopsik (Paris: Le Cerf, 1991).

24. Among many others, we may mention Joseph Gikatilla (thirteenth century), who explicitly distinguished the "genuine philosopher-sages" (*ḥakhmei ha-filosofim ha-'amitiyim*) who accept the "intellectual kabbalah" (*ha-ḳabbalah ha-śikhlit*) from other philosophers who must be opposed and rejected. The teachings of the former are to be taken without regard to their origin, just as we discard the peel when we eat fruit (Joseph Gikatilla, *Ginnat 'egoz* [Jerusalem, 1989], p. 233).

In a very different vein, Ḥasdai Crescas fought Aristotle on his own ground and, along with Christians and Muslims, was one of the medieval and Renaissance thinkers who helped dethrone Aristotelianism as the unchallengeable dominant science (Ḥasdai Crescas, *'Or Hashem* [Ferrara, 1995; repr. Jerusalem, 1970]; cf. Tony Lévy, *Les Figures de l'infini: les mathématiques au miroir des cultures* [Paris: Le Seuil, 1987]; Marc Tobiass and Maurice Ifergan, *Crescas: Un philosophe juif dans l'Espagne médiévale* [Paris: Le Cerf, 1995]).

Finally, the Maharal of Prague, a leading figure of the Renaissance, who was open to the science and philosophy of his own time, explicitly contrasted certain theses propounded by Maimonides and other "philosophers" to the doctrine taught by "the kabbalists"; he himself, as a "disciple of Moses," explained the coherence of that doctrine (see Chapter 2, n. 9).

25. Moshe Idel, "The Contribution of Abraham Abulafia's Kabbalah to the Understanding of Jewish Mysticism," in *Gershom Scholem's* Major Trends in Jewish Mysticism *50 Years After*, ed. Peter Schäfer and Joseph Dan (Tübingen: Mohr, 1993), pp. 117–143.

can always be piled up, one on top of another. The answers, though, must be empirical propositions (observed "facts") or logical propositions, or some combination of the two. In principle, these answers can always be criticized and called *in question*, because the ground of the truth of what we know, or think we know, is itself without ground. More precisely, "if the true is what is grounded, then the ground is not *true*, yet not false."[26] The underpinning of our certainty, the basis of what we know to be true because doubting it would be meaningless—because we have no *reason* to support such a doubt—constitutes a world-picture inherited from our acquisition of language like a sort of mythology. Wittgenstein saw this clearly.[27] Even if the base shifts, even if we encounter grounds for doubt, critical judgment will be exercised in the first instance on the basis of those preconceived notions, and not others, thus restarting the inquiry. In other words, in the best case the answer can lead only to a "yes, but . . ." sort of agreement, because the affirmation, narrative, or "primitive fact" on which the answer is built itself depends on the conditions of possibility. The latter, in turn, are subjected to criticism, so that the answer never puts an end to the inquiry. It starts it all over again, shifted by this rhetorical device of "yes, but . . . ," in which no consent is unconditional. It is this capacity for displacement that mythical narratives have in common with philosophical hypotheses, in that the latter always express an ontology to be questioned, explicitly or implicitly. Even if the form of the discourse is different, notably in its relation to the dream image, space, and time, the truth of the philosophical answer, like that of myth, is found not so much in its content as in its capacity to make one think, by provoking and guiding the new questions.

This "yes, but . . ." also implies a form of consent that, though only provisional, determines the conditions of the inquiry. Judgment can be suspended indefinitely only in the speculative and the academic; daily existence consists of actions, which are conscious or unconscious answers to questions that are themselves conscious or unconscious, practical solutions to the thousands of problems, great and small, posed by daily life.

26. Ludwig Wittgenstein, *On Certainty* (*Über Gewißheit*), ed. G. E. M. Anscombe and G. H. von Wright, trans. Denis Paul and G. E. M. Anscombe (Oxford: Basil Blackwell, 1969–1975), §205 (see below, n. 38). We accept some empirical and logical evidence not because it has been "proven"—that is, is solidly established on foundations that are impervious to doubt and criticism—but, on the contrary, because we have no good reason to doubt it. We accept the propositions that express this evidence because it would be unreasonable to doubt them. "But there isn't . . . a rule" about when doubt is unreasonable (ibid., §452). Furthermore, "there are cases where doubt is unreasonable, but others where it seems logically impossible. And there seems to be no clear boundary between them" (§454). In fact, when we learn to talk and think we form a set of beliefs, a sort of "nest of propositions" (§225), strongly dependent on how we proceed in our language games. This set constitutes a "world-picture" that we hold in the background of all our thoughts so long as we have no reason to modify it.

27. For example: "this is my hand," "I have spent my whole life in close proximity to the earth," and "the earth existed long before my birth" are elements of a world-picture that has been inherited like a sort of mythology.

> Nothing in my picture of the world speaks in favour of the opposite. But I did not get my picture of the world by satisfying myself of its correctness; nor do I have it because I am satisfied of its correctness. No: it is the inherited background against which I distinguish between true and false. The propositions describing this world-picture might be part of a kind of mythology. And their role is like that of rules of a game; and the game can be learned purely practically, without learning any explicit rules. It might be imagined that some propositions, of the form of empirical propositions, were hardened and functioned as channels for such empirical propositions as were not hardened but fluid; and that this relation altered with time, in that fluid propositions hardened, and hard ones became fluid. The mythology may change back into a state of flux, the river-bed of thoughts may shift. But I distinguish between the movement of the waters on the river-bed and the shift of the bed itself; though there is not a sharp division of the one from the other. But if someone were to say 'So logic too is an empirical science' he would be wrong. Yet this is right: the same proposition may get treated at one time as something to test by experience, at another as a rule of testing. (ibid., §§93–98)

Only skeptical philosophers escape the conditioning of "yes, but . . ." and only to the extent that they refuse to take their inquiry seriously when no answer has the rigor of the questions and cannot wield an affirmative power equal to the destructive force of the criticism. Otherwise even skeptics, as David Hume saw so well, are reduced to accepting the empirical value of habit in their practical judgments, even if their "academic skepticism" negates the ostensible reality of the cause-and-effect nexus.[28]

Of course, the form of the discourses, and their historical affiliation with a particular school, a particular century, and the social context of a particular zeitgeist, are sources of differences that only total confusion would obliterate and consign to oblivion. Philosophical systems are distinguished in particular by the greater or lesser place they accord to deconstructive critical inquiry and to the structures that are supposed to answer its questions. They are also distinguished, of course, by the nature and origin of the mythical elements, representative schemes and gnoseological intuitions that are part of what Jacques Schlanger calls their metaphysical structure.[29] But it is still customary to contrast, academically, the mythical narrative that speaks to the imagination with philosophy as a reasoned, analytical, and critical argument, initiated by Socrates in Greece and continued by Plato, Aristotle, and the schools that followed them.[30]

This contrast is welcome, because it is only too true that words can be used to say anything, including the greatest follies. Our imagination never finds it difficult to propose structures—some of them with a perfectly rational appearance—to plug up our ignorance and perhaps to ease our apprehension of the unknown. Nevertheless, from the very beginning (meaning Plato) philosophical reflection has been imbued with mythical elements that it could not or perhaps really did not want to discard, such as the mythical cosmogony of the *Timaeus*. These mythical elements have diverse sources. It may be, as in the *Timaeus*, a narrative handed down by tradition, or a narrative considered to be revealed truth of divine origin—what theologians later denominated the Holy Scriptures. But it can also be a properly philosophical structure, used by philosophers in their system to answer questions they have raised and whose facets they have studied in great depth. This is clearly the case with Plato's Myth of the Cave and his theory of anamnesis, used in the *Meno* to resolve the paradox of knowledge that would be impossible without prior acquaintance with what we are about to learn.

But it would not be difficult to find examples of this in all the great philosophers, because they are rarely content just to ask questions and insist on trying to answer them too. Such examples include Aristotle's *primum mobile* and agent intellect, Spinoza's unique infinite substance and its modes, Leibniz's monads, and Kant's transcendental object. Furthermore, Plato justified recourse to the imagination of "prophets" as a way of knowing by "divination," as long as we then call on reason to control its product.[31] In practice, it does not seem possible that theorizing, or explaining by means of visible causes, so-called natural

28. See David Hume, *An Inquiry Concerning Human Understanding*, Section V (ed. Charles W. Hendel [Indianapolis: Liberal Arts Press, 1955], pp. 54–68); Atlan, *Enlightenment*, on causality as proximity, pp. 190–191, and on the confusion of the "rational" with the "familiar-in-the-Western tradition," p. 123.

29. Jacques Schlanger, *La structure métaphysique* (Paris: PUF, 1975).

30. Incidentally, that this contrast is not limited to the Western philosophical tradition that began in Greece has been shown by Roger-Pol Droit in a study of India and its philosophical tradition, which developed against a different mythological and historical background (Roger-Pol Droit, *L'oubli de l'Inde: Une amnésie philosophique* [Paris: PUF, 1989]).

31. Plato *Timaeus* 71e–72e.

laws, or theories or models (as we call them today to reduce their scope), can manage without imagination. What we call the scientific method—that is, supervision by reason and verification of the observations of sense organs by diverse individuals and perhaps on the basis of cooperative experiments—has demonstrated its effectiveness, but only in the relatively limited domains in which it can be applied. Philosophy is rightly distinguished from the sciences in that it continues to raise questions for which the scientific method is inadequate or to which it cannot (yet?) furnish definitive answers. In proportion as the questions relate to objects or beings that are simultaneously complex and unique, such as a particular human being, or that tree over there, or the shape of that mountain, or "the universe," scientific theorizing is limited by the underdetermination of theories (or models) by facts.[32]

Philosophical inquiry has certainly benefited from the successive scientific revolutions, but the attempts at answers have not always eliminated images, whose function, like myth, is to show rather than to demonstrate. The borrowings, sources, and direct or indirect relations that create a family resemblance among philosophical systems, for all that they may be quite different and even contradict one another on essential points, usually derive from these mythical elements. The inherent property of myth is to lend itself to multiple interpretations in the form of analogies, metaphors, and the like. The same image can be integrated into different philosophical systems, though sometimes with contrary meanings. We often recognize these elements of myth that are common to different philosophies in the words that designate them and have different meanings for different authors. The notion or picture of the monad, in Giordano Bruno and Leibniz, goes back to Plotinus, although it carries a very different meaning for each of these authors. Spinoza was an expert in reusing words from the vocabulary of medieval Scholasticism, replete with theological images, but with new meanings that he was careful to specify in the definitions, axioms, and propositions of his *Ethics*. Elsewhere I have referred to this as "double philosophical language,"[33] conscious and openly proposed as rigorously as possible, and not to be confused with the doubletalk of the vicious or the delirious.

2. CRITICISM OF KANT'S *CRITIQUE*: METHODOLOGICAL DOGMATISM AND EMPIRICAL SKEPTICISM IN SALOMON MAIMON

The critical philosophy originated by Kant believed that it could avoid myth and dogmatic assertions yet retain a solid ground for rational knowledge of things, both the fact of this knowledge and its possibility. Do our judgments about descriptions of what we perceive correspond *in fact* to a reality that is independent of our perceptions? For example, is "the flame is blue" a fact? If so, when our understanding makes use of facts described in this way, associates them, and deduces from them an intelligible reality by application of its own categories, one of which is causality, what does the legitimacy of its operations rest on? Does the statement that "a flame heats" have a foundation *de jure*? Or is it only, as Hume maintains, an association, perhaps illusory, of two facts, that our habitual perception of first one and then the other produces in us? Kant's transcendental philosophy is supposed to answer these questions, which are posed in his *Critique of Pure Reason*. But as an early critic of the *Critique*, Salomon

32. See Atlan, *Tout, non, peut-être*, pp. 130–148; *SR* II, chap. 6, §4.

33. Ibid., pp. 195–202.

Maimon,[34] showed clearly, Kant's transcendental philosophy cannot avoid employing new "occult qualities,"[35] such as the notions of the transcendental object and of the thing in itself, which are unknowable but indispensable for answering the two initial questions: *Quid facti? Quid juris?*[36] Maimon, retracing each stage of Kant's progress, shows the extent to which Kant's answers to these questions are tainted by circular reasoning and beg the question. They simply pose all over again, although shifted, the very questions they are supposed to answer, producing insoluble aporias like those that Kant himself had analyzed in his criticism of the metaphysical notions of the existence of God, the immortality of the soul, and freedom. Maimon's critique of the *Critique* seems to have influenced post-Kantian thinkers, especially Fichte and Hegel, to a greater extent than is generally realized.[37] In any case, Maimon endeavored to resolve the problems he had identified in Kant in

34. Salomon Maimon, a true adventurer of thought, had an extraordinary biography and intellectual career, from a yeshiva in a poor town in Poland to the intellectual and bourgeois circles of Berlin that were enlivened by the philosophy of the Enlightenment. Supported for a while by Moses Mendelssohn, he became, after an intellectual quest during which he learned to think by immersing himself successively in Maimonides, kabbalah, Spinoza, and Leibniz, one of those who, according to Kant himself, best understood the Kantian method (see: Salomon Maimon, *Solomon Maimon: An Autobiography*, trans. J. Clark Murray [Urbana: University of Illinois Press, 2001 (London, 1888)]; or [abridgement of Murray but better English] *Solomon Maimon, An Autobiography*, ed. Moses Hadas [New York: Schocken, 1947]).

35. Martial Guéroult, *La philosophie transcendantale de Salomon Maimon* (Paris: F. Alcan, 1929), p. 138.

36. Here is Maimon's own description of the process that led him, in 1786, to write his *Versuch über die Transcendentalphilosophie* (1790), in which he provided a reading, enlargement, and finally a critique of the *Critique of Pure Reason*, which he carried further in his later works:

When I came to Berlin, Mendelssohn was no longer alive, and my former friends were determined to know nothing more of me. I did not know what to do. In the greatest distress I received a visit from Herr Bendavid, who told me that he had heard of my unfortunate circumstances, and had collected a small sum of about thirty thalers, which he gave to me. Besides, he introduced me to a Herr Jojard, an enlightened and high-minded man, who received me in a very friendly manner, and made some provision for my support. A certain professor, indeed, tried to do me an ill turn with this worthy man by denouncing me as an atheist; but in spite of this I gradually got on so well, that I was able to hire a lodging in a garret from an old woman.

I had now resolved to study Kant's *Critique of Pure Reason*, of which I had often heard but which I had never yet seen. The method by which I studied this was quite peculiar. On the first perusal I obtained a vague idea of each section. This I afterwards endeavored to make distinct by my own reflection, and thus to penetrate into the author's meaning. Such is properly the process which is called *thinking oneself into a system*. But as I had already mastered the systems of Spinoza, Hume and Leibnitz in this way, I was naturally led to think of a system that would synthesize them all [*Koalitionsystem*]. This in fact I found, and I gradually put it in writing in the form of explanatory observations on the *Critique of Pure Reason*, just as this system unfolded itself to my mind. Such was the origin of my *Transcendental Philosophy*.[*] Consequently this book must be difficult to understand for the man who, owing to the inflexible character of his thinking, had made himself at home merely in one of these systems without regard to any other. Here the important problem, *Quid juris?* with the solution of which the [*Critique*] is occupied, is wrought out in a much wider sense than that in which it is taken by Kant; and by this means there is plenty of scope left for Hume's skepticism in its full force. But on the other side the complete solution of this problem leads either to Spinozistic or to Leibnitzian dogmatism. (*Autobiography*, trans. Murray, pp. 279–280 [ed. Hadas, as far as *, pp. 104–105])

(In addition to Guéroult, *La philosophie transcendantale*, see also Sylvain Zac, *Salomon Maïmon: Critique de Kant* [Paris: Le Cerf, 1988].)

In a letter to Kant accompanying the first few chapters of the *Versuch*, in which Maimon lists the points where the system of the First Critique is inadequate (although he deems that work to be as classic and as irrefutable in its genre as is Euclid's *Elements* in its), he returns to these questions and to the difficulty of answering them without having recourse, in one case (*quid juris*) to a Spinozist notion of ideas of the understanding, and, in the other case (*quid facti*) to Hume's skepticism: "I define a new class of ideas which I call *ideas of the understanding* [*Verstandesideen*], which signify *material totality*, just as your ideas of reason signify *formal totality*. I believe I have opened the way to a new means of answering the aforementioned *Quid Juris* question. . . . [As for] the question, *Quid facti?* You seem to have touched on this, but it is, I think, important to answer it fully, on account of the Humean skepticism" (Maimon to Kant, April 7, 1789, in Immanuel Kant, *Philosophical Correspondence, 1759–99*, trans. Arnulf Zweig [Chicago: University of Chicago Press, 1986 (1967)], p. 134).

37. See Guéroult, *La philosophie transcendantale.*

order to ground the possibility of objective knowledge on reason, through recourse to Spinoza and his theory of understanding and to Leibniz, from whom he borrowed the concept of the "differential of consciousness."[38] The immanence of the infinite understanding in the finite understanding of each human being, along with the notion of

38. See below n. 202 and §12. Maimon used Leibniz's mathematical formalism to try to provide a Spinozist solution—ideas of the human understanding as part of the infinite understanding—to the Kantian problem of the ground and of the thing in itself. In this interpretation, the notion of the thing in itself is the result of attaining a limit, which makes them nonmensurable and unthinkable, because equal to zero, like the differentials *dx* or *dy*, as opposed to their ratio, *dy/dx*, which is finite and measurable. Later, however, Maimon renounced this *salto mortale* or "fatal leap," recognizing its failure. It was not only the analogy that was the problem, even though he had recognized the difficulty when he proposed it:

> The objections that can be made against the introduction into philosophy of mathematical concepts of the infinite are not unknown to me. In particular, because even in mathematics they are still subject to many difficulties, it might seem that I am trying to explain something obscure through something even more obscure. I dare say, however, that these concepts do in fact pertain to philosophy, that they were transferred from there to mathematics, and that the great Leibniz arrived at the discovery of the differential calculus through his system of monadology. A magnitude (*quantitas*) that is considered to have no size (*quantum*) is much stranger than the abstract quality of quantity. But these concepts, both in mathematics and in philosophy, are mere ideas that represent not objects but their mode of origin; that is, they are mere boundary concepts that can be approached but never reached. They are formed by a constant regression or decrease to infinity of the consciousness of an intuition. (Salomon Maimon, *Versuch über die Transcendentalphilosophie* [Berlin: Voss, 1790; repr. Darmstadt: Wissenschaftliche Buchgesellschaft, 1963], note on pp. 27–28)

He had come to realize that the ground provided no support for anything because it always assumes the idea it is supposed to ground.

These problems of transcendental philosophy remain unsolved to this day, which justifies now, as then, the species of relative relativism that combines academic skepticism and existential dogmatism, to which Maimon ultimately arrived. The questions about the ground of the truth of empirical propositions, of logical propositions, and of the legitimacy of their associations return as an inquiry that is always open and always insistent, in Wittgenstein's last notebook, *On Certainty*, which is an endless reflection about "I know . . .": for example, "this is my hand," or "that earth existed long before my birth," or "that 12 x 12 = 144."

The questions are the same, even if they do not employ the classic expressions *quid facti* and *quid juris*. But here we are dealing with the linguistic watershed of contemporary philosophy, which asks these questions about our knowledge and affirms that some propositions, respectively empirical and logical, are true because they are beyond doubt. What is this knowledge and how is it different from belief? The only conceivable answers merely take note of what we might call, in Spinozist language, the "given true idea," which precedes—both logically and in the process of learning language—the experience of error and of doubt. For "the reasonable man does *not have* certain doubts" (Wittgenstein, *On Certainty*, §220). For some items of evidence, it is the reasons for doubting them that must be grounded, rather than the reasons for believing them. What we know with certainty, we know, along with every reasonable person, because we have no reason to doubt it. In general, and contrary to the notion of an initial tabula rasa produced by methodical doubt, belief precedes doubt: "Doesn't one need grounds for doubt?" (§122). And "if the true is what is grounded, then the ground is not *true*, nor yet false" (§205). To be able to doubt, we must first have the experience of the true and then of the false. This experience is both individual and collective, because it takes place as we learn language. This is why knowledge expresses "*comfortable* certainty" (§357), satisfaction (§299), and membership in "a community which is bound together by science and education" (§298). (In Spinoza's language one might say, respectively, the joyful satisfaction of the *conatus* that increases its power of action and the shared common ideas of reason.) As such, primitive knowledge is related to "something that lies beyond being justified or unjustified; as it were, . . . something animal" (§359), and is also part of "the *natural law* which our inferring apparently follows" (§135). The last, as for Hume, causes us to believe in what is repeated, because "the principle that what has always happened will happen again (or something like it) . . . is not an item in our considerations" (ibid.). By contrast, the propositions we use to speak of them, like "what we call 'a mistake'[,] [play] a quite special part in our language games" (§196).

Here we find our mixture of Hume's academic skepticism and the dogmatism of Spinoza's "given true idea," i.e., the primitive evidence of the true, a sign of itself of which every human being has experience. When this evidence is known adequately, it is, for Spinoza, the "idea of God," whose essence implies existence. Most often, however, this knowledge is confused and mutilated by inappropriate uses of language in the service of imagination, including, and especially, as wielded by those given to discourses about God, i.e., "theologians." Whence, for Wittgenstein, the nonpropositional character of what *plays the role* of the ground of empirical and logical knowledge. For "giving grounds . . . come[s] to an end sometime. But the end is not an ungrounded presupposition: it is an ungrounded way of acting" (§110). And if I think about it as if I must speak of it, then "what I hold fast to is not *one* proposition but a nest of propositions" (§225), not all of which are necessarily formulated. That is, "our talk gets its meaning from the rest of our proceedings" (§229). All of this taken together constitutes a world picture, which we inherit "like a sort of mythology" (see above, n. 27).

the differential generalized to the activity of consciousness, allowed him to affirm both the unity and multiplicity presupposed by our cognitive activity; for example, the simultaneously associative and dissociative character of the cause-and-effect nexus.[39] But, criticizing his own criticism, he later concluded later that this is a "fatal leap" and doomed to failure: "I recognize," he wrote in 1792, "that in my first composition I attempted this fatal leap and tried to reconcile Kantian philosophy with Spinozism. Now, however, I am fully persuaded that this enterprise, natural to any independent thinker, is impractical and think rather of synthesizing Kantian philosophy with Hume's skepticism."[40]

In practice, however we picture the mechanism by which the data of experience are given as facts, after which the operations of the intellect transform them into the content of objective knowledge, the finite and piecemeal character of our understanding and of specific sciences means that this mechanism can never be conscious. As Martial Guéroult writes, with regard to Maimon, "the explanation of this mechanism can produce only a hypothesis, which is simultaneously an idea in the Kantian sense (because it relates to something that by our very nature transcends our knowledge) and a fiction in the Leibnizian sense (because it is a method of explanation, whose principle [the differential] cannot be conceived as metaphysically real or be realized without contradiction. Whether it is outside us or inside us the 'occult quality' is set as the ground of the phenomenon and, like the 'secret forces' of which Hume speaks, leads to skepticism."[41]

So at the end of the journey there is no escaping Hume's empirical skepticism, despite the Kantian and post-Kantian attempts to refute it. Note at once that this unavoidable skepticism is only "empirical" or, as Hume himself designated it, "academical philosophy."[42] It rules out neither the use of mathematics and the type of formal certainty that it produces nor the invocation

39. See Martial Guéroult's exposition of this question, as posed by Spinoza, as an aporia with no solution. On the one hand, the effect is posited as incommensurable and having nothing in common with its cause: "For a cause differs from a thing it causes, precisely in the quality which the latter gains from the former" (*Ethics* I 17, Note). Indeed, "a thing which is the cause both of the essence and of the existence of a given effect, must differ from such effect both in respect to its essence, and also in respect to its existence" (ibid.). On the other hand, he has already demonstrated, apparently contradictorily, that "things which have nothing in common cannot be one the cause of the other" (*Ethics* I 3).

In Guéroult's formulation, "the combination of commensurability and incommensurability between cause and effect is required for there to be a relationship between them: without incommensurability, effect and cause could not be distinguished; without commensurability, they would be totally separated and could not be linked." To put this another way, the contradiction is resolved by the immanent character of real causality, which implies that cause and effect are incommensurable only with regard to what the latter gains from the former, whereas they are commensurable with regard to the rest, because they have in common at least the fact of being causally produced as modes of substance. See Martial Guéroult, *Spinoza* (Paris: Aubier, 1968), vol. 1, pp. 286–295 (quotation on p. 290).

40. *Magazin zur Erfahrungsseelenkunde* (1792), Part 2, p. 143; quoted by Guéroult, *La philosophie transcendantale*, p. 138.

41. Ibid. The negative answer to both *quid facti* and *quid juris*, and the circularity that Maimon identified in the search for a transcendental ground of knowledge, are summed up in the dilemma that confronts him: "Either the fact itself [the use of the form of hypothetical judgments for empirical objects] is false and the examples given of it are based on an imagined illusion, in which case the categories are useless; or the fact is true in itself, but there is no principle that can make it knowable to us, and the categories remain, both before and after their laborious deduction and schematization, simple forms that cannot determine objects" (ibid., p. 139).

42. Hume, *An Inquiry Concerning Human Understanding*, Section XII. Total and unmitigated skepticism, in addition to annihilating itself the moment it is erected as a systematic principle, can also lead to a flight into the irrational or to faith in some revealed doctrine, as by Jacobi in his controversies with Kant and Mendelssohn. See: Immanuel Kant, "What Is Orientation in Thinking," pp. 235–249 in *Kant, Political Writings*, ed. H. S. Reiss (Cambridge: Cambridge University Press, 1991); Sylvain Zac, *Spinoza en Allemagne: Mendelssohn, Lessing et Jacobi* (Paris: Méridiens Klincksieck, 1989).

of habit in the rational action-oriented organization of our experiences. On the contrary, mathematics retains a privileged status because its objects are directly determined by our understanding and constructed by our intuition. It is only its relationship with experience that remains problematic if we would turn it into an ultimate ground. In this, Maimon's skepticism is particularly modern (or even "post-postmodern"). On the one hand, the linkage of formal truths, the real, and experience is not established from top to bottom, derived from a first principle, but from bottom to top, through examination of how rational thought acts on the specific facts within each individual branch of knowledge. This reveals the privileged status of the natural sciences to which mathematics can be applied, a privilege that comes down to the fact that mathematical law replaces the principle of causality. The positivists of the early twentieth century, reflecting on the nature of physical law, recognized the inadequate and magical—or even "superstitious"[43]—character of the principle of causality. We have seen how the search (whatever the cost) for the hidden causes of phenomena, which remains the rule in biology, sometimes makes it possible for the theoretical generalizations of this science (and its medical applications) to drift toward a magical thinking in which, allowing for minor terminological modifications, Paracelsus would have felt perfectly at home.[44]

On the contrary, the mathematical knowledge that constructs and determines its objects—and that accordingly must be distinguished from the statistical processing of observations that makes possible interpretive shifts at least as large as those toward magical thinking—allows the natural sciences to which it can be applied to determine their objects, too, and to set them up as objective reality. This skepticism, which is rather like the contemporary relative epistemological relativism,[45] means that undoubted objective knowledge is possible locally, even though there is no general principle that permits us to ground, a priori and in all circumstances, the truth of some knowledge. And thus, as Guéroult writes,

> the result is a doctrine that is far from systematic. . . . It is a synthesis [*Koalitionsystem*] in which we are preoccupied less with the intrinsic truth of ideas than with their intellectual profit; a sort of intellectual pragmatism . . . that is in harmony with the eclecticism of Maimon's early works. This idea that systems differ less in their content than in their degree of systematization [i.e., mathematization] ultimately produces not a stricter systemization of the elements, but an indefinite expansion of the framework of the doctrine; it makes it possible to combine in the same philosophy, with a sort of indifference, elements drawn from contrary theories, elements whose reconciliation is not realized but only suggested.[46]

This is why Spinoza's system makes it possible to provide a rational ground for knowledge of the objects of

43. See below, at n. 55.

44. See Atlan, *Enlightenment*; and Chapters 1 and 3 of this volume.

Some see the mathematization of biology, based on techniques for the analysis of dynamic systems, still in its infancy, as the next development in this science. Then—as in the physics of complex systems—not only the final causes of teleological judgment but also the efficient causes of the principle of causality would vanish from biology. See, for example: R. C. Strohman, "Epigenesis and Complexity. The Coming Kuhnian Revolution in Biology," *Nature Biotechnology* 15 (1997), pp. 194–200; Atlan, *La Fin du tout-génétique?*; idem, "Information génétique," *Encyclopaedia Universalis*, DVD-CD-ROM (1999); idem, "Epigenesis and Self-Organization: New Perspectives in Biology and Medicine," in *Conformational Diseases: A Compendium*, ed. B. Solomon, A. Taraboulos, and E. Katchalski-Katzir (Basel: Karger, 2001), pp. 291–297.

45. See: Atlan, *Tout, non, peut-être*, pp. 245–250; Hilary Putnam, *Reason, Truth, and History* (Cambridge, MA: Harvard University Press, 1981).

46. Guéroult, *La philosophie transcendantale*, pp. 140–141.

the world. His monism allows us to answer the questions of fact and of the legitimacy of the ground of our knowledge more satisfactorily than Kantian dualism does. Although absolute rationalism is possible, it is no more than a possibility whose realization can be perceived only dimly, empirically and locally, every time some manipulation of causes proves effective. Its truth clearly cannot be demonstrated rationally. The logical possibility of Spinozist rationalism leaves Humean skepticism intact.

Note that the juxtaposition of Spinoza's dogmatism with Hume's empirical skepticism is not as incongruous as it might seem; in both cases we are dealing with a methodical dogmatism that is quite different from the theological dogmatism that survives in Descartes and Leibniz. Unlike the God who, in a different way for each of these authors, "guarantees" the coherence of the system through his very inscrutability, the mythical element in Spinoza, which seems to ground the objective truth of his philosophy, is not so much the infinitely infinite unique substance, or even the hierarchy of attributes, immediate and mediate infinite modes, and finite modes, but the specific method of the *Ethics*. The geometrical method is the mythical element that, even though it too grounds nothing, displays the formal truth of its propositions *from within the system*, by means of their proofs, because "the eyes of the mind, whereby it sees and observes things, are none other than proofs" (*Ethics* V 23, Note).

From this perspective, Maimon's method can be compared to the original approach of the contemporary philosopher Gilbert Boss. Boss pursues philosophy as the quest for truth through systematic investigation of how two philosophies differ, in order to better understand this difference from within each of them. For him, even if "the choice of the authors to juxtapose is in principle quite free and needs to respect only one requirement, namely the philosophical value of the two doctrines,"[47] some conditions of historical and cultural context or of form and content are more propitious for the success of this project than are others. Because "the philosophies of Hume and Spinoza satisfy these conditions excellently,"[48] they can be the subject of the thousand-plus pages of his analysis and comparison.

What is more, Spinoza is easier to put together with Hume's skepticism and critical philosophy than other major precritical philosophers would be. The dogmatic or "metaphysical" character of the *Ethics*, in the pejorative sense of the positivists, is not as obvious as might be supposed from the historical context of its composition and the vocabulary of its propositions. For example, its proof of the existence of God withstands the Kantian critique, because in it the word "God" does not designate "a transcendent, ineffable, and empty entity about which one may wonder whether it may be nothing else than an idea in our mind," but is instead "reduced to the being of Nature" or "the infinite totality of all real things."[49] In practice, the ontological proof is vulnerable because it posits the existence of a concept as if it were a thing and deduces God's existence from the concept of his essence. But if God's essence is that of the totality of things that exist, it is the essence of that whose existence cannot be denied, because the word does not designate just a concept but real objects in their concrete existence. In brief, the tautological proposition "it pertains to the essence of all existing things to exist" is immune to the criticism of the

47. Gilbert Boss, *La différence des philosophies: Hume et Spinoza* (Zurich: Le Grand Midi, 1982), pp. 11–13.

48. Ibid., p. 15.

49. Guéroult, *Spinoza*, vol. 1, pp. 477–479.

ontological proof of God. (At most one can argue on the basis of the nonreferential character of the word "all"; but, as I have shown elsewhere,[50] this, like negation and the possible, is one of the "pseudo-concepts" that are indispensable for describing reality, along with the notions of existence, reality, and, of course, essence itself.) This is what Spinoza, in the *Treatise on the Emendation of the Intellect*, calls the "given true idea," from which Thought produces all of the ideas that constitute the infinite understanding, of which each human intellect is a part. This is also the anchor point of what can be called the myth of nature, with a content both ontological and gnoseological, that grounds his philosophy. As in Renaissance Platonism, "Nature, divine thought, produces its objects according to reasons, just as the mind of the geometer produces its figures from itself."[51] But this Spinozist nature, not only Thought but also Extension—it, too, divine (meaning active)—produces its objects according to the order of causes, which is the same as that of reasons. Whence the form of this "given idea," postulated only in the *Emendation of the Intellect* (§§33–42) and explained in the *Ethics* as an adequate idea of God, that is, of eternal and infinite substance, of which "the human mind has . . . an adequate knowledge" (*Ethics* II 47). Like the geometer, who experiences truth in the very act of understanding the properties of the figures as he draws them, the philosopher experiences truth in his own experience of intellection; except that the "given true idea" held by every human mind, inasmuch as it is a part of the infinite intellect, is obscured by imagination and the muddled use of language, which inevitably pushes human beings, who are "continually affected by external bodies," to associate "the name God with images of things that they are in the habit of seeing" (ibid., Note). This is the source, among other forms of inadequate knowledge, of the images and anthropomorphisms employed by the prophets, especially when they are amplified by theologians who erroneously believe them to be intellectual truth rather than an encouragement to virtue.

The qualitative and irreducible experience of this true knowledge that knows itself to be true for all eternity is that of mathematics.[52] Spinoza's monism, like Plato's but in a different way, guarantees its relevance for the adequate knowledge of the objects of nature—and not only of the entities of reason—as soon as reason is trained to recognize inadequate causes, the main sources of the errors that our imagination inevitably paints for us in order to explain the sequence of events.

More generally, Spinozism can be "reconstructed" (understanding this process of reconstruction in the sense advanced below, with regard to the nominalism of William of Ockham) by reading it through the lens of contemporary analytical philosophy and its criticism of language.[53] The classic and "dogmatic" character—chronologically precritical—of Spinoza's discourse does not prevent it from being understood in this way. On the contrary, not only does the analytical

50. Atlan, *Tout, non, peut-être*.

51. Marsilio Ficino, *Theologia Platonica*, quoted by Guéroult, *Spinoza*, vol. 2, p. 480.

52. Guéroult (*Spinoza*, vol. 2, pp. 417–487) shows how the "given true idea" is also the source of knowledge of the third type, or intuitive science, both the ultimate edifice of the *Ethics* and the constitutive experience on which the process of construction itself is based in the first and second parts of the book. "Thus the *Ethics* . . . grounds and legitimates the method by which it is possible. Nevertheless, the *Ethics* does not fall into a vicious circle, because it is the nature of true knowledge to know that it is true knowledge and to discover it itself the immutable ground on which it rests for all eternity" (ibid., p. 487).

53. J.-P. Brodeur, "Essai sur le discours spinoziste," thesis, Université de Paris–IX Nanterre, April 1975. I would like to thank Michel Pierssens and Robert Nadeau, of the philosophy departments of the University of Montreal and the University of Quebec in Montreal, who drew my attention to this work.

reconstruction fail to dissolve Spinoza's system as a metaphysics "devoid of factual content";[54] it actually highlights the links between the language of its ontology and the meta-language of its epistemology (the three types of knowledge), which relates to the truth conditions of the ideas and propositions that express it. Far from causing the system to break down, the analytical method is shaken by its confrontation with Spinoza's own analysis of language and must examine itself critically. The need for new inquiries in its own field then takes the form, inter alia, of the theory of reference. The relations among linguistic pragmatism, syntax, and semantics are seen to be more intricate than expected when a philosophical discourse, unlike a discourse in formal language, is necessarily both descriptive and axiological (like the identity of reality and perfection for Spinoza).

This sort of process leads us (like Maimon in his synthesis) to a diachronic association of the apparent classicism of Spinoza for whom, as for the Scholastics, there can only be knowledge by causes, with the postcritical analytical modernity of Wittgenstein, for whom "belief in the causal nexus is superstition."[55] This confrontation ultimately leads us to recognize (as did Wittgenstein himself) the necessity of some minimum dogmatism, like that of the "given true idea," which is inevitable even for those who entertain the illusion that they can avoid all dogmatism.[56] But as we have seen, this does not make it possible to refute Hume's empirical skepticism, in which the habits we acquire from our temporal existence and the timeless truths of mathematics coexist without ever being confounded.

54. A. J. Ayer, *Language, Truth and Logic*, 2nd ed. (London: Victor Gollancz, 1946), p. 41.

55. Wittgenstein, *Tractatus Logico-Philosophicus*, trans. D. F. Pears and B. F. McGuinness (London: Routledge and Kegan Paul, 1961), 5.1361. The philosopher and mathematician Jean Cavaillès seems to have undergone a similar process. In his last book, this active participant in and witness of the development of positivist philosophy in the first half of the twentieth century, who died too young, makes a sort of progression toward Spinoza, after obligatory detours through Leibniz, Kant, Bolzano, Carnap, and Husserl (Jean Cavaillès, *Sur la logique et la théorie de la science* [Paris: J. Vrin, 1997]). His quest for "a doctrine of science" leads him to "a philosophy of the concept" rather than of consciousness (p. 90), in which "the idea of the idea manifests its generative power" (p. 46). For Georges Canguilhem, Cavaillès's friend and posthumous editor, "it is because Spinoza's philosophy represents the most radical attempt at a philosophy without a *Cogito* that it was so close to Cavaillès, so present to him when he had to justify himself, whether the idea of his career as a Resistance fighter or the idea of the structure of mathematics" (Canguilhem, quoted by J. Cebestick in his postscript, ibid., p. 142). For Gilles-Gaston Granger, this last book, which Cavaillès wrote in prison, was "the postscript to the work already written," but also, above all, "the preface of the book that he was not able to write." Granger sees it as the extremely dense—too dense—summary of "a philosophical evolution" whose "atmosphere" struck him as "astonishingly Spinozist," especially its theme of the immanence of science as a whole in the operational activity of each of its parts. Cavaillès had arrived "at the threshold of this new 'philosophy of the idea,' which he was not allowed to elaborate"; this last book completed "what should have been only the first stage of his career. He still had to rewrite the *Ethics* (in the light of modern mathematical thought) to complete it fully" (G.-G. Granger, "Jean Cavaillès ou la montée vers Spinoza," *Les études philosophiques* 3 [1947], pp. 271–279).

56. See above, nn. 27, 38, and 53.

3. RECONSTRUCTION AND ACTIVE THINKING

Thus the only way for critical philosophy to totally escape myth and "occult qualities" is to renounce its ultimate ground or first principles, that is, to stop expecting definitive answers to its inquiry. Every explanatory answer incorporates myth. Even the so-called masters of suspicion cannot evade it: Marx, with the Golden Age that has returned as humanist messianism; Nietzsche, with Zarathustra and eternal recurrence; and, of course, Freud and Oedipus. They all employed myth to organize their inquiry and frame their answers.

The last attempt to erect a strict separation between myth and philosophy was made by the positivists of the Vienna Circle in the early twentieth century. Their program, modeled on mathematical physics, aimed at achieving the status of science by using only statements that could be decided logically or verified empirically. Any form of discourse whose statements did not comply with the rules of formal language, and thus could neither be verified nor refuted, were considered to be "devoid of factual content" and "non-sense" and were to be relegated to the obscure corners of metaphysics.[57]

The failure of positivist analytic philosophy to achieve this goal, coming on top of the inability of the great philosophical systems spawned by the Enlightenment to ground universal morality and politics on objective knowledge, spawned a certain pessimism about the usefulness—and even the point—of philosophy. The death of philosophy was regularly proclaimed, even though since Hegel, and then Heidegger, the announcement—in philosophical discourse—of its demise was belied by the author's own work. It was, in fact, just one more mythological way of speaking, as if the essential vocation of philosophy, like mythology, had always been to supply definitive answers, so that now, after these had been shown up as inadequate and always subject to further questioning, one could only ponder its disappointment—in philosophical discourse. It is true that "giving meaning" or "finding a meaning" by answering the eternal questions about being, essence and existence, good and evil, was the vocation of the great philosophers of antiquity, who engaged in both physics and metaphysics. Even then, however, the Skeptics and other cynics had no problem demonstrating the emptiness of the doctrines advanced, if they were taken as providing undoubted answers. The diluted or subjective relativism to which the baton has passed today is no reason to despair of philosophy (any more than in the past), on condition that we know how to distinguish the roles of myth and critical inquiry in what it teaches us.[58] As in the case of Salomon Maimon, empirical skepticism or relative relativism can coexist, although tenuously and with no claim to a synthesis, with elements of myth, rationalized and organized as if into a dogma. On the one hand, philosophy as critical inquiry retains its full value. It is no less necessary today than it was in the past if we would avoid the snares of new beliefs, or old ones that have been given a facelift. On the other hand, we readily admit that every philosophical current can contribute interesting and important answers, even though they are burdened, to a greater or lesser extent, with myth, as a function of the questions it asks and investigates. We learn not only to follow its inquiry but also to extend it and delve further into the answers it suggests. Myths, unlike articles of faith, are not made to be believed in literally.

Thus, going behind the history of ideas and leaping over the chronological sequence of philosophical systems, we forge a dialogue between the great philosophers of ancient Greece and the European Middle Ages and traditions that originated in the Middle East and Orient, associating them too with different levels of mythical images and with the logic of rational argument. For us, philosophizing can be something other than a simple historical catalogue of opinions that are

57. See, for example, Ayer, *Language, Truth and Logic*.

58. The ambit of this relative relativism, which I have defended elsewhere (Atlan, *Tout, non, peut-être*), includes several exponents of neopositivism and analytical philosophy, among them Hilary Putnam (*Reason, Truth, and History*) and Richard Rorty (*Philosophy and the Mirror of Nature* [Oxford: Basil Blackwell, 1980]).

more or less grounded, more or less reasoned and systemized. Through this diachronic dialogue we are finding and renewing, for our own benefit today, the experience of thinking.[59]

We can agree, then, that all philosophy is also an attempt to rationalize a set of myths borrowed from diverse sources, even though not all philosophers devote the same effort to this and not all are equally successful at it. Thus for us today, as for those who devoted themselves to this activity in the past, philosophizing is not only interpreting and commenting on the great texts of philosophy, whether historically, logically, or hermeneutically, but also actively thinking while engaged in these interpretations and commentaries. We have seen, with regard to Hume and Spinoza, how one can conceive of this activity, located at the interface between the philosophies brought into dialogue. With regard to a specific philosophical system, this is the task that Claude Panaccio calls "reconstruction." Panaccio applies it, with some success, to one of the major philosophical systems of the Middle Ages, the nominalism of William of Ockham.[60] As with Gilbert Boss's choice of Hume and Spinoza, this indicates (perhaps superfluously) that the success of the enterprise depends heavily on the choice of the philosophical system or systems that serve as the pre-text for the work of philosophizing. Both the reconstruction and the analysis of the differences clearly require, as a first step, interpretation and commentary and the deconstruction that draws on them. The richness of what is reconstructed obviously depends on that of the initial structure(s) and their ability to avoid being reduced to fields of hopeless ruins.

This having been said, there is no question of going backward and forgetting what critical and positivist philosophy have taught us. Our irreversible exposure to Kant's *Critique*, and to the critiques of the *Critique*, forces us to follow the strait and winding path of the intercritique. Assigning the timeless character of *philosophia perennis* to these works does not mean ignoring the history and geography of their production.

On the one hand, we cannot avoid reading the ancient texts, whether we want to or not, on the basis of the world that surrounds us here and now. Elsewhere I have discussed the anachronisms, confusions, and muddles of every sort that need to be avoided when we look for new meaning in ancient texts, tempted by the lure of grand explanatory syntheses of science and mysticism. On the other hand, this does not mean that our modern, critical and postcritical culture vanishes magically from view when we learn something from these texts, which express a mythological if not indeed magical—and certainly prescientific—rationality.[61] On the contrary, our contemporary scientific rationality is the unavoidable horizon or cultural backdrop of the various questions, inquiries, doubts, fears, and hopes at the origin of the problems of existence

59. Leibniz advised his friend François Mercure van Helmont, who taught him Jewish kabbalism—as he interpreted it through the lens of his Christian theology and Quakerism—to jot down his thoughts as they arrived and wait until later to organize them, if possible. In this he seems merely to have been describing his own way of proceeding and his own experience of thinking as a mixture of original inspiration and analyses of various existing texts, analyses necessarily shaped by the perspectives of these works, reinforcing one another even though they derived from different sources and seemed to follow contrasting paths (see Allison Coudert, *Leibniz and the Kabbalah* [Dordrecht: Kluwer, 1995], pp. 69–70). Below we shall encounter the role of this intellectual friendship in the formation of Leibniz's philosophy. All those with experience of thinking will readily recognize themselves in these descriptions, whether or not the subsequent attempts at systematic organization proved successful.

60. Claude Panaccio, *Les mots, les concepts et les choses: la sémantique de Guillaume d'Occam et le nominalisme d'aujourd'hui* (Montreal: Bellarmin, 1992). See also: Pierre Alféri, *Guillaume d'Ockham: Le singulier* (Paris: Minuit, 1989); A. de Libera, "Retour de la philosophie médiévale?" *Le Débat* 72 (1992), pp. 155–169.

61. Atlan, *Enlightenment*, passim.

and civilization that are made even more difficult because they are posed at the limits of our scientific and technological knowledge. Our experience of new objects, discovered or manufactured, and our new modes of existence raise problems that relate not only to "values" but also to the nature of reality; but the science and technology that spawn these problems provide no way to resolve them. To a large extent this explains the renewal of ethical inquiry, quite unimaginable half a century ago. The ancient depictions of life and death, of good and evil, both individual and collective—in short, of the human condition in its universal and particular aspects—still serve as sources of inspiration for the attempts to resolve these problems. The ancient myths are still present. The specter of the artificial man—whether a golem, Faust's homunculus, or Frankenstein's "monster"—has returned to haunt biotechnology.[62] Beliefs in the immortality of the soul or reincarnation are superimposed, helter-skelter, on the idea of cloning living individuals, in a delirious projection by a vulgarized genetics that has assigned the properties traditionally ascribed to the soul to mythological entities known as "DNA" and "the genome."[63] Finally, we saw in the previous chapter how, albeit more subtly, the frantic search for hidden causes reintroduces magical thinking into ostensibly scientific images, when the latter induce us to take as "true" and definitive the knowledge, which can always be improved, of partial causes. In this dialogue (or rather confrontation) of scientific and prescientific forms of reason we are well advised to return to the watershed moment when science was not yet clearly distinguished from magic. That can offer us a better vantage point on the tokens employed by these different forms of reason, while the advantage of three centuries of testing this modernity allows us to evaluate their results. The moment of the break between science and magic, between philosophy and theology—the seventeenth century in Europe—is the transitional period, the window through which we can learn something from the ancient philosophers without giving up the accomplishments of the *Critique* and, of course, with no mad and chaotic confusion.

62. See Atlan, preface to *Le Golem*; Dominique Lecourt, *Prométhée, Faust, Frankenstein: Fondements imaginaires de l'éthique* (Le Plessis-Robinson: Synthélabo, 1996).

63. See Chapter 1. See also: the [French] National Ethics Advisory Committee for the Life Sciences and Health [CCNE], "Reply to the President of the French Republic on the subject of Reproductive Cloning," Opinion 54 (Apr. 22, 1997), http://www.ccne-ethique.fr/docs/en/avis054.pdf; Henri Atlan, "Transfert de noyau et clonage: aspects biologiques et éthiques," *L'aventure humaine* 8 (1997), pp. 5–18.

4. EAST AND WEST: NEW PHILOSOPHICAL MYTHS AND THEOLOGICAL-POLITICAL CONFLICTS

At this transition point we find, for similar reasons but with different ramifications, Leibniz and Spinoza. The history of modern philosophy, which was first written and taught in the Europe of that age, is imbued with new myths about its origins, whose offshoots are evident in the ideological and political conflicts of the subsequent centuries. The continuing or renewed interest evinced by seventeenth-century scholars and philosophers in the natural magic of the Renaissance, which in the meantime had become an occult science of hermetic, kabbalistic, and alchemical inspiration, but from which the New Science was not yet totally divorced, raised the question of its ancient antecedents. The answers were heavily loaded with myth; the origin of these traditions, considered to be pre-Christian, were referred to as "oriental" (with all the ambiguity of this term), which may mean a "Semitic" East accessible to the West in its Christianized guise, or a pagan, "Brahmanist" or "Buddhist," Orient, initially held responsible for all evils. Much later the latter, too, was co-opted by the West, through an Indo-European Aryan myth that allowed the Christian West to submerge its Eastern/Semitic origins in a more distant and more "authentic" Orient.[64]

In this rearrangement of ambiguous terms, which was until recently the stuff of lectures and treatises on the history of philosophy, Spinoza seems to be the eye of the hurricane, motionless at the center of the stormy theological and political battles waged for or against a "Spinozism" that was always being redefined to suit the needs of the case. That Spinoza occupies a special place in the history of philosophy has been sufficiently underlined both by the modern philosophers who rejected him[65] and by those who, on the contrary, attempted in diverse ways to rework him in modern terms.[66] Because of his roots in European philosophy written in Latin, both Scholastic and Cartesian, and in republican and Protestant Holland of the seventeenth century, Spinoza stands at the defining moment of modern philosophy, breaking with Scholasticism but still linked to the most profound problems of ancient Neoplatonism and Stoicism. Thanks to his roots in Jewish philosophy written in Hebrew, within and outside the ex-Marrano Jewish community of Amsterdam, he occupies an analogous position, breaking with the contemporary rabbinic theologians and "Kabbalistic triflers" (*TTP*, chap. 9, p. 140), but still linked to the mythical "ancient Hebrews" with whom (and with St. Paul, too) he identifies when he defends himself against the accusations leveled by these "theologians" (Letter 21 [73]). Spinoza's Jewish and Hebrew sources have been studied extensively.[67] Some

64. For the philological aspects of this story see Olender, *The Languages of Paradise*.

65. These include Bergson, for whom "all philosophers have two philosophies: their own and Spinoza's" (letter to Léon Brunschvicg, 1927, in Henri Bergson, *Mélanges* [Paris: PUF, 1972], p. 1483), and Hegel, for whom "Spinoza is made a testing-point in modern philosophy, so that it may really be said: You are either a Spinozist or not a philosopher at all" (G. W. F. Hegel, *Lectures on the History of Philosophy*, trans. E. S. Haldane [Lincoln: University of Nebraska Press, 1995 (London: K. Paul, Trench, Trübner, 1892–1896), vol. 3, p. 283]).

66. Here we could cite, far from exhaustively, among contemporaries and only in France, Martial Guéroult, Robert Misrahi, Gilles Deleuze, Alexandre Matheron, Pierre Macherey, and J.-F. Moreau, for whom Spinoza is obviously not just one more philosopher like all the rest.

67. By virtue of its erudition, Harry A. Wolfson, *The Philosophy of Spinoza* (Cambridge, MA: Harvard University Press, 1934 [1962]), is perhaps the most extensive and exhaustive of these studies. But despite its subtitle, "Unfolding the Latent Processes of his Reasoning," the book offers a picture of Spinoza's philosophy that is grossly distorted in several essential points by the preconceptions of its author, for whom it can be reduced to a juxtaposition of sources. This often reduces to "allowing verbal analogies to take precedence over the meaning required by the analysis of the works, . . . substituting filing for reflection and association of ideas for intelligence" (Guéroult, *Spinoza*, vol. 1, p. 442).

A short article by Joël Askénazi, the French co-translator (with Jocelyne Askénazi-Gerson) of Spinoza's *Treatise on Hebrew Grammar*, opens interesting horizons while picking up the challenge of analyzing Spinoza's most explicitly "anti-rabbinic" work, the *Tractatus* (Joël Askénazi, "La parole éternelle de l'Ecriture selon Spinoza," *Les nouveaux cahiers* 26 [1971], pp. 15–39). He too invokes the idea, to which we shall return, of an infinite primordial wisdom, of which the Scriptures are only a particular expression, suited to the human beings who received and transmitted them. Spinoza could have been aware of a formally analogous idea—which he would have applied in his own way—of a Torah of *'azilut* (the world of "emanation"), of which the written Torah—the Sacred Scriptures—is only one realization, appropriate to our world. The primordial Torah is composed of the infinite combinations of letters and words of the written Torah and of the infinite meanings they spawn. This idea is found in a number of texts, some of which have been cited in the previous chapters. Askénazi mentions the introduction to Naḥmanides' Torah commentary, which Spinoza could certainly have known. Note that the similar idea of the primitive Torah as the "garment of the Infinite"—that is, that the infinity of combinations produced from its letters is the garment that makes it possible for us to perceive the Infinite, as well as the first movement by which the one and undifferentiated Infinite first differentiates itself in its game with itself (see Chapter 2)—is especially developed by the Lurianic kabbalah in the Neoplatonic version of Israel Sarug, which was circulating in Amsterdam in those days. Spinoza could also have been aware of this notion from the writings of Abraham Herrera or of Joseph Delmedigo, whose *Ta'alumot ḥokhmah*

of these studies, as we shall see, are handicapped by a need to offer post-factum justification of his excommunication by demonstrating the heretical and anti-Judaic aspects of his doctrines. Others, on the contrary, are animated by an apologetic or polemic desire to demonstrate his ties to medieval Judaism. In particular, the question of his links to the kabbalistic doctrines of the disciples of Isaac Luria, which were spreading in Europe and especially in Holland at the time, was raised soon after his death. When the *Ethics* became known, a comparison—benevolent or malevolent, but rarely neutral—with the content of the Latin works that introduced Christian readers to some of the main elements of the Lurianic kabbalah seemed to be de rigueur.[68] But that was only the beginning and one aspect of the history of Spinozism in Europe, generally fantasized as a cross between East and West, both of them somewhat mythical.

Thus we return to Hegel's remarkable expression: "Thought must begin by placing itself at the standpoint of Spinozism; to be a follower of Spinoza is the essential commencement of all Philosophy."[69] We must not be misled by this judgment; we must be aware of what sort of beginning this is, given that, for Hegel, his own system was the culmination of all philosophy as the unveiling of spirit. In Hegel's history of philosophy, this beginning is the mythical "Orient" of the "Western" philosophers and the role that the "Oriental" spirit plays in his own philosophy of history.[70] As was noted by Pierre Macherey,[71] this mythical Orient

(one of the first expositions of the Lurianic kabbalah published in Europe, incorporated and reworked by Knorr von Rosenroth into his *Kabbala denudata*) was in his library. (Although the book is in Hebrew, the title page also bears the Latin title *Sapientiae abscondita.*) See Paul Vulliaud, *Spinoza d'après les livres de sa bibliothèque* (Paris: Bibliothèque Chacornac, 1934), pp. 99–103; repr. in J. Préposiet, *Bibliothèque spinoziste* (Paris: Les Belles Lettres, 1973), pp. 339–343.

68. Building on the earlier works by Pico della Mirandola and Johann Reuchlin (fifteenth and sixteenth centuries), who had opened up this road and begun to translate and comment on the *Zohar* and Abulafia, and searching for an original tradition common to all the major religions, two Christian kabbalists, Christian Knorr von Rosenroth and François-Mercure van Helmont, seem to have played a decisive role. These two were the authors and editors, with their friend Henry More, of a work entitled *Kabbala denudata* (Sulzbach, 1677). For a summary of this volume see Gershom Scholem, *Kabbalah* (New York: Quadrangle, 1974), pp. 416–419. See also idem, "Die Erforschung der Kabbala von Reuchlin bis zur Gegenwart" (Pforzheim, 1970) [repr. in idem, *Judaica 3: Studien zur jüdischen Mystik* (Frankfurt a.M., 1973), pp. 247–63; and (in French) as "L'Etude de la Kabbale depuis Reuchlin jusqu'à nos jours," in *Le nom et les symboles de Dieu dans la mystique juive*, trans. Maurice R. Hayoun and Georges Vajda (Paris: Le Cerf, 1983), pp. 198–199].

Kabbala denudata soon inspired several authors to make more or less justified associations with Spinozism. They include Henry More and especially Johann Georg Wachter, whose initial reaction, in *Der Spinozismus im Jüdenthumb, oder, Die von dem heütigen Jüdenthumb und dessen geheimen Kabbala vergötterte Welt* (Amsterdam, 1699), was extremely negative. Seven years later, however, after making honorable amends and avowing his initial lack of understanding, he was unstinting in his praise, in his *Elucidarius Cabalisticus, sive, Reconditae Hebraeorum philosophiae brevis et succincta recensio* (Amsterdam, 1706). As for Leibniz, who, following his youthful involvement with the Rosicrucians, never lost interest in Christian, alchemical, and kabbalistic esoterism, he too was inspired in part by these texts and their authors. Although his philosophical work clearly cannot be reduced to these sources, his attitude toward Spinoza was strongly influenced by Wachter's high regard as well as by the knowledge he may have had of this "Hebrew kabbalah" through Knorr von Rosenroth, Helmont, and Wachter himself. (See: Georges Friedmann, *Leibniz et Spinoza* [Paris: Gallimard, 1946; 2nd ed. 1962; 3rd ed. 1974]; Scholem, *Kabbalah*, pp. 394–395 and 416–419; idem, *Abraham Cohen Herrera* [Jerusalem: Mossad Bialik, 1978] [Hebrew]; Coudert, *Leibniz and the Kabbalah.*) Leibniz's annotations of Wachter's book were published in 1854 by Foucher de Careil (see Alexandre Foucher de Careil, *A Refutation Recently Discovered of Spinoza by Leibnitz*, trans. Octavius Freire Owen [Edinburgh: T. Constable, 1855]), as part of his crusade against the "Oriental renaissance" in Germany (see below).

69. *Lectures on the History of Philosophy*, trans. Haldane, vol. 3, p. 257.

70. "The Oriental spirit is closer in its determination to the sphere of *intuition*, for its relationship to its object is an immediate one. But the subject is still immersed in substantial existence, and has not yet extricated or liberated itself from its original state of unmixed unity to attain subjective freedom" (Georg Wilhelm Friedrich Hegel, *Lectures on the Philosophy of World History. Introduction: Reason in History*, trans. H. B. Nisbet [Cambridge: Cambridge University

is perfectly suited to serve as the point of origin for Hegel's version of history. More precisely, it is the East of India and Buddhism as discovered and reinterpreted by Europe, variously adapted in the service of philosophical, religious, ideological, and political causes, with no concern for authenticity (or even, sometimes, for coherence). In his study of this astonishing phenomenon, Roger-Pol Droit has described this European invention of a fantasy Buddhism, centered on the cult of and quest for nonbeing.[72] The common denominator of all of the works, some of them quite learned, that contributed to the invention of the related myth of Indo-European Aryanism was the ostensible value attached to nothingness at the core of an unnatural religion that was not only inhuman but also absurd. Hegel played a role in the incorporation of this myth into the intellectual history of Europe in the nineteenth century. In an astonishing syncretism, he regularly associated his made-up version[73] of Spinozist substance with that of an equally fictive Buddhist Nirvana, "swallowing up in unconsciousness the unity with Brahma, nothingness."[74] But he had an excuse, because he was not alone; the same association was taken for granted, even by some of the greatest thinkers, including Kant, for whom the swallowing up of personality in the void was characteris-

Press, 1975], p. 199). But "the first form which the spirit assumes is that of the Orient" (ibid., p. 198) and "World history travels from east to west; for Europe is the absolute end of history, just as Asia is the beginning" (p. 197). This is why Hegel's philosophy, clearly European and the culmination of this history, makes it possible for the subject to emerge, in and through knowledge of universal reason, in which "everything depends on grasping and expressing the ultimate truth not as Substance but as Subject as well" (G. W. F. Hegel, *The Phenomenology of Mind*, trans. J. B. Baillie [New York: Harper and Row, 1967 (London: Macmillan, 1931)], p. 80). We should remember that, for Hegel, history is the self-constitution of spirit by objectivation of Universal Reason. This unfolding in time appears as the succession of different political regimes, which are also different moments of the advent of Mind or Spirit. But this progression takes place in space as well, moving from east to west, like the sun and its light. The difference is that in the West, aging is not, for Mind, the prelude to death, but, on the contrary, maturation and the accomplishment of the end of history as inscribed in divine providence. Thus, "the . . . scheme of historical phases contains a general survey of world history. Its aim is also to make the historical process intelligible in the light of the Idea and its underlying necessity" (Hegel, *Reason in History*, p. 196). America is outside the field of philosophical history, because it is "the country of the future, . . . a land of desire for all those who are weary of the historical arsenal of old Europe. . . . Prophecy is not the business of the philosopher" (ibid., pp. 170–171). Then Africa and the blacks, destined for slavery, are removed from the true theater of world history. What remains is the succession, historical and geographical, that begins with the "Chinese and Mongol Empire," continues with the "Hindu world," then Persia, Egypt, the Greek world, and finally the Roman world and Christianity. But this culmination is not truly achieved until after the lapse of several centuries, in the Teutonic world that begins with Charlemagne. This allows Hegel to fine-tune the chronology and place the Muslim world before this full realization of Mind! For him, only Christianity, thanks to the Incarnation, makes possible knowledge of God and of the ultimate goal of humanity (ibid., pp. 35–42). We must thus understand, building on Leibniz but better than he did, that the realization of Mind in history, "so rich a product of creative reason as world history," is a "theodicy," "a justification of the ways of God (such as Leibniz attempted in his own metaphysical manner, but using categories which were as yet abstract and indeterminate)" (ibid., p. 42).

71. Pierre Macherey, *Hegel ou Spinoza* (Paris: F. Maspero, 1979); see also Michel Hulin, *Hegel et l'Orient: Suivi de la traduction annotée d'un essai de Hegel sur la Bhagavad-Gita* (Paris: J. Vrin, 1979).

72. Roger-Pol Droit, *The Cult of Nothingness: The Philosophers and the Buddha*, trans. David Streight and Pamela Vohnson (Chapel Hill: University of North Carolina Press, 2003).

As shown by, inter alia, Edward W. Said (*Orientalism* [New York: Pantheon, 1978]), the Near East has been the victim of an analogous projective construct.

73. Guéroult, *Spinoza*, vol. 1, p. 466. Here we find a list of Hegel's most important misinterpretations of Spinoza's ontology and logic, warped in the process of their integration into Hegel's own philosophy and his demonstration of their ostensible inadequacies. In addition to the mistaken identification of substance with an abstract and empty absolute, in which the reality of objects and individuals is dissolved like that of the modes and attributes, we also find an analysis of the mistranslation of the famous expression *omnis determinatio negatio est*, scandalously hijacked in the Hegelian dialectic. For Guéroult, these interpretations inspired a whole line of idealist commentators, up to the present day, including those who later called for "a return to Kant" (ibid., pp. 462–468).

74. Quoted by Macherey, *Hegel ou Spinoza*, p. 27.

tic of the "*pantheism* (of the Tibetans and other oriental peoples); and in consequence from its philosophical sublimation *Spinozism* is begotten, which is closely akin to the very ancient *system of emanation* of human souls from the Godhead (and their final reabsorption into it). . . . But really this is a concept in which the understanding is simultaneously exhausted and all thinking itself has an end."[75] As we shall see, Emile Bréhier, in his *History of Philosophy*, continued this same derivation, in a less polemic but equally "obvious" manner, back to Plotinus (the emanation?), representing the pagan East, and the Stoics, for the "Semitic" aspect. (I leave it to readers, for the moment, to guess which of the two orients he associates with Spinoza.) Thus Hegel was not totally original in his treatment of these questions, although, for him, this image of a Spinoza bereft of thought and intelligence could hardly have been obvious. What was forced on Hegel and incorporated into his philosophy was, instead, the picture of an anomaly, of a thinker who was simultaneously modern, insofar as he was the heir of Descartes, and primitive, on account of his "Oriental theory of absolute identity."[76] Hegel depicted Spinoza "as a living fossil who, in the Low Countries and the height of the seventeenth century, introduced, under a Cartesian disguise, modes of thought that were pre-Christian, indeed pre-Hellenic; in short, oriental."[77] Hegel's "living fossil" is not far from the "monsters" of Bayle and Leibniz, to whom we shall return shortly.

All of this is found in the nineteenth century, in a coexistence of fascination and horror with this mythical combination, supposed to be the germ of a veritable mental illness, a monstrous anomaly of human thought, threatening to invade Europe and submerge the Christian West. In this specter Frédéric Ozanam, a French professor and the originator, with Lacordaire, of a leftist Catholicism with a social orientation, the founder of the Society of Saint Vincent de Paul and also the author of an essay on Buddhism, saw "the former enemy [that] has not changed." As Droit summarizes Ozanam's horror of the East, "the doctrine of the Orient [i.e., Buddhism] did no more than pursue the fundamental error of pantheism; it was 'the soul of paganism of all times and all places' [and] prolonged a 'tradition of error'. . . . For Ozanam, those were 'the same opinions' that were found from India to the mysteries of Thebes, from Alexandria to the Gnostics,

75. Immanuel Kant, "The End of All Things," in Immanuel Kant, *Religion and Rational Theology*, trans. and ed. Allen W. Wood and George Di Giovanni (Cambridge: Cambridge University Press, 2001), p. 228.

76. *Lectures on the History of Philosophy*, trans. Haldane, vol. 3, p. 242.

77. Quoted by Pierre Macherey, *Avec Spinoza: Études sur la Doctrine et l'histoire du spinozisme* (Paris: PUF, 1992), p. 189. See also above, n. 70.

To deny almost all human beings the consciousness of rational subjects is evidence of a certain arrogance, as if only European languages employed the first-person singular as a subject or object. It is probably to be explained by the "spirit of the age." But it is really the last straw, and quite sad, for thought to "swallow up" Spinoza's philosophy in this denial. The Spinozist human subject, obviously quite different from a transcendental subject, exists not one whit less as an individual with a body that is sufficiently complex for the idea of this body to be endowed with linguistic reflexivity. Every human being has this experience, if only in an inadequate (because limited) fashion, in his or her own individuality, and sometimes even adequately, through common notions of reason. In addition to Alexandre Matheron, *Individu et communauté chez Spinoza* (Paris: Minuit, 1969, 1988), a classic, the analysis by Laurent Bove (*La stratégie du « conatus ». Affirmation et résistance chez Spinoza* [Paris: Vrin, 1996]) depicts the *conatus* as a self-organizing affirmative power of Nature, a *causa sui* in each of its modes, autonomous or heteronomous depending on the degree of adequacy of its self-awareness. As such, this idea of the subject is especially suited to our modern theoretical models of our own experience of ourselves, both determined by and identified with our body and with the unconscious consciousness we have of it in its interactions with others. (See Atlan, *SR* II, chap. 7, on a physical theory of intentionality; and ibid., chap. 6, on a Spinozist perspective on evolution and the theory of action.)

from John Scotus Erigena to Spinoza. 'They have reappeared to the favor of the metaphysicians of Germany.'"[78] These ubiquitous condemnations of a theological and philosophical nature are clearly very different from those that later nourished racist theories. But the latter, which generally contradicted them, promoted the biologization of the Aryan myth and thus contributed new weapons to traditional Christian antisemitism even while continuing to associate the harmful effect of Buddhism with that of the Semites. Gobineau seems to have evaded the contradiction by considering Buddhism to be a degenerate form of primitive Brahmanism, which vigorously resisted it to the point of chasing it from "Aryan" India to find a new home among Chinese and Burmese, for whom it was good enough![79]

As for Spinoza and his naturalism, not everything in this muddle is obviously false. A better understanding of Spinozism than Kant's or Hegel's might also detect elements derived from the "Orient," but of a different nature: not the cult of nothingness and negation, which is totally foreign to the universe of the *Ethics*, but borrowings from Jewish, Stoic, and Platonic sources, modified and integrated into his post-Cartesian rationalism, in which pre-Christian antecedents can be made out. It is as if Spinoza, in "his combination of archaism and the avant-garde,"[80] "skipped over or rather under his time"[81] and is a timeless anomaly in a historical process that is supposed to obey the law of rational development. This unique characteristic is one of the elements that allow Spinozism to subsist in a perpetual present. It is a philosophy that "has never stopped being, always in the present, an object of fascination and reflection, in the sense that every century of European culture has reinvented it in some sense, modeling itself after its image thereof."[82] It is modern not only in its own age but even in our own, "probably because we project our own current fantasies on it."[83]

78. Droit, *The Cult of Nothingness*, p. 80, quoting F. Ozanam, *Essai sur le bouddhisme* (1842).

Times have changed. By an irony of history, not only has the myth of the quest for nothingness evaporated and Buddhism become better known; it has even become a religion, or a doctrine, lived and practiced by a growing number of Westerners. The Dalai Lama addresses the Western media almost as much as the Pope does. The chance of reading (but is it chance?) brought Roger-Pol Droit's book to my attention and made me aware of Ozanam's imprecations against Buddhism precisely the same week that the latter was beatified by John Paul II at Notre-Dame de Paris. Although the Pope took advantage of the occasion to request Protestant forgiveness for the Saint Bartholomew massacre (because the ceremony took place on its anniversary), he did not mention the missionary zeal of Europeans in Asian countries, which inflammatory remarks like Ozanam's justified and supported. I came away with the curious impression of having watched a televised match between His Holiness the Pope and His Holiness the Dalai Lama, conducted through the medium of the Blessed Frédéric Ozanam.

79. Ibid., pp. 104–118. See also below, n. 95.

80. Macherey, *Avec Spinoza*, p. 30.

81. Ibid., p. 189.

82. Ibid., p. 7.

83. Ibid., p. 6. Macherey would distinguish the permanence of this thought, *sub species aeternitatis*, from the timelessness of *philosophia perennis*, because

> not only does it produce, in a commemorative mode, identifiable and autonomous doctrinal effects; it also has the strange power of being in resonance with much of what we do, of mingling with what we read, as if it constituted, *sub species aeternitatis*, an intellectual structure whose permanence, far from being timeless, is defined by an unchangeable faculty of adaptation or adherence to the most determined forms of the present and of presence. Thus it is the opposite of *philosophia perennis*, which is a doctrine for all times: but a philosophy that is understood and acts in the present, here and now, without however closing itself up within the limits of any particular finite reality. (ibid.)

But isn't this a merely verbal distinction? Can one truly conceive of *philosophia perennis* as a closed system that remains frozen in itself in a static timelessness? As philosophy, it can be only the expression of active thought, which inspires the feeling that it is, as Spinoza says, the true philosophy. If so, its timelessness must be that of a presence that is constantly being renewed in history. Its species of eternity, like that of rational knowledge in the *Ethics*, is what we experience when we construct and discover geometric truths, which are also eternally present in the contrasting interpretations of successive generations of mathematicians and logicians. This should not make us believe, of course, that all interpretations are equally valid. As in the case of geometric truths, and of mathematics in general, the successive interpretations in their

But this presence in the present, permanently between past and future, also results from Spinoza's spatial position in a field composed of twin fantasies, between "East" and "West." We will see the myth operating in the opposite direction in the much more recent context of the twentieth century, when the "West in crisis" is devalued, and even more so after Europe's shameful behavior toward its own peoples and toward those who, at home and abroad, were the victims of its hegemonic empire. Spinoza was condemned then by some Jewish thinkers for having sold out to the West and for being a traitor to his own Oriental people.[84]

various contexts of application, including projections of "our fantasies of reality" (ibid.), do not always have the same value (or lack of value), even if only from the cumulative perspective that is also our own, because we can look back on all previous interpretations. Our perspective is that of the last comer and of the last word; as such it is always present, because the past of previous interpretations is concentrated in it, while the future does not yet exist.

Interpretations have succeeded one another, in fact, since the publication of the *Ethics* and its author's death. We have had "the materialist and atheist Spinoza of the eighteenth century, in which the rationalism of the Enlightenment recognized itself while opposing him," then "the intuitive and mystical Spinoza of the romantic nineteenth century," and finally "the theoretician and political Spinoza of the history of the twentieth century, split between the tendencies to liberalism and revolution" (ibid., p. 7). Macherey's *Avec Spinoza*, even more so than his earlier *Hegel ou Spinoza*, contributed to this revolutionary political reading in the 1970s, when many intellectuals still saw Marxism as the "impassable horizon," as Sartre put it. In this context it was important to create a philosophical link between dialectic materialism and Spinoza, one that was less idealistic and more revolutionary than that to Hegel. (In addition to the many authors more or less directly influenced in this direction by Louis Althusser, Antonio Negri's *The Savage Anomaly: The Power of Spinoza's Metaphysics and Politics*, trans. Michael Hardt [Minneapolis: University of Minnesota Press, 1991], is a fine and moving example of this revolutionary neoromanticism with a materialist and "scientific" inspiration.) Perhaps we are now moving toward a twenty-first–century Spinoza, the theoretician and practitioner of an ethics whose prize is the possibility of salvation through knowledge, generated by the natural sciences, of "common properties," accompanied by knowledge of what this knowledge does and does not know, so that the "intuitive science" of the particular can go beyond the previous without short-circuiting it. Pierre Macherey has recently embarked on an illuminating annotated literal reading of the *Ethics*, beginning with Part V, on freedom, in the tradition of Guéroult, who never got beyond the first two parts in his commentary—combining, as it were, the "radical idealism" that the latter preaches in his *History of the History of Philosophy* with the materialism of the former in the "order of reasons."

84. The reverse myth of the satanic or simply perverse West energized the campaigns for national liberation that accompanied decolonialization, as well as the search, after the genocide of the Second World War, for a forgotten "authentic" Jewish identity. As clearly seen by Gilles Kepel (*The Revenge of God: The Resurgence of Islam, Christianity, and Judaism in the Modern World*, trans. Alan Braley [University Park: Pennsylvania State University Press, 1994]), this campaign was the work of individuals who had attended university and had assimilated the myth of the decadent West from within the European culture that nurtured them. Unfortunately, far from being soothed by its political and cultural successes, this movement has flourished and proliferated. Today it is one of the engines of the "return to religion" that animates the fanaticism of Muslim Jihad, Jewish messianism, and also, by reaction, new fundamentalist Christian crusades.

5. THE INDIAN EAST AND SEMITIC EAST IN THE CONTEMPORARY HISTORY OF GREEK PHILOSOPHY

Before we come to this reversal, we find that until the start of the twentieth century the "oriental" character of Spinoza's philosophy was associated in university lecture halls with the orientalism of German Romanticism. This link, which seems to be more impartial and less fraught with ideology, is always presented as self-evident, an objective datum of the history of philosophy. Consider what Emile Bréhier, a leading French scholar of ancient and medieval philosophy, wrote in his 1928 study of Plotinus: "The German scholars who, during the last few years, have singularly increased our knowledge of the philosophy of India through their translations and commentaries did not fail to call attention to the affinity between certain Western thinkers and Indian thought. . . . Identity in the philosophy of Schelling, the union of the soul with God in the intellectual love in Spinoza, are conceptions closely related to the identity of the self with the universal being in Plotinus. They are to be found in the Upanishads."[85] Spinoza, as often, is mentioned here in passing, along with Schelling, who lends support for Bréhier's thesis about Plotinus' orientalism. For Bréhier, what was original about Plotinus' philosophy, in antiquity, was its combination of Greek rationalism with a mysticism that derives neither from the "Semitic East" of the salvation religions, as in Philo of Alexandria, for example, or from the Babylonian and Persian East, as in some of Plotinus' own disciples, but from the more distant Orient of India. For "the position of Plotinus with regard to Greek rationalism is analogous to that of Spinoza to Cartesianism. Spinoza, too, endeavored to solve the problem of eternal life and blessedness within a thoroughgoing and unreserved rationalism."[86] But this is not what sets Plotinus apart. On the one hand, "the fundamental problem which occupied the thought of Plotinus," from which Bréhier derives the rest, is: "Can rationalism have a religious significance? How can the soul exist, so to speak, in its own right? How can the problem of destiny retain a meaning in a universe the forms of which are graduated according to a necessary law of reason?"[87] On the other hand, this was also the problem addressed by other philosophical schools in the early centuries of the Christian Era. Not only had Greek philosophy been widely disseminated with Alexander's empire in Asia and Africa, it was also receptive to local influences in these same lands. Its center was no longer in Athens, but in Alexandria and the cities of Asia Minor, which were open to the Semitic and Babylonian cultures and to Hindu traditions.[88] This is why an

85. Emile Bréhier, *The Philosophy of Plotinus*, trans. Joseph Thomas (Chicago: University of Chicago Press, 1958), p. 123.

86. Ibid., p. 42.

87. Ibid., p. 41f.

88. The scope of these interchanges may explain why so many theologians and philosophers, searching for universalism, thought they had found the mythical sources of primordial, eternal, and universal wisdom in whatever was "ancient," "Egyptian," "Hebrew," "Brahman," etc. Plotinus himself taught that "these teachings are . . . no novelties, no inventions of today, but long since stated, if not stressed; our doctrine here is the explanation of an earlier and can show the antiquity of these opinions on the testimony of Plato himself" (*Enneads* V 1.8, trans. Stephen MacKenna, rev. B. S. Page, 4th ed. [London: Faber and Faber, 1969], p. 376). In the sixteenth century, Justus Lipsius, reviving the Christianized Stoic tradition of second-century Alexandria and rewriting the history of philosophy, traced it back to Adam's prelapsarian contemplative life. Pythagoras, who supposedly rediscovered it after it had been concealed, and in parallel to its desiccation by the Jewish schools of the talmudic era, was taken as matching "the traditional model of the wise and religious man, who studied in Egypt, in Babylonia, and with the Brahmins; Clement went so far as to say that he had been circumcised in the Judaic rite in order to better indicate the concord between his philosophy and Judeo-Christianity" (Jacqueline Lagrée, *Juste Lipse et la restauration du Stoïcisme* [Paris: J. Vrin, 1994], p. 24). "Among the ancient philosophers, Pythagoras of Samos is considered, on account of his wisdom and his piety toward God, to have surpassed all those who ever practiced philosophy. This Pythagoras, I say, not only patently knew our institutions and those of the Jews, but on many points he emulated and outdid them" (Justus Lipsius, *Manuductionis ad stoicam philosophiam* [Antwerp, 1604], I, 6). Leibniz, too, was aware of this tradition and could consequently assign to philosophy a unifying vocation that might help overcome religious schisms. He was quite explicit about this, notably in his *Preliminary Dissertation on the Conformity of Faith with Reason*, in which he condemned "this bad doctrine, . . . very ancient and apt to dazzle the common herd," of "a universal soul forming the ocean of all individual souls, and . . . alone capable of subsisting, whilst individual souls are born and die." But Leibniz did not trace this bad doctrine to the Neoplatonism of Plotinus, although he notes that "Plato's Soul of the World has been taken in this sense by some." It is rather the Stoics whom he is inclined to see as the root of those whom he calls "'Monopsychites,' since according to them there is in reality only one soul that subsists." He goes on to refer to an orientalist of his own day, Bernier, and the association of kabbalah with Spinoza, which he considered to be valid on the basis of the treatises of a "proselyte Jew" and of "a learned man who confuted" him (the latter being Georg Wachter; see above, n. 68):

Bernier observes that this is an opinion almost universally accepted amongst scholars in Persia and in the States of the Grand Mogul; it appears even that it has gained a footing with the Cabalists and with the mystics. A certain German of Swabian birth, converted to

amalgam of Greek rationalism with a physical and physiological cosmogony, in a system that gives meaning to human destiny (and the gods' as well), was a feature of the various Stoic schools as well. According to Bréhier, however, the latter expressed the Semitic influences of the salvation religions—Judaism for ancient Stoicism, and then Christianity. Plotinus' resistance to these influences illustrates, on the contrary, his openness to India, where the human soul and human intellect could blossom only by being based on a single reality, that of the single and motionless One, thanks to an experience of love, an experience that is both intellectual and ecstatic.[89]

In this context, it bears noting that Spinoza is naturally associated with the rationalist mysticism of Plotinus and the Indian identification of the self with the universal soul, in which individual destiny vanishes, rather than with the Stoics and the Semites and their salvation religions:

> There is a whole side to the speculation of Plotinus which is no less foreign to Hellenism than to the religions of salvation. And it is not the Hellene in him that protests against the idea of a divine providential activity which would deliberately exert itself in behalf of man. Hellenism can very well be reconciled with this piety and did reconcile itself with it as, for example, in the Stoics. It is in the name of an entirely different religious ideal that he protests.
>
> We sense in Plotinus the same resistance to accepting this idea which we feel in Spinoza and Schelling, who, like him and for similar reasons, rejected the traditional idea of the religion of salvation.[90]

Bréhier was not mistaken when he numbered Spinoza among those who reject the concept of a personal god and of a providence overseeing individual destiny. But taking into account what was known, both before and after him (though not without some misunderstanding), as Spinoza's materialism, he might have listed the classical Stoics along with Plotinus among his remote precursors. Clearly, though, Bréhier is one of those for whom Spinoza is a mystic and "idealist." The revival of the ma-

Judaism some years ago, who taught under the name Moses Germanus, having adopted the dogmas of Spinoza, believed that Spinoza revived the ancient Cabala of the Hebrews. And a learned man who confuted this proselyte Jew appears to be of the same opinion. It is known that Spinoza recognizes only [a single] substance in the world, whereof individual souls are but transient modifications. (Gottfried Wilhelm Leibniz, *Theodicy: Essays on the Goodness of God, the Freedom of Man, and the Origin of Evil*, ed. Austin Marsden Farrer, trans. E. M. Huggard [La Salle, IL: Open Court, 1985], pp. 78–79)

Finally, in a line running through Hegel, Schopenhauer, Nietzsche, William James, and others, the fascination with Oriental origins has survived among Western intellectuals, inspiring Herder, for example, who "at one time or another idealized the Hebrews, the Egyptians, and the Brahmans of India" (Olender, *The Languages of Paradise*, p. 31). Here one finds the same theological and political prize that Olender called the "cycle of the chosen peoples" (ibid., p. 37), reanimated by the rivalry between "Aryan and Semite: 'two twins' at the origin of civilization" (ibid., p. 30).

89. With regard to what Plotinus writes of "the word 'One' and the thing which it designates," "beyond being," untouched by numeration and of which nothing can be said, Bréhier notes that "Inge [in *The Philosophy of Plotinus*, 3rd ed. (London: Longmans, Green, 1929), vol. 2, pp. 107–108] has remarked that perhaps Plotinus utilizes the word 'One' only because the Greeks had no symbol for zero. He calls 'One' that which Scotus Erigena, in the *De divisione naturae*, would later call 'nothing' (*nihil*) and proposes that Plotinus used the word One only because the Greeks had no symbol for zero!" (*The Philosophy of Plotinus*, p. 157 n.)

> And this name, The One, contains really no more than the negation of plurality: under the same pressure the Pythagoreans found their indication in the symbol "Apollo" (α = not; πολλῶν = of many) with its repudiation of the multiple. If we are led to think positively of The One, name and thing, there would be more truth in silence: the designation, a mere aid to inquiry, was never intended for more than a preliminary affirmation of absolute simplicity to be followed by the rejection of even that statement: it was the best that offered, but remains inadequate to express the nature indicated. For this is a principle not to be conveyed by any sound; it cannot be known on any hearing but, if at all, by vision; and to hope in that vision to see a form is to fail of even that. (Plotinus, *Enneads* V 5.6; trans. MacKenna. p. 408)

90. Bréhier, *The Philosophy of Plotinus*, pp. 117–118.

terialistic interpretation, triggered by the Marxist wave that dominated French intellectual life after the Second World War, had not yet touched him. Furthermore, perhaps despite himself, he remained firmly within the traditional dichotomy, which goes back to Hegel, of Western philosophy that is Christian and consequently—but only for that reason—Judeo-Christian and "Semitic," versus Oriental (meaning Hindu) philosophy, which somehow infiltrated our Western civilization. This is what he seems to say in his conclusion:

> The new type of idealism created by Plotinus forms an independent and permanent force in the history of philosophy. I need not here even begin to tell its history. It would be appropriate to show how, in our Western civilization, its spirit has been manifested in the form of a philosophy which is both religious and rationalistic but still profoundly opposed to Christian thought.[91]

Next he describes this philosophy by alluding to the Spinozist rejection of free will as an "empire within an empire":

> Its essential trait, which has persisted through the centuries, is the affirmation of the complete autonomy of the life of the spirit. . . . The One, which is the very ground of this life, is absolute freedom. Freedom, in us, is not realized, consequently, as a spontaneity arising from nothing within an existing world, like an "empire within an empire," but through an increasingly intimate communion with the life of the universe.[92]

He had already stated his thesis, as clearly as possible, in his chapter on Plotinus' Orientalism, right after the passage quoted above, where he associates Spinoza and Schelling in their common resistance to the notion of divine providence:

> It is not because Spinoza was a Cartesian and a rationalist that he rejected the truths of the Christian faith, such as the freedom of the will or creation, with which Descartes came to terms completely. It is because he conceived the relations between the soul and the universal being in an altogether different way from a Christian.
>
> With Plotinus, then, we lay hold of the first link in a religious tradition which is no less powerful basically in the West than the Christian tradition, although it does not manifest itself in the same way. I believe that this tradition comes from India.[93]

Thus even in the West this religious tradition, which rebels against Christian thought and has a non-Western origin, has come to rival the power of the Christian tradition. Although Spinoza adopted this tradition, it cannot have a "Semitic" origin because "Semiticism" has long since vanished, fulfilled by the Christianity that replaced it. As with Philo of Alexandria, who was its product, Semiticism has been Hellenized and Westernized. This is why, in contrast to "India," it could not establish a philosophical and religious tradition distinct from Judeo-Christianity (i.e., Christianity). So in the tradition in which Plotinus is the first link we also find Spinoza, associated in the nineteenth century with the German *Lebensphilosophen* and Romantics like Schelling. But the Stoics, with whom we would have expected Spinoza to be grouped—if only because they too strongly rejected free will and because their asceticism, too, aimed at living their lives as part of nature—are relegated to the bygone past of their antiquity. On the one hand, they are not

91. Ibid., p. 196.
92. Ibid.
93. Ibid., p. 118.

"idealists"; on the other hand, because they too expressed an Oriental influence[94]—that of ancient Semitism—they vanished as it did, absorbed into the true philosophical and religious tradition of the Christian West.

All these stories of East and West, involving "Sanskritists and Semites"[95] as partners and rivals in the infiltration of Christian civilization, show the extent to which histories of philosophy are based on the philosophical myths inherited by the philosophers who write them. Now we should examine how this contrast of East and West can operate in the opposite direction, still with the focus on Spinoza, but denouncing him as "Western" and a traitor to his people, with anti-Judaism as the key to his philosophy.

94. In his monumental *History of Philosophy* Bréhier expounds the elements of Stoicism that cannot be reduced to Greek thought:

> In the place which the Stoics assigned to God, in the manner in which they conceived the relation of God to man and the universe, there were new traits that had never before appeared among the Greeks. [The Stoics' God is not the same as the] Hellenic God or the God of popular myth, as well as Plato's Good and Aristotle's Mind. . . .
>
> The God of the Stoics is neither an Olympian nor a Dionysus, [because he] lives among men and reasonable beings and . . . arranged everything in the universe for their benefit. His power penetrates everything, and his providence overlooks not even the slightest detail. . . .
>
> Here we have the Semitic idea of the omnipotent God who governs the destiny of men and things, and it is quite different from the Hellenic conception. (Emile Bréhier, *The Hellenistic and Roman Age*, trans. Wade Baskin [Chicago: University of Chicago Press, 1965], pp. 35–36)

This notion that Stoicism has a "Semitic-oriental" origin seems to be common; it can even be found in a history of Jewish philosophy by David Cohen, a disciple of Rabbi Abraham Isaac Hacohen Kook (see below). In the same context we also find a reference to legendary traditions that find Egyptian and Hebraic influences in Pythagoras and Plato (see above, n. 88). This would not leave very much that is truly Greek among the Greeks, aside from Homer, Socrates, and Aristotle. And even the last-named, according to Diogenes Laertius, was among those who maintained that philosophy—although the term itself was coined by Pythagoras—was born in a foreign land: "Some say that the study of philosophy originated with the Barbarians. In that among the Persians there existed the Magi, and among the Babylonians or Assyrians the Chaldaei, among the Indians the Gymnosophistae, and among the Celts and Gauls men who were called Druids and Semnothei" (Diogenes Laertius, *The Lives and Opinions of Eminent Philosophers*, trans. C. D. Yonge, Book I, Introduction).

95. These are the terms that a distraught Foucher de Careil employed to describe German intellectual circles of the nineteenth century: "The Buddha was holding school," and "Sanskritists and Semites were meeting evenings . . . to talk about Nirvana or Dyana" (Alexandre Foucher de Careil, *Hegel et Schopenhauer* [Paris: Hachette, 1962], p. 307; quoted by Droit, *The Cult of Nothingness*, p. 102). He viewed the Oriental renaissance in Germany as a menace both to the Christian West and to French Cartesianism, "a questionable mixture of nihilism, of Jewish—and thus anti-Christian—influence of racial superiority and of Kantian rationalism" (ibid., p. 101). Schopenhauer, too, whose "Buddhism" has a dubious relationship with authentic Buddhism, was not immune to the spirit of the age and its stereotypical images. Nevertheless, he did not see any connection between the Jewish Orient of the Old Testament and India, in which he saw the pessimism and true aspiration for nothingness that fit with his own philosophy. On the contrary, he considered the Jewish origins of Christianity to be a repulsive stench, the *foetor judaicus*, the Jewish reek of optimism and moral law, a fundamentally Jewish idea that had paralyzed and arrested Western philosophy (see E. de Fontenay, "La pitié dangereuse," in *Présence de Schopenhauer*, ed. R.-P. Droit [Paris: Grasset, 1989], pp. 77–88). Clearly he had Spinoza in his sights, but in association with everything that Schopenhauer detested about the West: the Bible, Descartes, Kant, Hegel, everything that, for him, breathes "an insipid Protestant and rationalist—or, more precisely, Jewish—optimism—all are the fault of "this all-pervasive *foetor judaicus*" (ibid.). As a contrast he offers the pure Orient, with its Buddhism and pessimism, which alone recognize, correctly, the misfortune of being born. "We can thus hope that one day Europe will be purified of all Judaic mythology, . . . [when] people of Asiatic origin from the linguistic branch of Japhet [*sic*] will take back the sacred religions of the homeland" (quoted by Droit, *The Cult of Nothingness*, p. 103). Completing the permutations of the three terms—Indian Orient (which would later become "Indo-European" or "Aryan"), Semitic Orient, and the West—taken two at a time, Jewish thinkers would later indict Spinoza as a philosopher who was excessively "Western," as compared to the East of an authentic Judaism to be rediscovered. Foucher de Careil had good reason for seeing the Buddhist study circles of Berlin as "a sect of Indianists—one that is flourishing today—that sees in Christianity nothing more than a product of India that was spoiled on the way, in Palestine" (quoted by Droit, ibid., p. 102), although he did not see this as inconsistent with the curious association of Sanskritists and Semites that so disturbed him.

But it was the same Foucher de Careil who published Leibniz's critical annotations of Wachter's second book, which presented Spinozism in a favorable light as an expression of the "secret philosophy of the Hebrews" (see above, n. 68). In his analysis of these marginalia, Georges Friedmann rightly considers them to constitute "a commentary on a work at second hand, important for the historian of Leibniz because of his interests and his mystical and kabbalistic preoccupations, but of scant value for understanding Spinoza." Furthermore, internal criticism of Leibniz's text shows how

6. SPINOZA'S "ANTI-JUDAISM"

The simplest way to unravel this tangled web begins with the contentious issue of Spinoza's excommunication. Historians have reconstructed the social, intellectual, and political environment of seventeenth-century Holland, in which the leaders of the Amsterdam Jewish community and the successive circles of the young Spinoza's Jewish, crypto-Jewish, and non-Jewish friends moved. There is a common but mistaken notion that Spinoza's excommunication by the Jewish community of Amsterdam was the result of doctrinal conflict with the Synagogue, analogous to Galileo's excommunication by the Church. But Spinoza was only twenty-five at the time and his philosophical works had not yet been written; he himself was still in the stage of learning and questioning. The circumstances of this excommunication, long shrouded in mystery, have now been almost fully clarified by I. S. Révah[96] and Henry Méchoulan.[97] Rather than a doctrinal conflict, it was a case of expulsion, a sort of exile, from a well-defined community, pronounced by its leaders, the notables and rabbis who bore primary responsibility vis-à-vis the Dutch authorities for its civil and religious order. The Dutch Protestant democracy allowed freedom of religion only within the officially recognized churches and synagogues, which were responsible for their members' civic behavior. Consequently, any transgression that involved the community administration, such as failing to pay the taxes and levies imposed on its members or publicly flouting the prescribed laws and orthodoxy, was condemned. Any overt opposition to the community's civic and religious organization was punishable as a disturbance of public order, which the community leaders were bound to uphold. Excommunication for a longer or shorter period of time was more of a social than a religious or doctrinal sanction. Even Rabbi Manasseh Ben Israel was placed under the ban, for twenty-four hours, when he came into conflict with the community notables.[98] This is why the community leaders were willing, time after time, to lift the excommunication of the resolutely heterodox Juan de Prado, Spinoza's friend and tutor, who repeatedly made honorable amends and was reintegrated into normal social life, though ultimately on condition that he leave Amsterdam for a distant Jewish community that would not have to answer for him to the Dutch authorities.[99] In this context, Spinoza's indefinite excommunication, employing an impressive ritual text that has survived, as well as his refusal to do anything to rejoin the community, must be understood as the price he had decided to pay, releasing himself from these constraints by leaving the community's jurisdiction. It was the only way, in that era, that he could think, study, and write freely, removed as far as possible from the political and religious turmoil of his century. We must not forget that in those days, when the Christian churches were torn by the theological and political conflicts that followed the Reformation, the

much "Foucher de Careil, led on by his own preconceived notions, exceeded scientific precision when he entitled the manuscript *A Refutation Recently Discovered of Spinoza by Leibnitz*" (Friedmann, *Leibniz et Spinoza*, p. 212). Thus we have good reason to suspect the authenticity of the knowledge and precision of judgment, both of Leibniz and his defender, concerning Hinduism, Buddhism, the kabbalah, and Spinoza, all wrapped up in the same package, through the abusive or polemic interpretations by these translators and expositors. As we have seen, many of these judgments, cleansed of their strident polemical excess, have become accepted facts and scholarly conventions in the history of philosophy. As such, they are always there to be taken up again, grossly or subtly, as the stakes of various theological and political battles that have been waged over the centuries.

96. I. S. Révah, *Des Marranes à Spinoza* (Paris: Vrin, 1995).

97. Henry Méchoulan, *Être juif à Amsterdam au temps de Spinoza* (Paris: Albin Michel, 1991).

98. Ibid., p. 55.

99. Révah, *Des Marranes à Spinoza*, pp. 374–375.

Jewish communities of the Diaspora were riven by the messianic upheaval of Shabbetai Ẓevi, inspired by popular kabbalah—an episode that had catastrophic social consequences for these communities, some of which, like that of Amsterdam, had scarcely recovered from the Inquisition and life as crypto-Jews.[100] In this context, the relationship between the content of the *Ethics*, conceived, like every major philosophy (if not more so) *sub species aeternitatis*, and traditional Judaism, with its doctrinal diversity, cannot be reduced to the specific circumstances of Spinoza's exclusion from his community. In particular, the hypotheses that most horrified and offended Christian theologians and philosophers in his published works—the absolute determinism of nature despite the imagined experience of free will and the total union of mind and matter in God/Nature—are expressed, sometimes in identical terms, in the writings of some rabbis of kabbalistic inspiration but undoubted Orthodoxy, both before and after him.[101]

It is the *Tractatus Theologicus-Politicus* that is most often held up as expressing a deep-seated anti-Judaism and that, for some, "represents one of the most violent attacks, not only against biblical ideas, but against the spiritual activity of the Jewish people in general."[102] For the author of these lines, Jacob Gordin, a representative of a dynamic segment of the European Jewish intelligentsia of the early twentieth century and one of the moving spirits of the revival of Jewish studies in France after the Second World War, "Spinoza played a fateful role in the decay of the Jewish soul."[103] Although Gordin recognized and admired the philosopher's greatness and power, he adopted the judgment of the Jewish Kantian Hermann Cohen about the man who "rejected his own religion, [who] made his own people contemptible," who committed "treason beyond human comprehension."[104] Gordin was unfortunately one of those modern Jewish thinkers who deemed it advisable to add their imprecations to those that Christian philosophers had been uttering for the previous three centuries. He did so by borrowing the accusation of Jew-hatred and reinterpreting a fine epigram of Nietzsche's,[105] which he quotes at the end of his 1935 article, "The Spinoza Case." We can benefit by trying to understand his motives.

100. On this subject, see Méchoulan, *Être juif à Amsterdam*, chap. 5; Gershom Scholem, *Sabbatai Sevi: The Mystical Messiah, 1626–1676*, trans. R. J. Zwi Werblowsky (Princeton: Princeton University Press, 1973). For a suggestive portrait of the trauma suffered by a typical European Jewish community during the rise and fall of the Sabbatean movement, see Isaac Bashevis Singer's novel *Satan in Goray* (New York: Noonday Press, 1955). For a critique of Scholem's hypothesis about the causal link between Sabbateanism and the Lurianic kabbalah, see Moshe Idel, *Kabbalah, New Perspectives* (New Haven: Yale University Press, 1988), pp. 258–260. Idel denies the existence of such a link, citing theoretical considerations about the nature of mass political messianism, which he contrasts with the theurgical strains of kabbalistic messianism already to be found in Abulafia and in the *Zohar* and with the Lurianic school of Safed and Jerusalem. He also takes issue with Scholem's thesis because of historical considerations—the gap of several decades between the emergence of Sabbateanism and the spread of Lurianic kabbalah in Europe in the classic version propounded in the writings of Ḥayyim Vital. At the height of the Sabbatean ferment, the latter was known only to a small circle of adepts in Jerusalem; well-read rabbis in Europe knew only the Neoplatonic version circulated by Israel Sarug and expounded by philosophical kabbalists such as Abraham Herrera, Joseph Delmedigo, and Manasseh ben Israel, with whom Spinoza could have been acquainted. Below we will see Spinoza's disinterest in the questions raised by the Sabbatean movement, including in his correspondence with Oldenburg about the possibility of the Jews' regaining national sovereignty in the Land of Israel.

101. Below we will encounter, in contrast to the opinions of Leibniz, Pierre Bayle, and most subsequent philosophers influenced by theology and ideology, the qualified response by Rabbi Kook in the early twentieth century. See also the traditional doctrines of absolute determinism surveyed in Chapter 2.

102. Jacob Gordin, "Les crises religieuses dans la pensée juive" (1946), in *Ecrits: Le renouveau de la pensée juive en France* (Paris: Albin Michel, 1995), p. 65.

103. Ibid., "Le cas Spinoza" (1935), p. 147.

104. Quoted by Gordin, ibid., p. 162.

In accordance with the state of Spinoza studies at the start of the twentieth century, Gordin believed that the key to the puzzle was to be found in the friendly relations between the young Spinoza, before and after his excommunication, and a group of Mennonites, a liberal and revolutionary Christian sect. Spinoza was supposed to have been influenced by this circle, which was imbued with the Christian heresy of the Socinians and related to the Gnosticism of Marcion, condemned by the Church in the second century, in part because of its Manichean dualism and rejection of the Jewish and Old Testament sources of Christianity. That is the alleged secret of Spinoza's doctrinal anti-Judaism. For Gordin, "there is no doubt that the work that reveals the true laboratory in which the propositions of the *Ethics* were concocted is the *Tractatus Theologicus-Politicus*, . . . which is one of the most violent attacks not only against biblical ideas but also against the spiritual activity of the Jewish people in general. This was caused by the personal problems faced by Spinoza, who found himself in a tangled web of spiritual and even financial submission to the representatives of the anti-Jewish spirit of the age, the milieu of the Quakers and Socinian Mennonites of Holland."[106]

Today, however, thanks to Révah, we know that Spinoza's youthful friendships were more diversified and that, while continuing to attend synagogue during his father's lifetime, he also frequented heterodox Jewish deists such as Uriel da Costa and especially Juan de Prado. During the first few years after his harsh exclusion from what had always been his family and social circle he seems to have been somewhat at a loss (though still close to Juan de Prado, also under the ban) until he decided to devote himself to the study of the sciences and philosophy and then to leave Amsterdam. He did continue to correspond with some of his Mennonite friends, now become disciples of Spinoza the philosopher, including Louis Meyer, Simon de Vries, and Jarig Jelles, who studied his works, attempted to ease the material life of the "hermit of Rijnsburg," and later supported the publication of his works.

Gordin's initial thesis is thus totally implausible, first of all with regard to the *Ethics*, in which it is hard to reconcile the monism of the unique substance and the illusory character of evil with Gnostic Manichean dualism; but also with regard to the *Tractatus Theologicus-Politicus* and the adaptation of his doctrine to the Judeo-Christian line commonly accepted by the public he addressed. Although Spinoza's correspondence demonstrates that he accepted the need to tailor the form of his discourse so that he would be understood by his readers, he does not compromise on what he holds to be essential, in any case at this stage of his career; namely, the three objectives that he set himself in the *Tractatus* and stated in a letter to Oldenburg (Letter 19 [68]): to combat the theologians' false opinions about the relationship between God and Nature; to persuade his

105.

Dem "Eins in Allem" liebend zugewandt,
Ein amore dei, selig, aus Verstand—
Die Schuhe aus! Welch dreimal heilig Land!——
Doch unter dieser Liebe fraß
unheimlich glimmender Rachebrand:
—am Judengott fraß Judenhaß!—
—Einsiedler, hab ich dich erkannt?

"An Spinoza" (KSA 11, Herbst 1884 28[49])

Lovingly devoted to the One,
An *amor dei*, blessed, for the Intellect—
Shoes off! What thrice-holy land!——
Yet under this love there ravened
an uncanny and glowing thirst for vengeance:
—the Jewish god devoured by Jew-hatred!
—Hermit, have I recognized you?

106. Gordin; "Les crises religieuses," pp. 65–66. See also: Negri, *The Savage Anomaly*; Jean-Pierre Osier, *Faust Socin ou le Christianisme sans sacrifice* (Paris: Le Cerf, 1996).

readers that he was not an atheist—i.e., that he believed in the true God and the true idea thereof, even if he seemed to be an unbeliever with regard to the false ideas about God and freedom; and to defend his freedom to practice philosophy by demonstrating that theology and philosophy are independent of each other. Should someone ever identify Gnostic elements in Spinoza's sources, they would certainly not be a Gnostic dualism derived from Marcion, but perhaps those that had infiltrated medieval philosophy through hermetic and kabbalistic Platonists, both Jewish and Christian, and that can be found in Giordano Bruno and Leibniz, for example.

Gordin dropped this thesis later; in his "Religious Crises in Jewish Thought," written right after the end of the Second World War, he did not develop the lines quoted above about the "laboratory" of the *Ethics* and Spinoza's "personal problems"—which allude to his 1935 "The Spinoza Case"—because "it is not this sad aspect of Spinoza the renegade that interests us here." Instead he expounds, briefly but explicitly, a different thesis that expresses, without justifying, Jewish religious thinkers' horrified judgment of the "renegade." He borrows the phrase "devilish irony," which Hermann Cohen had applied to the title of the *Ethics*, although Cohen, of course, like all Kantian religious moralists of practical reason and free will, was totally incapable of understanding its moral truth and depth. For Gordin in 1946, it was "Spinoza's almost satanic vigor, . . . because he is a true philosopher, perhaps one of the greatest," that explains how he has "played such a sinister role in the development of spiritual problems in Jewish thought, whose full weight we are still feeling today."[107] Here the stakes have changed. It is no longer, as in 1935, a question of restaging the excommunication ceremony, as if we were in Amsterdam in the seventeenth century.[108] In 1946 the order of the day was understanding the religious crisis in modern Jewish thought, whose origin was the intellectual emancipation that accompanied the social and political emancipation of European Jewry. Ever since Mendelssohn and his "unfortunate initiative" to introduce Spinoza to enlightened Jewish thought, even as he himself hesitated at the crossroads, there had been not only an "intellectual revolt" but also "a new awareness of the Western spirit," an attitude and awareness that, since that time, "have been eating away at the Jews' heart." This reflects a thesis that played an important role, both positive and negative, in the development of Judaic studies after the Second World War: Western culture is in crisis and authentic Jewish thought must recover its non-Western roots. Here we may recognize the theme of the return to the sources, whose slide into obscurantist fundamentalism, which would certainly have saddened Gordin himself, would follow within one generation. But for Gordin, who wanted to arouse awareness of the originality of Jewish tradition as a contrast to the post-Emancipation dominance of European culture, Spinoza played a negative role because he shuffled the deck. Precisely because he was a Western philosopher he was, like Plato and Aristotle, essentially foreign to Judaism. He set us off on the wrong track because of his Jewish origins and rabbinic reading and disrupted the Jews' mind and heart because of "his truly non-Jewish thought," camouflaged by "a false atmosphere of Judeo-Christian quasi-religiosity."[109]

107. Gordin, "Les crises religieuses," p. 67.

108. Although it would not be easy, he said, especially with regard to "a man supposed to be the classic defender of freedom of thought, . . . we must ratify or reject" the Amsterdam verdict. For "we have a moral duty to Jewish history, to ourselves, and to Spinoza's titanic thought, to accept this courage and this responsibility. *Benedictus* or *maledictus*, blessed or cursed? We must decide this question" ("Le Cas Spinoza," p. 147).

109. Gordin, "Les crises religieuses," p. 67.

Fifty years later, having rediscovered the non-Western sources of Jewish thought, encountering on the way many traces of the mutual fecundity of its relations with Greek philosophy, we can no longer share this judgment and even tend to reverse it. For us Spinoza, who is both "Oriental" (which means pre-Christian for Kant and Hegel) and "Western" as a philosopher, offers privileged access to the ancient wisdom that the philosophers in Greece and the rabbis in Babylonia and the Land of Israel pursued and sometimes found. Their universe was indeed pre-Christian and, as Hegel intuited, closer to the world of Indian philosophy and Buddhism than to that of theology. But the reasons for this were not what Hegel thought they were, which his ignorance of Buddhism and misunderstanding of Spinoza left him unable to analyze. We see this path as offering the promise of a universalism that is more real than the variety that theology traditionally reduces to the dimensions of the "monotheistic" universe. As for the trial in Amsterdam, taking account of the specific conditions in which the community lived, we have no reason to revisit it. *Pace* Gordin, we do not have to pronounce guilt or innocence, for all time, by examining, as if we were living in seventeenth-century Amsterdam, why Spinoza's "personal problems" led him to be a "renegade." The complex and difficult question of the relations between Spinoza's philosophy and Judaism must be reevaluated independently of his excommunication, which involved only that one Jewish community of Amsterdam. The content of his writings—especially the *Ethics* but the other works as well—makes it much more important to try to understand his philosophy and, through it, to think actively in a context that is as much Jewish as Greek, and perhaps Buddhist as well. We still need a study of the relations between the two "Eastern cultures" of Judaism and India. A first "summit meeting" produced a passionate report, despite its necessarily superficial and journalistic character, given the brevity of the event and its extravagant character.[110] To explain these relations there is no need to admit (but neither to exclude) the possibility of more or less mythical mutual influences in the remote past, alluded to in the *Zohar* and in the brief autobiography by Shabbetai Donollo in the preface to *Sefer Ḥakhmoni*, his commentary on *Sefer Yeẓirah*.[111] It is also

110. See Rodger Kamenetz, *The Jew in the Lotus: A Poet's Rediscovery of Jewish Identity in Buddhist India* (San Francisco: HarperSanFrancisco, 1994).

111. Donollo, a physician, rabbi, astronomer, and geometrician who lived in Italy in the tenth century, recounts his youthful wanderings when he was already a physician but eager to know and understand all the lore of his day, especially the "science of the stars and planets." Not having found this among the rabbis, he searched for it in the wisdom of the Greeks, the Muslims, and the people of Mesopotamia and India. After studying from books he embarked on a voyage to meet the wise men of the nations and learn this wisdom from them, until he found "a Babylonian sage named Bagdash," who not only was expert in the science of the stars and planets but also "knew how to calculate in order to truly understand what has been and what will be." It turned out that "all of his wisdom was in agreement with the *Baraita de-Shemuel* and all the writings of Israel, as well as with the books of the Greeks and Macedonians; but the wisdom of that particular sage was the most open and accessible"—unlike the aforementioned books, and in particular the *Baraita de-Shemuel*, "closed at the highest point" and to which no Jewish master had been able to provide him with the keys. Only after long study of astronomy and geometry with the Babylonian sage did he "apply his intellect to expound all the books that came to his hands, combined all their lore with that of the Babylonian sage and with his own studies, and wrote it down as a commentary in the book entitled *Ḥakhmoni*."

The ambivalent appreciation of the "Orient" as a place where the primordial wisdom has been preserved since time immemorial and developed in the form of the astrological and magical sciences of antiquity can be found in the Bible; first of all with regard to Cain and his descendants, founders of the first city and inventors of stockbreeding, music, and metallurgy, "east of Eden" (Gen. 4:16–22); then with regard to King Solomon, the wisest of men, whose knowledge "surpassed the wisdom of all the people of the east, and all the wisdom of Egypt" (1 Kings 5:10–11 [4:30–31]). The *Zohar* reestablishes the Abrahamic nexus of the transmission of this wisdom from Adam through the children of Abraham's concubines. Among these women, Keturah is none other (though under a different name) than Hagar, the mother of Ishmael who was banished from the family encampment, but whom Abraham took back after Sarah's death (*Zohar*,

possible that it is the Christian tradition that constitutes an exception to the universal nature of a wisdom that is less theological and more naturalistic. By a reversal of perspective fairly common in the history of civilization, after several centuries of successful evangelism the entire non-Christian universe came to be seen by this tradition as outside its catholicity, that is, in the strict sense, its self-proclaimed universalism.

Finally, one of the reasons most commonly invoked to explain the anti-Spinoza fury of some modern Jewish thinkers is that it has become conventional to view the *Tractatus Theologicus-Politicus* as the harbinger of critical and historical analysis of the Bible. This method, departing from the literal sense of the text, is understood to be intrinsically heretical, because it ostensibly destroys the foundations of traditional Judaism. What the *Tractatus Theologicus-Politicus* actually assails is a particular concept of scriptural revelation according to which the truth of the text is identical with its literal meaning. But divergence from this idea is not unique to Spinoza and can be found time and again within the rabbinic tradition. Defense of the literal meaning of the Bible does not coincide with defense of Jewish orthodoxy. In contrast to some Protestant denominations, a literal understanding of Scripture, which Spinoza mocks among those who would take it for philosophical truth, by no means constitutes the ground of rabbinic Judaism. On the contrary, the latter sees itself as rooted in the Oral Torah, a tradition of constantly renewed interpretations, and takes the text of the Written Torah as merely the medium of its formal continuity. Those who lack direct experience of the various forms of rabbinic exegesis[112] need only think of Maimonides, who had no compunctions about attaching an allegorical meaning to any biblical passage that contradicted the established scientific truths of his age, those taught by Aristotle. Similarly, the twelfth-century grammarian and Neoplatonist kabbalist Abraham Ibn Ezra identified several passages of the Pentateuch that could not have been written by Moses (Spinoza, incidentally, cites him as one of the precursors of his own method of exegesis).[113] Maimonides and Ibn Ezra have always been recognized as undisputed teachers of orthodox rabbinic Judaism, even though later rabbis have not always been in full agreement with them.

Genesis 133 [trans. Matt, vol. 1, p. 252]). The Bible informs us that Abraham took another wife, whose name was Keturah, and conveyed all his property to Isaac. "But to the sons of his concubines Abraham gave gifts, and while he was still living he sent them away from his son Isaac, eastward to the east country" (Gen. 25:6). The "East," obviously located somewhere east of Eden, is Babylonian or Indian rather than Semitic. The *Zohar* takes these "gifts" to be the Adamic lore in its lesser form of magic and sorcery, which the "people of the east" received from Abraham, whereas he transmitted the higher and purer form to Isaac, Sarah's son. Isaac in turn transmitted it to Jacob/Israel rather than to the latter's twin and rival Esau (whom the Talmud views as the ancestor of Rome), both of them sons of Rebecca. (This mythical history leaves the role of Ishmael, the son of Abraham and Hagar, up in the air. In practice, whether the plural "concubines" is to be understood literally or designates only Hagar-Keturah, by whom he had several other children, remains an open question; the *Zohar* cites the view of Rabbi Ḥiyya that the plural is meant literally, along with the contrary opinion.) This transmission does not seem to have been as simple as it might seem, because it was not until the age of Solomon and his extraordinary knowledge that the wisdom of Israel again included the science of celestial and terrestrial objects, until then reserved to the wise men and prophets of the Orient, such as Laban (Jacob's father-in-law) and the pagan prophet Balaam (see Chapters 1 and 2). It was this knowledge, which he too had to acquire in part through impure and dangerous commerce with demons, that gave Solomon his unmatched knowledge of natural objects as well as of the hidden meaning of the Torah precepts, such as the ritual of the red heifer (see *Zohar*, Numbers 19; Rashi on 1 Kings 5:10).

112. See Atlan, *SR* II, chap. 12.

113. With regard to the passages in the Pentateuch to which Ibn Ezra calls attention, Spinoza wrote that, in order to avoid being denounced as a heretic by "the Pharisees," "Aben Ezra, a man of enlightened intelligence, and no small learning, who was the first, so far as I know, to treat of this opinion, dared

not express his meaning openly, but confined himself to dark hints which I shall not scruple to elucidate, thus throwing full light on the subject" (*TTP*, chap. 8, pp. 120–121). "We have thus made clear the meaning of Aben Ezra and also the passages of the Pentateuch which he cites in proof of his contention. However, Aben Ezra does not call attention to every instance, or even the chief ones; there remain many of greater importance, which may be cited." Finally, he writes that an interpretation with which he disagrees "savours of the ridiculous; if respect for Aben Ezra allows me to say so" (*TTP*, chap. 9, n. 14, p. 272).

> More directly anti-Jewish are Spinoza's attacks on the Talmud and the Pharisees, to whom he attributes puerile and absurd interpretations, not to mention his mockery of the "kabbalists." It is this above all that has painted him as a Christianizing renegade or apostate, a man who placed his genius in the service of revenge against his people and religion, which he never stopped displaying as primitive and obscurantist, a fossilized national creed doomed to vanish. Clearly the facts have proven

In the introduction to his commentary on the Pentateuch, Abraham Ibn Ezra explains his method of exegesis as a mixture of looking for the obvious meaning that accords with grammar and the context, when this is possible and *plausible for a rational mind*, and recourse to exegetical traditions, notably from the Midrash, when it is not. Of the other four methods that he rules out, he rejects the literal reading of Scripture, which, following the Sadducees, refuses to associate the Oral Torah—the Pharisaic tradition—with the Written Torah, as the most reprehensible of all.

Naḥmanides, too, was not afraid to take a stand against earlier rabbinical authorities, such as Rashi and Maimonides, sometimes even more harshly than Spinoza, precisely on points of interpretation of the biblical text that he felt could not resist criticism. For example, he wonders (as have many) about the extraordinary life spans of several centuries that Genesis assigns to Adam and the early generations. He posits that human nature and the natural environment were different before the Flood. The validity of this explanation is not at issue here, no more than its literal or metaphorical meaning. What we want to underscore is the context of the critical interpretations of the text and the sometimes unbridled verbiage that accompanies them. Naḥmanides dismisses as so much hot air Maimonides' explanation that extraordinary longevity was the lot only of the handful of individuals whose names and lifespans are explicitly reported by the biblical text (commentary on Gen. 5:4). This critical and even disputatious attitude toward Maimonides (for another example, see Atlan, *Enlightenment*, pp. 118–119) has not kept him from being accounted one of the most important rabbinic commentators on the Bible, alongside Rashi, with whom, incidentally, he often disagrees.

Clearly, then, a certain latitude in the interpretation of Scripture was no revolution in rabbinic thought when the issue was harmony with the dictates of reason, whether in the sciences or in the "true Torah," meaning the esoteric wisdom of the kabbalah. We have already mentioned Maimonides, who, in his *Guide of the Perplexed*, said that he would have been willing to understand the account of Creation in a way that would be compatible with the Aristotelian postulate of the eternity of the universe, had he believed that theory to be "scientifically" demonstrated like Aristotle's other ideas, as he in fact did with the noncorporeality of God, although that was contradicted by the plain sense of many biblical verses. Yet the midrashic and kabbalistic literature are full of stories about "what preceded the creation of the world" (see, for example, Chapter 2, §11, nn. 89 and 98 and appendix). In order to harmonize the narrative of Genesis with this "emanatory" vision, it is often interpreted as the last stage in the process by which the infinite One produces the multiple and finite. (This process is clearly timeless, because time is one of its products.) The name *'Elohim*, usually taken as the subject of the verb "created" in Genesis 1:1—which, rendering the Hebrew word order literally, is "In the beginning (*bereshit*) created Elohim the heavens and earth"—is read instead as its object and thus as the outcome of this process. This is the tack taken, for example, by the *Zohar* (Genesis 15a), as if some primordial being named *Bereshit*, which denotes, inter alia, the *sefirah* of *Ḥokhmah* or Wisdom, "created Elohim [along with] the heavens and earth." In this reading the noun *'Elohim*, traditionally rendered as "God" despite its plural form, designates structured divinity, such as that of the "angels" or the "palaces." "This palace is called *Elohim*," explains the *Zohar*. Furthermore, if, in connection with an act of emanation, "*created* is written here, no wonder it is written, *God created the human being in his image* (Gen. 1:27)" (ibid. [trans. Matt, vol. 1, pp. 110–111]). For his part, the anonymous sixteenth-century author of *Ma'arekhet 'elohut* (The System of Divinity) explains that in Genesis the name *Elohim* designates the second *sefirah*, *Binah* or intelligence, also known as "the mother," through which the creative process takes place (*Ma'arekhet 'elohut* [Jerusalem, 1963], chap. 4). Moses Cordovero sums it all up in the equation, "In-the-beginning created God (Elohim)" = "Wisdom (*Ḥokhmah*) emanated Intelligence (*Binah*)" (Moses Cordovero, *Shi'ur ḳomah* [Jerusalem, 1994], chap. 59, p. 128).

Conversely, we should recall that the Copernican revolution was received as a brilliant discovery and celebrated as the victory of scientific truth by several rabbi-astronomers of the seventeenth century, notably (though not only) the disciples of the Maharal of Prague, just when the trial of Galileo was in progress (see the well-documented article by André Neher, "Copernicus in the Hebraic Literature from the Sixteenth to the Eighteenth Century," *Journal of the History of Ideas* 38[2] [1997], pp. 211–226). We may conjecture that a Spinoza living in Renaissance Prague would not have had to leave his community to practice philosophy freely—on condition, of course, of not publicizing it and certainly of not aspiring to write for the non-Jewish environment, still subject to its ecclesiastical authorities, who evinced greater or lesser tolerance depending on the place and the denomination.

him wrong on that account, while we had to wait almost three centuries for his improbable hypothesis of a return to Zion to become a political reality. But could he have imagined anything else? What do we know about the intellectual content of what he learned in the synagogue in his youth? His only image of the Talmud had to be the caricature painted by Christian scholars, a muddle of superstitions and useless quibbles. It is not certain that, during his few years of Jewish education, he would have had an opportunity to discover another and deeper dimension of talmudic exegesis. As for kabbalah, he probably became familiar with it later, in common with the educated men of his time who were nurtured by the alchemical-hermetic-kabbalistic traditions of the Renaissance, through the writings of Giordano Bruno, Marsilio Ficino, Pico della Mirandola, and others. The one enormous difference was that his knowledge of Hebrew gave him direct access to many of the original texts rather than having to read them in Latin translation. As Henry Méchoulan asks, "who are these Jews" whose preconceived notions he ridicules along with those of the "theologians" in the *Tractatus Theologicus-Politicus*? "The Hebrews, the Pharisees, or his contemporaries?"[114] We shall return to this point.

The anti-Jewish sentiments in the *Tractatus Theologicus-Politicus*, even if they express rancor and resentment, are not an end in themselves. They are "used, 'rationalized' so as to coincide with his readers' anti-Jewish prejudices."[115] It is there, in the Christian reading public for whom he wrote in Latin, rather than in a fundamental anti-Judaism, that we must seek the keys to Spinoza's language. We cannot know how he felt about his parental community after he had given it up as lost and placed his entire life in the service of "true philosophy." Nor can we know how he replied (or even if he did) to Oldenburg's intriguing request for his reaction to the fact that "everyone here is talking of a report that the Jews, after remaining scattered for more than two thousand years, are about to return to their country" (Letter 16 [33]). This must refer to the messianic adventure of Shabbetai Ẓevi, which originated in Turkey but reached all the Jewish communities of Europe. During its brief trajectory, this was the most important event in the lives of these communities, including the Amsterdam community. Oldenburg noted that "few here believe in it, but many desire it." For his own part, he added, "unless the news is confirmed from trustworthy sources at Constantinople, which is the place chiefly concerned, I shall not believe it." Nevertheless he wrote to Spinoza for his opinion as well as "to know what the Jews of Amsterdam have heard about the matter, and how they are affected by such important tidings which, if true, would assuredly seem to harbinger the end of the world." Not only did Spinoza not answer; the known correspondence with Oldenburg, which had always been amicable, soon broke off and was not resumed for ten years. No one knows whether there was any relationship of cause and effect between the apparently innocent question and the suspension of their correspondence.

Spinoza may have replied, indirectly and without emotion, in the *Tractatus Theologicus-Politicus*, written during that ten-year hiatus, when he evokes in passing (end of chapter 3) the possibility of the Jews' return to Zion as a moot hypothesis rather than as a concrete political reality. Spinoza behaved as if he was no longer interested in (or did not want to be interested in, or pretended not to be interested in) the future of the community and people from which he had been divorced, internally, after having been excluded from it socially. From that time on, what

114. Méchoulan, *Être juif à Amsterdam*, p. 147.
115. Ibid.

interested him was what he wrote about in his letters, before and after this decade, as if the messianic epic and its ruination, catastrophic for the Jewish communities that had placed their faith in it, had never taken place. His correspondence refers only to issues of physics and philosophy, about which his friend and admirer Oldenburg, secretary of the Royal Society in London, asked his opinion, as well as the reasons that had moved him to write the *Tractatus*:

> 1. The Prejudices of theologians. For I know that these are the main obstacles which prevent men from giving their minds to philosophy. So I apply myself to exposing such prejudices and removing them from the minds of sensible people.
>
> 2. The opinion of me held by the common people, who constantly accuse me of atheism. I am driven to avert this accusation, too, as far as I can.
>
> 3. The freedom to philosophise and to say what we think. This I want to vindicate completely, for here it is in every way suppressed by the excessive authority and egotism of the preachers. (Letter 30, trans. Shirley [not in O.P.])

The correspondence was resumed ten years later when Oldenburg thanked Spinoza for sending him a copy of the *Tractatus*, which he had just received. In the meantime, Spinoza had not been able to persuade his public either about the prejudices of theologians or that the charge of atheism leveled against him was false. Hence he asked Oldenburg to suggest how he could modify the *Tractatus*, by "point[ing] out the passages . . . which are objected to by the learned, for I want to illustrate that treatise with notes, and to remove if possible the prejudices conceived against it" (Letter 19 [68]). This is followed by his superb replies to Oldenburg, who had advised him "to give a clear explanation of what you think" on the most problematic issues, adding that "if you do this and thus give satisfaction to prudent and rational Christians, I think your affairs are safe" (Letter 20 [71]).[116]

So we should read Spinoza not as "a renegade Jew" but as the philosopher he wanted to be and was. His knowledge of Hebrew, which gave him direct access to basic texts of kabbalah and Jewish philosophy, certainly contributed to the originality of his philosophy. But his relationship with these texts and authors was intellectual, of the same nature as that entertained by his contemporary scholars and philosophers with the Jewish, Christian, and Muslim traditions of the Middle Ages and Renaissance.[117] We should not be looking here for a social, political, or religious relationship, or for a rejection of a such relationship. Obviously we cannot ignore the existence of a real and complex personal situ-

116. Spinoza's replies are Letters 21 (73) and 23 (75).

117. According to a thesis defended several years ago (M. Bedjaï, "Métaphysique, éthique et politique dans l'œuvre du Docteur Franciscus van den Enden. Contribution à l'étude des sources et des écrits de B. Spinoza," Université Paris I Panthéon-Sorbonne, 1989), Spinoza, like Descartes and Leibniz, was also familiar with the alchemical and hermetical tradition of the Rosicrucians. This supposedly constituted the core of his exchanges with van den Enden, his teacher of Latin and student of Hebrew. (I would like to thank Martine Israitel ["Un philosophe par le feu," *Nouveaux Cahiers* 124 (1996), pp. 43–44] for calling my attention to this work.) During the Renaissance, before its origins in Greek Gnosticism were discovered, this tradition was thought to be anchored in Egyptian esotericism. But those who were looking for an ancient, non-Aristotelian source for the New Science often associated it with Platonism and Jewish kabbalah as well. A rivalry of sorts sometimes prevailed between pagan Egyptian science and Hebrew science, as shown by the distance between Giordano Bruno, a magus of the sun although a Dominican, and Pico della Mirandola, a Christian kabbalist. Most often, however, they were closely associated as objects of study by the scholars and philosophers of the seventeenth century, as indicated by the biographies of Kepler, Newton, Leibniz, and others.

ation. Geneviève Brykman's study can enlighten us on this point.[118] But any attempt to reduce the content of an author's work to his or her biography—and in this case more than in others—is vain speculation. Baruch/Benedict's personal and intimate relationship with Judaism, complex and conflicted, at least at the outset (as far as one can judge), remains a part of his private life, with all its passions and emotions—precisely what Spinoza's philosophical efforts seek to liberate him from, along with his readers.

As for Spinoza's "Marranism," we need only read his writings, including his correspondence, to see how little his ideas about Scripture and about Christ could satisfy Christians, even "prudent and rational" ones, of his own century and the next. What is more, while he was writing the *Tractatus* he declined the offer of a chair at the University of Heidelberg, which would have provided him the best possible entrée to the world as a Marrano of his century. But, he said, he wanted to preserve his freedom to practice philosophy; delivering lectures open to all would limit him if he endeavored "to avoid all appearance of disturbing the publicly established religion" (Letter 54 [48]). The image of the Marrano, Christian on the outside but secretly preserving remnants of Jewish belief and practice, over several generations, suits him poorly. On the contrary, he left a community of former Marranos who had returned to classic Orthodox Judaism, as indicated by the activities and the writings of its rabbis. Perhaps the closest semblance to his social position is that of the modern "assimilated Jew." This model has become commonplace today, represented by a long series of great names—Marx, Freud, Einstein, Bergson, Simone Weil, and others—who maintained, each in his or her own way, complex relations with their ancestral people. Their common trait was not Marranism but the falling away from Judaism that accompanies acculturation and assimilation to the majority culture. We know that even individuals whose childhood was steeped, emotionally and linguistically, in a traditional Jewish environment may succumb to such acculturation. Until the Second World War, the intellectual experience of European university circles, with their modern laicization of Christian tradition, was the most effective route to this. It was probably the same evolution experienced by the young Spinoza, with a vast difference in the mechanism: in his time it was far from commonplace and could take place only in the dramatic context of conversion or excommunication.

Finally, it is hard to accept Méchoulan's peremptory judgment: after a clear explanation of the conduct of the Jewish leaders of Amsterdam, who had to maintain civil and religious order and had no choice but to excommunicate Spinoza, he considers him to be "a thinker who may have caused them—the Jews, in general—as much harm in a single book as centuries of Christian antisemitism."[119] But Méchoulan is confusing cause with effect. We have known centuries of Christian antisemitism followed by the "progressive" anti-Judaism of some proponents of the Enlightenment, succeeded in turn by the racism of the nineteenth and twentieth centuries; hence we cannot retain our composure in the face of criticism of certain modes of biblical interpretation. These critiques are embedded, in fact, in an exegetical agenda—which we can understand without sharing—that proclaims the demise of the Jewish people as its natural destiny, with the national religion of ancient Israel supplanted by Christianity, the "new Israel" with its uni-

118. Geneviève Brykman, *La judéité de Spinoza* (Paris: Vrin, 1972).

119. Henry Méchoulan, "A propos du *herem* de Spinoza," *Nouveaux Cahiers* 124 (1996), pp. 39–42.

versal vocation, before the anticipated reign of philosophical religion. We shall see below what Spinoza's "Christianity" was really like. But if we extract the *Tractatus* from the antisemitic context in which we have become accustomed to placing it, it is no more baneful to Judaism (and the Jews) than many rabbinical critiques of the views of other rabbis, deemed to be mistaken if not indeed inane.[120]

In an alternative history imagined by Rabbi A. I. Kook, to be discussed below, this is probably what would have ensued had Spinoza not been excommunicated and had he written his works in Hebrew, thereby continuing to develop as a Jewish teacher. It is high time to stop this demonization, "in the name of Judaism," of Spinoza, whose first agents, we must not forget, were Christian philosophers of the seventeenth and eighteenth centuries, including Pierre Bayle, Leibniz, and several champions of the Pantheism Controversy in Germany.[121] Spinoza's conflict with his ancestral community is a symptom; accusing him of a Christianizing "anti-Judaism" is tantamount to making the symptom the cause of the disease. The real malady is the sum total of the ambiguities that accompanied the political and cultural emancipation of the Jews in Europe. Spinoza's expulsion from Judaism during the years when he was discovering the learned world of Europe is a symptom, like the abandonment of Jewish belief and practice by the majority of European Jewish thinkers and scholars before the Second World War. It is a sign of a sort of internal colonization of the Jewish communities and of their failure to negotiate their inclusion in European modernity without having to pay the admission fee in the coin of the acculturation and Christianization of their intellectuals. But it is also, as in other cases of decolonization, a symptom of the European intelligentsia's inability to accept foreign cultures without grafting them onto the Christian philosophical and theological tradition, either directly or in the reactive mode of the deists and freethinking atheistic materialists.

120. The *Zohar* directs harsh words against those who are content with the obvious and plain sense of the biblical text. Those whose Scripture does not go beyond the surface meaning of the narrative fail to see that the garment hides the Torah so that it can be perceived by all; those who see only the precepts of the Law perceive only its body. Only those who search for the hidden meaning and study the wisdom of kabbalah discover its soul, in the expectation that in the future they will penetrate even more deeply and discover the soul of its soul. This is why those who insist—for whatever reason, including the danger of being misled—that there is nothing to look for in the written Torah and in the Talmud, beyond the stories and commandments, destroy the world and return it to the initial chaos, to *tohu-bohu* (*Zohar*, Numbers 152; see also *Tikkunei ha-Zohar* 43, quoted and commented on by Moses Cordovero in his *Or ne'erav* [see Ira Robinson, *Moses Cordovero's Introduction to Kabbalah: An Annotated Translation of His* Or Ne'erav (New York: Yeshiva University Press; Hoboken: Ktav, 1994)]).

In this text Cordovero also rejects the objections advanced by those who would proscribe the study of esoteric wisdom, because of the inherent danger of falling into error, by demonstrating, from a passage in *Sefer Habahir* (§68), the sublimity of these errors, which are themselves inherent in the love of wisdom: "As it is written [regarding wisdom] (Prov. 5:19), 'With its love you shall always err'" (*The Bahir: An Ancient Kabbalistic Text attributed to Rabbi Nehuniah ben HaKana*, trans. Aryeh Kaplan [New York: S. Weiser, 1979; repr. Northvale, NJ: Jason Aronson, 1995]). <Note that "err" is the normal sense of the Hebrew verb *shagah* employed here, and that the "delight/rapture/ravishment/infatuation" of translators through the centuries reflects a prejudice that wisdom and error are incompatible.> (On this verse and its erotic evocation of the love of wisdom, see also B *'Eruvin* 54b.) After carefully distinguishing these errors from heresy he cautions novice kabbalists against embarking on this lore without an adequate preparation in the Bible, its rabbinic commentaries, and the legal dialectics of the Talmud.

121. See Pierre-Henri Tavoillot, *Le crépuscule des Lumières: les documents de la "querelle du panthéisme," 1780–1789* (Paris: Le Cerf, 1995). See also: Alexis Philonenko, introduction to E. Kant, *Qu'est-ce que s'orienter dans la pensée?* (Paris: Vrin, 1988); Zac, *Spinoza en Allemagne*.

7. SPINOZA'S "PHARISEES" AND THE END OF PROPHECY

Spinoza was, however, mistaken about one major historical issue. Like most philosophers and theologians of his time,[122] he radically misunderstood the intellectual, social, and political dimension of the Pharisees and their project.

We cannot know whether Spinoza was merely repeating ideas accepted by his readers or whether he really did misconceive the true nature of the Pharisees' achievement, as expressed in the Talmud—where, in fact, the religious hypocrisy later attributed to them by their Christian adversaries is analyzed at length and its bearers denounced as "false Pharisees."[123] It is hard to answer this question, because we do not know the extent to which Spinoza ever had the opportunity and inclination to study the Talmud, and in particular tractate *Sanhedrin*, where we find diverse manifestations of the Pharisees' opposition to the power of the Sadducee priests.

This opposition was translated into action by the creation of new institutions—synagogues and schools—that constituted the core of one of the first comprehensive educational systems in the world, which culminated in the academies of *Talmud* (the word means "study") in the Land of Israel and in Babylonia. There, the most talented *talmidei ḥakhamim* or "disciples of the sages"—as the rabbis of that era referred to themselves—sharpened their minds in every department of their contemporary cyclopedia of theoretical and practical knowledge, with a universal objective and specific method. In place of the Temple

122. There were some rare exceptions, of whom Pascal is perhaps the most interesting.

Spinoza's harsh judgments about the Pharisees (e.g., in chap. 5 of the *Tractatus*) are evidence of a violent contempt, a combination of the remnants of his bitterness with the Jewish community that had excluded him and ready adoption of the stereotypes entertained by seventeenth-century Christian circles. This applies equally to his conception of the Mosaic law, which ostensibly nurtured hatred of the foreigner more than other legal codes do. In his condemnation of the Pharisees we can discern a retrospective projection of the habitual stereotypes about the Jewish communities of the period, which refused to accept Christianity. In support of his reconstruction of history he invokes two verses of Psalm 139 and a statement by Tacitus (of all "reliable" sources!) about "the character of the people and the obstinacy of their superstition" (*TTP*, chap. 17, p. 230), conveniently forgetting that the same Tacitus accused the persecuted Christians of his day of a hatred of humanity. Neglecting his own rule that Scripture is to be explained on the basis of Scripture, he also "forgets" the biblical exhortations to respect the stranger, notably "love your neighbor as yourself" (Lev. 19:18), repeated explicitly with regard to the stranger: "the stranger . . . shall be to you as the native among you and you shall love him as yourself; for you were strangers in the land of Egypt" (Lev. 19:34). On the other hand, he does not hesitate to quote the Gospels (Matt. 5:43) to the effect that the Hebrews were commanded to "Love thy neighbour and hate thine enemy" (*TTP*, chap. 19, p. 250)—a dictum that is not to be found in the Hebrew Bible, and which, furthermore, is a fairly common attitude among human beings. In fact, he is engaged in a retrospective fabrication in which, to persuade his Christian readers, he adduces as "proofs" what they considered to be solid evidence. Nevertheless, and still to satisfy the needs of his argument, he portrays the ancient Hebrew republic instituted by the Mosaic law as a near-perfect polity, the result of "Divine revelation to Moses" (*TTP*, chap. 17, p. 216). In fact, the only defect of this legislation—a flaw that, according to him, led to the destruction of that republic—was also revealed to Moses by God, thereby demonstrating that the mechanism of this destruction can be included simultaneously within the contexts of prophetic revelation and historical determinism, which is by definition ordained by God. This requires him to postulate that the Hebrews, in the covenant that established their theocracy, originally accepted the law of Moses without constraint, just as the citizens of a democracy leave behind their natural state without constraint. "It is, in fact, in virtue of a set covenant, and an oath (see Exod. 24:7), that the Jews freely, and not under compulsion or threats, surrendered their rights and transferred them to God" (*TTP*, chap. 17, p. 219).

But when it comes to the Pharisees, the objective is to set them up as the antithesis of Christ, who "was sent into the world, not to preserve the state nor to lay down laws, but solely to teach the universal moral law . . . —His sole care was to teach moral doctrines, and distinguish them from the laws of the state." Spinoza also affirms, providing no historical justification and this time allowing that the Hebrews were under constraint, that "the Pharisees, in their ignorance, thought that the observance of the state law and the Mosaic law was the sum total of morality; whereas such laws merely had reference to the public welfare, and aimed not so much at instructing the Jews as at keeping them under constraint" (*TTP*, chap. 5, p. 71). This permits him to reduce the Pharisees' project to the fact that they "continued to practice these rites after the destruction of the kingdom, but more with a view of opposing the Christians than of pleasing God" (ibid., p. 72).

This is the theological and political basis of Spinoza's "Christianity," as expressed in the *Tractatus*, which was meant for Christian readers and philosophers for whom the real issue of the historical Pharisees was totally foreign, as it may well have been for Spinoza himself. In this frame of reference, Christianity is the only conceivable response to the end of biblical prophecy. Anything else is unpardonable and absurd obstinacy whose origin must be the hatred of the stranger that the Jews are supposed to have inherited from ancient Hebrew law. What is more, as in the case of the antisemitism of contemporary peoples directed against their Jewish neighbors, diverse nations' hostility to the ancient Hebrew state is justified, for Spinoza, by the Jews' supposed prior hatred of them: "Nor was a general cause lacking for inflaming such hatred more and more, inasmuch as it was reciprocated; the surrounding nations regarding the Jews with a hatred just as intense" (*TTP*, chap. 17, p. 229). Here too he is retailing an analysis current in Christian circles about the stubborn Jews and projecting it onto the relations between the Jewish commonwealth and other states of antiquity. Spinoza may have helped entrench this idea, which has had a long life and is particularly prominent in certain Enlightenment authors, notably Voltaire. All the same, we must recognize that the ambient antisemitism had no need of Spinoza to integrate this notion into the spirit of the age.

In fact, the no-less entrenched idea that Hebrew civilization instituted by the Mosaic law was distinguished from other civilizations of antiquity by its fierce xenophobia, which it supposedly carried to an extreme because of the perverse and absurd (but effective) notion of the Chosen People, was itself a retrospective projection of the classic perception of the Diaspora communities as a "people accursed." These communities were supposed to bear sole responsibility for their condition, which they had inherited in some way from the excessive particularism of the Hebrew state of antiquity. It is not too difficult to justify this notion of chosenness, observing that all the peoples of antiquity defined their national identity in relation to their tutelary deity, in a reciprocal election of the people by their god; the Hebrews had no monopoly in this (see Atlan, *SR* II, chap. 10). But this means viewing Christian civilization from the outside and thus as less universal than it would like to believe itself to be. Similarly, there is nothing extraordinary about the centripetal organization of humanity, deriving from awareness of membership in one's own tribe or group of related tribes. There is no denying that this common situation was the starting point of an exceptional history, and we can even try to grasp its form (see Atlan, *Entre le cristal et la fumée*, chap. 11). This, for better or for worse, must be credited to the Pharisees and their project of reconstituting Judaism as a political, social, and religious organization suited to the condition of the scattered Jewish communities after the destruction of their state and the loss of their political and priestly sovereignty. Not only were there no longer kings, there were no prophets and priests, either; and no more audible god. In their place, as the vehicle of individual and collective salvation, "wisdom," including that of Solomon, the "wisest of men," was taken up as the model.

But for Spinoza, promoting the conventional contrast between universal and spiritual Christianity and particularistic and materialistic Judaism, the Pharisees had to be rigged out in the expected stereotypes, including their denunciation as "vilest hypocrites" (*TTP*, chap. 18, p. 241), a Christian tradition dating back to the Gospels. They are those whose "passion" impelled Pilate, seeking to appease them, to "consent to the crucifixion of Christ, whom he knew to be innocent" (ibid.). Similarly, Spinoza confounds—consciously or not?—the conservatism of the Sadducees with the Pharisees' openness to innovation. The former, the priestly aristocracy, monopolized the political power centered on the Temple and collaborated with the Romans. For them, the Mosaic law could be reduced to respect for the laws of the state, meaning the social and political organization they had instituted. The Pharisees, by contrast, organized into bands of "companions" who called themselves "students of the sages," launched a religious, intellectual, legal, and sociopolitical revolution that reinvented Judaism for the ages.

123. B *Sotah* 22b.

with its priests, which was no longer the center of Jewish life and had become a mythical reference to the rites of another age—that of the origins and that of the eschatological future—the text of Scripture became a formal reference to a lore that was held to be eternal, but whose content was to be uncovered through the study of the Oral Law and new insights (Hebrew *ḥiddushim* 'innovations, renewals') into it. Commentary or exegesis of Scripture, of the Written Law understood as the word of God in the literal sense of its plain meaning, was by no means the sole objective of this study. The ceremonial reading of the sacred text, with its obvious sense, was one of the new rituals intended to stimulate curiosity and interest in the study of wisdom. But knowledge was really to be found only in the Oral Law, postulated as being revealed to the sages just as the Written Law had been revealed to Moses. Henceforth this was the only true path to individual salvation. This is what is at stake in the talmudic debates in the eleventh chapter of *Sanhedrin*, where we find that the Sadducees and others rejected the Pharisees' innovation of assigning higher value to the Oral Law than to Scripture itself, which was supposed to be fenced around and contained within the ambiguous domain of the sacred and immutable. But the Pharisees had the temerity to affirm that the Torah—both written and oral—had come down from heaven into the world and consequently possessed the capacity to allow human beings to achieve salvation in this world and immortality in the next.

It is hard to say what Spinoza knew of these discussions, because we do not know the extent of his study of the Talmud. He does refer to the debate between the Pharisees and Sadducees about the resurrection of the dead—but only to demonstrate (correctly, as it happens) that the former, as "men expert in the law, summoned a council to decide which books should be received into the canon, and which excluded" (*TTP*, chap. 10, p. 155), depending on whether or not the content of these books could be harmonized with their own teachings. Consequently he does not refer to tractate *Sanhedrin* but to a different passage, in tractate *Shabbat*, which deals with the books that some sages had sought to exclude from the biblical canon. In any case, it is clear that the doctrines of the ancient Pharisees are not the subject of the *Tractatus Theologicus-Politicus*. Its focus is the status of Scripture, that is, the books of the Old and New Testaments, which alone possessed theological interest and religious value for the Christian readers, mainly Protestant, to whom the book was addressed.

When he diminished the vast enterprise of rabbinic Judaism after the destruction of the Temple to the single concern of preserving the xenophobia that supposedly always characterized the Hebrew monotheism taught by Moses, Spinoza was mythologizing history to make it serve his ends, employing a literary genre that is closer to the Freud of *Totem and Taboo* and *Moses and Monotheism* than to serious history. We recognize, of course, that neither of them was interested in historiography. Since Spinoza's time the complicated story of the birth of Christianity in the largely Hellenized Jewish milieu of first-century Palestine has given rise to a vast literature that relies in part on the critical method he himself inaugurated for the Old Testament, but now applied to the Gospels. A well-documented and original exposition can be found in the work by Paula Fredriksen.[124] She analyzes the construction of the multiform image of Jesus Christ through the successive texts of the New Testament as well as their image of the "Pharisees," who function as a foil for primitive Christianity, quite different from the historical Pharisees as they appear, inter alia, in their own writings.

In retrospect, it was the Pharisees' revolution that made it possible for the dispersed Jewish people to survive, despite the exile and in their exile. But it had more in mind than just proposing a political solution to the Jews' statelessness. Those who called themselves "Pharisees" (Hebrew *perushim*, "separated" or "distinguished" by their traditions of study and practice) constituted, along with the Sadducees, Essenes, and Zealots, the four groups that, according to Flavius Josephus, composed the Jewish people at the start of the first century.[125] By the end of the first century, after the emergence of the early Christian Church

124. Paula Fredriksen, *From Jesus to Christ: The Origins of the New Testament Images of Jesus*, 2nd ed. (New Haven, CT: Yale University Press, 2000).

125. Flavius Josephus, a participant in and historian of the Jewish war against the Romans (66–73 CE), describes these groups as follows: the Zealots or Sicarii who preached the revolt against the Romans were "a sect . . . having nothing in common with the others. Jewish philosophy, in fact, takes three forms. The followers of the first school are called Pharisees, of the second Sadducees, of the third Essenes" (Josephus, *The Jewish War*, trans. H. St. John Thackeray [LCL], II.118f. [= §8.2]).

and destruction of the Temple, the other three groups had disappeared. The Sadducees lost their raison d'être when the Temple ritual disappeared along with all traces of national sovereignty. The Zealots were crushed and massacred by the Romans. As for the Essenes, some merged into the new sect of Christians and others probably rejoined the Pharisees. The last-named are thus the sole ancestors of modern Judaism, totally Pharisaic, as it survived the crises of the first century. We know about them not only from Josephus but also, and even more so, from their own texts, which are the stuff of the Babylonian and Jerusalem Talmuds. They claim descent from a tradition of study and practice inaugurated five centuries earlier, at the end of the Babylonian exile, by those known as *soferim*, "writers" or "scribes." It is to these scribes, the most celebrated of whom was Ezra, that historical criticism attributes the definitive form of the biblical texts. But throughout this period, following the destruction of the First Temple and exile to Babylonia and then the return to Zion and construction of the Second Temple, in which the divine presence was less directly perceptible than it had been in the First Temple, they had to deal with the end of the mythological era and the disappearance of prophecy, which had been one of its manifestations.

Spinoza was perfectly aware of the importance of these phenomena. It is a great shame that he ignored this major dimension in the birth of rabbinic Judaism and seems to be aware only of the birth of Christianity. Spinoza's analyses in the *Tractatus Theologicus-Politicus* raise the question of the possible shift from prophecy to philosophy in both logical and historical terms, even if this history is revised to serve his purposes. These analyses are conducted from a Christian perspective on this transition, and this is their great failing from our modern vantage point; his reconstruction is derived almost exclusively from the Christian Bible, meaning the Old and New Testaments seen as a seamless whole, with the Old prefiguring the New. But it has the merit of posing the problem of the possible existence of at least two types of knowledge, each with the capacity to help human beings attain salvation—meaning happiness—although they are quite different from each other: prophetic revelation of the good and the bad, to encourage progress along the path of salvation by means of morality and religion; and the philosophical and scientific quest in which the true path to salvation is discovered by means of the natural light of reason and the wisdom that can accompany it. The historical record of the transition from one to the other, of the process by which prophecy—the form that mythology assumed in ancient Israel—disappeared and was supplanted by something else, which may be designated "philosophy," the love of wisdom, "study" (*talmud*),

This is followed by a long description of the doctrines, beliefs, and lifestyle of the Essenes (about whom we are beginning to learn more from the Dead Sea Scrolls). Finally, Josephus concludes his description of the philosophy of what he calls these three "sects" with brief notes that compare them to the philosophical schools of the Greco-Roman world:

Of the two first-named schools, the Pharisees, who are considered the most accurate interpreters of the laws, and hold the position of the leading sect, attribute everything to Fate and to God; they hold that to act rightly or otherwise rests, indeed, for the most part with men, but that in each action Fate co-operates. Every soul, they maintain, is imperishable, but the soul of the good alone passes into another body, while the souls of the wicked suffer eternal punishment.

The Sadducees, the second of the orders, do away with Fate altogether, and remove God beyond, not merely the commission, but the very sight, of evil. They maintain that man has the free choice of good or evil, and that it rests with each man's will whether he follows the one or the other. As for the persistence of the soul after death, penalties in the underworld, and rewards, they will have none of them.

The Pharisees are affectionate to each other and cultivate harmonious relations with the community. The Sadducees, on the contrary, are, even among themselves, rather boorish in their behaviour, and in their intercourse with their peers are as rude as to aliens." (Ibid., II.162–166 [= §8.14])

"science," or "Gnosis," is clearly broader than the account that can be afforded solely in the historical context of Christian theology. Nevertheless, and even though this was not its author's explicit intention, the *Tractatus Theologicus-Politicus* can also serve as an introduction to an analogous historical account of the same phenomenon, transposed into another historical and cultural framework, notably that instituted by the Pharisees in parallel with the Christian apostles and in competition with them, and to answer the same question.

The issue has been clearly analyzed by Alexandre Matheron in an entertaining reconstruction of his own, based on the *Tractatus Theologicus-Politicus* and the special status that Spinoza accorded to Jesus Christ.[126] We shall return to this.

After the end of the mythological period—that is, when inspired augurs and prophets were no more—the relationship to prophetic Scripture had to change. A straightforward reading of the text and meditation on it as "the word of God" were no longer adequate, because the experience of prophetic knowledge had disappeared. Henceforth the public reading of Scripture was a ritual that merely preserved the memory of something important in the past. Taken to the extreme, even understanding the words and sentences was no longer essential, because there was no longer any real experience of what they refer to. Inevitably, the rabbis[127] took possession of it and argued from (and especially around and alongside) the text. But the crucial thing is how one does this. Does one keep talking, beginning with God[128] and his word, as if it were still possible to have direct knowledge of the deity? Or does one alter the perspective and begin from the experiences of this world, in which God does not speak? In the former case, the prophetic Scriptures can serve as a source of inspiration, as if nothing has changed, even though they speak of a world and reality that no longer exist. Interpreting them in the terms of mundane experience serves as the foundation of a theological discourse that begins with God, as if the experience one can have of him, based on faith or reason, provides knowledge of the same deity as that of whom the prophets spoke. In the second case, Scripture furnishes a record of that former reality and serves as a pretext for glosses on different levels. The objective is not so much to understand the plain and obvious meaning of this text—because the main character of the epic can no longer be heard or seen—as it is to attach one or several meanings to current

126. Alexandre Matheron, *Le Christ et le Salut des ignorants chez Spinoza* (Paris: Aubier-Montaigne, 1971).

127. Spinoza deals with the status of the Apostles in chap. 11 of the *Tractatus*: were they prophets or rabbis? He concludes that they were prophets *of Christ* when they preached, with Christ in the role of the "mouth of God," that is, the physical means by which the word of God came to them. For the rest, however, they were rabbis or teachers; and, as theologians, already fallen into conflict over issues such as predestination and salvation by faith, as taught by Paul, versus salvation by works as taught by James. (Spinoza himself supported the latter position, if only for political reasons, as shown by the last chapters of the *Tractatus*.)

128. According to a report that Tschirnhaus gave Leibniz after a conversation with Spinoza (quoted in W. N. A. Klever, "Spinoza's Life and Works," in *The Cambridge Companion to Spinoza*, ed. Don Garrett [Cambridge and New York: Cambridge University Press, 1996], pp. 46–47), the latter told him that "most people's philosophy starts with creatures, Des Cartes started with the mind, he [Spinoza] starts with God." (It was during the same conversation that he spoke of "a sort of Pythagorical transmigration, namely that the mind goes from body to body" and added that "Christ is the very best philosopher.") But in the *Ethics* he starts from a different God: the idea of God/Nature (*Deus sive Natura*), both *natura naturans* and *natura naturata*, as he conceives of him adequately and understands him clearly and distinctly, thanks to reason and to knowledge of the third kind. This notion of God is given to all human beings, but most of them conceive of him in an inadequate, confused, and anthropomorphic fashion, including and especially the prophets, because their vivid imagination, the very faculty that makes it possible for them to be prophets, also clouds their intellect. Thus Spinoza begins not from God as he is known from the prophets but from the adequate idea of him he owes to his "true philosophy."

existence, as the continuation of the bygone past, to the extent possible, taking account of the inevitable break.[129] Spinoza, for his part, takes up the theologians' discourse in order to refute them—and notably one theologian whom he clearly studied in depth and who had confronted the same problem of the relationship between philosophy and prophecy. I mean, of course, Maimonides, the author of the *Guide of the Perplexed*, who prefaces his book with an address to readers that the young Spinoza cannot have been indifferent to.[130]

Spinoza begins by borrowing the categories of Maimonides' theory of prophecy, which, however, he modifies in two essential points. For both of them, a highly developed imaginative faculty is characteristic of prophets, magicians, dreamers, and visionaries, as well as statesmen. Developed intellectual faculties are the province of philosophers and scholars. For Maimonides, however, the scriptural prophets combined a strong intellectual capacity with their exceptional imaginative power and can consequently be deemed philosophers. Spinoza rejects this idea, with its implication that the prophets expounded philosophical truths in allegorical form. Furthermore, Maimonides assigned a unique status to Moses, whose prophecy had only the name in common with that of the other prophets. Moses "did not receive prophetic inspiration through the medium of the imaginative faculty, but directly through the intellect. . . . Moses did not, like other prophets, speak in similes."[131] He attained the highest form of intuitive theoretical knowledge possible for a human being and in this he was, even more than Aristotle, the very model of the philosopher.[132] Spinoza rejects this conception of Moses as an Aristotelian philosopher and replaces him with Jesus Christ the Spinozist philosopher. First he demolishes Maimonides' method of allegorical interpretation aimed at extracting philosophical truths from Scripture. To this end Spinoza invented the critical method

129. In the *Tractatus*, Spinoza develops a point of view that is original but not devoid of ambiguity. He pretends that the God who revealed himself to the prophets was the same as his own *Deus sive Natura*. The natural light of reason gives him an adequate idea of God, whereas the prophets had only a confused idea of him, clouded by imagination. His avowed objective is to demonstrate that the freedom to practice philosophy does no harm to religion, not because Scripture teaches philosophical truths but, on the contrary, because the goals pursued by prophets and philosophers are quite distinct: obedience to divine law for the former, comprehension of the true nature of God and of human beings for the others. "Theology is not bound to serve reason, nor reason theology" (*TTP*, chap. 15, p. 194). For this objective to be achieved, however, the reader, whether "philosopher" (*TTP*, preface, p. 11) or "prudent and rational Christian" (Letter 20 [71]), must be persuaded that these two paths, although different, nevertheless lead to salvation; in other words, that salvation by obedience is a possible path for those who are not philosophers—the unlettered. But if this solution is the same as that which philosophers achieve by means of rational knowledge and knowledge of the third kind, then prophetic revelation must share something, in its origins, with philosophical knowledge. There must have been an original unity of philosophy and prophecy before they diverged in their manifestations and goals, because they necessarily reveal the word of the same God and lead by different paths to the same beatitude and same eternal bliss, even though one group has the clear and distinct idea that accompanies the intellectual love of God and the others has only the confused ideas that are expressed by the prophetic texts and religious dogmas. This is the object of Spinoza's "moral certainty," which, in the absence of "mathematical certainty," relies on the authority of Scripture (*TTP*, chap. 15, p. 195). To ground this original and unique meeting, Spinoza retraces the path followed by Maimonides with the same objective, but totally deconstructs its method—at least in appearance—and displaces the original unity of prophecy and philosophy from Moses to Jesus, whom he always calls Christ or Jesus Christ.

130. In the introduction to the *Guide*, Maimonides analyzes several species of internal contradictions that may be present in a text. Some of these are intentional, surreptitiously introduced by the author to draw the attention of careful readers and allow them to grasp a point he did not want to state explicitly, because it was too difficult for the average reader to understand.

131. Maimonides, *Guide* 2.36 and 2.45; quotation from 2.36 (trans. Friedländer, p. 227).

132. See Shlomo Pines, "Spinoza's *Tractatus Theologicus-Politicus*, Maimonides, and Kant," *Collected Works of Shlomo Pines*, ed. W. Z. Harvey and M. Idel (Jerusalem: Magnes Press, 1997), vol. 5, pp. 660–711, esp. p. 670.

of reading the Bible—philological and historical analysis of the documents bequeathed by the prophetic tradition—and insisted that Scripture be understand only self-referentially. He has no trouble demonstrating that neither the Pentateuch nor the prophetic books can be read as philosophical texts without distorting them capriciously, a method that is unacceptable even when employed to make them agree with the dictates of reason. Remarkably, however, Spinoza employs precisely that method to place Jesus just where Maimonides located Moses. Maimonides, laboring mightily against the literal biblical text,[133] interprets the visions and voice of God revealed to Moses as allegorical descriptions of intellectual truths that God conveyed to Moses directly, without physical images or perception. Spinoza employs the same method to invent a Jesus Christ–Philosopher who teaches the universal religion through the Apostles. The latter occupy the role of prophets in relation to the discourses of Christ, which are equivalent precisely to the word of God, like the voice that Moses heard. As for Jesus Christ himself, he received the wisdom of God more than any other man, directly in his mind, by the intellect, and not through a physical voice as Moses had.[134] Jesus Christ serves as the voice or mouth to utter the word of God heard by the Apostles, who are the prophets of the universal religion that offers salvation to all human beings, even those who are not philosophers, whereas Moses and the prophets taught a religion suited only to the Hebrews.

8. THE SALVATION OF THE UNLETTERED IN THE *TRACTATUS THEOLOGICUS-POLITICUS*

This raises the question of the form of the *Tractatus Theologicus-Politicus* and the various contradictions or inconsistencies in it. One way to resolve these contradictions is to interpret Spinoza's text in a somewhat far-fetched way and assume that the author was just as aware of them as anyone else and left it for his readers to resolve them at their leisure. Alexandre Matheron does this with great flair when it comes to Spinoza's Christ and the salvation of the unlettered. Following Leo Strauss[135] and Shlomo Pines,[136] however, we can also view the contradictions as a way to prick the attention of alert readers and tell them, indirectly, what average readers might misunderstand were it stated explicitly. Maimonides says precisely this in the introduction to the *Guide*, where he devotes a long passage to several ways of assessing contradictions and alerts his readers to look for the contradictions he has subtly introduced into his treatise and to infer from them what the author could only hint at. Considering the role Spinoza assigned to Maimonides in the *Tractatus Theologicus-Politicus*, even if only to refute his core ideas, it seems at least plausible that he adopted the method of intentional contradictions suggested in the introduction to the *Guide*, just as he borrowed the form of the Maimonidean theory of prophecy in order to draw the opposite conclusions. Despite what one might think,

133. Maimonides, *Guide*, 1.35–36 and 1.46–48. We should remember Naḥmanides' vigorous objections to these interpretations. Like most kabbalists, he saw no need to do violence to the text in such matters, because he was not troubled by the use of bodily images to describe divine realities, meaning the *sefirot*. See Chapter 1; Atlan, *Enlightenment*, pp. 118–119; *SR* II, chap. 11.

134. See below, end of §13 and §14.

135. Leo Strauss, "How to Study Spinoza's *Theologico-Political Treatise*," in idem, *Persecution and the Art of Writing* (Glencoe, IL: Free Press, 1952), pp. 142–201. See also idem, "How to Begin to Study the *Guide of the Perplexed*," introduction to Moses Maimonides, *The Guide of the Perplexed*, trans. Shlomo Pines (Chicago: University of Chicago Press, 1963), pp. xi–lvi.

136. See Pines, "Spinoza's *Tractatus Theologicus-Politicus*."

this philosophical doublespeak is not dictated solely by politic caution. It aims to convey an unusual message more than to protect the author against criticism and condemnation. Neither Maimonides nor Spinoza hid his thought well enough to escape being banned. If we accept this hypothesis, it does not really matter whether Spinoza truly believed in the historical reality of Jesus Christ the Spinozist philosopher, any more than whether Maimonides believed in the historical reality of Moses and the prophets as Aristotelian philosophers. Each was trying, in his own fashion, to move a society still dominated by the confused recollection of the prophetic tradition to accept rational knowledge as a path to salvation. This is why the emblematic figure of the founder, traditionally considered to have a "divine" nature, whether Moses "the man of God" (*'ish ha-'elohim*) for the Jews, or Jesus Christ for the Christians, is also endowed with the attributes of the philosopher par excellence.

Matheron accepts the challenge of going as far as possible in the quest for consistency in Spinoza's various ways of referring to Jesus Christ and the simple dogmas of the universal religion he is supposed to have taught.[137] Belief in these dogmas, and joyful rather than compulsory obedience to the universal "divine law," which can be reduced to the practice of justice and charity—that is, to loving one's fellow as oneself—are the means by which the unlettered can achieve Spinozist salvation and beatitude. Matheron came to the conclusion that, for Spinoza, not only was Jesus Christ the most eminent Spinozist of all, he was even further advanced in knowledge of the third kind than Spinoza himself, inasmuch as he was capable of demonstrating "geometrically" that obedience to the universal religion—that is, the practice of justice and charity, received as the commandments of a God conceived in anthropomorphic fashion as a king who must be placated—guarantees the salvation of those who cannot truly know and love God with the intellectual love to which the empire of reason and knowledge of the third kind lead. Whereas Spinoza, to use his own terms, could have no more than a "moral certainty" about this, Jesus had mathematical certainty thereof. Matheron's strategy is to establish perfect coherence among the various expressions employed by Spinoza, in both the *Tractatus Theologicus-Politicus* and the *Ethics*, even though the language may support multiple interpretations, especially in the *Tractatus*. After eliminating several possible interpretations, Matheron reaches an astonishing conclusion that leads us, by an unexpected byway, to Salomon Maimon's notion of the kabbalah as expanded Spinozism, which we will consider below (§15).

137. *TTP*, chap. 14.

Matheron wonders how the idea that the unlettered can be saved by faith in an anthropomorphic deity and the obedience that follows from this faith is even possible in Spinoza's system, given that the author of the *Ethics* holds that philosophical knowledge, with the unconstrained virtue that accompanies it, is the only way to achieve salvation. Evidently we are to view such faith and obedience as a provisional moral code, a preparation for this knowledge, a form of teaching that can set one on the correct path. Spinoza seems to say as much in the *Ethics* when he states that "even if we did not know that our mind is eternal, we should still consider as of primary importance piety and religion, and generally all things which, in Part IV, we showed to be attributable to courage and high-mindedness" (V 41)—that is, before we attained the third kind of knowledge and the intellectual love of God. But this applies only to the philosopher who already tends to live under the sway of reason. These are "the dictates of reason" (IV 24) that can be employed to conceive of "a system [*ratio*] of right conduct, or fixed practical precepts [*dogmata*]" (V 10, Note). And these allow us to become accustomed to virtue, meaning conduct guided by the dictates of reason, by exercising our memory and imagination, because this is "the best we can do . . . so long as we do not possess a perfect knowledge of our emotions" (ibid.). All of this applies to the philosopher, and not to the unlettered who are targeted by religion, who cannot use their reason properly, whose ideas are for the most part inadequate, and who rarely go beyond knowledge of the first kind. Their entire life cannot bring them to Spinozist salvation, because that can be achieved only through knowledge of the third kind, which cannot do without knowledge of the second kind, attained through reason. How can Spinoza and Jesus Christ accept—one as a matter of "moral certainty," the other as a matter of "mathematical certainty"—that the unlettered can be saved and that obedience to religious precepts is the road that leads them there? The answer, according to Matheron, is that they seriously entertain the possibility of a Spinozist counterpart of the kabbalist theory of metempsychosis; the eternal essence of an individual's body comes into existence in response to external causes that make it exist for the limited duration of a human life, but it subsists as eternal when the body ceases to exist, after death. There is nothing to exclude that other external causes might bring it into existence several times, in several bodies, so that its progress along on the road to salvation advances bit by bit, over the course of several lifetimes. To support this unexpected conclusion Matheron cites, in addition to his own rigorous demonstration, Spinoza's remark to Tschirnhaus about the transmigration of souls, which the latter passed on to Leibniz along with other elements of the doctrine of the still-unpublished *Ethics*.[138]

And why not, after all? Transmigration of souls was in vogue in those years, to judge, inter alia, by the place it occupied for Leibniz's circle of friends and correspondents. There is a certain irony, nevertheless, in the fact that Leibniz himself, for all his openness to everything derived from occultism and especially from the "Kabbalah of the Hebrews," which he had learned from his friend van Helmont, refused to believe in metempsychosis. A strong defender of the latter's ideas, he nevertheless parted company with him on this point, which he saw as a possible source of conflict with the Church and its theologians.[139] By contrast, nothing in the eternal determinism of the *Ethics* rules out a Spinozist interpretation of the kabbalistic idea of the transmigration of souls as a means of *tikkun* or progressive "repair," from existence to existence, of the initial inadequacies of created finite beings who are unequally endowed with the capacity for knowledge that is indispensable for their freedom. We can also understand why, if this was truly part of Spinoza's doctrine, he was careful to keep it concealed, given the grave danger of magical or literal interpretations.

But another answer to the problem of the salvation of the unlettered, both simpler and more complicated, is stated explicitly in the last chapters of the *Tractatus Theologicus-Politicus*. There he explains the modality of obedience to the universal religion, that is, the specific content of the practice of justice and charity, otherwise known as the "divine law," not in the generality of the faith and beliefs that make it possible but in the details of its application. For Spinoza, who

138. See above, n. 128.

139. Coudert, *Leibniz and the Kabbalah*, pp. 56, 75, 81, and 84.

denied churches and theologians any authority in this domain, the political sovereign bears exclusive responsibility for interpreting and applying the universal religion in pursuit of the public weal. Persuading his readers of the wisdom of this model of theological and political authority—notably in light of the wars of religion that were tearing Christendom apart—was Spinoza's explicit objective in the *Tractatus*. Spinozist democracy must guarantee both the freedom to practice philosophy and religious freedom, meaning freedom to believe in one or another theological interpretation of the tenets of faith. But no church may impose its interpretation with regard to works, that is, the actual modalities of the practice of justice and charity so that the public good always takes precedence over the welfare of individuals. The salvation of both the unlettered and philosophers necessarily follows the path of obedience to the political sovereign in the practice of justice and charity within a Christian state, which is itself subject to the "catholic" universal religion. It was in order to persuade his "philosophical reader" (*TTP*, preface, p. 11) and "prudent and rational Christians" (Letter 20 [71]) that Spinoza concocted his image of Christ and his universal religion, which are equally compatible with philosophical knowledge and with the beliefs of the unlettered, although interpreted differently. Of course these different interpretations are incompatible as a matter of theory, because one is adequate and corresponds to the truth of the intellect, whereas the other is confused and the product of ignorance. But both may lead to the same practice of justice and charity, under a common political authority; works are the only road to salvation, whether they are inspired by religious obedience (a majority of citizens) or by philosophical knowledge (the small minority).

This brings us back to deciphering the doublespeak of the *Tractatus Theologicus-Politicus* and to the stance one takes as to its theoretical or political coherence, despite the apparent self-contradictions of Spinoza's doctrines. Matheron himself has to read behind the seven dogmas or postulates of universal religion, which Spinoza attributes to Jesus Christ–Philosopher, in order to demonstrate that the statement of belief in an anthropomorphic God who rewards, punishes, and forgives, adapted to the masses' capacity to understand, masks "neutral" terms in which the philosopher can be found. He is too cautious and too rigorous not to express his fear that, in this search for full theoretical consistency between the *Tractatus* and the *Ethics*, his proof is not wholly satisfactory. Similarly, with regard to the presentation of Jesus Christ as the greatest philosopher of all time, he considers in passing the possibility that Spinoza, in his own reading of Scripture, applied the allegorical method he denounced when employed by "theologians." If we keep in mind the Maimonidean model that Spinoza rejects, in order to extend to the "universal" Christian state the quasi-perfection of the ancient Hebrew state, within the limits of its particularism, we see that for Spinoza the figure of Jesus Christ plays the role that Maimonides ascribed to Moses. We can understand also that for the citizens of a Christian state obedience to the Christian "divine law" permits salvation, inasmuch as this obedience can be freely given and without constraint, as a matter of "devotion"—echoing Spinoza's depiction of the Hebrews' voluntary submission to the Law of Moses as the founding covenant of their theocratic state.[140] The problem he raises is the same as that posed by Maimonides: How one can read the prophetic Scriptures in a world from which prophecy has departed and in which the only possible knowledge is that provided by reason and sensory experience—a world in which

140. See above, n. 122.

Scripture can be understood only as an imperfect witness of a vanished experience of knowledge, an experience we can no longer have, so that we cannot even know the sequence of natural causes that made it possible? This is why, as Shlomo Pines maintains, Spinoza's criticism of Maimonides can deepen our understanding of Maimonides' thought and, in particular, its real or apparent weaknesses and inconsistencies.[141] This applies especially to the issue of prophecy as a natural phenomenon that we can accept as having existed in the past, even if we do not know the causes that produced it or those that made it vanish.[142] It is in this context that the problem of the interpretation of Scripture can be addressed to both of them. Spinoza departs from his predecessor even more strongly because the solution he wanted to offer (already quite heterodox for a Christian public) owes nothing to the particularism of the religion of the Hebrews, historically and theologically outdated, for himself as for his readers. Moreover, if the greatest philosopher, on whom Maimonides modeled his Moses, was Aristotle, this could not be the case for Spinoza in his own construal of Jesus Christ. The latter must have known "true philosophy"—which had to be Spinozism.

141. See Pines, "Spinoza's *Tractatus Theologicus-Politicus*," p. 660.

142. In several forums I have suggested that the ritual use of hallucinogenic plants played an important role in the origin of this phenomenon and the diverse forms of revelation of the sacred, shamanic and other, that can be recognized under other skies and in other civilizations. As for the disappearance of prophecy and mythical consciousness in general, it seems to correspond to a progressive but relatively rapid evolution of human knowledge and its capacity for interior dialogue, achieved through the internalization of the voices and images formerly interpreted as originating elsewhere (see above, §1, nn. 13 and 15; and Chapter 2, n. 70). Neither Maimonides nor Spinoza ever ventured this type of hypothesis; but the latter's critical reading of the former allows us it to ascribe to both of them a naturalist conception in which the word of God has an intellectual form for scholars and philosophers and a pictorial form for prophets. The Talmud describes the end of prophecy as a progressive phenomenon (B *Sotah* 48b). One of its first manifestations was the suspension, in the time of David and Solomon, of the oracular practice of inquiring of the Urim and Tummim. This practice was abandoned as soon as it was found to be ineffective. In addition, the disappearance of prophecy was a collective phenomenon; according to their colleagues, some of the talmudic sages had the capacity to be prophets—the *Shekhinah* or Divine Presence rested on them—but their generation was unworthy (ibid.). Later, only a few rabbis retained the capacity to "understand God," that is, to correctly interpret prophetic visions and voices.

9. THE PHARISEES AND THE BIRTH OF JUDAISM

Nevertheless, not only is this the same problem that Maimonides faced; the latter was merely rephrasing in Aristotelian terms an issue that, even before the birth of Christianity, had already confronted the Pharisee scholars, whose heirs, the "rabbis" (i.e., teachers), filled the vacuum left by the passing of the prophets and priests. Moreover, their practical solution to the problem of the salvation of the unlettered was the same and involves the law of the community. This has major implications for understanding the means of this salvation. If the path to salvation, in its collective dimension, rests on a social organization that promotes the salvation of individuals by enabling those endowed with the necessary capacities to access it, that path cannot be understood as one that *necessarily* leads to this goal. In this it differs from the royal road of philosophy, where adequate knowledge of God necessarily leads to the intellectual love of god and to "blessedness." Obedience to the law of the community, even when unforced, is a necessary but not sufficient condition. We shall return to this point.

Maimonides' problem with these teachers whose authority he was bound to accept also related to the "true philosophy" that was his reference point. For him this was Aristotle; but the talmudic sages had created their own schools in a more eclectic philosophical environment influenced by Stoicism and Neoplatonism, and even by Epicurus, and still imbued with Mazdaism and Gnosis.

The Pharisees, like the philosophers of antiquity but in their own way, had to break with the mythological way of thought they had inherited while maintaining their position as its heirs. Unlike most medieval philosophers and theologians, who explicitly rejected mythology—and all the more vigorously because it was pagan—the Pharisees proclaimed themselves the heirs of the Law of Moses and successors to the prophets, even as they established a new doctrine (the Oral Law) and new institutions to teach it. Thanks to their efforts, the original heritage was preserved so well that it could be transformed and adapted to the new objective reality in which rational and discursive thought occupied an ever-larger place in understanding the world and applying the law. As for Aristotle,[143] *sophia* and *boulê*, wisdom and cogitation (Hebrew *ḥokhmah* and *sevarah*), complement and eventually replace prophecy and divination as ways of knowing what can be known and of articulating moral law. The Pharisees invented the very concept of Sacred Writ, a term that is ambiguous because applied to a text that serves only as a formal reference, fixed once and for all time, for an interpretive discourse that is always in motion. The multiple meanings of this discourse—still known as the "Oral Law" even after it was committed to writing—intentionally overturn the fixity of any ultimate meaning of Sacred Writ, because the explicit content of prophetic revelation, the essential meaning behind the text, has vanished from the field of possible experience.

143. *Nicomachean Ethics* 1142b.

The history of the Jewish people cannot be understood if we fail to assess this transformation. We are so accustomed to viewing the "People of the Book" in a historic continuity that begins with Moses and the Exodus that we overlook the break represented by rabbinic Judaism. It was in fact a new religion, supplanting the ancient Hebrew creed that had carried on, sometimes for better and sometimes for worse, after the golden age of Solomon's kingdom and then survived the destruction of the state, first by Nebuchadnezzar and later by Vespasian. From this perspective, primitive Christianity and talmudic Judaism can be seen as rival responses to the institutional, social, cultural, and moral crisis that could have wiped out the legacy of the Hebrew civilization constituted around the law of Moses and the prophets of Israel. As Spinoza underscores, the Mosaic law was originally adapted to the Hebrews' sovereign national existence in the Land of Israel. Hence the destruction of their state undermined the sociocultural foundations of the Jewish people. By all rights they should have disappeared, like other peoples who lost their sovereignty. But the destruction of the Hebrew state, like that of every state, had both external and internal causes: military conquest and subjugation by more powerful empires, on the one hand, and the social, moral, and religious decadence of its institutions, royal and priestly, on the other. It could survive only through an institutional transformation that would provide new life, detached from that of a sovereign nation in its own land, to what could be perceived as the core of this heritage, in the new cultural and historical context.

But what was this essential core? What were the radically new conditions that impeded the continued normal transmission of this heritage?

The core was the civilizational model and the prophetic message embodied in the Mosaic law and the prophetic books. The new development was the end of prophecy as the unique source of knowledge underlying this message and model. Too often we forget this decisive human factor in the history of all ancient peoples: the end of the age when men and women had a special relationship with the godhead, however they referred to this link, and the dawn of the philosophical era in which the logos spoke only through the mouths of human beings themselves—the beginning of a period when, in Spinozist terms, the word of God could be heard and understood only in the natural light revealed to human thought and not in the light of prophetic revelation.

From this starting point, preserving the core of the heritage required an answer to the only question that still mattered: What elements of the prophets' teaching can be understood and retained in a world without prophets? Christianity's answer was the message of justice and love preached by the prophets of Israel, embodied in each clime by the formula appropriate to that society, "speaking as a Greek to the Greek, a Jew to the Jew," with a minimum of dogmas and beliefs to bond the faithful together, and notably belief in the divinity of Christ, the man-god and god-man with whom one can have communion. For the rabbinic Judaism introduced by the Pharisees, the answer to the same question was a legal code inspired by the Law of Moses, focused not on the organization of a sovereign state but on the practice of justice and charity by individuals joined in communities defined and constituted by adherence to this law.

In this project, the Pharisees arranged things so that the new structure they devised would not be perceived as disconnected from what had been destroyed, notably with regard to the relationship to the Law. As Francis Schmidt has shown, they kept alive "the system of thought of the Temple," an evident paradox, given the undoubted break with the Temple praxis. In particular, whereas the rituals conducted in churches seek to reproduce certain rites previously limited to the Temple in Jerusalem, delocalizing and imitating them—the vestments of priests and bishops are based on those of the High Priest in Jerusalem, the altar and incense are at least partially modeled on those of the Temple—there is no attempt, not even partial and approximate, to replicate the Temple service in the synagogue, which is, literally, a house of assembly (*beit keneset*) or house of study (*beit midrash*). In fact, this is no paradox at all, because it is precisely thought about the Temple, study and reflection about a practice that no longer takes place because it has no place:

> But the main point is elsewhere. There is a monument of inexpressible beauty round which is constructed the passage from before 70 to after 70, and which ensures the continuity of Judaism beyond the break of 70–135, in spite of all the setbacks: the Mishnah. Because in it are inscribed the Temple, its architecture, its grand liturgies. In this respect several treatises—such as the one entitled *Tamid* in the fifth order called *Qodašîn*–'Holy Things'—can be read as the parts of a *Liber Ritualis* of the Temple. That in the second century, when the fire of the altar is definitively extinguished and the Temple razed to the ground, the compilers of the Mishnah would judge it necessary to conserve these traditions and these ceremonial laws from now on cut off from every institutional anchorage, is a fact [that] has given rise to numerous questions.[144]

Some see this curious association of the minute study of vanished rituals with images, beyond time and space, that soon became legendary and miraculous, as the creation of a new "memory" oriented toward a projected restoration in a longed-for future. But things are not so

144. Francis Schmidt, *How the Temple Thinks: Identity and Social Cohesion in Ancient Judaism*, trans. J. Edward Crowley (Sheffield: Sheffield Academic Press, 2001), pp. 264–265.

simple, because that future is eschatological, itself beyond time, whereas the future longed for in history is pictured quite differently. The most striking example of this is the paradox of Maimonides' emphasis on the sacrificial rite. Eleven centuries after the destruction of the Temple and the suspension of sacrifices, Maimonides' commentary on the Mishnah and his subsequent mammoth compilation of Jewish law, the *Mishneh Torah*, go way beyond the norm in the time and effort devoted to codifying the various categories of sacrifices, as if they were offered regularly, treating them in the same detail as all the other Torah precepts whose practice and application was still the daily experience of the Diaspora communities. This same Maimonides, in his *Guide of the Perplexed*, held that the sacrificial ritual had been ordained as a sort of compromise with the generally accepted customs of the idolatrous peoples of antiquity.[145] We might suppose him to hold, then, that sacrifices would not be necessary in a different environment and will be even less important in the Messianic age, when knowledge of the true God will fill the earth. (Many rabbis disagreed.) But the paradox of Maimonides' attention to the practical details of the sacrifices in his legal code, as if they were commandments to be actually performed, is no paradox when we recall that rabbinic Judaism after the destruction of the Temple is defined by its attitude toward the *study* of the Temple ritual.[146]

145. In Maimonides' view, a primitive people, ignorant of philosophical truth, could not have understood and accepted the total suppression of the sacrificial cult by a Law whose chief objective is to induce the fear and love of God. Consequently the Torah had to prescribe sacrifices in minute detail, with a didactic end suited to the spirit of the age. But the ritual was permitted only within the Temple precincts and its meaning was displaced from the astral cults to worship of the true God, with rational knowledge of the deity as its objective.

146. Here we encounter the principle, expressed repeatedly, that study of the Torah, which is both law and wisdom, is the ultimate value. With regard to the several types of sacrifices (Heb. *ḳorbanot*, literally "what is brought near [to the altar]"), one of the most explicit talmudic formulations of this principle is the following:

> R. Isaac said, "What is meant by the verses, 'this is the law (*torat*) of the sin-offering' [Lev. 6:18] and 'this is the law of the guilt offering' [Lev. 7:1]? [They teach that] whoever engages in the study of the law of the sin offering is considered to have offered a sin offering, and whoever engages in the study of the law of the guilt offering is considered to have offered a guilt offering." (B *Menaḥot* 110a)

Maimonides takes up these words and drives the point home at the end of his commentary on that same tractate:

> Whoever studies the Torah is deemed to have offered a burnt offering, a meal offering, and a sin offering. The sages said with regard to scholars who engage in the study of the rules (*halakhot*) of the sacrificial service, that they are considered [to have merit] as if the Temple were built in their day. This is why it is appropriate for men to study the details of the sacrifices and to deliberate about them, rather than saying that these are matters for which there is no use in the present time, as most people do. (commentary on M *Menaḥot* 13:11)

As for the physical rebuilding of the Temple, as a component of the return to Zion always hoped for and sometimes attempted, Rashi explicitly notes its eschatological and ahistorical character in his comment on a talmudic passage that deals with the modalities of this reconstruction (B *Sukkah* 41a). He writes that there can be no advance plans for the construction of the Third Temple, neither political nor architectural, because, when the time comes, it will descend from heaven fully built. The Third Temple will not be identical in every architectural detail to the Second Temple, just as that was not an exact copy of the First Temple. But in the absence of prophets, who alone could "see" and prescribe the particulars of the design, no planning is possible and the Third Temple must descend from heaven, where it is already assembled and waiting. Most later rabbinical commentators accept Rashi's interpretation. Here we can see, along with Schmidt, "the projection of an idealized towards an ideal future" (Schmidt, *How the Temple Thinks*, p. 265).

This idea is invaluable today for averting the messianic and political delusions that urge construction of the Third Temple in Jerusalem without further delay. It teaches us that "human hands"—political planning—must not be involved in the realization in history of the Messianic aspiration. That must remain in the "hands of YHWH," as stated by the biblical verse quoted by Rashi in his commentary on B *Sukkah* 41a: "You made a place for your abode, YHWH, your hands established a sanctuary, Adonai" (Exod. 15:17; see also Rashi on this verse).

There are some people today who claim Maimonides as an advocate of the opposing view, namely, that the religious obligation to build the Temple is always in force, if the political circumstances and balance of power permit it. According to this thesis, the idea that the Temple will descend from heaven fully built is merely one possibility and refers in any case

This indeed constitutes a new system. In the Mishnah,

> the great work of the Sages from the end of the Second Temple, . . . there is codified and transmitted the "new system" that makes the thinking of the Temple go out beyond the very boundaries of the Temple. After 70, from being sacrificial Jewish society is transformed into a non-sacrificial society. But the thinking of the Temple inscribed in the Mishnah, normative, commented on in the Talmuds, studied in the Houses of Study, applied to all the actions of daily life, has at the same time the power to ensure the everlastingness of Judaism from the time before to the hereafter of the Temple.[147]

In Chapter 5, when we turn to the oracular practices of the Temple and assess their value for a world without prophecy, a world in which the random is no longer sacred and the lottery can no longer serve as a means of divination, we will consider an example of this strategy by which the talmudic sages maintained a theoretical continuity with the Temple mode of thought while totally breaking with it in practice.

In the two new and henceforth rival systems, apostolic Christianity and rabbinic Judaism, the prophetic message of justice and charity remained the path to salvation, in this world and in eternity, but was understood in different ways. For one, it is salvation by faith and by obedience to what Spinoza calls the divine law or universal religion, without specifying the means by which this obedience becomes joyous and unconstrained (yet is still effective), an indispensable condition if it is to lead to salvation, that is, to happiness and blessedness. For the other, the path of salvation is obedience—here too ideally joyous and unconstrained—to a divine law that is potentially universal but is concretely defined in a particular process. The study of this law as the path of salvation is its most important obligation and makes it possible for all to reach the goal, to the extent that their natural capacity and circumstances of life permit.

10. THE BOOK OF ESTHER

One way to measure the magnitude of the transformation is to analyze the fascinating though relatively little known role that the talmudic sages assigned to the story of Esther, and the book that recounts it, in traditional Jewish thought.

The story is set "in the days of Ahasuerus," one of the successors of Cyrus on the throne of the Persian Empire, where the descendents of the tribes of Judah and Benjamin, exiled after the destruction of the First Temple by Nebuchadnezzar, lived in scattered communities. After his conquest and annexation of the Babylonian Empire, Cyrus authorized the exiles to return home and rebuild the Temple. This period—the last years of the First Temple, the Babylonian exile, and the return to Zion and construction of the Second Temple—is also the time of the last prophets and the great transformation. This culminated, several centuries later, in the synagogue, the rabbinic institution that came to be the center of Jewish life in the Diaspora, in place of the priestly Temple, which had been the heart of the ancient Hebrew state. The book of Esther is the first biblical

only to "the heavenly Jerusalem." Maimonides' authority can be invoked to support this thesis and the radical theological and political projects derived from it because Maimonides included the obligation to rebuild the Temple, along with the injunction to offer sacrifices, in his enumeration of the 613 commandments. But this is tragically to forget that, for Maimonides perhaps more than for any other commentator and codifier, studying the Temple ritual is tantamount to performing it, satisfies the obligation to do so, and may even have greater value.

147. Schmidt, *How the Temple Thinks*, pp. 265–266.

text that applies to the descendants of the survivors of that state—the kingdom of Judah (Hebrew *yehudah*)—the global designation *yehudim*, originally "Judahites" but now also "Jews," including those who are not members of the tribe of Judah. The transformation was completed following the Romans' destruction of the Second Temple. During the siege of Jerusalem that preceded the destruction and the ensuing centuries of exile, a leading Pharisee sage, Rabbi Johanan ben Zakkai, negotiated his own escape from the city and surrender to the Romans in return for the establishment of a house of study in the small town of Yavne, located between Jerusalem and the Mediterranean. This founding act of Judaism as we know it today, distinct from ancient "Hebraism," continues the story that began several centuries earlier "in the days of Ahasuerus."

On the surface, the plot of the book of Esther concerns something quite different. It centers on the plan to exterminate all the Jews in the Persian Empire, concocted by Ahasuerus' chief minister, one Haman, who is motivated both by a lust for their wealth and a desire for personal revenge against Mordecai, a prominent hanger-on at the royal court. The pretext exploited by Haman to persuade the king is that the Jews are a refractory people with their own peculiar customs, different from those of the rest of his subjects. Ultimately, however, his wicked design is frustrated and converted into the Jews' triumph over their enemies, thanks to Queen Esther, whom Ahasuerus had married following a "beauty contest" and who had concealed her religion from him.

The first thing we observe is that the narrative prefigures a process that would be repeated again and again during the course of Jewish history in the Diaspora. But there is more. The book enjoys unique status in the ritual tradition instituted by the rabbis as the only biblical book, aside from the five that constitute the Pentateuch or Torah, that must be written on a parchment scroll for public reading. Whereas the ritual reading of the Torah is carried out section by section, each week of the year, the reading of the book of Esther is one of the obligations associated with the festival of Purim, a sort of Jewish carnival, which celebrates the events of this story and their happy conclusion. Thus the book of Esther, along with Pentateuch, is Sacred Scripture par excellence. It is meant to be read aloud and in public, but physical contact with it "renders the hands impure." Its sacred character places as it were a taboo on the scroll in which it is scrupulously copied word for word. By definition, the Sacred Scriptures render hands that touch the bare scroll impure.[148] The talmudic tractate *Megillah* reports the Sages' deliberations about Esther's inclusion in the biblical canon. R. Judah states, in the name of his teacher Samuel, that the book of Esther is not sacred as a "writing," but only as a "reading." Consequently "it does not render the hands impure" and does not have to be covered by a case or wrapper. Several generations earlier, however, Rabbi Simeon had held that the book of Esther, like Ruth and the Song of Songs, "renders the hands impure."[149] In any case, it is the only book whose public reading is conducted according to a ritual modeled on the public reading of the Torah. The Talmud contrasts "Sacred Scripture" to the wisdom books, notably Ecclesiastes, which "does not make the hands impure because it is the wisdom of Solomon."[150]

148. B *Shabbat* 14a. The very notion and term "Sacred Scriptures" (*kitvei ḳodesh*) seems to have originated in this context. Originally there was only the Law of Moses or the book of this or that prophet. The idea of Sacred Scriptures (or, more commonly in Hebrew, the "Reading" [*miḳra'*]) first appeared with the definition of the biblical canon, which differentiates it from texts or oral discourses that could be collected and written down but do not have the status of Scripture, fixed once and for all time.

149. B *Megillah* 7a.

150. Ibid. <I.e., it is the product of human contemplation rather than divinely inspired.>

The observances of the Purim festival and the public reading of the *megillah*, the "scroll" of Esther, are analyzed and discussed in the talmudic tractate of the same name, *Megillah*. In the rabbinic concept of Sacred Scripture, as contrasted to the Oral Law (meaning the exegetical tradition recorded in the Talmud), all the other books of the Bible—the Prophets and the rest of the Hagiographa—have a status intermediate between Scripture and received tradition. Originally, only the Five Books of Moses had the status of written law and could serve as the formal reference and source of legitimacy of their own interpretation. In some fashion, however, the rabbis elevated the book of Esther to almost the same plane of Sacred Scripture par excellence. Although the written Torah retained its status as the constitutive legal code, the book of Esther was made into the founding narrative of the new "Jewish" reality and the new religion, "Judaism," and marked their entry onto the stage of history. What is so special about this book, as distinct from all the other books of the Bible, that it rates it such a status? Those others texts, too, served as sources of inspiration, through the constant reading and study, both and public and private, instituted by the Pharisees as the Oral Law at the heart of the social organization of Jewish communities.

Reading the book of Esther makes its idiosyncratic quality readily evident: no name of God appears in it. At the same time, it seems to be a sort of archetype or paradigm of the next two millennia and more, until the twentieth century, of the history of the scattered Jewish communities: "a certain people, scattered and dispersed among the other peoples in all the provinces of your realm, whose laws are different from those of any other people and who do not obey the king's laws" (Esth. 3:8) and who are fit only to be annihilated and despoiled of their wealth. The plot comes within an inch of success; then the cycle starts all over again, with communities that are once again flourishing but dependent on the sovereign who shelters them, until a new crisis threatens their destruction—time and again down through the generations. Paradoxically, in the book of Esther as in later cases, the Jews are accused of separatism, even though they are already far advanced on the road to assimilation. One indication of this assimilation, among others reported in the books of Ezra and Nehemiah (which also relate to the Persian period), is Esther's own name, derived from the pagan goddess Astarte or Ishtar, and given her as a matter of course to replace her Hebrew name, Hadassah.[151] Of course, persecution itself impedes assimilation, which merely starts the process all over again.

The Pharisees presented the account of these peripateias as a historical book, the foundation of a Sacred Scripture without God, in which the word of God can neither be seen nor heard, alongside the Prophetic books that convey the core of the foundation myths. There is not a single mention in it of the main actor of the Pentateuch, designated by several names, both singular and plural, that theologian-translators have rendered by various designations for "God," according to their conception of him in terms of the Greek *theos* and the Latin *deus*, inherited from Greco-Roman mythology.

This peculiarity of the book of Esther is explicitly noted in the Talmud,[152] where the rabbis justify the radical departure from all the rituals prescribed by the Law of

151. Esther 2:7; B *Megillah* 13a.

152. B *Ḥullin* 139b. See also B *Yoma* 29a: "Esther is the end of all miracles"—more exactly, the last of those recounted by Sacred Scripture. Hanukkah, instituted to commemorate the Maccabees' victory several centuries later, also involves a miracle (a small cruse of oil that kept the Temple candelabrum lit for eight days, although in nature its contents should have sufficed for only a single day). But, concludes the Talmud, the story of Hanukkah was not meant to be written; unlike Esther, it was not incorporated in the biblical canon.

Moses with the observation that Moses himself had foreseen a time when God would "hide his face." Applying the sort of wordplay characteristic of talmudic exegesis, the book of Esther is offered as the fulfillment of this prophecy, in both content and form, inasmuch as the name Esther is very close to the Hebrew for "I will hide."[153]

The talmudic sages' institution of Esther as Sacred Scripture, whose reading is formally on a level that makes it resemble the Pentateuch, is one of the strongest symbols of the import of their enterprise. In keeping with their usual practice, they assigned the new ritual obligation to read the Megillah in public, like so many others, to the oral tradition conveyed "to Moses from Sinai." Because the Law of Moses prescribes that nothing be added to or subtracted from it, it was conventionally allowed that all of the sages' exegetical innovations, past, present, and future, had been transmitted to Moses from Sinai; although, the sages admitted, he might sometimes have been hard put to recognize them.[154]

Their project saw itself as the continuation of the ancient story of Israel, recounted in the written Torah, but creating the new and different content of the Oral Torah, suited to a world in which prophets and their god(s) were no more than a confused memory. As part of this effort the sages coined new terms to designate the hidden God of whom they now had to speak: the Place (*ha-maḳom*) and the "Sacred blessed is he" (*ha-ḳadosh barukh hu'*, normally rendered as "the Holy One, blessed be He"), in the Talmud and midrashic literature; the "Infinite" (*'ein sof*) or the "Infinite that is blessed" (*'ein sof barukh hu'*), the "Name" (*ha-shem*) or the "glorious Name" (*ha-shem ha-nikhbad*) or, simpler still and more radical, "He, may he be blessed" (*hu' yitbarakh*), in the kabbalistic literature; not to forget the feminine noun *Shekhinah* 'divine presence', formerly visible, now exiled, hidden and scattered inside individuals, so that whether or not "She" and "He" are united depends on human action.

As in the story of Esther, the initial experience of what can be lived and understood—Spinoza's "given true idea"—can no longer be that of God and divine revelation to the prophets. In the best case, reflection and practice may perhaps lead to knowledge of this hidden God by some other path, on which knowledge of God is confounded with knowledge of nature and where, as for Maimonides and the kabbalists we have mentioned—as well as for Spinoza in the *Ethics*—knowledge of good and evil is confounded with that of the true and the false and with the experience of the glorious life of eternal blessedness. This is the program launched by the Pharisee sages who, in opposition to the Sadducee priests, broke with the idea that the Mosaic law was valid only for regulation of national sovereignty. They insisted, rather, that it is also a means of salvation, both individual and collective.[155] This

153. Deuteronomy 31:18. <The verbs in the Hebrew text there, *hastēr 'astīr* 'I will surely hide', are almost homophonous with *'estēr* 'Esther'.>

154. A talmudic legend (B *Menaḥot* 29) depicts Moses sitting in the back row at a lecture by Rabbi Akiva, totally at sea. At the end of the discourse, to his astonishment, he hears the teacher affirm that all these unfamiliar notions were transmitted "to Moses from Sinai" (see Atlan, *SR* II, chap. 12). The same principle is applied here: "The Holy One, blessed be He, showed Moses [on Sinai] the minutiae of the Torah, and the minutiae of the Scribes, and the innovations that would be introduced by the Scribes; and what are these? The reading of the Megillah" (B *Megillah* 19b).

155. See the long debates reported in the last chapter of Tractate *Sanhedrin* about whether or not the Torah can lead to salvation. The sages (Pharisees) compete to show off their exegetical skill, capping biblical verses that allude to eternal salvation. Despite their attempt to show that the Torah promises "a share in the world to come" and the "resurrection of the dead," their opponents have no trouble demonstrating the novelty and arbitrary nature of these interpretations for those who hold to the plain sense of the biblical text. The Pharisees can only suggest the poverty of a

necessarily includes the possibility of the salvation of the unlettered, without being limited to it. Salvation through obedience to the law, illuminated by study based on reason and tradition, leaves open the possibility that the unlettered (the *'am ha'areẓ* or "people of the land") may gradually achieve the status of the sage (*ḥakham*)—whence the designation *talmid-ḥakham* or "student-sage" for one who devotes the major part of his life to searching for truth by means of such study.

The attitude of the Pharisee sages is thus just the opposite of what Spinoza imputes to the Jews, namely, that if they "were at a loss to understand any phenomenon, or were ignorant of its cause, they referred it to God" and never to any mediate or particular cause (*TTP*, chap. 1, p. 21).[156] The special status they accorded to the book of Esther indicates, on the contrary, that they perceived the political history in which they were caught up as a sequence of particular causes in which, unlike the prophetic narratives, the invocation of God makes no contribution to explaining or describing the occurrence of some event. Indeed, as we shall see in Chapter 6, with regard to the unusual use of the word *ḳeri* in the sense of "chance" and its curious linkage to God in the Bible, it is the meaning that human beings attach (or do not attach) to it that makes it possible for their history to be seen as guided or as merely a succession of random happenings. Human history, of which the book of Esther is a paradigmatic narrative, is merely a sequence of specific causes and effects that engage the passions of men and women. The fabric of history, its "plot," is shaped by Eros or desire, which powers the three main engines: the allures of glory and power, of money, and, of course, of sex. Spinoza's remark takes on added significance when we realize that it is borrowed, word for word, from Maimonides,[157] who employed it to justify his allegorical reading of the biblical text and attributed this propensity to assign everything directly to God to the form and style of the prophetic text. Note that the same applies to all myths, in which human beings are manipulated by gods or supernatural heroes who possess them. Spinoza reverses Maimonides' observation by assigning this tendency to the Jews in general, as if their excessive piety deprived them of the capacity to employ their reason. He also dismisses the Jews of the time of St. Paul and the Apostles as having "despised philosophy" (*TTP*, chap. 11, p. 164). To be logically consistent he must have excluded from this generalization at least St. Paul himself, whom he accuses of excessive (and poor) philosophizing. We may surmise that he also excluded Maimonides, inasmuch as he does him the honor of stopping to examine and criticize his work. Curiously, he refers to the twelfth-century Maimonides as "the first among the Pharisees who openly maintained that Scripture should be made to agree with reason" (*TTP*, chap. 15, p. 190). Jumping over eleven centuries, he forgets Philo of Alexandria (whom he quotes elsewhere) and the talmudic sages who were well-versed in Greek philosophy and science, even if they demurred at it.[158] As for the ancient Pharisees, whose

Torah reduced to a national code and its lack of suitability to the new world that is emerging under their eyes. "You have falsified your Torah yet it has availed you nothing," says R. Eliezer ben R. Jose to the Samaritans, who, like the Sadducees, denied that the "resurrection of the dead [can be proven] from the Torah" (B *Sanhedrin* 90b).

156. Spinoza offered a similar explanation in his *Compendium of Hebrew Grammar*, but without the pejorative connotation, of a verb form, peculiar to Hebrew, that "relates an action to its main cause." This is the *hiphil*, which is a causative or indicates that something fulfills its function. There he does not speak of religious devotion; nor is God necessarily and directly the main cause that makes things happen. Rather, the principal cause may fit into the nexus of mediate and particular causes.

157. *Guide of the Perplexed* 2.48.

158. For example, B *Pesaḥim* 94b; see Atlan, *Enlightenment*, pp. 265–271.

milieu produced St. Paul (the student of Rabbi Gamaliel), Spinoza always treats them with the conventional disdain, although, as we have seen, they were among the first to confront the need to transmute the prophetic discourse in which all comes from God into a dialogue of wisdom and intelligence.

In Chapter 5 we will analyze one element of this transformation, which involves precisely the relationship between chance and destiny. The oracular chance of the lottery, through which some effect worked by the deity can be known, is replaced by the chance of fortune, meaningless and indifferent, first introduced in the story of Esther, where the date of the planned massacre is chosen by lot; whence the name Purim ("lots") given to this festival, which is the paradigm for a story produced at random by absurd and grotesque events, in which everything is turned into farce, a catastrophe can be the occasion for deliverance, and good and evil are muddled in inebriation and carnival masks. Just as the "God-less" scroll of Esther is the twin of the Torah scroll, in which God's names are ubiquitous, so too the Purim lottery twins the lottery of the Day of Atonement, by which the High Priest, officiating in the Temple, designated the scapegoat before his once-yearly penetration of the Holy of Holies. We might say that, after the destruction of the Temple and suspension of the ritual, the new history, without a visible God, without prophet or augur, without enchantment or spells, was instituted under the sign of the tragicomic farce of Purim.

II. SALVATION FOR ALL AND A PROVISIONAL MORAL CODE

In reply to a potential convert who demanded that he summarize the entire Torah while standing on one foot, Hillel, the head of one of the two major Pharisee schools, stated the famous injunction that we must not do to another what we would not like to have done to ourselves. Then he added: "As for the rest, go and learn."[159] In fact, if the divine law can be reduced to this Golden Rule of universal morality—or, as Spinoza says, to the practice of justice and charity—we still face the question of the path to be taken in order to arrive there: that is, of the rules of conduct most appropriate for guiding the unlettered on the path of salvation by study, in the absence of prophetic visions. For, "since the Temple was destroyed, prophecy has been taken from prophets and given to fools and children."[160] In other words, it has not vanished completely but now serves as a memory of a magical but irrational world, that of childhood or of unbridled passion and imagination. We all start out as children, if not mad, but we can stay that way for only a limited period, longer or shorter, and to a greater or lesser extent. The doctrine of salvation through knowledge must take into account the fact that everyone begins in ignorance and that this salvation is played out for everyone within the limits of finite existence. It is in this respect that the wise man and the fool share a common destiny, as Solomon observes in Ecclesiastes.

On the other hand, knowing that one is ignorant is already the first step on the road to knowledge. The philosopher knows that his ignorance is inversely proportional to his capacity to assess what he does not know. Thus the salvation of the "ignorant" is the salvation of each and every human being, inasmuch as we are always only relatively wise and relatively ignorant. Here the double language of the philosopher-teacher is justified internally, as it were, independent of any prudent interest of blame or dissembling. Esoteric language that can be understood differently by

159. B *Shabbat* 41a.
160. B *Bava batra* 12b.

the wise man and by the unlettered has a pedagogical function, in that makes it possible for all to advance from their present situation, from what they already know, even confusedly, toward what philosophy and the experience of wisdom may perhaps bring them to know.[161]

The rabbis, engaged in their project of reorganizing Jewish society, faced a double constraint: the end of prophecy (and the consequent need to replace it with knowledge and wisdom) and the loss of political sovereignty. A social organization oriented toward the salvation of individuals, of both the wise and the unlettered, living under the laws of other sovereign states, must accept this sovereignty, even though imperfect and inadequate, in accordance with the talmudic principle that "the law of the sovereign authority is the law."[162] Here the ancient national sovereignty survives only as a memory that is both historical and mythical: historical, because it refers to a past that really existed; mythical, because the memory of that past refers to the mythological age of origins when prophetic revelation gave meaning to life, because God allowed himself to be seen and heard.[163] This leads to a doctrine of salvation, simultaneously individual and collective, promoted by a new ritual that forges an imaginary reality to be superimposed on the daily experience of a world that, although devoid of prophecy, remains potentially intelligible, and that each individual discovers in keeping with his or her abilities, with varying speed and varying clarity, at the end of childhood. The provisional moral code instituted in this way orders the social and individual rules and norms that guide human desire toward greater freedom instead of toward more onerous servitude. Even if we are willing and determined to follow Hillel's dictum, at the outset we do not necessarily know what is good and what is bad and cannot be certain what we would or would not want to have done to us. We know very well, through our memories of past experience and through imaginative projection, that something good in the present, here and now, may be evil in the future, and that, conversely, present evil may yield future good. Consequently, what we would not want others to do to us, because of our past experiences, may not be as repugnant for

161. In Proverbs (26:4–5), King Solomon advises, first, "answer not a fool according to his folly, lest you be like him yourself," and then, seeming to contradict himself, "Answer a fool according to his folly, lest he be wise in his own eyes." The Talmud (B *Shabbat* 30b) resolves the paradox with the notion that one need not answer a fool according to his folly unless his question involves Torah. As we saw above (Chapter 2, n. 112), Rabban Gamaliel applied this principle using ambiguous terms, in order to instruct a questioner in keeping with his intellectual capacity. Note that this same page of the Talmud contains the rabbinic discussion, which Spinoza cites explicitly, along with the talmudic reference, about the inclusion of Ecclesiastes and Proverbs in the biblical canon (*TTP*, chap. 10, p. 155).

162. <B *Gittin* 10b et passim.>

163. This is how we should understand the otherwise astonishing statement "whoever lives in the Land of Israel may be considered to have a god, but whoever lives outside the Land may be regarded as one who has no god" (B *Ketubbot* 110b). Clearly the reference is neither to the god of the philosophers nor the god of the theologians, but rather to the god of the Bible, or, more precisely, to how he is remembered, as associated with the land of the ancient Hebrews and their social organization. The talmudic passage underscores this link by means of a verse from Leviticus that states explicitly, as the peroration of a collection of statutes, "I am YHWH your god, who brought you out of the land of Egypt to give you the land of Canaan [so as] to be your god(s)" (Lev. 25:38). Rashi quotes a version of the midrash, only slightly weaker, that paraphrases the verse as follows: "I am god for those who live in the Land of Israel. Anyone who leaves is considered to worship idols." Today we see only too well what religious nationalism can do with such statements, lifted from their context, forgetting that they apply in particular to major Jewish communities whose greatest teachers were accordingly, like the others, "godless" or "idolators." Not to mention Moses himself, who, as is known, never entered the land of Israel and thus never had concrete experience of his own laws, which were meant to regulate the society in that land. The commentary on the verse is perfectly consistent with the fact that it comes at the end of a pericope that contains the laws of the sabbatical year, the Jubilee, the emancipation of slaves, the prohibition against lending at interest, and so on, all of which are part of that social legislation.

someone else whose formative experiences were different. In brief, the law of what we want is always ambiguous. Our past experiences make it impossible to content ourselves (in the manner of animals and infants) with the pursuit of immediate satisfaction as if that were the same as the search for pleasure and flight from pain. We know that postponing pleasure may avoid a greater pain and that, conversely, accepting pain may be the means toward greater pleasure. On the other hand, the satisfaction of desire cannot be postponed in all circumstances and indefinitely. When, how, and to what extent one should defer the satisfaction of desire constitutes a strategy that is usually expressed in a set of rules and standards we have designated the "second level of ethics"—the first being the almost instinctive quest for immediate pleasure and the equally automatic flight from all present pain.[164] This set of rules is most often a product of the history, geography, and language of each society, that is, of its culture in the broad sense, or what Cornelius Castoriadis called the "social imaginary." So even if we allow that the Golden Rule is universal, it is empty when not accompanied by its modes of application, as invented and instituted by each society, based on its own history and specific cultural, linguistic, religious, legal, and political traditions. To put this another way and speak like Spinoza, if the divine law and religion taught by Christ and the prophets are universal, their application necessarily requires some organization of the political sovereignty that regulates the concrete life of each nation. The political sovereign must enact legislation that stipulates how to love one's fellow as oneself and how to practice justice and charity, in a manner best suited to attaining the public weal. Spinoza holds that all must submit to it, both philosophers and fools, even if, for good or bad reasons, the rules decreed by the sovereign are deemed to be flawed.[165]

This social and collective aspect of the *Ethics* introduces additional problems that bring the question of the salvation of the unlettered back to the surface. The philosopher, living under the empire of reason, can understand that it is better to obey the laws of his city than to defy them, even if he deems these laws to be bad. This adequate knowledge of the social reality in which he finds himself living makes him know and love God along with the sometimes unbearable conditions—disease, frustration, and so on—of human existence. But what about the unlettered? How can they can find their salvation in obedience to divine law if that requires submission to a restrictive and oppressive sovereign and is the source of grief more than of joy? How can a forced and joyless obedience be related to Spinozist salvation, which is happiness?

To resolve these problems, we must closely examine the status Spinoza assigns to a provisional moral code that applies equally to philosopher and fool. Note, to begin with, that the last two propositions of the *Ethics* deal precisely with this provisional moral code. The conclusion seems to draw back from what came before. *Ethics* V 41, quoted above, which deals with piety and religion, returns to the subject of the initial propositions of Part V, before we have encountered the mind that knows itself to be eternal and the third kind of knowledge. It is with regard to this question that Spinoza, in the notes to the last two propositions (41 and 42), paraphrases Solomon's dictum (Eccles. 2:13) that wisdom is better than folly and explicitly contrasts the modes employed by the wise man and the fool to advance toward blessedness. Fools, unlike the wise, are crushed by piety and religion, which they experience as a burden and form of slavery; they can

164. See Atlan, *SR* II, chap. 2.
165. See his insistence on this point, *Tractatus*, chaps. 19 and 20.

conceive of salvation only after death, as the "reward for their bondage, that is, for their piety, and religion." But Spinoza employs these very same words, piety and religion, with regard to both the wise and fools. He has already furnished precise definitions of these terms, in keeping with the rule of his double language and the geometrical method of the *Ethics*: "Whatsoever we desire and do, whereof we are the cause in so far as we possess the idea of God, or know God, I set down to *Religion*. The desire of well-doing, which is engendered by a life according to reason, I call *piety*" (*Ethics* IV 37, Note 1). But the content of piety and religion, as a provisional moral code for the philosopher embarked on the road to wisdom, is not the same as it is for the unlettered. For the former, "so long as we are not assailed by emotions contrary to our nature, we have the power of arranging and associating the modifications of our body according to the intellectual order" (*Ethics* V 10). This reinforces "the mind's power, whereby it endeavours to understand things" (ibid., Proof). This doctrine instructs us "to frame a system of right conduct, or fixed practical precepts, to commit it to memory, and to apply it forthwith to the particular circumstances which now and again meet us in life, so that our imagination may become fully imbued therewith, and that it may be always ready to our hand" (ibid., Note).[166] As a result, we can conceive and experience little by little that "blessedness is not the reward of virtue, but virtue itself" (*Ethics* V 42). In this way the philosopher learns to "rejoice in this divine love or blessedness," because he increases in wisdom and is "conscious of himself, and of God, and of things, by a certain eternal necessity" (ibid., Note), which we can experience through the second, rational kind of knowledge. By contrast, the ignorant man, too, requires a doctrine of piety and religion, but he can conceive of it only confusedly, as passive submission, motivated by hope and fear, living "as it were unwitting of himself, and of God, and of things" (ibid.).

In other words, it is by knowledge of what he does, adequate comprehension of the causes that determine him, and the true satisfaction of the soul that flows from it that the wise man is superior to the ignorant, even if both can be guided to the same works on the road of salvation, namely, love of one's fellow, justice, and charity—as long as these works are carried out under the tutelage of the political authorities. For we must not forget, as the last chapters of the *Tractatus Theologicus-Politicus* indicate, that in every particular situation such actions always entail (though for different reasons) submission to the universal religion as it is concretely interpreted by the political authorities in pursuit (at least in principle) of the collective welfare. Although all are free to interpret their faith as they may, depending on their temperament, there is no question of leaving the specific content of these actions to be defined by churches, which are in any case divided and blinded by their theological aberrations.

All of this means that the provisional moral code to which they submit does not guarantee salvation, neither for philosophers nor for the ignorant. Philosophers achieve it only if their preparation is effective and enables them to advance beyond the second kind of knowledge to

166. To understand how this provisional moral code is fully integrated into Spinoza's ontology and epistemology and not a mere veneer or appendage, we may refer to Nicolas Israel, who shows how vigilance and presence of mind make it possible for the sequence of rational thought to penetrate duration and the mechanism of the passions, at certain opportune moments or occasions for action. See: Nicolas Israel, "La présence d'esprit: *Ethique* V, 10, sc.," *Philosophique* (1) (1998), pp. 79–97; idem, *Spinoza: le temps de la vigilance* (Paris: Payot, 2001); idem, "La distinction entre la durée et le temps dans l'œuvre de Spinoza," Ph.D. dissertation, Université Paris I Panthéon-Sorbonne, 1998.

the third kind and the intellectual love of God. There is no assurance that this difficult journey will be successful. The case of the unlettered is even more difficult. Their submission does not ensure happiness, neither in this life nor in some future reincarnation (if we accept Matheron's interpretation).

For Spinoza, then, a provisional moral code is a necessary but not sufficient condition for achieving salvation. Consequently it is no longer indispensable to search for the mechanism by which a provisional moral code, whatever it may be—and notably obedience to religion—can guarantee Spinoza's form of individual salvation. In fact, it does not. It is merely a prerequisite. Spinozist salvation, attainable only by means of the third kind of knowledge, will be achieved (or not) as a function of whether additional enabling conditions are met. There is no certainty of happiness or Spinozist blessedness, neither for philosophers who live under the empire of reason nor for the ignorant, whether in one life or in several. The hypothesis that Spinoza believed in the transmigration of souls is not required to make his theory of the salvation of the unlettered work, if we accept that the theory states only the necessary conditions for salvation and not a mechanism that necessarily leads to it.

What is more, the collective aspect implied by submission to political authority, whatever its nature, makes it possible to conceive of a different side of Spinozist salvation. There is another way, in addition to (and not necessarily incompatible with) reincarnation, for individuals to be "saved," even if they are not conscious of this salvation in their present life and thus never achieve personal experience of their own happiness and blessedness.

The progression traced out by the *Ethics*, from imaginative knowledge of the first kind through rational knowledge of the second kind and then on to the third kind of knowledge, is not limited to the individual's advance along the path marked out by Spinoza; it is also a collective process in which each individual can progress to the extent that his or her social and political environment permits. Conversely, this environment is more favorable when it has been instituted by a larger number of wise persons, or at least by persons who are under the empire of reason.[167] Every stage of the journey from bondage to freedom depends on internal factors, constituted by each person's intellectual capacities and emotional history, and on external factors, which may be more or less favorable and which themselves depend on the advancement of knowledge as a collective enterprise. The result is that every individual advance, however small, contributes to the collective evolution, which in return favors further individual progress. In this perspective, the effect of the provisional moral code is both individual and collective. Hence we are not limited to understanding Spinoza's doctrine as a theory of individual salvation only, valid exclusively for those rare philosophers who are endowed with the necessary and sufficient capacities to complete the journey. The distinction between the philosopher and the unlettered is not absolute.

In fact, the philosophers to whom the *Ethics* is addressed are only apprentice philosophers,[168] whom Spinoza takes by the hand to guide onto the path of freedom. The ignorant, too, are only relatively uninformed—even though the ideas produced by their minds are more inadequate than

167. This, incidentally, is Matheron's analysis in his first published work (based on his dissertation): Alexandre Matheron, *Individu et communauté chez Spinoza* (Paris: Editions de Minuit, 1969).

168. <Precisely *talmidei ḥakhamim*, disciples of the sages or student-sages, who aspire to "grow up" to be sages themselves.>

adequate—inasmuch as they too are endowed with reason, which, by definition, is compounded from the common notions shared by all human beings. Thus even if the provisional moral code does not function the same way for everyone, it contributes to the salvation of all. Put another way, individuals can find their "salvation" through participation in the collective progress, at the stage they have achieved, if they can find there, even partially, the form of joyous acquiescence that is the sign of the blessedness that can be fully attained only through the intellectual love of God. This is why it is so important that the social and political organization promote rather than impede this progress for the greatest number. It is also the root of the objective stated by Spinoza in the introduction to the *Tractatus Theologicus-Politicus*: drawing attention to "the freedom granted to every man by the revelation of the Divine Law" and concluding "that everyone should be allowed freedom of judgment and the right to interpret the basic tenets of his faith as he thinks fit, and that the moral value of a man's creed should be judged only from his works. In this way all men would be able to obey God wholeheartedly and freely, and only justice and charity would be held in universal esteem . . ." (*TTP*, preface [trans. Shirley, p. 393]). This "freedom granted to *every man* by the revelation of the Divine Law" is certainly not the freedom of the salvation, "as difficult as it is rare," of the end of the *Ethics*, but is its necessary condition and prerequisite.

As for each individual's actions—the manner in which people practice justice and charity—they may vary according to his or her circumstances and temperament. The provisional moral code, meaning the *habituation* essential for advancing along this path of salvation for all, may take diverse forms, including some that are religious and not at all philosophical, such as prayer,[169] as well as external rituals and ceremonies, on condition that each individual "mind his own business" and not "disturb public peace and quiet."[170] Chapter 5 of the *Tractatus Theologicus-Politicus* demonstrates that rituals and ceremonies are contingent on particular situations and are not

169. Letter 34 (21) to Blyenbergh, a devout Christian who was seduced by Spinoza's thought but had problems with what he saw as its moral and religious consequences. He could not understand that the difference between the just and the wicked had to do with the extent of their knowledge and understanding and with "the love of God, which proceeds from the knowledge of God" (Letter 32 [19]), if "both the virtuous and the vicious execute God's will" (Letter 35 [22]). Spinoza had already explained to him that this did not set them on the same plane, because "the wicked lack the love of God, which proceeds from the knowledge of God, and by which alone we are, according to our human understanding, called the servants of God. The wicked, knowing not God, are but as instruments in the hand of the workman, serving unconsciously, and perishing in the using; the good, on the other hand, serve consciously, and in serving become more perfect" (Letter 32 [19]). For Blyenbergh, trapped in his anthropomorphic conception of God, to believe that we depend strictly on God for our actions and that "the godless serve God with their actions as much as the godly do with theirs" empties of meaning "anxious and serious meditation aimed at making ourselves perfect. . . . We deprive ourselves of prayer and aspiration toward God, by which we have so often felt that we received extraordinary strength" (Letter 33 [20]; trans. Curley, pp. 368–369). Spinoza replies to this by emphasizing that "our understanding offers our mind and body to God freed from all superstition" (Letter 34 [21]). His opinion is not harmful, because "it offers to those, who are not taken up with prejudices and childish superstitions, the *only means* [emphasis added] for arriving at the highest stage of blessedness," namely, understanding and knowledge. But, he confesses, his own understanding, like every man's, is limited. Consequently he is unable "to determine all the means, whereby God leads men to the love of Himself, that is, to salvation" (ibid.). This is why he does not "deny that prayer is extremely useful to us," inasmuch as it is one such means—leaving us to infer, perhaps, that for him it is indeed a "prejudice" or "childish superstition."

170. *Political Treatise*, 3.10. Nevertheless, the fifth chapter of the *Tractatus* implies that the historical utility of rituals and ceremonies has always been related to the social organization, inasmuch as they are irrelevant to personal salvation. This is clear with regard to the political organization of the "government of the Hebrews" and the rites and commandments instituted by the law of Moses. But it is also the case for Christian rites, which are merely "external signs of the universal church . . . ordained for the preservation of a society" (*TTP*, chap. 5, p. 76).

indispensable for the "divine law," which consists merely of loving one's fellow and practicing justice and charity. In certain circumstances, however, some rituals, such as circumcision for the Jews and the queue for the Chinese (*TTP*, chap. 3, p. 56), may prove effective for supporting individuals in a specific social environment that is relatively conducive to their advancement. What is more, "the intellectual knowledge of God . . . has no bearing on . . . faith, and on revealed religion," because human beings cannot take "his nature as it really is in itself"—as grasped by that intellectual knowledge of God—for "a set rule of conduct [*certa vivendi ratione*] nor take [it] as their example . . . [for instituting a rule for] the practice of a true way of life [*ad veram vivendi rationem instituendam*]" (*TTP*, chap. 13, trans. Shirley, p. 513). In the absence of philosophical knowledge, they found such an example or model, and the rule that follows from it, in prophetic revelation. This "set rule," valid for all, necessarily has a didactic content different from that described in the *Ethics* (V 10, Note) as permitting philosophers who have embarked on the path of salvation to train themselves to live according to the dictates of reason. Although the practical actions are the same—love of one's neighbor, justice, and charity—the means for reaching them differ, because philosophers apply the precepts of reason, whereas the ignorant obey the law ordained by their creed. The seven articles of the "Catholic, or universal, religion" that Spinoza enunciates, which cannot "give rise to controversy among good men," are a shared minimum foundation whose practical consequences are valid for all, even if those who adhere to them are in thrall to error and hold superstitious beliefs about the nature of God. It is sufficient that these few dogmas, the necessary and sufficient "universal religion," constitute the "only . . . dogmas as are absolutely required in order to attain obedience to God, and without which such obedience would be impossible" (*TTP*, chap. 14, p. 186). But just as different individuals may understand these dogmas in different ways, the appropriate means by which one society comes to such obedience are not the same as those appropriate for another society and must take account of its social and political organization. Ancient Hebrew society offers one example. The minute regulation of all aspects of daily life, prescribed by the Mosaic law, promoted a social organization that was both democratic and theocratic, originally almost perfect, in which individuals' piety enabled them to obey the divine law in an unconstrained manner, which is the minimum and indispensable condition for (nonphilosophical) "salvation" through obedience. This underlies Spinoza's insistence that individuals must submit to the political authority with regard to their actions, including in the domain of religion, although they may think and interpret their faith as they wish (and can), unconstrained by any authority (*TTP*, preface and chap. 19). Note that for Spinoza this rule had a broader scope; philosophers, too, must submit to the political authority in all circumstances, even if they deem it imperfect or corrupt. In such cases (notably that of tyranny), philosophers must simply wait for the government to fall. In practice, because a tyrannical authority can maintain itself only by force, the wrath of the masses will doom it to destruction in the uprising that its oppressive rule is sure to trigger, sooner or later.

This inevitably poses a thorny question for us, who have witnessed the new zeniths of oppression and bloody dictatorships of recent history. To what extent must we obey tyrannical authority and unjust laws? If human beings under the empire of reason may never rebel against the constituted authority,[171] the Spinozist philosopher seems to be bound to wait patiently for the popular revolt and perhaps to associate himself with it; but he may not instigate it directly and even less may he attempt to lead it, although his thinking may serve as its catalyst.

Here we are far removed from the timeless serenity of the *Ethics* and its path to happiness through knowledge, on which Spinoza wishes to be our guide. But this is because the context of the *Tractatus Theologicus-Politicus* is essentially political, as indicated by its preface and last chapters. It must take into account the fact that most human beings do not live under the empire of reason, although the common ideas of reason are by definition accessible to them, at least in principle. In contrast to the philosophical objective of happiness and the intellectual love of God that leads to it, here the goal is merely to regulate society so that the collective environment is conducive to the cultivation of philosophy by those with the capacity for it. This necessarily involves compromises and misunderstandings, some of which may even be helpful; here, though, unlike on the path of wisdom, success can never be guaranteed. In fact, the failure would seem to be total, to judge by how political Spinozism was received by the Christian societies to which it was addressed. All the same, less than three centuries after his death, something very close to the democratic Christian state he wanted Holland to be (though located on the opposite shore of the Atlantic) became the greatest power in the world, for better and for worse: for better, judged by the freedom of thought and speech guaranteed by the United States Constitution; and for worse, judged by the proliferation of fundamentalist sects, televangelists, and other creationist movements, all of them inspired by the "free" interpretation of Scripture. It is a task for historians to evaluate the role played by Spinoza's writings in the drafting of this constitution, which refers to God and the Bible, by men descended from the Protestant churches of Europe. Not far from New Amsterdam, which became New York, in Philadelphia, where the Declaration of Independence and Constitution were adopted, the Quaker William Penn had dreamed of creating a state that would be based on religious freedom and love of one's fellows.[172]

171. This is implicit in *Ethics* IV 37, Notes 1 and 3. But it is stated quite explicitly in the third chapter of the *Political Treatise* (§§5, 6, and 9). Finally, in the *Tractatus*, Spinoza envisions only one exception to this rule—purely theoretical and quite inconceivable outside the context of biblical antiquity—which rescues the coherence of his theory of prophecy and reconstruction of the origins of Christianity. Anticipating the question of "by what right, then, did the disciples of Christ, being private citizens, preach a new religion" not approved by the local authorities, he replies that they did so "by the right of the power which they had received from Christ against unclean spirits" (*TTP*, chap. 19, pp. 250–251). Here he refers his readers to the end of chapter 16, where, he says, he showed "that all are bound to obey a tyrant, unless they have received from God through undoubted revelation a promise of aid against him." Backtracking to chapter 16, we find that he offers the example, recounted in the book of Daniel, that "of all the Jews in Babylon, there were only three youths who were certain of the help of God, and, therefore, refused to obey Nebuchadnezzar. All the rest, with the sole exception of Daniel, who was beloved by the king, were doubtless compelled by right to obey, perhaps thinking that they had been delivered up by God into the hands of the king" (*TTP*, chap. 16, p. 212).

We might expect Spinoza to leave the door open to the possibility that a philosopher, too, might refuse to obey a tyrant, because he is assured of divine assistance by "undoubted revelation," which may derive with equal plausibility from natural light as from prophecy. Daniel, the exception within the exception, might even suggest a model for this. In fact Spinoza rejects this idea, which he relates to the world of prophecy and myths of origin, which no longer pertain (if they ever did) to our human world: "Let no one take example from the Apostles unless he too has the power of working miracles. . . . We must therefore admit that the authority which Christ gave to His disciples was given to them only, and must not be taken as an example for others" (*TTP*, chap. 19, p. 251).

172. New Amsterdam and Amsterdam seem to have had much more in common than their name. According to Méchoulan, in Spinoza's day the latter was "the laboratory of money in its modern guise and of freedoms in their diverse appearances" (Henry Méchoulan, *Amsterdam au temps de Spinoza: Argent et liberté* [Paris: PUF, 1990], p. 8). It is difficult to avoid seeing this as a prefiguration of the American dream, in both its dimensions.

In parallel, the descendants of the Diaspora communities organized by the heirs of the Pharisees managed to survive centuries of exile—sometimes golden, sometimes painful—and of persecution, in Christian countries and Islamic lands, before they attained the capacity, both physical and moral, to reestablish their national sovereignty with the help of these same countries.

If we allow that Christ and his apostles on the one hand, and the Pharisees on the other, were philosophers who, moved by concern for the salvation of the unlettered, were also political leaders, we can understand that their successes were obtained at the price of centuries of warfare that was increasingly "total" and of massacres that increasingly knew no bounds.

Learning the lesson of this history, we would be well advised to remember that a confusion of roles is always dangerous. The "vivid imagination" that animates both politicians and prophets does not promote cultivation of the intellect, but the intellectual faculties of the philosopher and scholar are insufficient to guarantee the "happiness" of the people or of the "masses" (meaning the "unlettered") if the latter do not want it.

We would be well advised today to try to go beyond the *Tractatus Theologicus-Politicus* and its project of a Christian democracy. For our own societies, the method of the unfinished *Political Treatise* and its more general democracy are probably more appropriate. As for individual salvation and blessedness, the *Ethics* shows the way, even if each individual is left to his or her own devices and background in order to advance along it as far as possible.

12. LEIBNIZ AND KABBALAH: MATHEMATICS AND THEODICY

The complicated history of Spinozism in Europe actually begins with Leibniz. Leibniz, whose interest in alchemy and kabbalah has become increasingly evident from the study of his unpublished manuscripts and correspondence, violently condemned Spinoza's pantheism—precisely because he himself could be accused of it—as a detestable opinion and "monstrous doctrine" that confounds God with created beings and holds physical objects to be part of the divine.

> It is utterly true that Spinoza abused the Cabala of the Hebrews. And a certain person, who converted to Judaism and called himself Moses Germanus, followed his perverse opinion, as is shown in a refutation in German written by Dr. Wachter, who knew him. But perhaps the Hebrews themselves and other ancient authors, especially in the East, understand the proper meaning. Indeed, Spinoza formulated his monstrous doctrine from a combination of the Cabala and Cartesianism, corrupted to the extreme. He did not understand the nature of true substance or monads.[173]

But Leibniz's violent opposition to the "monstrous" combination of Kabbalah and Cartesianism is nevertheless marked by a certain ambivalence and merits brief examination here.

It was the time when, as Pierre Bayle wrote in his *Dictionnaire* (1695–1697), "Spinozist" meant "all those . . . who have hardly any religion and who do not do much to hide this,"[174] even if they had never read anything by Spinoza.

173. *Die Philosophischen Schriften von G. W. Leibniz*, ed. C. I. Gerhardt (Berlin, 1875–1890; repr. Hildesheim: Olms, 1962), vol. 3, p. 545 [English translation from Coudert, *Leibniz and the Kabbalah*, p. 77].

174. Pierre Bayle, *Dictionnaire historique et critique*, s.v. Spinoza [English: *Historical and Critical Dictionary: Selections*, trans. Richard H. Popkin, with the assistance of Craig Bush (Indianapolis: Hackett, 1991), p. 301].

Bayle dismissed Spinoza as an atheist because of his pantheism, which could be summarized by the formula "God is everything and everything is God." Since his time, this seems to have been the almost unanimous verdict of both anti-Spinoza theologians and Spinozists

The unrestricted search for truth outside established Christian theological doctrines, which had in any case been rent by schisms since the Reformation, produced *The Book of the Three Great Impostors*,[175] whose author, Christian Kortholt, was deemed sufficiently authoritative to be cited by Bayle in connection with Spinoza. (The other two "great imposters" were the Englishmen Thomas Hobbes and Lord Herbert of Cherbury, a Christian stoic whose *De Veritate* was detested by the theologians.) Bayle refers to the philosophy of the *Ethics*, even more so than the thesis of the *Tractatus Theologicus-Politicus*, as "the most monstrous hypotheses . . . [and] the most diametrically opposed to the most evident notions of our mind."[176] This "monstrous character" comes from the fact that, for Spinoza, "there is only one being, and only one nature; and this nature produces in itself by an immanent action all that we call creatures."[177] This, maintains Bayle, leads to a materialist pantheism that is, in fact, a worse atheism than the pagan doctrines of the authors of antiquity. For "the most infamous things the pagan poets have dared to sing against Venus and Jupiter do not approach the horrible idea that Spinoza gives us of God, for at least the poets did not attribute to the gods all the crimes that are committed and all the infirmities of the world. But according to Spinoza there is no other agent and no other recipient than God, with respect to everything we call evil of punishment and evil of guilt, physical evil and moral evil."[178] We see that the key to this verdict is the origin of evil in a world that, as the natural sciences had begun to demonstrate, was determined by impersonal mathematical laws. Leibniz himself faced this question and endeavored to answer it through a philosophical doctrine that, unlike Spinoza's, claimed to reconcile the new mechanical and determinist science with the spiritualist theology of the dominant European traditions—what is called (though clearly in error) the "Judeo-Christian tradition." Three centuries later we are still asking the same question; but the constant development of the sciences, including the life sciences, increasingly reveals the vanity of attempting such a reconciliation. This is why the contrast of Leibniz and Spinoza is pertinent for us; it involves two systems whose questions are very close but whose answers are violently opposed. We see this close opposition in the ontological, gnoseological, and axiological aspects of their philosophy. Spinoza's unity of the infinite substance and its modes corresponds to Leibniz's infinity of monads.[179] The monism of body and mind, extended to

as well. Without raising the issue of the definition of atheism implied by this formula, we may note that many kabbalists of the most rigorous religious orthodoxy have employed very similar formulas, as we shall see. In this they were following an obscure rabbi who seems to have exerted major influence, Joseph of Hamadan (thirteenth century), who employed this formula literally: "He is everything and everything is He" (*Fragment d'un commentaire sur le livre Beréchit, attribué à Rabbi Joseph de Suse, la capitale*, Fr. trans. C. Mopsik [Paris: Verdier, 1998], Hebrew p. 5, French p. 51).

175. *De tribus impostoribus magnis liber* (Kiel, 1680).

176. Bayle, *Dictionnaire historique*, trans. Popkin, pp. 296–297, 300.

177. Ibid., p. 301.

178. Ibid.

179. In his proof of the unity of substance (which is not the same as his proof of the unity of God, introduced, almost in passing, in the second note to *Ethics* I 8), Spinoza considers and rejects only the hypothesis of a finite number of substances. The possibility of an infinite number of substances is not mentioned, but neither is it formally ruled out by the proof. It is precisely this possibility that Leibniz elaborates in his theory of monads. Spinoza never contemplates it, probably because of difficulties entailed (and which apply to Leibniz's monad) by the conception of substance as simultaneously one and many, which would return us to the mystery of the unintelligibility of the nature of God. On the contrary, having first advanced a proof of the existence of one infinite substance endowed with one or several attributes, which is the cause of itself and of its modifications, the definition of God as substance constituted by an infinity of attributes allows Spinoza to prove his oneness, not as that of a single element in a countable set that might include multiple elements, but as the one and unique that can be conceived only as part of no class or set. The sequence of these proofs eliminates false ideas—anthropomorphic and others—about God and

all existing bodies (Spinoza's "mechanical animism"), corresponds to Leibniz's teleological animism in which the vitalist theories of the following centuries would be perfectly at home.[180] Finally, Leibniz's preestablished harmony—between final causes and free will in the mental world, and efficient causes in the mechanical operation of physical objects—miraculously expressed in his theory of motion and dynamics, where the differential and infinitesimal calculus proves its value, corresponds to Spinoza's determinism, devoid of free will and final causes, timeless and absolute, of which geometry offers an experience, with the possibility of liberation through the knowledge, equally timeless, of how these determining causes act in our body and mind. The only areas where the asymmetry is striking, and in Leibniz's favor, are dynamics (the theorem of "live forces," whose mathematical formulation would later be reinterpreted as an expression of kinetic energy in the laws of conservation) and mathematics, where the German philosopher (along with Newton) has been credited by posterity as the father of the differential and integral calculus. But the vast contrast between these irreversible advances in mathematical physics, on the one hand, and the spiritualist and vitalist philosophy within which these discoveries were made, on the other, produces a strong temptation to sever them and separate the wheat from the chaff, despite their author's affirmation that one requires the other.

Today we employ differential and integral calculus without referring to Leibnizian metaphysics (or any metaphysics at all). The same holds for the conservation of kinetic energy. In both cases we are simply splitting apart, as has often occurred in the history of science, the content of some well-established scientific knowledge from the more or less contingent conditions of its discovery. If we want to understand Leibniz's philosophy, though, we cannot make this separation, because the scientific aspects and metaphysical aspects of his thought were intimately intertwined.

In his *Discourse on Metaphysics*, when he corrects Descartes' error with regard to the conservation of the quantity of motion (mv), he expounds the need to "have recourse to metaphysical considerations in addition to discussions of extension if we wish to explain the phenomena of matter."[181] He advances extremely simple arguments to demonstrate that the "force" conserved is proportional to the square of the velocity and not to the velocity.[182] He calls this quantity, mv^2,

makes the true nature of his essence and existence—that is, of *natura naturans* and *natura naturata*—intelligible. On the logic of this sequence and the order of the arguments in these proofs, see Guéroult, *Spinoza*, vol. 1, chaps. 1–6, and pp. 10–11 of the introduction. Guéroult quotes Frege on the being of the one as that of a proper noun, "which excludes all multiplicity, . . . above all number, including the number one": "the one (*Eins*) that is neither property nor predicate nor susceptible of a plural, unlike unity (*Einheit*), which is the name of a concept (*Begriffswort*) and can take the plural" (ibid., p. 157, n. 59).

180. Renée Bouveresse (*Spinoza et Leibniz: L'idée d'animisme universel* [Paris: Vrin, 1992]) highlights the stark contrasts in the usage of the word "animism." The same holds for the expression "spiritual automaton," which Leibniz borrows from Spinoza while giving it a totally different meaning. Finally, recall that the "parallelism" of body and mind, often invoked to refer to the union of body and mind in their effects, as described in *Ethics* II 7, was invented by Leibniz, who applied it to Spinoza in order to make it easier to distort his doctrine and bring it closer to his own parallelism.

181. *Discourse on Metaphysics*, §18 (in Gottfried Wilhelm Leibniz, *Discourse on Metaphysics; and, The Monadology*, trans. George R. Montgomery [Buffalo, NY: Prometheus Books, 1992; first pub. Lasalle: Open Court, 1902]).

182. This in a section that bears the title, "An example of a subordinate regulation in the law of nature which demonstrates that God always preserves the same amount of force but not the same quantity of motion: against the Cartesians and many others":

> I suppose . . . that it will take as much force to lift a body weighing one pound to the height CD, four feet, as to raise a body B weighing four pounds to the height EF, one foot. . . . [But] it has been proved by Galileo that the velocity acquired by the fall CD [four feet] is double the velocity

the "live force," as opposed to the "dead force" that is the velocity or quantity of motion.[183] He held this principle of conservation, still known as the "theorem of live forces," to be a universal law. In fact, if we extend it to include potential and other forms of energy discovered later (heat, chemical, electrical, etc.), it is fully compatible with the principle of the conservation of energy, with the quantity conserved equal to the integral of the quantity of motion *mv*, that is, $\frac{1}{2}mv^2$.[184] These fundamental principles of dynamics, along with differential calculus, are Leibniz's contribution to modern science. We might think, accordingly, that the metaphysical and theological arguments that accompany his explanations and demonstrations could be stricken from Leibniz's philosophy, as relics of the prescientific age, whereas his scientific writings are marked by a mechanistic thought that is free of metaphysics, inasmuch as he always precisely defines magnitudes by means of the mathematical equations and laws that govern their relations. In fact, this is not possible. Not only does Leibniz himself frequently insist on just the opposite, even in his last works;[185] the coherence of his thought—both physical and metaphysical—is evident in the major axes of his philosophy. On the one hand, we have the pre-established harmony in "two realms, that of efficient causes and that of final causes," where "souls act according to the laws of final causes through appetitions, ends, and means [and] bodies act according to the laws of efficient causes or motions."[186] On the other hand, there is a distinction between the principle of sufficient reason as a principle of reality and the principle of contradiction as a principle of possibility, which are the two great principles on which "our reasonings are grounded."[187]

With regard to the pre-established harmony, Leibniz was confident that Descartes, had he realized that what is conserved is mv^2 rather than *mv*, would have arrived at a different notion of the relationship between soul and body and "would have come upon my system of pre-established harmony."[188]

acquired by the fall EF [one foot], although the height is four times as great. . . . We can see therefore how the force ought to be estimated by the quantity of the effect which it is able to produce, for example by the height to which a body of [a] certain weight can be raised. This is a very different thing from the velocity which can be imparted to it, and in order to impart to it double the velocity we must have double the force. (ibid., §17)

183. The terms have varied. Leibniz referred to the velocity as *conatus*, by which he meant an instantaneous value, whereas he used *impetus* for a quantity of movement in which velocity times mass was integrated over the time of displacement (see M. Guéroult, *Dynamique et métaphysique leibniziennes* [Paris: Les Belles Lettres, 1934], pp. 30–47). Whatever the terminology, however, for him the magnitude conserved is always mv^2 and not *mv*, as the Cartesians thought. The equation of work or kinetic energy with $\frac{1}{2}mv^2$ and its association with other types of energy in the principle of the conservation of energy, the first law of thermodynamics as developed in the nineteenth century, clearly support Leibniz against Descartes with regard to what is conserved in motion.

184. See ibid., as well as the notes by E. Boutroux in G. W. Leibniz, *La monadologie* (Paris: Delagrave, 1880), §80, pp. 185–186.

185. See, for example, §11 of the *Discourse on Metaphysics*, headed "That the opinions of the theologians and of the so-called scholastic philosophers are not to be wholly despised," and the end of §18: "This reflection is able to reconcile the mechanical philosophy of the moderns with the circumspection of those intelligent and well-meaning persons who, with a certain justice, fear that we are becoming too far removed from immaterial beings and that we are thus prejudicing piety."

186. *Monadology*, §79 (trans. Robert Latta [London: Oxford University Press, 1898]).

187. Ibid., §§31–39.

188. "Descartes recognized that souls cannot impart any force to bodies, because there is always the same quantity of force in matter. Nevertheless he was of opinion that the soul could change the direction of bodies. But that is because in his time it was not known that there is a law of nature which affirms also the conservation of the same total direction in matter. Had Descartes noticed this he would have come upon my system of pre-established harmony" (*Monadology*, §80). The square of the velocity does not indicate any direction, unlike velocity, which is a vector. Changes of direction do not change the sum of the live forces and are thus compatible with the properties of matter. For Descartes, by contrast, they could be attributed only to the action of the soul on the body. This is because, faced by the difficulty of conceiving of the union of soul and body

despite the impossibility of any link between the two substances, thought and extended matter, "Descartes wished to compromise and to make a part of the body's action dependent upon the soul" (G. W. Leibniz, *Theodicy, Essays on the Goodness of God the Freedom of Man and the Origin of Evil*, trans. E. M. Huggard [La Salle, IL: Open Court, 1985], §60, p. 156). However,

> two important truths on this subject have been discovered since M. Descartes' day. The first is that the quantity of absolute force which is in fact conserved is different from the quantity of movement, as I have demonstrated elsewhere. The second discovery is that the same direction is still conserved in all bodies together that are assumed as interacting, in whatever way they come into collision. If this rule had been known to M. Descartes, he would have taken the direction of bodies to be as independent of the soul as their force; and I believe that that would have led direct to the Hypothesis of Pre-established Harmony, whither these same rules have led me. For apart from the fact that the physical influence of one of these substances on the other is inexplicable, I recognized that without a complete derangement of the laws of Nature the soul could not act physically upon the body. (ibid., §61)

In other words, the discovery of the theorem of live forces made it possible to be more Cartesian than Descartes and to conceive of a total physical separation of the two substances. But this is only to better defend the idea of a "metaphysical communication . . . which causes soul and body to compose one and the same *suppositum*, or what is called a person" (§59).

For Leibniz, the live force or absolute force expresses more than the geometrical property of extension, inasmuch as it is defined by its future effect, namely, the capacity of a body to be lifted to some height while producing work. The live force, unlike the dead force (pure velocity), expresses a spontaneity at the origin of its future effect in the movement of the body. This future tendency contained in matter soon leads to a spiritualization of the universe through the theory of entelechy, in which the substantial primitive forces of scholasticism are associated, as in the monad, with the effects of efficient causes and of the spontaneity of final causes (*Monadology*, §18; Guéroult, *Dynamique*, pp. 186 and 203–210). This is because "one must admit that there is something in the body that is not magnitude or velocity, unless we would deny the body all power of action" (Leibniz [to Bayle], *Die philosophischen Schriften*, ed. Gerhardt [Berlin, 1875], vol. 3, p. 48). This is why Leibniz is sure that Descartes, were it not for his error about the quantity of movement, would have discovered the system of pre-established harmony.

As for the principle of sufficient reason, it was later adopted as a pillar of modern rationalism and confused with a principle of causality: which may be Cartesian, if, "concerning every existing thing, it is possible to ask what is the cause of its existence,"[189] and nothing exists without a cause; or Spinozist, if "there is no cause from whose nature some effect does not follow" (*Ethics* I 36).[190] In practice these are three distinct principles, of which the most classical, Descartes', is analytical, working back from effects to causes, and is the most vulnerable to Hume's critique. Spinoza's applies to the production of objects; here our knowledge is synthetic and proceeds, as in geometry, from cause to effects. The principle of sufficient reason, often confused with the principle of causality in its analytic or synthetic form, holds that adequate knowledge of efficient causes and/or of their mathematical expressions is sufficient to provide the reason for natural phenomena, so that recourse to final causes is superfluous. For Leibniz, though, efficient causes always depend on final causes. The latter express God's will in his choices among all the noncontradictory possibilities of real events, which correspond to "the best" and which computation of maxima or minima might be able to reconstruct. The principle of contradiction makes it possible only to determine what is false, because self-contradictory, or true (but only possible), because noncontradictory. On the other hand, it is by virtue of the principle of sufficient reason that "we hold that there can be no fact real or existing, no statement true, unless there be a sufficient reason, why it should be so and not otherwise, although these reasons usually cannot be known by us."[191] This is because facts are usually contingent and do not stem from geometric necessity. Their necessity—why things are the way they are and not some other way—is to be found only

189. Descartes, *Meditations on First Philosophy*, Second Replies, "Axioms or Common Notions," in John Cottingham, Robert Stoothoff, Dugald Murdoch, and (vol. 3) Anthony Kenny, eds. and trans., *The Philosophical Writings of Descartes* (Cambridge: Cambridge University Press, 1984), vol. 2, p. 116.

190. See also Axioms 3 and 4.

191. *Monadology*, §32.

in God: "God is sufficient."[192] This necessity differs from geometric necessity in that it is only moral necessity, which does not constrain God himself. Indeed, it "inclines" rather than necessitates; this makes it possible to rescue free will, both divine and human. Even if we happen to discover natural laws with their mathematical expression, this does not mean, for Leibniz, that they automatically express geometric necessity, as we have seen with regard to the theorem of live forces.[193]

Finally, we should remember that in the *Discourse on Metaphysics* the treatment of live force and the quantity of movement are immediately followed by an exposition of the "utility of final causes in physics." This section (§19) offers one of the clearest and most persuasive presentations, at least superficially, of the argument that would demonstrate the existence of God from the "obviously" purposeful organization of living beings. This is where he takes up the cudgels against the "new philosophers" (whom we recognize as Descartes and Spinoza) "who pretend to banish final causes from physics, . . . as though God proposed no end and no good in his activity, or as if good were not to be the object of his will." Next, criticizing the advocates of a purely mechanical universe (thus we may add Hobbes and Gassendi to the list), who admit only necessity and chance to their theories of physics, he employs an argument that is both theological and rational and invokes piety and the principle of sufficient reason as well: "I advise those who have any sentiment of piety and indeed of true philosophy to hold aloof from the expressions of certain pretentious minds who instead of saying that eyes were made for seeing, say that we see because we find ourselves having eyes."[194] David Hume delivered what should have been a knockout blow against this argument, which has nevertheless enjoyed new life as the "argument from design."[195] He saw it as the illegitimate use of a cause that is disproportionate to its effect.[196] Leibniz too seems to accept this rule that the cause must be proportional to the effect, when he insists that

192. Ibid., §39.

193. See above, n. 188.

194. *Discourse on Metaphysics*, §19.

195. In *The Origin of Species*, Darwin explicitly refers to William Paley, a major nineteenth-century representative of this notion. But the passage, which might suggest some convergence of ideas between the two, attributes to natural selection, rather than to some plan of the Creator's, the property of always seeking the good of creatures: "Natural selection will never produce in a being anything injurious to itself, for natural selection acts solely by and for the good of each" (Charles R. Darwin, *On the Origin of Species by Means of Natural Selection*, 1st ed. [London: John Murray, 1859], p. 201). In fact, Darwin's objective is to propose a mechanism for evolution—random variation and natural selection—as the basis of a theory that can serve as an alternative to Paley's natural theology. On the subsequent avatars of the argument from design and its ambiguities, see D. Lecourt, *L'Amérique entre la Bible et Darwin* (Paris: PUF, 1992).

196. We can summarize the argument from design and Hume's counterargument, addressed to those who "have acknowledged, that the chief or sole argument for a divine existence (which I never questioned) is derived from the order of nature" (David Hume, *Inquiry*, Section XI [ed. Hendel, p. 145]). Just as, from observation of a footprint in the sand, we necessarily induce the existence of a human being who passed that way, the order of nature should lead us to infer the existence of a supreme being who is its intelligent author. In the first case, however, we already have an experience of human beings, independent of the observed footprint in the sand; what is more, we have already had several experiences of the causal relationship between human beings walking on sand and their footprints. In the second case, by contrast, our only experience of the supposed cause is its effect, that is, the orderly structure of nature itself. What is more, this is a unique phenomenon, the totality of nature, and must therefore have a unique cause and the causal relation must be inferred without any repetition. Hence we can infer no more than the observed order, other than arbitrarily, in the supposed cause of what we observe.

> When we infer any particular cause from an effect, we must proportion the one to the other, and can never be allowed to ascribe to the cause any qualities, but what are exactly sufficient to produce the effect. . . . If the cause be known only by the effect, we never ought to ascribe to it any qualities, beyond what are precisely requisite to produce the effect: nor can we, by any rules of just reasoning, return back from the cause, and infer other effects from it, beyond those by which alone it is known to us. . . . The cause must be proportioned to the effect. (ibid., §11)

"the effect should correspond to its cause."[197] But he also makes a remarkable about-face, as with his principle of sufficient reason, in that he always subordinates the logical and mechanical efficacy of efficient causes to the metaphysical efficacy of final causes. As we see in the *Monadology*, it is the equality of full cause and entire effect that makes the conservation of live force a principle that is both logical and metaphysical. This conservation expresses the equality of the full cause and the total effect of work in motion, because the latter is the cause of the work done but also the effect of the work that produced it.[198] For him this is a truth that is both logical and metaphysical in that, although it can be expressed mathematically, it cannot be conceived of by mathematical thought alone. Leibniz includes and inverts this rule of equality in his principle of sufficient reason, which, for him, requires more than just a search for efficient causes. This is even clearer from the observation of living beings, where it is by virtue of the principle of sufficient reason and the rule of equality of cause and effect that "all those who see the admirable structure of animals find themselves led to recognize the wisdom of the author of things."[199]

Kant imposed some order on this confusion by establishing fundamental distinctions between types of final causes: not only external versus internal final causes, but especially a distinction between the latter (the only type that can be subjected to the teleological judgment required by the vitalism that dominated the life sciences at the time) and formal final causes, as exemplified by the properties of geometric figures. The latter are not really final causes, in the sense of a temporal phenomenon that is determined by its ultimate form or by an image thereof.[200] Nevertheless, in the same passage of the *Critique of Judgment* Kant endeavors to attenuate Hume's blow indirectly by defending the legitimacy of teleological judgment in the life sciences, ultimately going as far as the idea of a "physical theology," although only from the perspective of reflective judgment and non-authoritative reason. As a result, natural theology continued to flourish into the nineteenth and twentieth centuries, notwithstanding the efforts by Lamarck and Darwin to replace it with mechanistic theories of evolution. Despite the anthropomorphism, the persuasive force of the idea that "the eyes were made for seeing" is such that finalism, whether theological or simply vitalist, continued to dominate the life sciences until the start of the twentieth century, long after it had disappeared from physics. Only modern biology, with its basis in physics and chemistry, seems to have been able to send it packing by providing purely mechanical explanations of living phenomena.

In physics, at least, today we no longer have any reason to accept (not even in mathematical physics) Leibniz's distinction between abstract geometrical necessity and a concrete physical necessity that, despite its mathematical formulation, would express some metaphysical reality, such as a spiritual aspect of substance or God's wisdom. Interpreting Leibniz in

197. *Discourse on Metaphysics*, §19.

198. Guéroult, *Dynamique*, p. 43.

199. *Discourse on Metaphysics*, §19.

200. The mathematician Albert Lautman (*Essai sur l'unité des mathématiques* [Paris: UGE, 1977]) adopted this distinction with regard to the principles of maximum and minimum. Their necessity—formal and unintentional—is like that of the geometrical properties of a shape. *Pace* Leibniz, these principles do not express in themselves a temporal and intentional end that could serve as evidence of final causes. This important distinction is still often unappreciated, although it allows us to understand the difference between mathematical optimizations that can be applied in physics, biology, and other sciences, and which do not necessarily imply the existence of internal or external final causes, on the one hand, and the postulated internal ends of teleological thought, which always discovers the final causes of Leibniz's sufficient reason, on the other (see Atlan, *Enlightenment*, pp. 164–165).

a mechanistic or logical manner, the fact that the physical quantities of velocity, live force, and so on play a part in laws only through their mathematical definitions suffices for them to exclude not only all occult qualities but also any necessary reference to a metaphysics that might be deduced from them and give physical entities a concrete meaning. Leibniz's contributions to modern science—the differential calculus and the theorem of live forces—can and must be separated from his metaphysics, leaving the key elements of the latter (pre-established harmony and monadology) as relics of a bygone past associated with their author's theological concerns. As Guéroult shows, however, these interpretations of Leibniz, which distinguish his dynamics—mechanistic and scientific in the modern sense—from his spiritualist metaphysics, are really post-factum projections of concepts and a vocabulary that have meaning only in the light of the Kantian critique. Certainly this post-Kantian "reconstruction" of Leibniz is legitimate for locating his dynamics in the history of modern physics and mathematics. But there is no doubt that it deforms an overall perspective on his philosophy, both its development over a half century and the form it achieved in his last works, including the *Essays on Theodicy* and the *Monadology*.

We have considered his treatment of live force and his interpretation of Descartes' mistake. As for the differential and infinitesimal calculus, its directly metaphysical meaning for Leibniz is at least clear.[201] Far from being a simple computational algorithm and no more, as we consider it to be today, for him it proves the presence of infinite substance in every point, in every element of the universe, however unique it may be. The mathematical analysis of concrete objects can be extended to the continuum, for it is no longer limited by the discrete character of classical arithmetic and geometry. Thanks to differential calculus, the fullness of the whole is found in every part, however tiny.[202] This rationalization of each part or each state of a process, which results from the law of the whole that is present in each part thanks to its differential, is in principle a property that is both physical and cognitive. It refers equally to objects as states of substance and as states of the consciousness that knows them. This is why "a state of consciousness is fundamentally just a differential of

201. In this he joins a long tradition of mathematicians, philosophers, and physicists who could not separate their reflections on the mathematical infinite from their cosmological and metaphysical intuitions. See, for example: Tony Lévy, *Figures de l'infini: Les mathématiciens au miroir des cultures* (Paris: Le Seuil, 1987); Françoise Monnoyeur, *Infini des philosophes, infini des astronomes* (Paris: Belin, 1995).

202. Note that this metaphysical use of the infinitesimal calculus is more sophisticated and more profound than the mere invocation of a *sense* of the infinite, for example by Pasteur, who, rebelling against the rampant positivism that surrounded him, wanted "to serve the spiritualist doctrine, quite forsaken"; a moderate positivism, like that of Renan, who replied to him, sufficed to refute it (see Pasteur's address on the occasion of his admission to the Académie française, and Renan's reply, in Louis Pasteur, *Oeuvres* [Paris: Masson, 1939], vol. 3, pp. 326–351).

In the *Essays on Theodicy*, with regard to the question "of physical evil, that is, of the origin of sufferings," Leibniz held that it had

> difficulties in common with that of the origin of metaphysical evil, examples whereof are furnished by the monstrosities and other apparent irregularities of the universe. But one must believe that even sufferings and monstrosities are part of order; and it is well to bear in mind . . . that these very monstrosities are in the rules, and are in conformity with general acts of will, though we be not capable of discerning this conformity. It is just as sometimes there are appearances of irregularity in mathematics which issue finally in a great order when one has finally got to the bottom of them.

It is in this context, which could not be more explicit, that Leibniz proceeds to show how mathematics, notably the differential calculus, supports and illustrates his metaphysical principles:

> That is why I have already in this work observed that according to my principles all individual events, without exception, are consequences of general acts of will.
>
> It should be no cause for astonishment that I endeavour to elucidate these things by comparisons taken from pure mathematics, where everything proceeds in order, and where it is possible to fathom

consciousness, and if consciousness is the entire universe in little, then a state of consciousness is the entire consciousness in little."[203] We can understand the seductive attraction of this theory for Salomon Maimon, who, in the Kantian moment of his intellectual development, saw it as the way to overcome the inadequacies he had found in the transcendental philosophy of the *Critique of Pure Reason*. Invoking the differential of consciousness and the adequate ideas that are part of Spinoza's Infinite Understanding, Maimon for a time believed that he could ground the legitimacy of every perception on the presence in it of all substance, from the point of view of the perceived as well as from that of the perceiver.

The history of Leibniz's treatment by historians of philosophy is quite remarkable if viewed from the perspective of what Martial Guéroult calls the history and philosophy of the history of philosophy.[204] For all practical purposes, the nonscientific and metaphysical elements of his philosophy have assumed increasing importance as historians of science move further away from the positivist scheme of the linear progression of the three states of humanity, with a strict separation between its scientific culmination and the earlier states, mythological and then metaphysical. In the positivist era, at the start of the twentieth century, Leibniz was seen (by Louis Couturat and Bertrand Russell, for example) as a mechanic and logician, whereas his metaphysics tended to be unknown or was neglected as a superficial expression of a bygone past. In the post-positivist 1930s, Guéroult himself rehabilitated the profundity and authenticity of the metaphysical and theological dimensions of Leibniz's philosophy. Finally, as we shall see, an increasing number of contemporary historians have demonstrated how Leibniz, like Newton, drew on alchemical, kabbalistic, and even Gnostic sources in his studies, both philosophical and scientific.[205] This also seems to have been the fate of most of the scholars and philosophers of the seventeenth century who are taken as precursors of modern rationalism. But rationalism in the sciences developed on its own, independent of philosophy, with its own rules that also evolved over time, pursuing ever-greater technical and operational efficacy and deferring indefinitely the discovery of the ultimate explanation of phenomena.

From a vision of the history of philosophy that is wholly oriented toward its present end and culmination in science, for which it was only a preparation and auxiliary, we come back to a more timeless vision, in which the coherence and the "eternal" stakes of each system receive their full due, although the circumstances in which it was elaborated are not ignored. In his unfinished second work, the monumental *Spinoza*, Guéroult's reflections about the history of philosophy

them by a close contemplation which grants us an enjoyment, so to speak, of the vision of the ideas of God. One may propose a succession or series of numbers perfectly irregular to all appearance, where the numbers increase and diminish variably without the emergence of any order; and yet he who knows the key to the formula, and who understands the origin and the structure of this succession of numbers, will be able to give a rule which, being properly understood, will show that the series is perfectly regular, and that it even has excellent properties. One may make this still more evident in lines. A line may have twists and turns, ups and downs, points of reflexion and points of inflexion, interruptions and other variations, so that one sees neither rhyme nor reason therein, especially when taking into account only a portion of the line; and yet it may be that one can give its equation and construction, wherein a geometrician would find the reason and the fittingness of all these so-called irregularities. That is how we must look upon the irregularities constituted by monstrosities and other so-called defects in the universe. (Leibniz, *Essays on Theodicy*, Part 3, §§241–242 [pp. 276–277])

203. Guéroult, *Dynamique*, p. 194.

204. Idem, *Dianoématique* (Paris: Aubier Montaigne, 1979).

205. In addition to the authors cited below, see also, for example, R. E. Butts, "Leibniz's Monads: A Heritage of Gnosticism," *Canadian Journal of Philosophy* 10 (1980), pp. 7–62.

proclaim a reborn interest in which the practice of philosophy is once again recognized as an activity of thinking, constantly being renewed and brought up to date by its interactions with the major philosophical systems. Although these systems have been inherited from the past, they bear within themselves—as we have seen with regard to the attempts to reconstruct or analyze their differences—a capacity for thought and for inspiring thought that transcends, though it does not ignore, the historical circumstances of their invention.

Although, in retrospect, Spinoza—unlike Pascal, Descartes, and Leibniz—made no contribution to this science in its infancy, his philosophy emerges rather better than theirs. Thanks in part to the original and radical use he made of his sources, he could conceive of an absolute rationalist monism without having to burden himself with the reality of final causes, other than those that our finite condition imposes on our imagination. Consequently his philosophy retains not only its coherence but also its ethical relevance vis-à-vis contemporary scientific models in which the elements of the universe are entities subject to determining causes, albeit in diverse ways as a function of the complexity of their principles of organization. As we saw in Chapter 2, this adequacy across the centuries—Spinoza's contemporary relevance—as well as that of some of his sources can be recognized only by accepting a more intricate and adult conception, nontheological and less "pious" in the doctrinal sense, of human freedom.

13. LEIBNIZ AND SPINOZA: SCIENCE AND PHILOSOPHY IN THE EUROPE OF THE CHURCHES

As we have seen, Leibniz's philosophy per se is more difficult to accept in the modern scientific context—despite his major contribution to the latter—than that of Spinoza, who seems to have recognized straightaway the implications of the impersonal mechanistic universe of which the scientific revolution of his time offered only a glimpse.

It is remarkable, though, that this contrast appears against a backdrop, implicit in Spinoza, explicit in Leibniz (and in other major figures of the scientific revolution, such as Kepler, Boyle, and Newton, following in the footsteps of Paracelsus, Bruno, and other alchemists), that links the first buds of modern science with the kabbalistic, hermetic, and alchemical pre-science of the Renaissance. The stakes are the same, then as now: the origin of evil and the possibility of averting it, in an ordered world in which we discover the efficacy of the logos, perceived as thought in action, as language in formulas, and as the activity of nature.

Galileo's excommunication by the Catholic Church was recent history, and theological politics continued to be strongly interwoven with science and philosophy.[206] For Leibniz, as for Bayle and later for Kant, it was a time when no philosopher wanted to be accused of Spinozism, lest he find himself in serious trouble

206. The theological and political context is that of the wars of religion triggered by the Reformation, dealing with the truth of Scripture and the power conferred by its interpretation. This explains why first Hobbes and then Spinoza, in their pursuit of a natural philosophy and a rational political morality, could not avoid questions about the status of Scripture. One might think, following the salutary efforts of the Enlightenment and Kantian philosophy over the next two centuries, that this issue has lost its relevance today. We see, however, that it rebounds as if nothing had happened, perhaps because of the exaggerated hopes produced by these philosophies with regard to the capacity for freedom and happiness. Every fundamentalist doctrine of our own age asserts, in its own manner, that it alone possesses the revealed truth and can transmit the unique meaning of the miraculously recorded word of God—found in the Bible for some, in the Koran for others.

with the ecclesiastical and political authorities. As for "Hebrew Kabbalah," it could be perilous to mention that, too, even in its Christianized version, as shown by the misadventures of van Helmont, imprisoned by the Inquisition in Rome on charges of Judaizing. Van Helmont and his colleague Knorr von Rosenroth, both of them friends of Leibniz and his mentors in kabbalah, were the representatives, in seventeenth-century Germany, of the Christian hermetic/alchemical/kabbalistic tradition of the Renaissance, of which Pico della Mirandola was perhaps the most notable example, and the counterparts of Robert Fludd and Henry More in England. Christian Knorr von Rosenroth is known as the chief author of *Kabbala denudata*,[207] a Latin translation of excerpts from the *Zohar* and exposition of the themes of important texts of the Lurianic kabbalah, including passages from *'Emeḳ ha-melekh* (Valley of the King).[208] Their goal was to disseminate these ideas in order to promote the reunion of the Christian denominations, which had been sundered by doctrinal questions, and to promote a universal religion in which these issues would be resolved rationally, in harmony with the new natural philosophy and physical sciences that were beginning to take shape. They also hoped to encourage the conversion of the Jews by showing them the "true" meaning of their own tradition so that they would accept this universal religion, which could only be Christianity. They thought that the "Jewish Kabbalah," which they interpreted through the lens of Christian motifs, was the closest survivor of the ancient tradition conveyed to Moses, the true *prisca theologia* that the Jews themselves had forgotten.[209] Leibniz seems to have been well versed in both the religious and the political aspects of this project, even if he demurred at one or another point of doctrine, such as metempsychosis. That was very important in the thought of van Helmont, who had helped translate a treatise on this doctrine, attributed to Isaac Luria, that was incorporated into *Kabbalah denudata*. Leibniz, for his part, seems to have never missed an occasion to wax ironic about this part of van Helmont's system, which he nevertheless esteemed highly, "even though it was bristling with unintelligible paradoxes" and even though he said that he had not "understood his arguments and proofs . . . because some of these dogmas are based on the traditions of kabbalistic Jews rather than on indisputable arguments."[210] In general, Leibniz—like van Helmont and his friend Henry More—was searching for a middle road between Cartesian metaphysical mechanics and the spiritualism of the "enthusiasts and mystics," in order to "provide a reasonable explanation of those who have assigned life and

207. See above, n. 68.

208. Naphtali ben Jacob Elḥanan Bacharach, *'Emeḳ ha-melekh* (Amsterdam, 1953; rev. ed., Jerusalem, 1994). This is one of the earliest systematic expositions of the kabbalah of Isaac Luria, as interpreted by Israel Sarug, before the interpretation of Ḥayyim Vital began circulating outside the closed circles of Safed and Jerusalem (see above, n. 100). In *'Elim* and *Ta'alumot ḥokhmah* (1629), followed by *Novelot ḥokhmah* (1631), Rabbi Joseph Solomon Delmedigo, known as the "physician of Candia," had expounded concepts of Lurianic kabbalah in the medical and scientific context of the age, notably the Copernican revolution, in which Delmedigo, who studied astronomy with Galileo himself, was involved personally. According to Gershom Scholem, Bacharach drew heavily on Delmedigo (Scholem, *Kabbalah*, pp. 394–395). Leibniz could have had some familiarity with the content of his book, through Knorr von Rosenroth's *Kabbala denudata*; it is almost certain that Spinoza knew Delmedigo's work, inasmuch as his library included a copy of it (see above, n. 67).

209. See Coudert, *Leibniz and the Kabbalah*. For example, they perceived the Holy Spirit and Christ in the "first wisdom" of the *Zohar* and the *Adam Ḳadmon* (primordial man) of the Lurianic kabbalah. In this they were continuing the tradition of Pico della Mirandola, in contrast to Giordano Bruno, who preferred the heliocentric hermetic tradition, despite its pagan origins.

210. Quoted by Anne Becco, "Leibniz et François-Mercure van Helmont: Bagatelles pour des monades," *Studia Leibnitiana, Sonderheft*, vol. 7 (1978), pp. 119–142.

perfection to everything."[211] As Anne Becco noted, for van Helmont and Leibniz "theology could not be separated from philosophy; its foundation is rational. The *magia naturalis* provides the underpinning for a rational and philosophical theology."[212]

Several authors, notably Becco, cited by Georges Friedmann[213] and more recently by Allison Coudert,[214] have insisted on the Neoplatonic, Gnostic, and kabbalistic influences in Leibniz's later philosophy, that of the *Monadology* and the *Essays on Theodicy*. Van Helmont had revived Leibniz's earlier interest in kabbalah, applied now in his search for a "universal characteristic of numbers" as the rational and divine language of nature. In a recent study, Suzanne Edel shows how Leibniz employed this form of thought, which he discovered in Jakob Böhme, Pico della Mirandola, and the combinatorics of Reuchlin and the school of Abulafia, in his project to construct a universal language, an "alphabet of ideas," which would make it possible to find and develop the cognitive and creative virtues of the lost Adamic language to which the true kabbalah, the "science of signs," still bore witness.[215] The *Discourse on Metaphysics* was an incomplete and inadequate expression of his philosophy, unable to offer a reasonable alternative to Descartes' mechanistic dualism. But by going back to his earlier lectures, reanimated by his conversations with and about van Helmont, he found the inspiration and terminology for his theory of monads. This propounded a nondualistic solution to the problem of the relations between body and mind and the division of matter. The Leibnizian monad made it possible to conceive of "units" that partook of the "single" being of the unique substance, of extension endowed with life, as well as of "a complex, complete, infinite, and multiple unit, to infinity," thanks to the arithmetic properties of combinatorics that underlie "magical mathematics," and of the infinite units and infinitesimals of calculus. To return to Becco, "science and magical pre-science were authoritatively joined."[216] For those aware of this intellectual environment, in which the entrance to modernity was also the exit from the Renaissance, his idea of an infinite single substance, both one and many, necessarily evokes a certain analogy with the undifferentiated infinity (*'ein sof pashuṭ*) of Lurianic kabbalah, whose first "garment" consists of the letter-numerals of the primordial Torah, which, in their in-

211. Ibid., p. 135. This is why Leibniz defended van Helmont against the charges of eccentricity and "enthusiasm"; for both van Helmont (who in the meantime had become a Quaker) and himself, "Enthusiasts have this in common with Libertines, that both say things against reason" (quoted by Coudert, *Leibniz and the Kabbalah*, p. 56). Nevertheless, Leibniz was aware of his friend's difficulties formulating his ideas and expressing them clearly. He was ready to assist him in this task, to the point of having ghosted a text on Genesis attributed to van Helmont (see Becco, "Leibniz et François-Mercure van Helmont," p. 126).

212. Becco, "Leibniz et François-Mercure van Helmont," p. 126.

213. Friedmann, *Leibniz et Spinoza.*

214. Coudert, *Leibniz and the Kabbalah.*

215. Suzanne Edel, "Métaphysique des idées et mystique des lettres: Leibniz, Böhme et la Kabbale prophétique," *Revue de l'histoire des religions* 213(4) (1996), pp. 443–466. See the next note for the relations between these studies and Leibniz's mathematical and scientific investigations.

216. Becco, "Leibniz et François-Mercure van Helmont," p. 140. To complete our treatment of this topic, it is important to consider Leibniz's interest not only in alchemy but also in the universal language of numbers and letters, which, though artificial, functions as a natural language to name things as they are, on the model of the primeval language employed by Adam to name objects and creatures in Genesis (see Edel, "Métaphysique des idées"). His mathematical work seems to have been conducted in the same context as his inquiries into this language or universal characteristic of numbers, which he called "true Kabbalah," in contrast to the "magic of letters," which he dismissed as a "vulgar species of Kabbalah" (ibid.). The former was successful and make Leibniz one of the founders of modern science; but the latter seems to attach him to the occult past of Renaissance magic. Things are probably not so simple, however. His interest in "the wonders that could be discovered through numbers, letters, and a new language, which some called Adamic and which Böhme called the language of nature" (ibid.), had been stimulated by his readings in kabbalah, directly or through Jakob Böhme, as

finite combinations, constitute the origin of the infinity of possible meanings later found in the infinity of possible states of the "head that is not known" (*resha' delo' 'ityadda'*),[217] a stage in the process of differentiation and of the expression of knowing and knowable wisdom. In any case, Leibniz, a brilliant physicist and mathematician as well as a philosopher and theologian, identified important elements in this tradition and transformed and incorporated them into his own project to reconcile science and theology.

Spinoza played quite a different role, even though he too rationalized elements of myth borrowed from diverse sources. Contending with the same intellectual environment,[218] he found his path determined in part by two particularities, one cultural and the other social. On the one hand, he had direct access to some of the ancient kabbalistic texts, as shown by the catalogue of his library and occasional quotations from them in his works. These include his reference in the *Ethics* (II 7, Note) to "those Jews who maintained that God, God's intellect, and the things understood by God are identical" and to "ancient Hebrews" in a letter to Oldenburg, where he draws on St. Paul to defend his position about God and nature against the charges of atheism occasioned by the publication of the *Tractatus Theologicus-Politicus*.[219] On the other hand, he occupied a unique social position vis-à-vis the established religions. He had an overt dispute with the Jewish community of his time, from which he had been excluded, and needed long years to overcome his resentment and go beyond polemic in order to achieve the serenity required by his own philosophic method.[220] But he never joined any Christian

well as by his relations with van Helmont. Van Helmont, too, was looking for an alphabet of nature and thought he had found it in the structure of Hebrew. Leibniz, for his part, looked for kabbalah, or the science of signs, not only in the arcane recesses of the Hebrew language but also in any language "in which words are correctly understood and employed"; he attributed to natural languages—notably, like Böhme, to German—properties that brought them close to the language of nature. He valued Böhme's approach, which he considered to be a continuation of the true rather than the vulgar kabbalah. He also appreciated van Helmont's work, to the point of ghosting on his behalf a commentary on the first four chapters of Genesis and translating the preface to that work (which does seem to have been written by van Helmont himself), in which the alphabet of nature is taken up again (see Becco, "Leibniz et François-Mercure van Helmont," the section headed "Leibniz the real author of a forgery by van Helmont"). There too, however, as in his mathematical endeavors, Leibniz looked for a tool to support rational thought and help him discover the laws of nature. He seems to have given combinatorics—that of Raymond Lully to begin with and later of Abraham Abulafia, with Pico della Mirandola between them—an intermediate status between the mysticism of letters, which he rejected, and "modern" calculus, the science of magnitudes and numbers. The language of nature was supposed to provide an "alphabet of ideas" whose associations and combinations, like mathematical computation, would make it possible not only to say but also to rationally reconstruct the hidden divine order of objects in nature. Although only the mathematical branch of his enterprise was crowned with success, we can see the other side of his quest for a "universal characteristic" (or symbolic language) as an ancestor of the calculus of propositions of modern logic, from which the formal languages of computers emerged. Can we perhaps see the extent to which computer-based simulations have now supplanted mathematical analysis and differential and integral calculus, when those cannot be employed because of the complexity of the systems studied, as the culmination of Leibniz's quest, some three centuries later?

217. See Chapter 1, n. 53, and §11.

218. This is the same challenge taken up by Father Mersenne, better known as Descartes' correspondent, in his *La vérité des sciences contre les sceptiques or Pyrrhoniens* (1625), ed. Dominique Descotes (Paris: Champion, 2003). He envisioned and discussed three attitudes toward the new thought expressed by nascent modern science. Of these three—those of the skeptic, of the alchemist, and of the Christian philosopher—he commended the third, because it rests on confidence in the pragmatic truth of the sciences, with no strong link to metaphysics and theology, unless it is the a priori confidence that "God does not deceive us." This modus vivendi between science and Catholic dogma came to dominate French philosophy, whereas German idealism, in its Protestant milieu, continued to look for a Leibnizian synthesis, even after Kant.

219. See Spinoza, Letters 19–23 (68 and 71–75). "I . . . [agree] with Paul, and, perhaps, with all the ancient philosophers, though the phraseology may be different; I will even venture to affirm that I agree with all the ancient Hebrews, in so far as one may judge from their traditions, though these are in many ways corrupted" (Letter 21 [73]).

As we shall see, this reference to more or less mythical "ancient Hebrews" and their corrupted traditions can be understood as a

church, despite his intimacy with the Mennonite collegians for a few years after his excommunication.[221]

Nevertheless, in the *Tractatus Theologicus-Politicus* and his correspondence, and even in the *Ethics* (IV 68, Corollary), Spinoza is careful to associate Christ with the more or less mythical figures of Adam, the patriarchs, the Hebrew prophets, and Moses; he accepts, at least as "conclusions on the subject . . . drawn solely from Scripture" (*TTP*, chap. 1, p. 14), the possibility that God spoke to Moses in a "real" or "supernatural" voice, and to the Apostles "through the mind of Christ" (p. 19). "God revealed Himself to Christ, or to Christ's mind immediately" (chap. 4, p. 64), for "Christ was not so much a prophet as the mouthpiece of God" (ibid.). "God made revelations to mankind through Christ as He had before done through angels—that is, a created voice, visions, etc." (ibid.). In this same chapter 4, "Of the Divine Law," he develops at length what he had announced in chapter 1 concerning prophecy: "The voice of Christ, like the voice which Moses heard, may be called the voice of God, and it may be said that the wisdom of God (i.e., wisdom more than human) took upon itself in Christ human nature, and that Christ was the way of salvation" (chap. 1, p. 19).

But none of this prevents him from keeping his distance from the literal meaning of what he has just said (on the contrary, he does so several times): "Those doctrines which certain churches put forward concerning Christ, I neither affirm nor deny, for I freely confess that I do not understand them" (ibid.). He does not deny them, but neither does he understand them, because they are not derived from Scripture itself. Nowhere in the Bible has he read "that God appeared to Christ, or spoke to Christ,"

veiled allusion to *prisca theologia* (primordial philosophy); as we have seen, the Christian kabbalists of his time—and Leibniz after them—traced this back to the authentic teachings of Moses and primitive Christianity, which they looked for in the "Kabbalah of the Hebrews."

These curious references, along with some passages in the *Tractatus*, raise difficult problems of interpretation. Gershom Scholem, in *Abraham Cohen Herrera* (p. 26), discusses at length the plausibility of several authors' conjectures about the possible influence of Neoplatonic kabbalistic texts on Spinoza's thought, through the intermediacy of Amsterdam rabbis, notably Manasseh ben Israel. But we must be careful not to be misled by purely verbal associations while forgetting that Spinoza uses old terms with new meanings. In another reference to the "Jews" or "Hebrews," he seems to be borrowing the old Aristotelian-Maimonidean formula of God who is "knower, known, and knowledge." But this reference is a comment on his own anti-Maimonidean notion, leading to the conclusion that material extension is an attribute of God's, just as thought is. To demonstrate its ancient pedigree he affirms that "this truth seems to have been dimly recognized by those Jews who maintained that God, God's intellect, and the things understood by God are identical" (*Ethics* II 7, Note). It would clearly be a mistake to see this as a reference to the Maimonidean thesis of the creative intellect, even "dimly," given that elsewhere he fiercely criticizes this idea (*Ethics* I 17, Note; see also Guéroult, *Spinoza*, vol. 1, pp. 275–277). He employs this formula in its immanentist and absolutely monist context; hence the Jews in question must be looked for rather among the literal interpreters of the biblical text, who are not troubled by the theological question of God's incorporeality, or among kabbalists who borrowed the Aristotelian formula in a different context. They include Moses Cordovero in the sixteenth century and, later, the hasidic master Shneor Zalman of Liadi (eighteenth century), in his *Tanya*, as well as his opponent, Rabbi Ḥayyim Volozhiner, in *Nefesh ha-ḥayyim* (1824). Spinoza may have been alluding here to concepts of an immanent deity, a sort of amalgam of Neoplatonic emanation and Stoic materialism, as found in kabbalists such as Abraham Ibn Ezra and Judah Halevi, in the Middle Ages, and especially in Moses Cordovero, closer to his own time. The Neoplatonic hierarchy of the One and of Being was thus associated with the Stoic corporeality of God that had always horrified the "theologians" (including Plato, Aristotle, and, of course, Maimonides) but toward which early kabbalists such as Naḥmanides and then Cordovero adopted a more nuanced stance (see Chapter 2; *SR* II, chap. 11; Askénazi, "La parole éternelle de l'Écriture selon Spinoza"). Note too that Henry More considered the corporeality of God to be "mystical materialism" and one of the most suspect elements that Spinoza had taken from the Kabbalah (see Scholem, *Abraham Cohen Herrera*, pp. 54–56).

220. On this evolution see Brykman, *La judéité de Spinoza*.

221. In addition to the references already cited, notably I. S. Révah (*Des Marranes à Spinoza*), there is an abundant literature on this question and on the role this episode played in Spinoza's life; see Negri, *The Savage Anomaly*.

but only—and this is a significant difference—"that God was revealed to the Apostles through Christ; that Christ was the Way of Life, and that the old law was given through an angel, and not immediately by God; whence it follows that if Moses spoke with God face to face as a man speaks with his friend (i.e., by means of their two bodies) Christ communed with God mind to mind" (ibid.).

We can imagine the perplexity of his correspondent Oldenburg, a favorably disposed reader who pressed Spinoza to clarify, among other things, his "opinion concerning Jesus Christ, the Redeemer of the world, the only Mediator for mankind, and concerning His incarnation and atonement" (Letter 20 [71]). To this, as well as to requests that he clarify his position on God's immanence and on miracles, Spinoza replied by making his meaning perfectly clear: "I will speak on the three subjects on which you desire me to disclose my sentiments, and tell you, first, that my opinion concerning God and Nature differs widely from that which is ordinarily defended by modern Christians" (Letter 21 [73]). This is followed by several additional clarifications, which he later developed further, at Oldenburg's request, in another letter. With regard to the immanence of God and Nature he based himself, as we have seen, on St. Paul, ancient philosophers, and the ancient Hebrews. As for "the doctrines added by certain churches, such as that God took upon Himself human nature, I have expressly said that I do not understand [them]." In fact, it is no incapacity of his that keeps him from understanding, but the absurdity of the Incarnation taken literally, which seems to him "no less absurd than would a statement that a circle had taken upon itself the nature of a square" (ibid.).

So when he demurs at the doctrines of "certain churches" it is not to agree with those of others, especially given that he emphatically disassociates himself from the beliefs of "modern Christians," too. As Guéroult observes rightly, this means that he distances himself from everything that is not the authentic doctrine of primitive Christianity, which he reconstructs on the basis of his reading of the Gospels and the Apostles. His "modern Christians" actually include all Christians of the various denominations established since the time of Jesus: Catholics, Lutherans, Calvinists, and others. Contrary to his own method of drawing on his knowledge of Hebrew when reading Scripture, including the Gospels, even that of John—"although John wrote his gospel in Greek, his idiom was Hebraic" (Letter 23 [75], trans. Shirley)—the churches, like his correspondent, measuring "oriental phrases by the standards of European speech," have made a travesty of the primitive doctrine. "Christians interpret spiritually all those doctrines which the Jews accepted literally" (ibid.).[222]

14. SPINOZA'S "CHRISTIANITY," ANCIENT PHILOSOPHY, AND ETERNAL WISDOM

How should we understand, then, what Spinoza writes about his beliefs concerning revelation and the nature of Christ? Spinoza's "Christianity" has been the subject of a vast literature, in which many conjectures, some more and some less persuasive, have been advanced.[223] We have already seen some elements of this. Let us try to home in on and, as much as possible, extract from behind the philosophical doublespeak a coherent vision of this "Christianity"

222. Guéroult, *Spinoza*, vol. 1, pp. 585–586. We should note this reference to "oriental phrases," which Spinoza, too, associates with a supposed authentic primitive doctrine. This association, which we have seen to be widespread, certainly plays a role in Spinoza's alleged "Christianity."

223. See, for example, Sylvain Zac, "Le problème du christianisme de Spinoza" (1957), repr. in his *Essais spinozistes* (Paris: Vrin, 1985), pp. 105–117. The analyses by Martial Guéroult, and, more recently, by

whose metaphorical meaning could fit naturally into the Spinozist doctrines of the *Ethics*.

We obtain a better understanding of the terminology employed in the *Tractatus Theologicus-Politicus* and the letters when we realize that, in *Ethics* IV, Spinoza uses "the spirit of Christ; that is, . . . the idea of God" (Prop. 68, Cor.), to designate allegorically the "idea of God in Thought," which he had previously formally defined (*Ethics* I 21–23) as the immediate infinite mode under the attribute of Thought and defined elsewhere as "absolutely infinite understanding" (Letter 66 [64]).

This assimilation of "Christ or the spirit of Christ" to infinite understanding (of which every idea is part, including the idea of a human

Robert Misrahi (notes and comments in his French translation of the *Ethics* [Paris: PUF, 1990], especially pp. 444–446 and 448–449), attempt to reconcile the philosophy of the *Ethics* with the text of the *Tractatus*. Nor can we ignore Alexandre Matheron's *Le Christ et le Salut des ignorants chez Spinoza*, which casts new light on frequently neglected aspects of Spinoza's thought. First of all we must remember that one of the main objectives of the *Tractatus* was to persuade readers, almost all of them Christians, that the Old and New Testaments speak of a prophetic religion, originally limited to the requirements of the Hebrews' national life in their land under the monotheistic theocracy instituted by Moses, and then extended to the pagan world and universalized by the Christian Apostles. This prophetic religion is clearly not identical with the philosophical religion of which he speaks in the *Ethics*; and this is where he takes issue with the "theologians." The origins and intelligibility of these two types of religion are different: in the former case, they derive from the light of prophetic revelation, which appeals to imagination and faith and aims at the salvation of the greatest number through obedience; in the latter case, from natural light, which appeals to reason, and acceptance of the truth—for which geometry provides the model—that human beings under the empire of reason cannot fail to recognize as such. The moral code derived from them is the same, however. One can accept the authority of Scripture because of the purely moral certainty that it stems from the prophets, without having any adequate knowledge of the true nature of God, which can be procured only by philosophical certainty (see *Tractatus*, chap. 15). This is how Spinoza hopes to persuade his readers that, contrary to the prejudices and rumors that circulate about him, it was not his intention in the *Tractatus* to "militate against the practice of religious virtue" (Letters 19 [68] and 21 [73]).

Put another way, the *Tractatus* is essentially an educational text with social and political goals, intended to pave the road for the type of thinking essential for science and philosophy and directed at a predominantly Christian reading public for whom, in the wake of the Reformation and wars of religion, interpretation of Scripture had assumed both theological and political importance.

Measured against his philosophy per se, Spinoza's "Christianity" or "Judeo-Christianity" is a cloak or mask, socially and politically necessary (see M. Francès, note on chap. 1, in *Traité de l'autorité politique*, trans. Madeleine Francès [Paris: Gallimard, 1994], p. 145). Inquiring about the traditional theological vocabulary of both the *Tractatus* and the *Ethics*, Robert Misrahi raises the question more pointedly: "Why are these traditional concepts found in a system as modern and subversive as the *Ethics*? Why are they found with a sense that is so radically different from their former meaning?" (Misrahi, notes on the *Ethics*, pp. 458–459). His first answer invokes "social prudence" and "philosophical prudence." He also observes, however, that the *Ethics* furnishes the key for a radical transposition of these expressions into specifically Spinozist concepts. As such, the camouflage of Spinoza's language is not a sign of deceit, falsehood, or bad faith, no more than that of Descartes, the "masked philosopher." Not only do these keys lay bare the precise meaning that the author assigns to these words, with no real attempt at dissimulation; the very fact that he employs such concepts while transposing them into his own philosophy reflects back on them what Spinoza believed to be their true meaning. "The profound truth of these concepts merits being expressed, because in practice they respond to a fundamental and decisive problem, that of blessedness and salvation. Spinoza endeavors to return to philosophy problems and concepts over which religion thought it had a monopoly" (ibid.).

These observations relate to terms such as substance, God, love, and blessedness, which go back to Scholastic theology. But what about Spinoza's references to biblical characters pregnant with myth, such as Adam, the patriarchs, Moses, and, of course, Christ? Does Spinoza provide us with the key to the transposed and radically different meaning he attached to them? Although the *Tractatus* and letters are relatively explicit—notably in chapter 17 with regard to Moses, the inspired legislator of the ancient Hebrews—these keys are to be found in the *Ethics*, with its systematic definitions and propositions. With regard to Christ, a comparison of the letters to Oldenburg, cited above, with the note to *Ethics* IV 68 is particularly illuminating (as suggested by Francès at the start of her note, and as noted by a number of commentators, including Guéroult and Misrahi). In the latter Spinoza discusses the hypothesis that human beings are born free. He then refers, with great brevity, to the biblical myth told "by Moses in the history of the first man." This narrative demonstrates, in the metaphor of the loss of this freedom, that the hypothesis is necessarily false. Spinoza goes on to note that "this freedom was afterwards recovered by the patriarchs, led by the spirit of Christ [about which the Mosaic works breathe not a single word]; that is, by the idea of God." The puzzle of the *Tractatus* and the letters can be clarified now: Spinoza employs the figure of Christ to designate what he defines in the first part of the *Ethics* (21–23) as the "idea

body that constitutes the mind of each human being) allows us to understand the otherwise curious statements in the *Tractatus* (chapters 1 and 4) about the adequate character of the content of the revelation through Christ: it is adequate because it is identical to the content of infinite understanding itself, unlike the content of the Apostles' conception thereof, acquired through imagination—that is, as for all the prophets, in words and images.

We cannot avoid wondering, though, what this rationalization of mythical and theological images meant for Spinoza himself. What advantage does he derive from associating the figure of Christ with the infinite understanding of God, as he defines it in the *Ethics* in terms of substance and the system of attributes and modes? Why does he find it necessary to make these mythical images philosophically intelligible in the *Ethics*, a self-sufficient philosophical work that, unlike the *Tractatus Theologicus-Politicus*, should not have to refer to any scriptural tradition? Some scholars, as we have seen, view it as a purely verbal concession—like the use of the word "God" for Nature—and as a compromise with the religious and political authorities, given that, even in democratic Calvinist Holland, a certain prudence was warranted where they were concerned. Although this may be true, the explanation is not sufficient, because Spinoza's definitions of God and of Christ were poorly chosen if their sole object was to persuade readers of the author's Christian orthodoxy. In fact, as we know, they did not protect him against charges of atheism and irreligion. On the contrary, to judge by his correspondence with friends he felt he could trust and who wanted to help him, where there is no cause to doubt his sincerity, he seems to have wanted to defend these notions vigorously. By integrating them into the system of the *Ethics* and giving them a rational explanation, he exemplifies his statement to Oldenburg that his ideas on these issues are those of the ancient philosophers and of the authentic traditions of the ancient Hebrews (among whom he includes Paul) before they were modified. The figure of Christ, a fully spiritualized symbol, that emerges from all this—from the *Tractatus Theologicus-Politicus*, his letters, and the *Ethics*—which he contrasts to the authentically human figure of Moses, is far removed from that of orthodox Christian theology. It brings to mind, instead, a number of images that hark back to primitive Christian Gnosticism,[224] whose interactions with the sources of the kabbalah are well known.

of God," that is, the immediate infinite mode under the attribute of thought, which, as he explains elsewhere, is "absolutely infinite understanding" (Letter 66 [64]). This is how he can speak, somewhat vaguely, of "Christ" or the "spirit of Christ" and of the man Jesus Christ as an unsurpassed paragon of morality. He can then affirm that "the wisdom of God (i.e. wisdom more than human) took upon itself in Christ human nature, and that Christ was the way of salvation" (*TTP*, chap. 1, p. 19). He can also tell Oldenburg, without contradicting himself, that this incarnation must not be understood literally: "I do not think it necessary for salvation to know Christ according to the flesh: but with regard to the Eternal Son of God, that is the Eternal Wisdom of God, which has manifested itself in all things and especially in the human mind, and above all in Christ Jesus" (Letter 21 [73]). This does not mean that God assumed human form. God's wisdom or God's infinite understanding is not God, but an infinite mode of thought caused by God.

Finally, we must mention certain points of tangency between Spinoza's Jesus Christ and the Moses of some rabbis. We have seen that Maimonides made Moses the greatest of all philosophers, as Spinoza does with regard to Jesus Christ. We should also recall a kabbalistic concept, mentioned above (Chapter 2, n. 89), that takes Moses to be "knowledge itself, that is, knowledge as the root of souls, from where all souls were drawn." This calls to mind Spinoza's notion of Christ as eternal and infinite understanding, the sum total of all individual eternal minds.

224. See, for example, Elaine Pagels, *The Gnostic Gospels* (New York: Random House, 1979, 1989). M. Tardieu notes the assimilation of Moses to the Logos, in a text ascribed to Simon Magus; whereas Jesus, the "common fruit of the pleroma," came to console the Valentinian *sophia*, wandering and in tears, "abandoned to the sorrows and passions of this world," and turns this sorrow into the "'material substance' of

There is another possible explanation, too, suggested by a reading of Spinoza's works in the context of the intellectual environment in which he wrote. Like Hobbes before him and Leibniz after, Spinoza endeavored to ground universal morality on reason, as uncovered by geometry and mechanics. It was taken for granted, however, that this moral code had to be the same as that preached by Christianity, the only religion that merited consideration, because of its exclusive domination of the spirit of the age. Adopting the Christian lexicon was not just a concession to or compromise with the authorities. Given that the Christian religion, despite its schisms, had acquired total hegemony over European social and intellectual life, it must contain a deep truth, concealed behind the distortions imposed on it by the theologians. Unlike Leibniz, Spinoza was not trying to harmonize science and theology in a philosophical system that combined the two. He did assert, however, that he could reconcile them with regard to moral and social behavior; this required a doctrine about the status of biblical revelation:[225] to wit, the

the universe." So too, "the joy of Isis, who carries in her the seeds of the world," in Plutarch, corresponds to "the laugh of Sophia, who gives birth to light," in a Gnostic text (M. Tardieu, "The Gnostics and the Mythologies of Paganism," in *Mythologies*, ed. Yves Bonnefoy, Eng. ed. Wendy Doniger, trans. Gerald Honigsblum et al. [Chicago: University of Chicago Press, 1991], vol. 2, pp. 677–680, on pp. 679–680). We might compare this joy and laughter with Wisdom's delight and rejoicing before the creation of the world (Prov. 8:30–31).

225. See S. Zac, *Spinoza et l'interprétation de l'Écriture* (Paris: PUF, 1965).

According to the *Tractatus*:

> . . . it is one thing to understand the meaning of Scripture and the prophets, and quite another thing to understand the meaning of God, or the actual truth. This follows from what we said in Chap. II. We showed, in Chap. VI, that it applied to historic narratives, and to miracles: but *it by no means applies to questions concerning true religion and virtue* [emphasis added]. . . . We have now shown that Scripture can only be called the Word of God in so far as it affects religion, or the universal Divine law. . . . For from the Bible itself we learn, without the smallest difficulty or ambiguity, that its cardinal precept is: To love God above all things, and one's neighbour as one's self. (*TTP*, chap. 12, pp. 170–172)

Spinoza thus presents, as the essence of the divine law, the commandment to love God, which immediately follows the *Shema*, the ritual proclamation of God's unity (Deut. 6:4–5) in traditional Judaism, supplemented by Hillel's famous dictum (B *Shabbat* 31a) that reduces the entire Torah to the injunction "do not do to your fellow what is hateful to you," which is an Aramaic contrapositive of Leviticus 19:18. On the other hand, the Talmud itself expounds the difference between prophecy and wisdom and holds that the latter is superior to the former, insofar as their relationship to truth is concerned: "A wise man is superior to a prophet" (B *Bava batra* 12a). The model is the wisdom of King Solomon, "the wisest of all men" (1 Kings 5:11 [4:31]), whose writings, notably the book of Proverbs, Spinoza sees (like his own works) as a celebration of the fruits of understanding and knowledge, which express natural light rather than prophetic gift (*TTP*, chap. 4).

Thus "true religion" and "real virtue" are taught, although by different paths, by both philosophical religion and the "word of God" preached by the prophets and transmitted in Scripture. The "councils [that fixed the scriptural canon], both Pharisee and Christian, were not composed of prophets, but only of learned men and teachers. Still, we must grant that they were guided in their choice by a regard for the Word of God; and they must, therefore, have known what the law of God was" (*TTP*, chap. 12, p. 171). We must recall that the "word of God" "signifies that Divine law treated of in Chap. IV; in other words, religion, universal and catholic to the whole human race" (p. 169; see also the entire passage that follows, through "we can thus easily see how God can be said to be the Author of the Bible: it is because of the true religion therein contained, and not because He wished to communicate to men a certain number of books" [p. 170]).

This claim to universality, ascribed to the primitive Judeo-Christianity taught by the Bible, is a classic datum of European thought from St. Augustine until the nineteenth-century historians of religion, for whom the Christianization of humanity was a return to the most authentic sources of human nature (see Olender, *The Languages of Paradise*, pp. 90–92). Today this claim is probably the most questionable element of Spinoza's theory of the relations between Scripture and "true religion," unless we give it a mythical content and expand this notion of universal primitive religion to include the roots of all religious traditions, biblical and other. This is the sense in which every nation and civilization can be considered to be chosen by its god, insofar as it is associated with this common origin in its own particular way (see Atlan, *SR* II, chap. 10). All of these traditions—not just the third classical monotheistic faith, Islam, but also the major traditions of India, Tibet, China, and Japan, derived from other scriptures like the Vedas and Upanishads, as well as all "animist" pagan creeds—express, in a form that is corrupted to a greater or lesser extent, a common religion, a mythical original element of human nature, universal in its existence but differentiated in the detailed laws that flow from it (see Atlan, *Enlightenment*, chap. 8; *SR* II, chap. 2). As we have seen, Spinoza himself, in the *Ethics*, seems to draw on this tradition as a myth of origin that can illustrate various points of his philosophy, rather than as a true history, given that he speaks of it with regard to what was "signified by Moses in the

scriptural texts are quite different from all others and express, in metaphorical, confused, and anthropomorphic fashion, corrupted remnants of the elements of an adequate knowledge (and thus of philosophical truth) revealed long ago, in a mythical past, and, in a fashion just as mythical, to Moses and to Jesus Christ.

As a result, we may see in Spinoza's recourse to these images his own version of an idea that was typical of his age, that of a *prisca theologia* or *prisca sapientia*,[226] a primitive philosophy that some (notably Paracelsus and Bruno) had looked for in alchemy and hermeticism, while others (such as Leibniz and his associates) focused their search on Jewish kabbalah. These circles, rather like the ancient Gnostics, often interpreted the person of Christ, sometimes assimilated to the *Adam Ḳadmon* (primordial man) of the kabbalah,[227] as well as the incarnation and resurrection, in symbolic and allegorical fashion, as totally spiritual, as Spinoza himself writes in his correspondence. In this context, equating "Christ or the spirit of Christ" (with no real distinction between the two) with "eternal wisdom" or "the infinite understanding of God," an infinite mode of substance, was a perfectly natural way of rationalizing the ostensibly concealed meaning of the ancient dogmas. As we have seen, all of this was also part of the longstanding problem of the transition from mythology to philosophy, that is, in the biblical context, of the movement from prophetic revelation to the light of reason.

This does seem to be what Spinoza explains to Oldenburg in his letters, where he refers first to St. Paul and then to "all the ancient philosophers, though the phraseology may be different" (Letter 21 [73]), when he wants to enlighten his correspondent with regard to certain enigmatic—and potentially dangerous—aspects of the *Tractatus Theologicus-Politicus*.[228]

It is therefore probable that Spinoza, like Leibniz, and like Descartes (perhaps) before him,[229] was faced by the same alternatives that seem to have dominated the spirit of seventeenth-century Europe. Confronted by the new objective science, which was distancing itself from its theological, magical, and alchemical roots, every philosopher had

history of the first man" (*Ethics* IV 68). It is hard to believe that he considered this "history" to be a historical account.

226. Note that, for Spinoza, the idea that the philosophy he constructed is the expression, in the language of his time, of a perennial philosophy that could have been expressed, in other times and other places, in different forms—though the content would remain the same—is compatible with his own theory of knowledge. Human understanding, as part of the infinite understanding, produces adequate ideas that are the same as those produced by the infinite understanding. It is thus normal that these adequate ideas can be produced whenever and wherever the circumstances permit human beings to gain illumination from their own intellect.

227. The mediate infinite mode under the attribute of extension is what Spinoza calls *facies totius universi* ("the 'face' of the whole universe") in the same text (Letter 66 [64]), where he refers to infinite understanding as an infinite mode under the attribute of thought. This curious expression, which his correspondent was expected to understand, probably refers to the well-known doctrines of the universe as a being with human form, animated by the World soul; that is, as a macrocosm that reproduces the structure of the human microcosm. Robert Misrahi believes that Spinoza derived this expression from Jewish kabbalistic texts he might have read, notably Herrera's *Puerta del Cielo* (see above, n. 219), while giving it a new meaning as part of his own system (Misrahi, notes on the *Ethics*, p. 349). Herrera uses the Lurianic term "faces" (*parẓufim*)—originally human faces—for the organized structures of the infinity of worlds.

228. One of the passages of the *Tractatus* in which Spinoza's thought is hard to decipher behind his dialectic (in addition to those already mentioned) is the statement that the difference between the Old and New Testaments is purely circumstantial, for they deal with the same religion, and that "the catholic religion (which is in entire harmony with our nature) was [not] new except for those who had not known it." This may be intelligible if we take it as an allusion to his mythical primitive Judeo-Christianity, which embodied the single truth of the true religion that has since been hidden behind ecclesiastical dogmas and theological disputes (*TTP*, chap. 12, p. 170).

to be "Christian" in some fashion or another, open both to science and to the moral code of the "true religion," or be consigned to the heretical categories of libertine skeptic or eccentric alchemist. To put it another way, it is quite likely that for Spinoza, too, the only universal religion that could be grounded in philosophy was a Christianity purged of superstition and of the errors of the theologians. For him, though, unlike Leibniz, this attitude, even if real and not merely tactical, could not carry decisive weight in the elaboration of his philosophy. The theoretical reconciliation of science and theology certainly was not one of Spinoza's objectives. His goal was to lead his readers down the path of true philosophy, which he knew to be true just as he knew that "the three angles of a triangle are necessarily equal to two right angles" (Letter 60 [56]). Nor could he entertain a compromise solution that would sacrifice any part of his monism, which was deterministic and nonteleological, even though it evoked Stoic and kabbalistic notions[230] that the established religion denounced as heretical. This means that when Spinoza and Leibniz confronted the same difficulties created by the inadequacies of Cartesian ontology, they took very different stands vis-à-vis the seventeenth-century zeitgeist. On the one hand, like Leibniz after him, the Spinoza of the *Ethics* undertook the elaboration of a new philosophy that reorganized elements derived from the ancient philosophical tradition. These elements were combined with Cartesianism, whose vocabulary and method Spinoza adopted, but whose dualist ontology, recourse to an unintelligible God, and consequent model of man were philosophically unacceptable to him.[231] On the other hand, in this project, having left the polemics of the *Tractatus Theologicus-Politicus* behind, he does not allow himself to be distracted (as Leibniz would be) by any concern for the immediate religious and political implications of his doctrine; this dispassion was facilitated by the fact that he belonged to no Church and was no longer affiliated with the Synagogue.

We can also understand why Leibniz, facing the same problems a few years later but occupying a very different social and cultural position, adopted an ambivalent attitude toward Spinoza, as Georges Friedmann has shown in detail.[232] He could understand him and sometimes even agree with him on certain points, such as the possibility of ethics in a totally deterministic world.[233] But he found it quite important to mark himself off from Spinoza, especially in public, and to avoid at all cost the accusation of pantheism to which his monadology and theodicy were always open.[234] This probably explains the violence of his denunciations, including that against "the monstrous doctrine[,] . . . a combination of the Cabala and Cartesianism, corrupted to the extreme."[235] For he too stood at grave risk of being charged with concocting a similar blend of kabbalah and Cartesianism.

229. See above, n. 218.

230. On the dignity and divinity of matter in certain streams of Jewish kabbalah, as well as absolute determinism, see above, Chapter 2.

231. See, for example, the preface to *Ethics* V.

232. Friedmann, *Leibniz et Spinoza*.

233. Both Friedmann (ibid., pp. 236–243) and Coudert (*Leibniz and the Kabbalah*, pp. 99–111) compare this theme in Leibniz's thought to the kabbalistic doctrine of *tikkun 'olam* ("repair of the world"), which Leibniz could have known through van Helmont. There too, however, Leibniz kept his distance from Spinoza; he evokes his "pre-established harmony" to reconcile freedom (divine and human), in the sense of the exercise of free will, with absolute determinism.

234. This obsession with maintaining a safe distance from the "monstrous doctrine," which could nevertheless not be ignored, continued to haunt European philosophy, especially German idealism and its offshoots, throughout the eighteenth and nineteenth centuries.

235. See above, n. 173.

15. EXPANDED SPINOZISM AND LIMITED KABBALAH

We can accept Leibniz's view of Spinoza's philosophy as a combination of kabbalah and Cartesianism, but with a positive take on its "monstrous" character, quite different from what it meant for Bayle and Leibniz. Seen in long retrospect and in the context of the spirit of his age, Spinoza can be seen as a "hopeful monster," as the biologist Richard Goldschmidt referred to the role of certain mutations in the evolution of a species.[236] The radically innovative and anticipatory character of his philosophy may cast Spinoza in the role of a mutant in the history of philosophy, a "monster" as compared to the norm for his century but in fact the herald of the modern age. In the same sense, Robert Misrahi speaks of the "modern and subversive" character of the content and language of the *Ethics*, where "subversive" is clearly meant positively, like the "monstrous" character of hopeful mutants.[237]

More precisely, it is Salomon Maimon's apparently paradoxical judgment that is best suited to a comparative reading of the *Ethics* against some major texts of Lurianic and post-Lurianic kabbalah, reinterpreted in the context of classical and modern philosophy. Maimon, whose career we reviewed briefly above, was especially well equipped for such a reading. His conclusion is just as astonishing as it is pithy: Kabbalah, he observed, "is nothing but an expanded Spinozism."[238]

We might have expected him to note a kabbalistic influence on Spinoza somewhat more prosaically, in the fashion of other authors. Instead he inverts the sequence, anachronistically, so that Spinozism is the origin from which "kabbalah" developed and expanded. Without concerning himself about this question of chronology, which clearly is not a problem for him, Maimon justifies his judgment as follows:

> In fact, the Cabbalah is nothing but an expanded Spinozism, in which not only is the origin of the world explained by the limitation [*Einschränkung* 'contraction', that is, *ẓimẓum*] of the divine being, but also the origin of every kind of being, and its relation to the rest, are derived from a separate attribute of God. God, as the ultimate subject and ultimate cause of all beings, is called Ensoph (the Infinite, of which, considered in itself, nothing can be predicated). But in relation to the infinite number of beings, positive attributes are ascribed to Him; these are reduced by the Cabbalists to ten, which are called the ten Sephiroth.
>
> In the book, *Pardes*, by Rabbi Moses Kordovero, the question is discussed, whether the Sephiroth are to be taken for the Deity Himself or not.

236. Richard Goldschmidt, "Some Aspects of Evolution," *Science* 78 (1933), pp. 539–547; repr. in idem, *The Material Basis of Evolution* (New Haven: Yale University Press, 1940), pp. 390–393. Goldschmidt was one of the first biologists to challenge the neo-Darwinian orthodoxy of continuous evolution by the accumulation of small mutations. As a developmental biologist in the 1930s he understood that large mutations that affect the process of embryonic development and produce monsters could be invoked to explain the appearance of the new models of organization that underlie the emergence of new species. His theories are extremely relevant today, because they introduce into evolutionary theory the notions of a "system of chemical reactions" and of global behavior organized in time and space—ideas that we are starting to better understand today, thanks to the identification of developmental genes. Goldschmidt perceived that these global modifications of the total *activity* of the genome, which may result, for example, from changes in the rates of certain chemical reactions, rather than individual mutations progressively accumulated and selected, may be what introduce a change that is viable—though not necessarily adaptive, in the sense of optimal—in the spatiotemporal structure of development; i.e., the dynamic organization that characterizes a species.

237. In *Ethique*, trans. Misrahi, p. 458.

238. *Solomon Maimon: An Autobiography*, trans. Murray, p. 105.

> It is easy to be seen, however, that this question has no more difficulty in reference to the Deity, than in reference to any other being.
>
> Under the ten circles I conceived the ten categories or predicaments of Aristotle, with which I had become acquainted in the Moreh Nebhochim [*sic*],—the most universal predicates of things, without which nothing can be thought.[239]

In other words, in kabbalah Maimon detected the idea, which he deemed Spinozist, of an infinite God who contracts in the infinite essences that are the attributes. What is more, if we disregard the *number* of *sefirot* (which, according to Maimon, varies as a function of certain considerations[240]), then the hierarchy in the production of objects and their essences, in and by the one and undifferentiated Infinite, which yields the multiple finitude of what exists, does evoke Spinoza's system in which absolutely infinite substance produces, under each of its attributes, first its immediate infinite modes, then its mediate infinite modes, and finally its finite

239. Ibid. Wolfson mentions Maimon's remark, in his Hebrew commentary on the *Guide of the Perplexed*, that Spinoza's point of view "agrees with the opinion of the Cabalists on the subject of *Zimzum*" (*The Philosophy of Spinoza*, vol. 1, p. 395). In practice, we are dealing with the classic issue, especially prominent in Neoplatonism, of how finite multiplicity can be produced by an undifferentiated and infinite One. The theory of the *zimzum* (contraction) of the infinite offers a metaphor for this. This is what Abraham Herrera writes in his introduction to the kabbalah, the *Puerta del Cielo*, also quoted by Wolfson: "From an infinite power, it would seem, an infinite effect would necessarily have to follow. . . . In a certain manner God had contracted His active force and power in order to produce finite effect" (ibid., pp. 394–395, quoting *Sha'ar ha-shamayim* [i.e., *Puerta del Cielo*] V 12).

Wolfson rightly underscores the difference between Spinoza's immanentist model and Herrera's theory of emanation, in which the contraction by which the Infinite limits himself must be an act of his intelligence and his will. On the other hand, it is not so obvious, *pace* Wolfson, that all kabbalists understood the *zimzum* in this voluntarist and theist sense rather than in an immanent sense very close to Spinoza's god, who creates and acts by the free necessity of his nature. The classic expression "it rose into his thought [*or* will] to create the world," which introduces several midrashic passages, is quoted in the Talmud, and is repeated in various kabbalistic texts (including Ḥayyim Vital's *'Ez ḥayyim*), raises the question of *where* his thought rose from, if all that "existed" was the one and undifferentiated Infinite. Clearly it could have risen only from the world itself, which, though still uncreated, coexisted with the Infinite as a necessity or desire to be created, as if it had been produced for all eternity by a timeless and impersonal intentionally. Moses Cordovero (*Shi'ur ḳomah*, chap. 60) analyzes this phrase, along with other anthropomorphic expressions in the midrash that he strips of their temporal form, although all of them refer to "*before* the creation of the world," as a sort of initial project that the Creator was *then* forced to modify in certain points (see above, n. 113; also Chapter 2, nn. 89–99 and appendix). In most texts, moreover, this Infinite that contracts into itself is represented by the *spatial* image of an infinite and undifferentiated light (*'or 'ein sof ha-pashuṭ*) or of an expansion (*hitpashṭut*, from the same root as *pashuṭ* 'simple, undifferentiated') *within which* the creative contraction takes place, the source of limits and the origin of differentiation. Thus Henry More could view kabbalah as "mystical materialism," because it describes divinity in terms of *res extensa*. According to Gershom Scholem (*Abraham Cohen Herrera*, pp. 54–55), Spinoza could have been aware of More's correspondence with Descartes on this subject. From this perspective too, and contrary to Wolfson's analysis, Maimon's remark is reasonable. It leads us to distinguish, as Scholem does, among the kabbalists themselves, differentiating those, including Herrera, who hold a clearly theist position that Spinoza rejected from those whose stance, fundamentally immanentist and pantheistic, is sometimes concealed behind a superficial theism (see above, n. 219, and below, n. 261).

240. The *sefirot* are sometimes grouped into seven or thirteen *middot* ("measures"), five *parzufim* ("faces"), four *'olamot* ("worlds"), etc. They can even be reduced to two, from which the others are deduced by operations of substitution and combination, as demonstrated by Rabbi Shlomo Eliaschow in the first chapters of his *Book of Knowledge*, quoted at length in Chapter 2 of this volume (S. Eliaschow, *Leshem shevo ve-'aḥlamah. Sefer ha-De'ah* [Piotrkow, 1912; new ed. Jerusalem, 1976], "Explanations of the World of *Tohu*, 1.1–3, pp. 11–45; see also Henri Atlan, "Formalisme de la Kabbale et concepts scientifiques," *Encyclopédie des religions* [Paris: Bordas, 1988], pp. 336–337). These are the two categories—conjunction and disjunction—that make it possible, as a matter of logic, to think. They are designated, respectively, *ḥesed* and *gevurah* (plurals *ḥasadim* and *gevurot*), which, depending on the context, may be translated respectively as "love," "grace," or "generosity" and as "might," "rigor," or "heroism," but also as "expansion" and "limitation," "indifferentiation" and "differentiation," or even "male" and "female."

modes.[241] On the other side, it is apparently the system of the ten *sefirot* as an image of these essences, of the relations they maintain among themselves, and of the particular essences thus produced, that constitutes this non-Spinozist "expansion." Maimon suggested understanding this system in light of the Aristotelian categories.[242]

241. For Alexandre Matheron (*Individu et communauté chez Spinoza*), this analogy applies even to the details of the structure of the "sefirotic tree," in which the ten *sefirot* are organized in three aspects and arranged in three columns (left, right, and middle). Matheron uses this arrangement to demonstrate and analyze a recurrent structure in the logical organization of the propositions of the last three books of the *Ethics*, with their proofs and notes. This structure copies the three-column sefirotic tree described by Gershom Scholem in his *Major Trends in Jewish Mysticism*. What is more, Matheron detects the same arrangement in the Spinozist state of the *Political Treatise*, in which he finds "ten kinds of institutions. The structure assembled in this way . . . can be set in one-to-one correspondence with the sefirotic tree of the kabbalists" (p. 344). Matheron presents detailed tables and diagrams of the specific analogies between each of the ten *sefirot* and these ten institutions (pp. 620–621), recognizing that some of them (e.g., between Sovereign and *Keter* ["crown"], Consults and *Ḥokhmah* ["wisdom"], Justices and *Din* ["judgment"], Territorial institutions and *Malkhut* ["kingdom"]) are more obvious than others. Similarly, other diagrams of the sefirotic tree illustrate the structure of the groups and subgroups of propositions in the third (pp. 616–617), fourth (pp. 618–619), and fifth (p. 622) parts of the *Ethics*. Responding to criticism of these analogies, Matheron, in the preface to the second edition of his book, notes that what he "said about the sefirotic tree" (aside from the structure of the Spinozist state) concerns only the physical order of the propositions in the last three books of the *Ethics*, not their content. He had not intended "to turn Spinoza into a kabbalist; to take an analogous example, to say that our students' dissertations have a threefold structure does not mean that their authors believe in the Trinity" (p. iv). Nevertheless, in addition to this formal analogy, Matheron does seem to find traces, in Spinoza's ontology, of a cosmic drama of the shattering of the vessels (p. 30 n. 27), which, in Lurianic kabbalah, follows the contraction (*ẓimẓum*) and self-limitation of the Infinite. Despite the strictures of Kant, Hegel, and the other critics of Spinoza's "orientalism," the individuality of object and beings is real for him and cannot be dissolved in the grand Whole of substance, whose modes they are. Hence the separate individuals can be in conflict with one another. Furthermore, although their essences are eternal, their finite existences in time succeed one another and cause one another; but they also destroy one another. An individual's existence ends only because of external causes produced by other individuals, because the *conatus* of each makes it "endeavour to persist in its own being" (*Ethics* III 6). This leads to what Matheron calls the "drama of separation," which is the origin, inter alia, of the "human drama" in which universal concord can be born only from discord and from the infinite conflicts that oppose individual existence. This is another way of envisioning the origin of evil and destruction in the structure and dynamic of divine power. In this division Matheron sees Spinozist modes; in the conflict of conations he finds a "drama of separation, whose ontological root, as affirmed by an entire kabbalistic current, must be sought in God himself" (p. 30). Referring, then, to the structure of the sefirotic tree he locates this root on the left side; more precisely, in the "attribute of Rigorous judgment" (the *sefirah* of *gevurah* or *din*).

Curiously, though, Matheron, continuing his interpretation of Spinozist ontology, distinguishes Extension from Thought, because only in the former can this separation be a source of division and conflict. In his analogy with the structure of the *sefirot*, he is tempted "to place Spinozist Extension on the left side, that of separation, and to attach Thought to the right side, which is that of unity" (p. 30 n. 27).

This is an unexpected interpretation if we remember that Extension, as an attribute, is no more divisible than Thought and that the modes of the latter, namely ideas, are at least as separate as the bodies to which they are attached, can come into conflict with one another, and, especially, are inadequate. Hence the source of division lies no more in Extension than in Thought. It is located, rather, in the transition from substance to the finite modes, by the intermediacy of the infinite modes. This is how it can be found, as in some kabbalistic currents, *in God himself*, that is, in the self-limiting Infinite. If we are looking for the distinction between Thought and Extension in the structure of the sefirotic tree, rather than between the left and right sides, we should note that most kabbalistic writers distinguish the three *sefirot* of the "head" (with the evocative designations "Crown" or "Will," "Wisdom," and "Intelligence") from the seven *sefirot* of the "body," which are often associated with the six spatial directions and with matter. The origin of separation may also be represented in Thought, in its left side, that is, in Intelligence (*binah*), which analyzes, distinguishes, and defines, although it is always joined to its right side, Wisdom (*ḥokhmah*). At the most one can say that the union of these two sides is "permanent" in the "head," through the "knowledge above," but only intermittent in the "body," through the "knowledge below." But we are already no longer in the root of things, because, in the scheme of the infinite sequence of worlds, the distinction between above and below, or head and body, or mind and matter, is only *relative* (see above, Chapter 2). These worlds produce one another, from top to bottom, as an infinity of sefirotic trees. It is hard to see how a comparison with the two attributes of Spinozist substance could be appropriate here.

242. In this sense, kabbalah could be seen as expanded Aristotelianism, too. In his eagerness for knowledge and understanding, the young Salomon Maimon, confronted by the hermetic character of the kabbalistic lexicon, soon arrived at the methodological conclusion that "for the understanding of Kabbalah the framework of some speculative system is required, be it Spinozism, Maimonides' philosophy, or astral magic" (Moshe Idel, *Hasidism: Between Ecstasy and Magic*

Finally, kabbalah does include irrational and antiphilosophical currents that insist on their inaccessibility to reason—an "expansion" that is even greater, outflanking all philosophy.

With regard to the coherence of the *Ethics*, we can readily understand that this "expansion" is in fact a diminution; or, rather, a plus that is unfortunately a minus.[243] But things may not be so simple; "expanded Spinozism," when applied to kabbalah, is a paradox. It pretends to designate a doctrine that comments on and interprets Spinoza, when this ostensible expansion preceded Spinoza in time. This would mean that Spinoza's system is in fact a contraction of kabbalah, to be understood either as a limitation of its field of view or as a chemical concentrate or precipitate that purifies and makes visible what was formerly in solution. Wachter, incidentally, says this in so many words when he tries to show that Spinoza expounded part of the kabbalah in systematic form.[244] With regard to its kabbalistic sources, then, Spinoza's thought is both a shortening of its horizon and a reinforcement of its coherence. In practice, anticipating the methods of modern science, which carefully limit its objects, Spinoza restricts his objects of investigation to those that can be thought rationally, even though he also had to include the confused thought of the imagination and of waking dreams. Inverting the description cited above, the doctrine of the *Ethics*, vis-à-vis its sources, is a minus that is really a plus. By reducing the number of places examined, it also limits those where the examination goes astray.

In fact, this can be said not only of kabbalah but of all of Spinoza's sources. We might say that Neoplatonism is an expansion of Spinozism; and so are Stoicism, the philosophy of Maimonides, Cartesianism, and so on. This casts new light on the question of his "sources." Rather than dealing merely with the historical question of the influences of a particular system, we can now also have exegeses, interpretations that develop, more or less adequately, certain points of Spinoza's thought, even if, chronologically, they preceded the publication of the *Ethics*. Like every interpretation, each of them simultaneously adds and subtracts; in this they resemble the multiple interpretations elaborated since

[Albany: SUNY Press, 1995], p. 40). In his autobiography, Maimon recounts how he wondered, when he first began studying kabbalistic texts, about the validity of a method of interpreting Scripture in which the system of *sefirot* was used as an analytical template. (This is a reference to Gikatilla's *Sha'arei 'orah*, each of whose ten chapters explains a series of words that allude to one of the ten *sefirot*.) Although he certainly seems to have appreciated the ludic aspect of this, he did not see how one could avoid falling into the excess of an overworked imagination, in the absence of any rational control of these interpretations.

In this he was repeating the critique of Spinoza, who mocked the dreams and madness of "some kabbalists" and rabbinic techniques of interpreting Scripture (*TTP*, chap. 9). As the founder of the historical and critical method of biblical exegesis, Spinoza could only laugh at the traditional methods that depart from the plain and obvious sense and use the text as a pretext (see Atlan, *SR* II, chap. 12). We should note, moreover, that Abraham Ibn Ezra, whom Spinoza drew on in the *Tractatus* (see above, n. 113), was also part of the Neoplatonic kabbalistic tradition. It is only in the past few decades that literary criticism has begun to do justice to hermeneutic methods that read certain texts in an active and creative fashion and disregard historical research about what the presumed author of the texts might have meant. After three centuries, the historical-critical method no longer has a monopoly on rationality. But it remains nevertheless a defensive barrier against uncontrolled use of allegorical or "hidden" meanings that are taken literally and confounded with the plain sense of the text. As always, it is not the proliferation of possible meanings that is the problem, but the confusion of domains and levels of meaning that produces a mad delirium. This, incidentally, is suggested by the talmudic story of the four sages who penetrated the "garden" of esoteric knowledge, one of whom emerged insane (B *Ḥagigah* 14b).

243. See Zac, *Salomon Maimon*, p. 12.

244. "In the preface of his third book, the *Elucidarius cabalisticus*, published in 1706, Wachter apologized for having misunderstood the true thought of Spinoza, 'an author whom I confess to have understood very wrongly or perhaps not at all at that time [of the first book], both because of his subtlety and because of the vulgar prejudices spread against him'" (Friedmann, *Leibniz et Spinoza*, p. 209).

the publication of the *Ethics* with the deliberate intention of making sense of or developing its doctrines. Even if we set aside the accusations of atheism, pantheism, or deism (all of them "theological" categories that were quite foreign to him), the hasty conclusions and the wild enthusiasms with their obvious misinterpretations, we also find that each of them has its positive and negative sides. This is even clearer in the case of great philosophers, such as Leibniz, Kant, and Hegel, not to mention our own contemporaries, for whom Spinoza may be only an indispensable point of departure that their own thought is supposed to be able to transcend. As we have seen, the "necessity" of going beyond Spinoza is generally rooted in their own inadequacies and misinterpretations rather than in Spinoza's thought itself. It is in this sense, where every interpretation is an "expansion" that is both more and less than what it interprets, that we can say that each of Spinoza's main sources is itself an "expansion of Spinozism."

Thus we can readily accept Salomon Maimon's assertion that kabbalah is "expanded Spinozism," noting, however, that kabbalah, like Neoplatonism and Stoicism, is not a homogeneous doctrine but a collection of works that share a common terminology even though often differing considerably on essential points. Some of these works clearly have nothing in common with Spinoza, if only because their authors refuse to consider them to be philosophy. What is more, some kabbalistic authors were preoccupied with elaborating a doctrine that would be in harmony with the dominant Aristotelianism of their day and with Maimonides, its chief advocate.[245] Others, like Naḥmanides and Ibn Ezra, seem to draw instead on Neoplatonic and perhaps Gnostic sources. Finally, some stress the operational aspects of practices that resemble, in fact if not in intention, the pagan magic of antiquity and the natural magic of the Renaissance. They may be the targets of Spinoza's attacks on the "dreams of the Kabbalists," in the context of traditional biblical exegeses whose methods he criticizes in the *Tractatus Theologicus-Politicus*.[246] Note in any case that, even in this context, he approvingly quotes Abraham Ibn Ezra, whose biblical commentary combines a concern for grammatical rigor with a traditional rabbinic inspiration that is not devoid of kabbalistic sources. Finally, almost all Jewish kabbalists were deeply involved in the rabbinic mainstream; far from discarding its legal aspect, they provided it with a new dimension of meaning.[247] Scholem refers to these diverse doctrines as "trends" in Jewish mysticism. This designation is not fully accurate if we understand mysticism as the opposite of rationality. As I have shown elsewhere,[248] some of these doctrines express a rationality that is better suited to modern thought than is the scholastic and Aristotelian rationality of the currents of Jewish philosophy derived from Maimonides and traditionally referred to as "rational." Turning Spinoza into an oddity of the history of this tradition (as he is in the history of philosophy) would imply that his philosophy, too, represents one of these trends. If so, we could characterize it in several ways. The most important is clearly the Cartesianism of the language

245. See above, n. 23.

246. See above, n. 242.

247. Scholem, *Kabbalah and Its Symbolism*, chap. 2, "The Meaning of the Torah in Jewish Mysticism," pp. 32–86. One work, among many, in which the legislative and kabbalistic aspects of rabbinic Judaism are joined in a synthesis of the *nigleh* (overt meaning) and the *nistar* (arcane meaning) is Isaiah Horowitz's *Shenei luḥot ha-berit*, which, precisely on this account, has been very popular ever since its publication in the seventeenth century (see Chapter 1, n. 30). The introductory section of this work, *Toledot 'adam*, a book in its own right, has recently appeared in an annotated English translation: Miles Krassen, *The Generations of Adam* (New York and Mahwah: Paulist Press, 1996).

248. Atlan, *Enlightenment*, chap. 3; and "Formalisme de la Kabbale et concepts scientifiques."

and the rational order of the proofs, which produce what Guéroult calls "mysticism without mystery."[249] As for the absolute monism of the unique substance, in which Hegel sees an "oriental" influence, it ends up as the "mystical materialism" denounced by Henry More.[250] As compared to the traditions of Western rationalism, still inspired by Christian theology, either directly or as a reaction, here, as in kabbalistic texts, we are dealing with elements of a thought that can be understood only as a product of myth and mysticism, with connotations of irrationality that are not necessarily justified, as I have sought to demonstrate in my *Intercritique of Science and Myth*.[251]

But this dual singularity of Spinozism also has the effect, suggested by Maimon's curious formula, of freeing the issue of sources from the chronological straightjacket and setting it in a timeless interpretive system. Instead of quarrying Spinoza for elements borrowed from kabbalah, Neoplatonism, Stoicism, Descartes, and other sources, we can consider these doctrines to be "expansions" of his thought in the sense of interpretations, and even exegeses, with the pluses and minuses of all exegesis. In the normal course of the world, a philosopher is interpreted by those who come after him. But a timeless philosophy can also be illuminated by earlier doctrines, with, as always, distortions in this or that direction. Spinoza's dialogue with the philosophies on which he drew is located in this timeless realm, and "in eternity there is no such thing as when, before, or after" (*Ethics* I 33, Note 2). Ironically, this is identical with a major principle of the rabbinic exegesis of the biblical text, for all that he mocks it in the *Tractatus Theologicus-Politicus*: "There is neither before nor after in the Torah."[252] According to this principle, when we are dealing with the Torah, meaning wisdom, and not merely history, then—as in eternal philosophy—explanations and interpretations follow an order that is not necessarily the chronological sequence of the incidents reported by the text.[253] This is how we can employ Spinoza's sources, including certain doctrines of philosophical kabbalah, as if they were exegeses and interpretations whose content and scope can be understood only by virtue of the underlying doctrine—Spinozism—on which they build.

The partisans of a particular religious orthodoxy may dispute this way of reading and studying texts with the help of "outside" philosophers. Let us say only that doing so is part and parcel of an ancient tradition of rabbinic orthodoxy, which can be traced from Philo of Alexandria, who employed Greek philosophy in order to interpret the Bible, through Maimonides, who read the Bible and Talmud as if they were commentaries on Aristotle, and on to Rabbi Abraham Isaac Hacohen Kook, who drew on Plato to understand certain points of kabbalah.[254] With Rabbi Kook, one

249. Guéroult, *Spinoza*, vol. 1, p. 9.

250. See above, nn. 219 and 239.

251. Atlan, *Enlightenment*.

252. See, for example, Rashi on Genesis 35:29 and on Numbers 9:1, drawing on B *Pesaḥim* 6b. The medieval tosafist Rabbenu Peretz, in the name of his teacher Rabbenu Jeḥiel, cited a tradition that God mixed up the order of the sections of the Torah because, had they been given in the same order as actual events, human beings could have employed the text to create whatever they wanted. See the marginalia in the Steinsalz edition of the Talmud, ad loc.

253. A more recent philosophy could resolve these ancient problems or make traditional difficulties evaporate. But it can also create new ones, because of presuppositions that may lead it into dead ends. The return to the "sources" as interpretive systems sometimes makes it possible to recover and continue. This was the method employed by Salomon Maimon, who held that philosophical systems must be used alongside one another, and not against one another, ahistorically, in order to extract the grains of truth that all of them contain. This is how, becoming a Kantian near the end of his life, after having first been a Spinozist and then a Leibnizian, he could demolish the splendid edifice of transcendental philosophy with the help of Spinoza, Leibniz, and Hume.

254. In this he seems to have been reviving a legendary tradition that was developed by Abraham Cohen Herrera and Joseph

of the last kabbalistic masters of the twentieth century, the circle is closed, because he wrote freely about Spinozism without viewing it as heretical or as an essential contradiction of his own current of thought. Quite the contrary, he found in it the absolute monism of his tradition. This did not prevent him from justifying Spinoza's excommunication by the Amsterdam community, for reasons of pedagogy rather than doctrine. But in a dialectic typical of his thought, these arguments bolstered his evaluation of Spinozism as a doctrine that could have occupied an important place in the rabbinic tradition. His argument is worth looking at. We should remember, to begin with, that the reasons for Spinoza's excommunication are not to be found in the finished doctrine of the *Ethics*, which the leaders of the Amsterdam Jewish community could not know, since the work had not yet been published or, most likely, even thought out. The impetus for the excommunication, as we have seen, probably derived from social problems associated with the specific situation of this community and its reaction to nonconformist behavior.

Rabbi Kook imagined what might have ensued had the community not excluded Spinoza from the Jewish people, whatever its reasons. Then, he thought, Spinoza would have written in Hebrew and been recognized as an important chain in the rabbinic tradition. He was glad, nevertheless, that things had worked out as they did, because he found Spinoza's mode of expressing this tradition to be dangerous. He saw it as "an aristocratic system fed by dry reasoning," in which reflection about "Divine 'essence,'" transmuted into "a subject of professional research," risks being transformed into the invocation of an "alien and pagan god." Had it been written in Hebrew, employing the vocabulary of the Torah and halakhah, Rabbi Kook believed, it would have imperiled the spiritual and national status of Israel. In other words, had Spinozism been presented as part of the rabbinic tradition—quite conceivable from a doctrinal perspective—it might have consigned to oblivion "'the Divine ideals' inherent in the spirit and practice of Judaism," in which God cannot be known in himself, but only through his relationship with the world and with human beings.[255]

There is a clear tension here between Rabbi Kook's positive evaluation of Spinozism, based on his own doctrines,[256] and his concern as rabbi and educator responsible for the community. Quite naturally, his disciple David Cohen devoted several pages to Spinoza in a work on prophecy and rational thought, one major section of which is devoted to a historical survey of the relations between kabbalah and philosophy. In a chapter titled "Infinite Extension and Infinite Thought (as

Delmedigo, who saw Platonism and pre-Aristotelian Greek philosophy as offshoots of ancient Jewish wisdom.

255. See Tzvi Yaron, *The Philosophy of Rabbi Kook*, Eng. version Avner Tomaschoff (Jerusalem: World Zionist Organization, 1991), pp. 46–48.

256. Rabbi Kook's is a philosophy of life, expressed in a body of work that is original in both its form and its underpinning, full of explicit or allusive borrowings from various currents of kabbalah. In its complex dialectic between inspired poetry and science, emotion and reason, the individual and the collective, we often find a desire to satisfy conflicting needs that correspond to diverse human experiences to which he attaches value, despite their differences; but he pays for this in the coin of linguistic paradoxes and tensions, as in his discussion of Spinoza. (See, in addition to Yaron's book, Henri Atlan, "Etat et religion dans la pensée politique du Rav Kook," in *Israël, le Judaïsme et l'Europe*, ed. J. Halpérin and G. Levitte [Paris: Gallimard, 1984], pp. 32–64.) According to the late philosopher and polymath Yeshayahu Leibowitz (personal communication), Rabbi Kook rendered a similar verdict with regard to the kabbalah itself: although it is true Torah, it exposes one to the peril of idolatry. Rabbi Kook justified his attachment to Maimonides, despite the differences in their philosophy with regard to this true Torah, by Maimonides' capacity to serve as a bulwark against this danger.

attributes of God): Intellectual Love," he relates these essential elements of Spinoza's ontology to talmudic and kabbalistic texts with clear pantheistic connotations, in which God is at the same time the "place" (*makom*) of the world and the content of the world. Despite his admiration for the "philosopher of Amsterdam," as he calls him, Cohen demonstrates the "heretical" character of his thought, distinguishing an orthodox panentheism ("everything is in God") from a heterodox pantheism ("God is in everything").[257]

Of course, the absolute monism of thought and matter, in which infinite Extension is an attribute of God, is fully compatible with the kabbalistic philosophical tradition to which Cohen belongs. He quotes several traditional interpretations of God's name *Hamakom* (literally "the place") as designating infinite *space*, notably in the work of Ḥasdai Crescas.[258] But it is what he calls Spinoza's pantheism ("God is all") that he considers to be the root of his heresy. For, according to the Talmud and Midrash, God "is the place of the world, but the world is not his place."[259] Nevertheless, "no place is empty of him," which entails that, at least in some sense, "he fills the entire world." As Cohen notes, this is stated explicitly by Ḥayyim Volozhiner in *Nefesh ha-ḥayyim*, who writes that "He surrounds [Hebrew *sovev*, which also has the sense 'causes'] and fills all the worlds" and explains the several meanings in which these two aspects must be understood.[260] Cohen refers to these doctrines, as well as similar ones expounded by other rabbinic texts of undoubted orthodoxy, including the *Tanya*, the foundational tract of intellectual Hasidism (Chabad), in which he finds clear pantheistic tendencies. Where this differs from Spinoza is in its recognition of an infinite that is *outside* and distinct from the world (which it encloses and causes) in addition to the infinite that fills it; that is, of transcendence associated with the immanence. Nevertheless, from the various kabbalistic texts that Cohen quotes it is clear that the difference is not so clear-cut as pretended by their authors, who refer only to formulaic articles of faith and theological dogmas. We really have to do with disagreements about words whose definitions are certainly confused, given that several kabbalistic philosophers have defended totally immanentist hypotheses—even though they too are suspect, according to Cohen, of teaching erroneous doctrine.[261] He mentions a liturgical "Song of Unity," attributed to the tenth-century scholar and philosopher Saadia Gaon, that is recited in some synagogues but banned in others: "Cause of all, filling all, with nothing above you and nothing below; being all you are in all, there is no place unfilled by

257. David Cohen, *Kol ha-nevu'ah: Ha-higgayon ha-'ivri ha-shim'i* (Jerusalem: Mosad Harav Kook, 1970), pp. 116–129. Cohen understands Spinoza's thought, in a curious and totally erroneous fashion, as an "idealism of negation" (p. 119), which he compares to Kant and Berkeley, although we know, on the contrary, that Spinoza's philosophy is in every respect affirmative and positive. Perhaps he, like so many others, was misled by Hegel's modification and tendentious interpretation of Spinoza's formula that "all determination is negation" (see above, n. 73).

258. Ibid., pp. 120–123.

259. <*Gen. Rabbah* 68:9 (et passim).>

260. Cohen, *Kol ha-nevu'ah*, p. 123.

261. To begin with, radically pantheistic expressions can be found in the writings of many kabbalists recognized and accepted by orthodoxy; for instance, "He is everything and everything is He" (Joseph of Hamadan [see above, n. 174]); "everything is in him and he is in everything" (Meir Ibn Gabbai); and others cited by Gershom Scholem in his chapter on "The Kabbalah and Pantheism" (*Kabbalah*, pp. 144–152). In addition, as Scholem writes, "the pantheistic tendencies in this line of thought are cloaked in theistic figures of speech, a device characteristic of a number of kabbalists" (ibid., p. 147). In practice, the issue of whether the kabbalah is pantheistic does not seem to have been posed in so many words before the publication of Wachter's study of Spinoza, which aimed to lay bare the kabbalistic sources of Spinoza's pantheistic heresy. Ever since, historians and commentators have disagreed about this question, which is in fact inappropriate and poorly phrased, because "much depends here . . . on the defini-

you."[262] Similarly, the numerological equivalence of *'Elohim* and *ha-ṭeva'* 'nature' had been used to introduce the concepts of *natura naturans* and *natura naturata* (*ṭeva' maṭbia'* and *ṭeva' muṭba'*) in a sense very close to that given them by Spinoza.[263] Spinoza's thought cannot be characterized simply as pantheism, in the common theological sense, any more than as acosmism, as is sometimes said. The subtle relationships between infinite substance and its attributes and modes, both infinite and finite, cannot be reduced to these muddled categories. Without constituting an "outside" or a transcendence vis-à-vis the world and nature, the infinity of infinite attributes (in addition to the Thought and Extension that constitute God or Substance, but whose essence or nature cannot be comprehended by human beings, who are only body and thought) postulates the infinity of "God/Nature" as essentially different from what we can picture through our human nature, although we can have a rational conception of its existence and absolute unity. Even the affirmation of absolute determinism and the negation of free will, which do not rule out (quite the contrary!) true freedom and the intellectual love of God, sometimes presented as *the* Spinozist heresy, can be found, at least by implication, in various orthodox authors,[264]

tion of a concept which has been employed in widely different meanings" (ibid., p. 144).

Among kabbalistic writers themselves, the correlative of this point was treated in a more technical fashion: Is the substance or "essence" (*'aẓmut*) of the one and undifferentiated Infinite identical with that of the *sefirot* that emanate from it and, through them, with that of the multiplicity of objects and beings in our world? More specifically, the question is focused (notably by Moses Cordovero in his *Pardes rimmonim*) on the nature of the first *sefirah*, *Keter* ("crown"), which is itself sometimes designated "Infinite" (*'ein sof*). Is it of the same nature as the undifferentiated Infinity from which it proceeds, or is it only an expression of the infinite creative power? That power is depicted as the first "garment" (*malbush*) of the Infinite, which is manifested through an infinity of realities that are progressively less "refined" (*ẓaḥẓaḥot*) and more differentiated in their existence. The garment itself, which is the subtlest and closest to the "body" of the Infinite and makes visible what it covers, is none other than the primordial "Torah," taken as a fabric woven of the letters of the alphabet and the infinite meanings potentially contained in their combinations, in the combinations of their combinations, etc. Interpreting these questions—and the mythical images used to answer them—in theological terms of pantheism and theism, or even of immanence and transcendence, produces rampant chaos. We can always find expressions that seem to support either position equally well. When we analyze their context, however, we notice that some ostensibly "pantheist" expressions—such as "God is in everything and everything is in God"—go with a "theist" discourse in which the accent is on the break represented by creation ex nihilo, through which the One, in an act of free will, calls the worlds and their multiplicity into being. We also observe, however, that some "theistic" expressions that stem from the standard orthodox theology of God the creator in fact conceal a pantheistic vision in which, to use the classical Neoplatonic formula, "all comes from the One and all returns to the One."

According to Scholem, Cordovero offers the most specific and detailed analysis of this question; he demonstrates the extent to which that kabbalist was torn between these two positions and resolved the tension by means of the formal distinction between the unity of essences, in which the Infinite is always present, one and identical in all things, and the multiplicity of existences, which involves only creatures and is contained in the Infinite only in a "hidden" or "potential" fashion. We also find in Cordovero the unmistakably panentheist formula "God is all that exists, but not all that exists is God" (Scholem, *Kabbalah*, p. 150, quoting *Sefer 'Elimah* 24d). The theological debate, whose character is unsuitable for the formal system on which it is projected, may recall the issue of the identity or difference, for Spinoza, of *natura naturans* and *natura naturata*. As we saw above (n. 113), the name *'Elohim*, the *creator* God of Genesis, is also taken to represent *created* (although divine) nature, as the object of the verb "created" in the first verse of Genesis. Finally, we should recall that Spinoza attributed the panentheist declaration that "all things are in God" to, among others, "all the ancient Hebrews" (see above, n. 219).

On the formula, "God is in everything and everything is in God," and other explicitly pantheistic and immanentist expressions of various kabbalists, including Abraham Abulafia and Joseph Gikatilla, and their influence on eighteenth-century hasidism, see also Idel, *Kabbalah, New Perspectives*, pp. 144–146 and 153–154; idem, *Hasidism*, pp. 17–18.

262. Cohen, *Ḳol ha-nevu'ah*, p. 124.

263. Ibid., p. 125.

264. As we saw in Chapter 2, with regard to the theses of Shlomo Eliaschow, one current of thought among the disciples of the Gaon of Vilna employed traditional doctrines drawn from the Talmud and Midrash to carry the affirmation of absolute determinism to its ultimate limit.

notably Ḥasdai Crescas, one of the only two Jewish philosophers (along with Abraham Ibn Ezra) to whom Spinoza relates favorably. In brief, from Cohen's book we gain a better understanding of the place that the *Ethics* would have occupied in the corpus of rabbinic literature—controversial on account of its terminology, no doubt, but probably no more so than many other works—had it been written (as in Rabbi Kook's alternative history) in Hebrew and had Spinoza not separated from his ancestral community before writing it. We can also imagine that, in these circumstances, he might still have published extensively in Latin, for the greater benefit of philosophy, just as, some centuries earlier, the eleventh-century rabbi and Hebrew poet Solomon Ibn Gabirol wrote a work in Arabic that, in Latin translation and attributed to an otherwise unknown Muslim author named "Avicebron," was incorporated into the philosophical tradition of Christian Europe.[265]

Whatever the case, we can certainly endorse Maimon's curious and ambiguous description of kabbalah as "expanded Spinozism." For us this means going back and forth without ever merging the two extremes, getting to understand Spinoza better with the help of this philosophical tradition on which he drew and applied in his own way but also coming to understand these texts in the light of Spinoza's thought, with which they share a contemporary and "eternal" value.[266]

265. Solomon Ibn Gabirol (Avicebron), *The Fountain of Life = Fons Vitae* [trans. Alfred B. Jacob] (Stanwood, WA: Sabian Society, 1987); see above, n. 19.

266. The permanent relevance of Spinoza's thought, constantly renewed in its successive and contradictory interpretations (see above, n. 83), can be found, *mutatis mutandis*, in the series of interpretations of the kabbalah, which "determined the enduring interest it enjoyed in various eras" (Scholem, *Le nom et les symboles de Dieu dans la mystique juive*, p. 196). In "The Study of Kabbalah from Reuchlin to Our Own Days" (ibid.), Scholem enumerates the successive fundamental notions of kabbalah held by European circles whose marginality depended on the age, from the time of Johannes Reuchlin and Pico della Mirandola, in the fifteenth century: an original adamic revelation and tradition going back to Moses and the Patriarchs, a secret Christian doctrine, Spinozist pantheism, Gnostic doctrine, theosophy in the manner of Jacob Böhme, and so forth. In succession to the myths and legends that, until the Renaissance, it shared with Alexandrian hermeticism, its association with the supposed atheism of "Spinozism" and then with Gnosis introduced it into the history of European philosophy in the eighteenth century through its influence on incipient German idealism. Within the rabbinic corpus per se it was the domain of "arcane lore" (*ḥokhmat ha-nistar*), whose circulation varied from age to age but was always restricted to the elite, those best suited and most worthy in terms of their intellectual capacity and moral qualities. Only in the twentieth century, when Scholem turned kabbalah study into an academic discipline, did the outside world come to perceive it as part and parcel of traditional Judaism. In general, though, it has been viewed as a heterodox and marginal excrescence, some sort of magical and superstitious occultism or, in the best of cases, as an esoteric mysticism wholly inaccessible to rational understanding. Like Spinozism, it led one to little-traveled paths, sometimes crossing the traces of that "wild anomaly" of European philosophy (to use Negri's lovely label for Spinozism), with the same combination of marginality and new relevance. Unfortunately, despite the explicit strictures and warnings pronounced by some rabbis, its longstanding historical and sociological association with religion has exposed kabbalah to the perils of enthusiasm and credulity, the customary nourishment of sects and cults.

CHAPTER 5

THE DESACRALIZATION OF CHANCE: FROM ORACULAR LOTTERY TO THE INDIFFERENCE OF THE RANDOM

1. CHANCE AND CASTING LOTS

Chance has not always been synonymous with an absence of meaning or with the random in the sense of indifferent. Quite the contrary: in magical thought it often serves as an oracle by which the deity is made manifest. Casting lots makes it possible to overcome a lack of knowledge and to reach a decision in situations of uncertainty. Far from being indifferent, the lottery reveals hidden things and gives voice to destiny at the same time as it determines it. In other words, casting lots is a technique of divination, a way of making the gods or demons speak and of gaining access to the knowledge of hidden things, inaccessible by the normal paths of experience that employ the senses, reason, and imagination.

There are many examples of this. The Chinese tradition of divination, the I Ching, with its elaborate system of interpretation, is based on casting sticks; their configuration on the ground is interpreted as providing indications about the past and future life of the person who threw them. The mystery that throwing the sticks at random makes them fall in a pattern that relates specifically to the person interrogating his fate certainly cannot be understood in any "clear and distinct" causal context. On the contrary, our ignorance of the mysteries of the future and of destiny seems to be redoubled by the lottery procedure, which is itself based on ignorance of the causes that produce a particular configuration. It is this increase in ignorance that seems to play the role of exorcism or conjuration: the mystery of casting lots conjures up the mystery of destiny. The lot literally "speaks" to us. The same holds for many other techniques of divination still in use all over the world, including the West, such as fortune-telling by means of tarots and other sets of cards.

This was also clearly the case in antiquity. Hence it is not astonishing that the casting of lots is employed in many circumstances, some of them dramatic, that punctuate the narratives and rituals of the Hebrew Bible. We may recall the ritual of the Day of Atonement, when the scapegoat is selected at random from two identical animals, one of which is consecrated to YHWH while the other is sent to "the devil," bearing all the sins of the community.[1] Later, in the story of Esther and the festival of Purim (the word itself means "lots"), we find Haman casting lots to determine the appropriate day for massacring the Jews. As a matter of etymology, the French word *sortilège* ("spell, charm, enchantment") derives from the use of lots in methods of sorcery and divination; the Hebrew word *goral* designates both the object (stone, slip of paper, or what have you) used in the lottery procedure and "destiny" in general. We see that our destiny is decided by a throw of the dice—like the image of "God playing at dice" that Einstein rejected in his criticism of the essential and irreducible indeterminacy of subatomic physics. But precisely this relationship is inverted and exorcised in the oracular procedure; contrary to appearances, the lottery does not determine destiny. Rather, destiny is determined by hidden causes that we cannot know but which the lottery discloses.

We encounter this ambivalence about the chance nature of lotteries, a respectful fear of the oracle of destiny combined with suspicion of the arbitrary and absurd character of such a mode of knowledge, throughout the Bible and especially in its exegesis by the rabbis.

There we find a juxtaposition of these two visions, that of the random lottery that links magical thought with the sacred and that of a desacralized randomness whose role in the management of human affairs is a makeshift one endeavors to reduce to a minimum. This desacralized randomness is asserted definitively in the Talmud, especially in legal affairs, foreshadowing the modern uncertainty that is itself the successor to "the chance and fortune" of the ancient Greeks. For Aristotle, as we know, a chance event does not lack a cause, but only a purpose or *final* cause. Among his famous four causes, it is the final cause that truly explains a phenomenon by revealing what it means. When two individuals meet, even if they take advantage of their encounter to deal with their mutual affairs and even if the meeting is produced by obvious efficient and material causes, it nevertheless remains fortuitous if it is not the result of some final cause, which may take the form, for example, of an interest of at least one of them to instigate their encounter. Also, although it may seem paradoxical today, one of his words for chance is *automaton*,[2] in the sense of an event or object that comes of itself, with no final cause or *telos*, intentional or otherwise, to give it meaning. Thus the robot-automaton, with its contemporary connotations of rigorous mechanical determinism, goes back to the term that designated chance and fortune for Aristotle. What has happened is simply that, in the interim, the notion of causality and causal explanation has been transformed. In parallel, the cognitive implications of relying on chance and lotteries have changed their meaning.

The desacralization of chance was completed in the seventeenth century, with the advent of rational mechanics and the principle of sufficient reason, which for all practical purposes reduced Aristotle's four causes to one: the efficient cause. No event can take place unless it has an efficient cause, and knowledge of that cause suffices to explain it. This idea has been diversified since then; the cause may be replaced by a set of causes or, especially, by a causal law whose generality and abstraction are greater than the effect it produces and explains. Nevertheless, the mechanical explanation no longer invokes the final causes in nature that formerly

1. See below, n. 30.
2. Aristotle *Physics* 2.

gave meaning to events. Since the eighteenth century, such final causes have found asylum—temporarily, and only for the idealist heirs of Leibniz and Kant—in the attributes of the human soul and in the suprasensible universe of moral law and freedom. Today, randomness is no longer merely the lack of a purpose that would provide meaning. This is because in our world, determined by physical laws and in which final causes no longer have a place, *every* natural event that is not produced by human art and planning lacks an end and purpose. The only conceivable end is the formal one of maximum or minimum, in mathematical physics; but this is just another way of expressing mechanical causality. Modern randomness refers to what is produced without a cause that is known, or even knowable, and whose isolated incidence accordingly cannot be predicted. The only possible prediction of such events—and even it is not always available—is a statistical forecast, which no longer relates to the occurrence of single events but only to the set of events we construct by adopting a perspective that makes them indistinguishable from one another.

Underlying this method is the law of large numbers. Much ink has been spilled in the attempts to give this empirical law a theoretical basis, from Pascal and Leibniz through the statistical interpretations of quantum physics. According to this law, if a random phenomenon—for example, casting lots, where the outcome cannot be predicted because we do not have adequate knowledge of all the causes that produce it—is repeated many times, we can predict the results as an average value for all these repetitions. The more times we cast lots, the more precise the predicted average; or, to put it a different way, the closer to certainty is our knowledge of the result for all of these many events taken together.

In other words, thanks to the law of large numbers, our lack of knowledge of the causes of a phenomenon or an individual event is replaced by knowledge of the mean of a set of phenomena that are supposed to be identical. Statistics and the computation of probabilities constitute a powerful method for taming and controlling chance, but their fundamental hypothesis is that all events or all individuals that make up a set of random phenomena can be taken as mutually indistinguishable. This is why the most spectacular applications of this law are in statistical thermodynamics, where we can postulate almost infinite collections of identical molecules. But similar results can be obtained for human mass behavior, such as visitors to a public place, public transport, traffic, opinion polls, and so on. This is possible, of course, only if we take a large number of individuals into account. The causes of individual actions are multiple and unknown to the observer-statistician; but, as with molecules, the mean behavior of the whole is known with great precision. In the human case, each individual knows—or thinks she knows—the causes of her actions and even thinks she decides freely, perhaps after deliberation. Nevertheless, the behavior of the group is determined, inasmuch as it obeys a statistical law that describes and makes it possible to predict it.

Some have seen this as a possible solution to the theological problem of human freedom in a world that is subject to the determinism of an omnipotent and omniscient God.[3]

3. See, for example, Ely Merzbach, "Random Processes in Rabbinic Texts" (Hebrew), *Higgayon* 1 (1989), pp. 31–38. The author, a mathematician at Bar-Ilan University, provides extensive documentation of various aspects of chance and fate in biblical and talmudic literature. It is not certain, however, that his proposed definitions are relevant in the contemporary scientific context where he would locate them, if only because he does not distinguish between the final causality of intentional human actions and the efficient causality of physical phenomena (unlike the kabbalist Joseph Gikatilla, who made this distinction

Unfortunately, rather than offering a solution it actually does away with the problem, because it requires that we treat human beings as if they were molecules in a thermodynamic system. Alternatively, it is merely one way—and probably the simplest—of denying the reality of free will, leaving only our subjective illusion of it as individuals.

Elsewhere[4] I have analyzed how we can conceive of a linkage between the behavior of individuals, in varying degrees of enslavement and autonomy, and that of the group produced by the aggregation of their individual behaviors.

We have also seen how, following some streams of the rabbinic tradition, we can imagine an absolute determinism in which free will is only an illusion, which is, however, necessary and thus real, insofar as it is one of the givens of human situations and human nature.[5] As such, and in keeping with the ideas in the *Ethics*, this imaginary free will, which we conceive in our ignorance of the determining causes, is to be distinguished from the freedom revealed to us by progress mastery of Spinoza's form of knowledge, which is both physical and intellectual.

In all cases, we are far from the oracular notion of casting lots to reveal the truth of future events in advance. Today, casting lots means acknowledging that we cannot know causes (or causal laws) and accepting, by convention, that the elements of a set of events cannot be distinguished from one another and are indifferent to whatever predictive knowledge we may have of the causes of their individual occurrences. Whether this ignorance of causes is temporary and associated with an inadequacy of our cognitive abilities, or permanent and associated with some essential indeterminacy, is then merely a metaphysical question about the nature of chance: Is it the chance occasioned by ignorance or is it essential chance?

This is clearly the case with the chance of an isolated fortuitous event, classically defined as the intersection of two independent causal series. Their encounter cannot have a known cause, inasmuch as the two series would then no longer be independent. Since the seventeenth century, when Fermat and Pascal invented the calculus of probabilities, with all its subtleties, the nature of mathematical chance has become increasingly complex. In the wake of the advances of mathematical physics and the study of dynamic systems in the nineteenth and twentieth centuries, chance has become an increasingly sophisticated object of theory. The opposition between random and determined, between chance and ignorance of causes, is no longer so simple today. The computation of microscopic indeterminacies in physics produces a quasi-deterministic knowledge of macroscopic laws. Conversely, the more recent discovery of deterministic chaotic states demonstrates that deterministic dynamic systems, defined by known equations, may behave in a stochastic fashion that we cannot distinguish from what might be produced by a lottery. Finally, the computer programs that generate random sequences by deterministic procedures—for example, the decimal sequence of pi—show that chance can be produced even if we are not ignorant of causes. For programmers, however, a long random series is defined by its "incompressible" character, that is, by the fact that it cannot be generated by any program

back in the thirteenth century; see Chapter 6). In general, it is extremely difficult to assimilate the oracular function of the lottery in the mythical and prophetic context of the Bible to its purely operational role in the modern world. As we shall see, the talmudic discussions on this subject are particularly interesting because they occupy the seam between a world that is full of the divine presence and a world in which the gods no longer speak.

4. Atlan, *Tout, non, peut-être*, chap. 1, "Le théorème de von Foerster–Dupuy."

5. See Chapter 2.

shorter than itself. This means that a long series produced by a "random-number generator" is not random (in this sense), except for those who do not know how it was produced. But this generator is a program that is usually much shorter than the series itself, which could, in theory, extend to infinity.

Still, the standard of the fortuitous and the random is the lottery, a situation of a priori equivalence and indifference in which prior knowledge of which lot will be selected is impossible.

Whatever the metaphysical status of mathematicized chance and its role in physical laws, the effect on our knowledge of things and on our management of human affairs based on this knowledge is the same. Modern desacralized randomness is the polar antithesis of the divinatory lottery. When we cast lots, it is to reach a verdict in a situation of a priori indifference and equivalence where a choice must nevertheless be made. Casting lots or rolling dice no longer brings to light a hidden knowledge and the "correct" choice that stems from it, but only the partners' consent to the selection process, *despite* its arbitrariness. The decision is not the result of knowledge that the oracle is held to reveal but of an agreement reached by convention, because we have no better way to decide in the absence of such knowledge. To put it another way, it is a makeshift to which we must resign ourselves when we have no other method to hand.

In this context, the invocation of chance appears to be the very negation of any attempt to manage our affairs on the basis of rational knowledge. It is surrendering without a fight, renouncing action, submitting to the lots however they fall out, not because they are held to articulate our destiny and the meaning of our existence but, on the contrary, because they offer the most definitive and irrevocable expression of the absurdity of our condition. One of its "highest" forms is Russian roulette, in which the absurdity and a priori lack of meaning are not only recognized (that might be the "beginning of wisdom") but are in fact actively sought out.

We can understand, in this context, why the Talmud considers games of chance to be an archetypal destructive activity, irrational in the sense of mad and suicidal, to the point that those who engage in them are automatically disqualified as witnesses in judicial proceedings, like all persons suffering from dementia.[6]

Thus we see the evolution from the sacred chance in which the oracle speaks to the destructive chance that asserts and institutes the reign of the arbitrary. But it is a short path between the two, because in both cases it comes down to giving up any attempt to understand: in the former, overwhelmed by the presence of the mystery that explains everything, even if we do not know how; in the other, felled by the absurdity in which nothing is explained as long as not everything is explained. For a thinking being, giving up the effort to render the world intelligible is neither more nor less than giving up on life, if mind and body are really one and the same thing and the life of one is the life of the other. Nevertheless, the same effort produces both the most astonishing discoveries and the wildest illusions when in our impatience we cannot accept the unexplained and prefer any explanation, however crazy, to ignorance of causes and a randomness of events. To put this another way, mastery of the random, even if desacralized as it is today, remains a knotty problem for our relations with objects. When must we accept the verdict of chance, and when not? In what conditions can the verdict of such an arbiter be other than arbitrary?

6. See Atlan, *Enlightenment*, p. 377 and n. 8.131.

This question applies even more acutely in the liminal situations where lots are cast to uncover a guilty party or—what amounts to the same thing—to designate a sacrificial victim. René Girard has made important contributions to the analysis of the subtle mechanisms of the rite of the scapegoat and its foundational role in the social regulation of the violence of all against all.[7] But Girard's theory leaves out the specific dimension the rite sometimes receives from the lottery that selects a living and conscious victim. Elsewhere[8] I have shown how Girard, in the second part of his *Things Hidden Since the Foundation of the World*, makes repeated use of an unsupported postulate that is belied precisely by analysis of mythical narratives in which the victim, assumed to be guilty, is designated by lot. We can review his thesis briefly. Girard takes it for granted that the scapegoat mechanism can function only in a state of ignorance. Its participants must believe in the victim's guilt or at least responsibility. They must not know that they are merely offering a sacrifice to a ritual in which it is their own violence that is channeled in pursuit of social concord. The mechanism would not work were it exposed, as in the relatively unsophisticated Freudian interpretation that mere awareness of a repressed conflict can keep it from exerting its neurotic effects. For Girard, the gospel of Christ's love is just such a revelation, begun in the Old Testament and completed in the New. But the mechanism, once unmasked, can no longer function. Modern societies can either return to the apocalyptic violence of all against all or transcend this violence through Christian love; no other option is available.

7. See especially: René Girard, *Violence and the Sacred*, trans. Patrick Gregory (Baltimore: Johns Hopkins University Press, 1977); idem, *Things Hidden Since the Foundation of the World*, trans. Stephen Bann and Michael Metteer (Stanford, CA: Stanford University Press, 1987).

8. Henri Atlan and J.-P. Dupuy, "Mimesis and Social Morphogenesis: Violence and the Sacred from a Systems Analysis View Point," in *Applied Systems and Cybernetics*, ed. G. E. Lasker (New York: Pergamon Press, 1981), vol. 3, pp. 1263–1268; Henri Atlan, "Founding Violence and Divine Referent," in *Violence and Truth: On the Work of René Girard*, ed. Paul Dumouchel (Stanford, CA: Stanford University Press, 1988), pp. 192–208.

But it has not been proved that awareness of the unconscious mechanisms of this rite suffices to eliminate its effects, especially those that permit social cohesion. As in the case of the Freudian interpretation, awareness may lead to a transformation of these mechanisms—by a sort of sublimation—rather than their disappearance, if it turns out (as Girard himself demonstrates in the first part of his book) that they are indispensable for maintaining social harmony.

2. A REREADING OF JOSHUA 7:1–19

At first sight, certain mythical narratives about lotteries, such as the biblical story of Jonah, seem to support Girard. Thanks to its oracular power, the casting of lots really can designate the guilty party; those who employ the procedure believe in good faith in the guilt of the person on whom the lot falls. Even in the book of Jonah, however, this belief needs to be reinforced and confirmed by his confession. What would have happened had Jonah challenged the result of the lottery, instead of confessing and encouraging the sailors to throw him overboard so as to still the anger of his god? But this dialectic of lottery and confession can be found in another and even more dramatic biblical story and especially in the Talmud's reading of it. The dialectic is further complicated by the fact that biblical and talmudic law attach no value to a suspect's confession as a way of establishing his guilt; only testimony about material facts, by witnesses who must also prove their own credibility, is admissible.

Our case involves an episode in the Israelite conquest of Canaan, under the leadership of Joshua. This war, unique in Jewish history and law,[9] was subject to the law of "proscription" (*ḥerem*), which required that everything associated with Canaanite culture be destroyed. This included an absolute ban on taking booty; doing so would be a violation of the covenant between Israel and its god. We learn that "Israel has sinned [collectively]" (Josh. 7:11), inasmuch as some plundered items have been concealed among their belongings; consequently they were defeated in battle and "the hearts of the people melted and became as water" (Josh. 7:5).[10] The covenant has been violated: "The Israelites will not be able to hold their ground against their enemies, . . . for they have [themselves] become proscribed" (Josh. 7:12). The only way to lift the proscription from the people of Israel is to identify the guilty party, locate the hidden spoil, and destroy it.

It is at this juncture that YHWH, the god of Israel, instructs Joshua as to the procedure to be followed. First he is to examine each tribe until one of them is "caught" by YHWH; then each of the families in this tribe until one of them is "caught"; then each of the households of this family; and, finally, within the household that has been "caught" in its turn, each of the men—until one of them is "caught" and thereby designated as the guilty party, to be burned with all his possessions. The successive "captures" are generally interpreted as involving one or several oracular practices.[11] Whatever the method, it is a lottery that plays the decisive role.

9. The rabbinic tradition holds that this war was unparalleled in the history of Israel (see, for example, Maimonides, *Mishneh Torah*, *Hilkhot melakhim*, on the several categories of wars that the Israelites/Jews might have to wage and the laws applying to each). The conquest of the land was to be accompanied by a total proscription (*ḥerem*) of the inhabitants, who were to be killed, and of their property, which was to be destroyed. This injunction appears repeatedly in the Bible as a command that the God of Abraham, Isaac, and Jacob gave to the tribes of Israel when they were about to take possession of the land that had been promised to their ancestors. But it is not to be applied, in any fashion, to any war subsequent to this founding event. A divinely enjoined total annihilation, it is more closely related to a Maoist "cultural revolution" than to "ethnic cleansing" aimed at preserving the homogeneity of a family or tribe. (Such homogeneity no longer existed in any case after the exodus from Egypt, because the Israelites led by Moses and then by Joshua consisted not only of the twelve tribes descended from the sons of Jacob but also of the "mixed multitude" of former slaves of all nationalities who were liberated along with them. What is more, the precept of extermination applied only to the "seven nations" that inhabited the Promised Land and not to others "from a distant country," as reflected by the story of the Gibeonites in Joshua 9.) A reading of the text suggests that the decadent Canaanite civilization was condemned for its own vices; the Israelites who, according to the Law of Moses they received in the wilderness, are coming to supplant them, are the means for blotting out all traces of their culture. This is the source of the proscription of plunder. As we see, the Israelites found it difficult to comply with this ban.

10. The narrative is prefaced by a laconic indication of the cause of the events to follow: "The Israelites, however, violated the proscription: Achan son of Carmi son of Zabdi son of Zerah, of the tribe of Judah, took of that which was proscribed, and YHWH was incensed with the Israelites" (Josh. 7:1). Within the narrative itself, however, the cause remains hidden until the climax, which is the discovery of Achan's guilt and his subsequent execution. Readers do not know whether the Israelites' insubordination was limited to Achan or was more general and extended to other individuals, unknown to Joshua and the people as a whole. In that case, Achan plays the typical role of the scapegoat, taking the guilt of the collective on himself. The Talmud analyzes this ambiguity at some length (B *Sanhedrin* 43b).

11. For example, the classical commentator David Ḳimḥi (Toulouse, twelfth-thirteenth century), writing on Joshua 7:13, associates the "capture" with the Ark of the Covenant, which designates a group or person by pulling them or him to it and freezing them in place. This could be interpreted as the effect of suggestion working on the guilty person, who gives himself away as a result of the terror produced by the ordeal. According to another interpretation offered by Ḳimḥi, the tribes were assembled in front of the oracle of the *Urim ve-Tummim* (see Exod. 28:10; see also Chapter 3) on the breastplate of the high priest, which was inlaid with twelve precious stones bearing the names of the twelve tribes. One of these gems darkened to designate the tribe among which the malefactor should be sought. The identification of the family, household, and finally the guilty individual then proceeded by the drawing of lots from an urn. Rashi, too, explains that the guilty party was designated through a lottery.

Achan, the guilty part (or at least *one* of the guilty parties),[12] is designated by this procedure. But Joshua is not content with this. He asks and even beseeches Achan to "pay honor to YHWH, the god of Israel, and make confession to him" (Josh. 7:19). Achan does so.

This text is the focus of a particularly rich and illuminating passage in the Talmud.[13] The Mishnah prescribes that those condemned to death should be encouraged to confess their crime before their execution. As mentioned above, under talmudic law confessions carry no weight for a court to render a verdict of guilt. The attempt to obtain a confession before execution could be interpreted as meant to make it easier for the judges to sleep at night or to "pay honor to the god of Israel" from which their authority proceeds. The Mishnah itself understands it as motivated by a concern for the condemned man, inasmuch as the death that expiates his crime will render him "innocent" again; in addition, he will be accounted a righteous person through his role in reinforcing the glory of God. The story of Achan is cited by way of illustration. But the Talmud raises a number of objections, notably about the procedure of trial by lottery, given that only the testimony of credible witnesses is normally admissible. Here the Talmud rewrites the entire story in the form of a dialogue it places in the mouths of the main characters.

In this retelling, a sort of dramatic midrash, Joshua begins by asking YHWH to tell him who is guilty, so that the Israelites can expiate the sin. To which YHWH answers that he is not an "informer"[14] and commands him to conduct a lottery.

From this we can understand that the lottery procedure does not express the word of God (at least not directly), since if it did so he would still be cast in the role of an "informer." It is hard to find any justification for this idea in the biblical text, unless it is the dialogue in verses 7–10, before YHWH makes Joshua aware that a sin has been committed and tells him how it can be expiated. Joshua bewails the military defeat and the weakness of the people. He and the elders prostrate themselves before the Ark of the Covenant and pray that Israel may again enjoy the support of its god. The divine response is that prostration and prayer are not appropriate here; Israel has sinned, and the only possible atonement is through the death of the guilty party. Two different logics seem to be superimposed here: one of dialogue and prayer, in which the god of Israel is personalized in some fashion, and one of an impersonal mechanism of crime and punishment. It is clearly the personal god who, according to the Talmud, does not want to be an informer, inasmuch as the lottery will set the atonement mechanism in motion.

Now, however, the talmudic retelling has Achan voice a fundamental objection to the lottery procedure. "How," he is supposed to have demanded of Joshua, "can you condemn me solely on the basis of a lottery? If I were to cast lots between the two leading men of the generation, yourself and the High Priest Eleazar, it would necessarily fall on one of you." In other words, the lot must fall on someone, and this has nothing to do with his possible guilt. The alternative is to suppose that all are guilty and that the lot merely designates one person, arbitrarily, so as to avoid having to condemn everyone. In either case, there is no justice here, but only the arbitrary designation of a scapegoat in a situation of equivalence, where all are equally guilty or equally

12. See end of n. 10.

13. M *Sanhedrin* 6:2; B *Sanhedrin* 43b.

14. The Talmud employs the Latin word *delator*. The same word is used in the Midrash and *Zohar* to designate the Accuser, otherwise known as "Satan" or "Sammael." In the ritual of the Day of Atonement, a lottery is employed to designate which of two goats will be offered in atonement. (See n. 29.)

innocent. Responding to this objection, Joshua (still according to the Talmud) asked Achan to confess, in order to "pay honor" to the god of Israel who enjoined that the lottery take place; that is, to justify a posteriori the casting of lots and to avoid casting aspersions on this method, because it will later be employed to apportion the land among the tribes.

The sage Ravina further discredits Joshua's procedure, suggesting that his request for a confession was merely a trick, intended to make Achan believe that he could escape death by admitting his guilt. If so, the trick is evidence that Joshua himself had strong reservations about condemning Achan solely on the basis of the verdict rendered by the lottery. As the passage continues, other rabbis amplify the scale and diversity of Achan's hidden crimes, which would justify his punishment in any case.

This story, which at first seems to be paradigmatic of the procedure of condemnation and execution, turns out to be a unique event set in exceptional circumstances and certainly not to be taken as a model. There are two good reasons for this, which we have already noted and which Maimonides expounds in his commentary on the Mishnah. First, guilt can be established only by the testimony of reliable witnesses, and not by oracles and lots.[15] Second, an individual may not testify against himself, which means that a suspect's confession cannot be used as the basis for conviction.[16]

So we are brought back down to the solid ground of talmudic law, which admits neither oracles nor divine intervention and where the Torah "is not in heaven." We still face the question of whether some theological meaning attaches to the story of Israel and its collective responsibility, in which the people's defeat is punishment for hidden individual sins that human tribunals cannot know about. This question is debated at length in the pages of the Talmud that follow our Mishnah, in which the sages seek to reduce the field of this collective responsibility to a minimum.

As for the issue of chance and its value as a source of oracular knowledge, these pages of the Talmud seem to capture the process of the transition from one mental universe to another, from the world of mythological or prophetic thought to that of modern desacralized thought. It is a transition from a world in which God speaks, states his view, and intervenes in human affairs by means of various manifestations (his "names" or his messengers, angels or gods), to a world in which there is no god, except perhaps in the consciousness of those who make him exist by defining themselves as his servants. In this world devoid of oracles and prophets, casting lots is just as arbitrary as rolling dice. Such arbitrariness is clearly unacceptable when justice must be rendered. It can still function for decisions that must be made in situations of indifference or an equivalence of the arguments, such as apportioning the land among the tribes when nothing supports, a priori, one allocation rather than another.

Why, then, this roundabout path, if that is where we will get to anyway? The answer is precisely what Joshua was seeking when he begged Achan to confess: to avoid totally discrediting the lottery procedure. Randomness and chance must still be taken into account, even in a desacralized

15. Num. 35:30; Deut. 17:6 and 19:15–19.

16. Finally, we must add that talmudic and subsequently rabbinic law always strives to avoid capital punishment and looks for any reason to doubt the veracity of the testimony heard. Application of this law made the death penalty almost unheard of, in total contrast with the literal sense of biblical law. We see that the latter was implemented through an interpretative lens that totally transformed the experience.

world where the authority of the prophet has been replaced by that of the judge, and knowledge through divination has given way to experience, memory, and rational argument. But why?

What role or roles may remain for the recourse to chance, such that we need to preserve it? The example of the allocation of the land among the tribes is instructive, because it is discussed elsewhere in the Talmud.[17] There we observe, though in another way, the same evolution from the sacred chance of divination to the conventional randomness of a world without oracles. The point of departure is the biblical text that commands that the division of the land be effected "according to [*lit.* by the mouth of] the lot" and "*only* by lot" (Num. 26:55–56). But the commentary is already part of a myth of origin, because this allotment of the land is the founding act of the establishment of the people of Israel in the Promised Land. The talmudic account tells of a miraculous procedure, in which the *Urim ve-Tummim* reveal to the high priest the name of the tribe whose lot is about to be drawn from the urn even before the lot is drawn.

The Talmud interprets "by the mouth" as referring to interrogation of the *Urim ve-Tummim*, which functioned as the "mouthpiece of destiny."[18] As for the expression "only by lot," it designates a lottery that is mute and double-blind, with the names of the tribes inscribed on lots contained in one urn and the names of the territories in a second urn. Yet the two procedures yielded the same result. According to this midrash, the high priest, wearing the *Urim ve-Tummim* on his breast, attuned his mind to the *ruaḥ ha-ḳodesh* or sacred spirit (one of the levels of prophecy) and announced the name of the tribe that was going to be drawn from the first urn and the territory that would be drawn for it from the second urn. Miraculously, the twin lottery coincided with his prediction. Why this curious procedure? What need was there for both a prophetic oracle and a lottery, with the risk that they might not coincide? The answer, according to the Talmud, is that the mutual reinforcement of the two procedures was meant to reassure the tribes that the distribution was fair. Otherwise there might have been disputes and contention, given the different populations of the various tribes and the disparate geographical locations and natural resources of the territories assigned them.[19]

So either procedure might have been deemed insufficient and questionable if used alone. In fact, the Talmud goes on to compare the allocation of territories "in this world" with another division of the land, alluded to by Ezekiel (Ezekiel 48), that will take place "in the world to come." This second, eschatological division will repeat the original one; but this time it will be perfect, because all the territories will be identical in the wealth and diversity of their natural resources. Furthermore, this ultimate apportionment will be made by God himself: "those are their portions, the word of YHWH" (Ezek. 48:29). According to the commentary on this talmudic passage by Rabbi Samuel Edels (known as the Maharsha, 1555–1631), the first division "by the mouth of the lot" is acceptable because it is temporary and affects only this world; but that of the future, eternal world will be made directly "by the mouth of YHWH."

Here, as in the story of Achan, we encounter the duality of the unmediated divine word vis-à-vis the prophetic word revealed by oracle and lottery. In both cases, for the Talmud,

17. B *Bava batra* 121b–122a.

18. B *Bava batra* 122a. The *Zohar* (Leviticus 101b) applies this expression to the lottery employed to select the scapegoat on the Day of Atonement, again to mean that the lottery "spoke" (see below, n. 30).

19. Rashbam on B *Bava batra* 122a. <"Rashi's commentary" on this tractate is actually by his grandson Samuel ben Meir, known as Rashbam.>

oracle and lottery coexist as ways to render a verdict between rival parties, following, incidentally, the aphorism of Proverbs: "The lot puts an end to disputes and decides between powerful contenders" (Prov. 18:18).

Thus the lottery is recognized as an indispensable makeshift in situations of doubt and indecision; but the context of the biblical epiphany in which it was employed forced the Talmud to superimpose an oracular character on it. Now we can understand the sages' interpretation of that imagined stichomythia between Achan and Joshua, namely, that it reflects the need to avoid discrediting the lottery as a means for revealing God's glory, despite its arbitrary nature and the impossibility of deeming it a means to attain reliable knowledge in criminal law.

3. RUTH AND BOAZ: CHANCE AND DESTINY

An episode in the book of Ruth points to an intermediate status between oracular chance and random chance. This passage, enriched by the midrashic glosses on it, paints a complex relationship between the chance of events and meetings, on the one hand, and the theme of *ḳeri* as randomly spilled semen, on the other. First we should review the plot (or *'alilah*) of the book. The Moabite woman Ruth chooses to link herself to the people and god of Naomi, her mother-in-law. Naomi, who had left the land of Israel ten years earlier, because of a famine, is now returning home penniless and bereft of her husband and two sons, one of whom had been Ruth's husband. Ultimately Ruth is "saved," in the sense that she avoids the slavery to which her poverty could have reduced her, because a rich relation, Boaz, marries her. Their great-grandchild is King David, the ancestor of the dynasty from which the messiah will be born.

According to the text, the first meeting of Ruth and Boaz takes place by chance. During the grain harvest, Ruth goes out to glean the ears that the harvesters leave for the disadvantaged of society—the poor, foreigners, orphans, and widows—as the law prescribes.[20] "She came and gleaned in a field, behind the reapers; and, *as chance would have it*, it was the piece of land belonging to Boaz" (Ruth 2:3). The Hebrew phrase, with its internal accusative, reinforces the fortuitous nature of this: *va-yiḳḳer miḳreha*, literally "her chance/destiny (*miḳreha*) came by chance (*va-yiḳḳer*) [to] Boaz's field." Subsequently Boaz arrives, learns that the unfamiliar young woman is Naomi's daughter-in-law, and takes her under his protection. The invocation of chance here is unexpected. In contrast to the story of Esther, the god of Israel is never absent from the narrative that describes the origins of the royal house destined to rule over Israel, in history and at the end of history. Moreover, right after she chances on Boaz's field, but before he sees here, he greets the harvesters with the expression, "YHWH is with you," to which they reply, "may YHWH bless you" (Ruth 2:4). Thus God is present, but so is chance. Finally, the word *va-yiḳḳer* 'came by chance' is precisely that employed in the book of Numbers (23:4) to designate the chance arrival of a prophetic message to Balaam, the heathen prophet, in contrast to *va-yiḳra' 'elav* 'he called to him' (Lev. 1:1), used when Moses is the recipient. As we have seen, the verb *va-yiḳḳer*, appropriate to pagan prophecy and derived from the same root as *miḳreh* 'fortuitous event', is assimilated to the "chance" of the scattered drops of sperm, the "random sparks" of postbiblical literature.[21] The midrash on the book of Ruth cannot allow such a fine occasion for bringing up

20. Lev. 19:9–10 and Deut. 24:19.
21. See Chapter 3, n. 66.

these associations to pass by unnoted. For Rabbi Johanan, the expression *va-yikker mikreha* suggests that Ruth's beauty was so physically arousing that "whoever saw her had a seminal emission (*ḳeri*)."[22] Another midrash[23] justifies this interpretation to some extent by establishing that the word *va-yiḳḳer*, which is used for the descent of the prophetic spirit on Balaam, connotes impurity, as in the idiom *miḳreh laylah* 'nocturnal event' (Deut. 23:11), which rendered a man impure for participation in the Temple ritual.

This makes slightly more sense if we recall the status of the inhabitants of Moab in the biblical typology. Balak, the king of Moab, hired Balaam to curse Israel at the end of their forty years of wandering in the wilderness, in order to avert their threat and halt their advance toward the Promised Land. After this approach failed (because Balaam, although a pagan, was a true prophet), Balak almost succeeded in detaching the Israelites from their god by sending the Moabite women to seduce them and entice them into idolatry (Num. 25).[24] One consequence is an absolute ban on intermarriage with and conversion ("entering the congregation of Israel") by Moabites, "to the tenth generation . . . forever" (Deut. 23:4 [3]). Although talmudic law interpreted this ban as applying only to males and not to Moabite women, it is clear that the latter were associated with an inglorious episode in the history of Israel. To complete the picture, we should recall that, according to Genesis, their eponymous ancestor Moab is the product of Lot's incestuous night with his older daughter.

Ruth's history inverts all of this. Not only does Ruth the Moabite enter the congregation of Israel by attaching herself to Naomi; she even marries Boaz and becomes the ancestor of the royal house of Israel. The book of Ruth recounts the secret nocturnal history of this reversal. The story that culminates in the glory of King David begins under the sign of incest, chance, and impurity—the sign of death. Here too we encounter the ambiguity of these sparks of *ḳeri*, the ambiguity of life and death in Eros, expressed now in the duality of chance and providence. The chance of the encounter is not yet the absurdity of blind and indifferent destiny; but neither is it the sacred chance by which the divine meaning is revealed directly. More precisely, chance is still a form of divine revelation, incomplete and limited, as in dreams or when addressed to pagans, designated by the word *va-yiḳḳer* (ויקר), a truncated form, stripped of the final aleph, of *va-yiḳra'* (ויקרא). But chance is no longer directly perceived as such. It is only after the fact that we may perhaps be able to assign it this function and give it this meaning. When the event takes place, it is simply the product of random chance. It is to exorcise this godless chance that Boaz and the harvesters greet each by invoking the name of YHWH. The midrash notes that this form of greeting was a relatively new custom at the time, one that seems to violate the ban on speaking God's name in vain. This innovation, along with the institution of the annual public reading of the book of Esther, is listed by the midrash among the new rituals that the earthly tribunal instituted and the heavenly tribunal ratified,[25] despite the a priori ban on adding to or subtracting from the Law of Moses.[26] The need to employ the divine name as a greeting derives from the natural tendency to forget it in a world that can be compared, at least in this respect, to the godless world of the book of Esther.

22. *Ruth Rabbah* 4:4.
23. *Gen. Rabbah* 74:7.
24. Numbers 25.
25. *Ruth Rabbah* 4:5. With regard to the annual public reading of the book of Esther, see Chapter 4, §10.
26. Deut. 4:2 and 13:1.

Both in the story of Ruth, with its glorious conclusion, and in that of Esther, whose end is merely survival in exile, the sexual outflanks the order of the divine law on all sides. However, inasmuch as the sexual, too, is divine "at its source," it makes some contribution to this order. Alongside the unveiled divinity that is named, personified in the individuals who hear it and receive it, sex, as Eros and Hermes, is the vehicle in history of a hidden divinity, that of chance and *ḳeri*, the nocturnal event that is both spermatic and prophetic.

We find another example of this ambiguity of the divine, as applied to the permitted and forbidden, in the law of levirate marriage, similar to the case of Ruth and Boaz, and in Naḥmanides' astonishing interpretation of it. When a man dies without issue, the widow is consecrated to his brother and must have a child by him, to carry on the name of the deceased.[27] What would otherwise be prohibited incest between brother-in-law and sister-in-law becomes an obligation after the husband's death. According to Naḥmanides:

> The ancient wise men who were prior to the Torah knew of the great benefit in marrying a childless dead brother's wife, and that it was proper for the brother to take precedence in the matter, and upon his failure to do so, his next of kin would come after him, for any kinsman who was related to him, who would inherit his legacy would derive a benefit from such a marriage. . . . Now when the Torah came and prohibited marrying former wives of certain relatives, it was the will of the Holy One, blessed be He, to abrogate the prohibition against marrying a brother's wife in case he dies childless [so as to permit levirate marriage], but it was not His will that the prohibition against marrying a father's brother's wife or a son's wife or similar wives of relatives be set aside.[28]

From this we might get the impression that God and his Torah are not in total agreement and that the institution of levirate marriage is a sort of compromise between them! Naḥmanides is commenting here on the story of the levirate union of the widowed Tamar—first to her brother-in-law Onan and then to her father-in-law Judah (Gen. 38:8–30)—whose offspring, Pereẓ, is the ancestor of Boaz and thus of King David. At the end of the book of Ruth, Boaz stands in for the brother-in-law, being a close enough relative of Ruth's first husband for his marriage with Ruth to be deemed her "redemption," but sufficiently remote that their union is not banned by the law. The permitted and the forbidden are always very close to each other, as if God and his law were pursuing objectives that are not always convergent.

As in the case of the lottery that convicts Achan, in its talmudic version, the divine can be revealed in different fashions, some of them mutually contradictory. The god of Israel who spoke directly to the prophets, or the god of the sages after the end of prophecy, is not to be confused with the divinity behind the oracle. The latter neither calls nor speaks; nor does it even allow itself to be conceived in some intelligible fashion.

The chance of the encounter between Ruth and Boaz lies in this brief interlude when chance, already desacralized, still preserves the memory of a time when it was providence. Projected into the future, it leaves open the possibility of grasping, after the fact and for those who reconstruct it, the meaning of a completed story.

27. Deut. 25:5–10. In fact, a special divorce procedure—exceptional for the Bible, mandatory today—is provided for a case in which the surviving brother does not want to consummate the marriage with his widowed sister-in-law.

28. Naḥmanides on Genesis 38:8, in *Commentary on the Torah*, trans. Charles B. Chavel (New York: Shilo, 1971–1976), vol. 1, pp. 499–470.

4. DESACRALIZED CHANCE AND THE PATHS OF KNOWLEDGE

Could chance play a specific role that needs to be preserved, as if it were an oracle by which truth speaks, given that we know it is only the acceptance of the arbitrary nature of a choice in a situation of ignorance and indifference?

Perhaps we can find an answer in the dramatic ritual of the scapegoat on the Day of Atonement, in which the high priest cast lots to determine which of two identical goats would be sacrificed on the altar in the normal way and which would be "dispatched to Azazel" (Lev. 16:8),[29] meaning it would be sent tumbling off a precipice in the wilderness, carrying with it in death all the sins of Israel. This dualism, unique in Jewish ritual, has provoked many questions in the monotheistic context of the traditional reading of the Bible. Long after the disappearance of the oracular *Urim ve-Tummim*, when the First Temple was destroyed, and then the end of prophecy, it continued to be practiced with full solemnity each year in the Second Temple. After that, too, was destroyed, and down to the present day, an account of this rite is recited in all synagogues, once

29. <The ritual is the source of the English "scapegoat" ("scape" in the sense of "sent away") and the French "bouc émissaire," which represent the Vulgate's rendering of the Hebrew *śa'ir la-'aza'zel* as *capro emissario*.> According to Rashi and Ibn Ezra, "Azazel" is a mountain in the wilderness. For Naḥmanides, who endeavors to uncover the arcane meaning (*sod*) of this ritual and quotes the midrash known as *Pirḳei de-Rabbi Eliezer*, Azazel is another name for Sammael, the angel of death, and designates "that 'prince' [power] which rules over wastelands" (trans. Chavel, vol. 3, p. 220). The *Zohar* (Leviticus 101b–102a) identifies him with the Satan of the book of Job and, more generally, with the accusing angel or informer (the Aramaic text uses the Latin word *delator*, like the talmudic passage, cited above, in which YHWH declines to be an informer and requires Joshua to employ lots to determine the guilty party).

According to the arcane meaning that these texts make explicit, the function of the ritual is to compel evil and destruction (which result from human misdeeds and in strict justice ought to annihilate the guilty) to reintegrate their original source, where they are united with the good and the construction of the world. Then it becomes possible to make amends for crimes other than by the death of the guilty party; repentance and forgiveness are conceivable even if the reality of the transgressions and crimes, as well as their consequences, has not been annulled. Azazel is the name of an angel, derived from *'ez*, another Hebrew word for "goat." The synonymous *śa'ir* (used for "goat" in this passage of Leviticus—one *śa'ir* for YHWH and one for Azazel) also designates a category of demons ("fauns"), as found in the next chapter of the biblical text (Lev. 17:7; see Chapter 3, n. 73). <The common root of both, and also of Seir as a cognomen of Esau, is *śe'ar* 'hair'; goats, demons, and Esau are all covered with hair.>

But as Naḥmanides reminds us (so does the *Zohar*), its homograph *śe'ir* designates the prince of the Edomites, the descendants of Esau, whose heritage is the sword and war, just as YHWH is the principle of the people of his twin, Jacob-Israel, the innocent or perfect one (Heb. *tam*) who lives in tents. Naḥmanides plays on the word *'avonotam* 'their sins', the faults of Israel that are carried away by the scapegoat, breaking it down into *'avonot tam* 'the sins of the innocent one', i.e., Jacob. That is, these sins can be atoned for and pardoned because in some sense they were committed in good faith, in innocence, and their prime cause is ultimately the Creator, who "forms light and creates darkness, makes peace and creates evil" (Isa. 45:7; see Chapter 2).

Thus casting lots between the first cause of evil (YHWH), where it is not evil, and the proximate cause (Azazel or Sammael or the Satan of the book of Job) indicates that they are not differentiated in their common source. As the *Zohar* states here, in certain circumstances they can even be taken as "equals," even though the latter is the "slave" of the former. This is the necessary condition for satisfying the "left side," that is, justice, which is good, but also bad in its accusatory function.

Here too, then, we encounter chance as an ambiguous mode of knowledge, in which demons and angels are interchangeable, as we saw in Chapter 3. The festival of Purim ("lots"), which commemorates the vicissitudes of the story of Esther, reproduces this foundational lack of differentiation; the day selected by lot for the planned massacre is transformed into a day of deliverance. For the canonical code of Jewish law, the *Shulḥan 'arukh* (*O.Ḥ.* 695, "Laws of Purim"), the wine consumed at the festival meal that commemorates this event is supposed to produce a stupor in which we can no longer distinguish between "the curse of Haman [the executioner] and the blessing of Mordechai [the designated victim, saved in extremis]." This obligation is derived from a statement by the talmudic sage Rava (B *Megillah* 7b), which is followed by an anecdote that is itself marked by the humor of the holiday:

> Rabbah and R. Zeira joined together in a Purim feast. They became intoxicated and Rabbah got up and cut R. Zeira's throat. The next day he prayed on his behalf and revived him. Next year [Rabbah] said, "Come, let's have the Purim feast together." [R. Zeira] replied: "A miracle does not take place every time."

a year, on the Day of Atonement, and constitutes one of the most solemn moments of Jewish ritual. Here the role of the lottery is clear; in some fashion it restores the unity that the opposition between god and devil seems to shatter. The lottery means that the choice between these two goats is indifferent because, "somewhere," the two goats are only one, just as god and the devil are only two aspects of the unity "above."[30] By the same token, the apportioning of the land among the tribes must not destroy the unity, whether of the people or of its land—the unity represented first by the portable sanctuary and then by the Temple of Jerusalem, the seat of the Ark of Covenant that contained the two tablets of the Law of Moses.

In this desire to preserve the verdict of the lottery, even in the absence of an oracle and despite all the ambiguities we have found in the talmudic commentaries, we see how the role of chance is taken into account by knowledge and action.

Chance, as ignorance of causes and recognition of the impossibility of making a choice with full knowledge of the state of things, is an opening, in contrast to the closure of a prepackaged meaning or of the desire

It may not be irrelevant that, according to another talmudic passage (B *Sanhedrin* 65b), Rava, who prescribed intoxication on Purim, created an artificial man (a golem) and sent it to Rabbi Zeira (the victim in this anecdote), only to have the latter reduce it to the dust from which it had been formed (see Chapter 1).

Perhaps this means that the humor of the game is another way of referring to the randomness of fate, which combines the a priori indifference with the destiny we perceive only in retrospect. But we are invited to draw a clear distinction between two types of laughter and play, invoked by apparently contradictory verses in Ecclesiastes. The first, Ecclesiastes 2:2, praises laughter (*śeḥok̩*, which can also mean "game" or "play"); but this praise is ambiguous, because the adjective *meholal* 'praised', here attached to laughter, can also be understood to mean that laughter or play is "mad" (see Rashi and Ibn Ezra on the verse). According to another verse, however, "anger is better than laughter" (Eccles. 7:3). Similarly, joy or gladness (*śimḥah*) is praised in one verse (8:15), but another (the second half of 2:2) disdains it as useless. The Talmud resolves the apparent contradictions:

> "Anger is better than laughter" refers to the anger that the Holy One, blessed be He, displays to the righteous in this world, which is better than the laughter which the Holy One, blessed be He, laughs with the wicked in this world. " . . . Laughter is to be praised" refers to the laughter that the Holy One, blessed be He, laughs with the righteous in the world to come. "Then I commended joy" refers to the joy of performing a precept. "And of joy, . . . what use is it"—this refers to joy [that is] not associated with the performance of a precept. (B *Shabbat* 30b)

In short, when confronted by the ironies of fate, in which good and evil and life and death are mixed, humor may be either mockery or the tragic laughter that triumphs over destiny. The former can be compared to the mad addiction to games of chance, which banishes its players from the community of those whose testimony is admissible under rabbinic law (see Atlan, *Enlightenment*, pp. 375–378). The latter, perhaps like Zarathustra's, is what constructs the future world in which God plays with the righteous, just as he played with Wisdom before and during the creation of the world (Prov. 8:30 and *Gen. Rabbah* 1:1).

30. See the previous note. This mixture of contraries has not prevented later interpretations that attach a retrospective meaning, despite everything, to the result of these lotteries. According to the Talmud (B *Yoma* 39a), the two goats were placed on either side of the high priest. He put both hands in the urn; the lot picked by his right hand was for the goat on the right, and that by his left hand for the goat on the left. But another meaning was extracted from the procedure: if the goat designated for YHWH was on his right side (and that for Azazel on his left), the people were walking in the path of God; and conversely. (In the first case, the lottery conforms to the normal order of things, with YHWH on the right and Azazel on the left; in the second case, the order is inverted.) Similarly, as in the miraculous interpretation of the parceling out of the land, in which chance and the oracle confirmed each other, the passage from the *Zohar* cited above (Leviticus 101b) reintegrates the verdict of the lottery with the oracular pronouncement. In the determination of the tribal territories, Joshua's lots "spoke"; on the Day of Atonement, the lot "rose of itself" into the hand of the high priest to designate which of the two goats, that on the right or that on the left, would be sent to Azazel. This interpretation is, incidentally, the Talmud's justification for the requirement that the lottery be truly blind and not biased by the high priest's desire to designate the goat on his right for YHWH. The lots in the urn had to be well mixed so that he could not "guide" his hands in the urn while drawing out the lots (B *Yoma* 39a).

Consequently it is by indirect knowledge, "by the mouth of the lot," that the "informer" or "accuser" is designated, and not directly by the mouth of YHWH, although the former is certainly in the service of the latter, as in Joshua's dialogue with YHWH, in which the latter declines to serve as an informer. For Rabbi Judah Ashlag, in his *Sullam* commentary on the *Zohar* (Numbers 239b and 240a), this makes perfect sense when we realize that oracular chance, "which speaks by the mouth of the lot," designates, in the architecture of the *sefirot*, "wisdom of the left side," which is both that for which atonement is required and the means of obtaining it (see Chapter 3).

for rationality at any price. The impatient illusion of total knowledge is at least as dangerous as the arbitrariness of flashes of inspiration and their interpretations.

Here we learn that the arbitrariness of chance may be preferable to the arbitrariness of a choice, just as ignorance recognized as such is preferable to false knowledge, or atheism to idolatry; in all these cases, knowledge that there are no causes that we really know—knowledge of our ignorance—is preferable to invocation of hypothetical causes, partially or totally hidden, as if we knew them. For in the latter case we designate these hidden causes by different names, such as those of partial causes inappropriately generalized, or by "God" as the First Cause, and then employ them, as magicians do, to explain everything straightaway, overcoming our impatience with an imperfect knowledge in which "truth springs up from the earth" (Ps. 85:12)[31] and then grows slowly and laboriously.

Understanding that chance has a role to play even when it can no longer serve as an oracle means recognizing that we cannot act as if our knowledge of causes is complete.

31. See Chapter 3, §15; Atlan, *Enlightenment*, chap. 9.

In other words, desacralized chance retains its value as a *theoretical possibility* of knowledge that is (still?) hidden. Unlike the oracle, it does not pretend to have any truth value or any utility as a means of *discovering* this hidden knowledge. But it does make it possible to transform ignorance into a possible opening to the future, without reducing this ignorance in the slightest or in any way guaranteeing the truth or correctness of a decision dictated by the outcome of the lottery, when all else fails, and in despair of ever achieving full knowledge.

The parties' agreement to submit to such a verdict, like that of the tribes who divided up the land, makes it possible to make progress by means of a method, that, as we have seen, also forges a kind of unity among the parties, of the sort that at least concludes the dispute and puts an end to the contest. As an opening to indeterminate possibility, chance makes it possible to break out of the confinement of incomplete meaning and knowledge that think themselves perfect. The arbitrariness of the illusion of knowledge is replaced by the conventional arbitrariness of the lottery, even if it entails constructing a new meaning, partial and provisional, based on that arbitrariness. This is why a lottery, unlike sworn testimony, is binding only because of the parties' consent and not because of its ostensible truth. As such it can serve to set things in motion when progress is otherwise impossible—but on condition that only things, and not human beings, are at stake—sharing out property, which is by definition exchangeable, impermanent, and subject to devaluation and revaluation.

As for the scapegoat, the victim chosen by lot, it is allowed a role in regulating violence, but with no illusions that anything can be known about its culpability, because the guilt in play is that of society as a whole and, at the limit, inherent in the nature of things and in the world as created. It is precisely this recognition that is expressed by the use of chance in the ritual, shifted first from human sacrifice to animal sacrifice and then to a mere narration of the animal sacrifice.

5. THROWING DICE

Neither ontological nor oracular, chance is what is left when everything else fails. The broken language of Stéphane Mallarmé's "A Throw of the Dice Will Never Abolish Chance" may be the best expression of the dialectic of this despair and the hope that, despite everything, can emerge from it. We all read Mallarmé's poem as best we can, picking up some hints of his meaning from the typography.[32] In it we encounter disillusionment and nostalgia—nostalgia for divinatory dice that could speak out and abolish chance, for the nonsense of events when we perceive portents and hidden meanings in them. . . .

> a simple insinuation into silence, entwined with irony,
> or the mystery hurled, howled, . . . the virgin index
> there AS IF

But

> the lucid and lordly crest of vertigo . . .
>
> the disillusionment of one who knows that
>
> **NEVER**, EVEN WHEN TRULY CAST IN THE
> ETERNAL CIRCUMSTANCE OF A SHIPWRECK'S
> DEPTH . . . **WILL ABOLISH.**

Perhaps resigned to

> his petty reason, virile . . . at the moment of striking, . . .
> a rock a deceptive manor suddenly evaporating in fog
> that imposed limits on the infinite.

And yet, despite everything, there is hope for a different transformation of chance.

> IT WAS THE NUMBER, . . . WERE IT TO HAVE
> LIGHTED, IT WOULD BE, worse no more nor less
> indifferently but as much, **CHANCE.**

It was still no more than a new mad outburst of

> the plume, . . . to the summit faded by the same neutrality of abyss
>
> NOTHING . . . WILL HAVE TAKEN PLACE . . . BUT THE PLACE . . . in this
>
> region of waves, in which all reality dissolves
>
> EXCEPT at the altitude PERHAPS, as far as a place fuses with, beyond, . . . A CONSTELLATION

Not as a sign, however, of "a final account in formation," but as waiting and doubting, a meditation in which the dice are not thrown, but thought: a constellation, then,

> cold with neglect and desuetude, . . . attending, doubting, rolling, shining
>
> and meditating before stopping at some last point that consecrates it
>
> All Thought expresses a Throw of the Dice.

32. Stéphane Mallarmé, *Un coup de dés jamais n'abolira le hasard* (Paris: E. Bonniot, 1914); repr. in *Igitur. Divagation. Un coup de dés* (Paris: Gallimard, 1976), pp. 409–429. <English translation, including the prefatory note quoted below, by A. S. Kline, www.poetryintranslation.com/PITBR/French/MallarmeUnCoupdeDes.htm. In the penultimate line, "à quelque point dernier qui le sacre," Kline's "that crowns it" has been modified to conform to Prof. Atlan's emphasis on *le sacré* and the sense of the present chapter. For the entire translation (unmodified), in the original layout, see the Appendix to this chapter.>

We should pay attention to the author's notice to readers that the poem is to be read aloud, with the white space indicating silence that hints at what cannot be said; as well as his regret that this prefatory note must be read and expressing his preference that it be forgotten, because "it tells the knowledgeable reader little that is beyond his or her penetration." But he is also worried that it may confuse the uninitiated:

> The "blanks" indeed take on importance, at first glance; the versification demands them, as a surrounding silence, to the extent that a fragment, lyrical or of a few beats, occupies, in its midst, a third of the space of paper. . . . The variation in printed characters between the dominant motif, a secondary one and those adjacent, marks its importance for oral utterance and the scale, mid-way, at top or bottom of the page will show how the intonation rises or falls.

Later Mallarmé deemed it sufficient to announce, succinctly, that "this tale is addressed to the intelligence of the reader who dramatizes it" (introduction to *Igitur ou la Folie d'Elbehnon*)—a tale that includes dice, shadows, and the abolition of chance (ibid., pp. 41–66). As in the last verse of the poem, every thought always expresses the mixture of meaning and the absurd.

APPENDIX: STÉPHANE MALLARMÉ, "A THROW OF THE DICE WILL NEVER ABOLISH CHANCE," IN THE ORIGINAL TYPOGRAPHICAL LAYOUT (ENGLISH: A. S. KLINE)

A THROW OF THE DICE

NEVER

EVEN WHEN TRULY CAST IN THE ETERNAL
CIRCUMSTANCE

OF A SHIPWRECK'S DEPTH

Can be
only
the Abyss
raging
whitened
stalled
beneath the desperately
sloping incline
of its
own wing
through an advance falling back from ill to take flight
and veiling the gushers
restraining the surges
gathered far within
the shadow buried deep by that alternative sail
almost matching
its yawning depth to the wingspan like a hull
of a vessel
rocked from side to side

THE MASTER beyond former calculations
where the lost manoeuvre with the age
rose
implying that formerly he grasped the helm
of this conflagration of the concerted
horizon at his feet
that readies itself
moves and merges
with the blow that grips it
as one threatens fate and the winds
the unique Number which cannot be another
Spirit
to hurl it
into the storm
relinquish the cleaving there and pass proudly
hesitates
a corpse pushed back by the arm from the secret
rather
than taking sides

a hoary madman
on behalf
of the waves
one overwhelms the head
flows through the submissive beard
straight shipwreck that of the man
without a vessel
empty
no matter where

ancestrally never to open the fist
clenched
beyond the helpless head
a legacy in vanishing
to someone
ambiguous
the immemorial ulterior demon
having
from non-existent regions
led
the old man towards this ultimate meeting with probability
this
his childlike shade
caressed and smoothed and rendered
supple by the wave and shielded
from hard bone lost between the planks
born
of a frolic
the sea through the old man or the old man against the sea
making a vain attempt
an Engagement
whose
dread the veil of illusion rejected
as the phantom of a gesture
will tremble
collapse
madness
WILL NEVER ABOLISH

AS IF
A simple *insinuation*
into silence *entwined with irony*
or
the mystery
hurled
howled
in some close *swirl of mirth and terror*
whirls *round the abyss*
without scattering
or dispersing
and cradles the virgin index there
AS IF

a solitary plume overwhelmed
untouched *that a cap of midnight grazes or encounters*
and fixes

in crumpled velvet with a sombre burst of laughter
that rigid whiteness
derisory
in opposition to the heavens
too much so
not to signal
closely
any
bitter prince of the reef
heroically adorned with it
indomitable but contained
by his petty reason virile
in lightning

anxious
expiatory and pubescent
dumb *laughter*
that
IF
The lucid and lordly crest *of vertigo*
on the invisible brow
sparkles
then shades
a slim dark tallness *upright*
in its siren coiling
at the moment
of striking
through impatient ultimate scales *bifurcated*
a rock
a deceptive manor
suddenly
evaporating in fog
that imposed
limits on the infinite

IT WAS *THE NUMBER*
stellar outcome

WERE IT TO HAVE EXISTED
other than as a fragmented agonised hallucination

WERE IT TO HAVE BEGUN AND ENDED
a surging that denied and closed when visible
at last
by some profusion spreading in sparseness
WERE IT TO HAVE AMOUNTED

to the fact of the total though as little as one
WERE IT TO HAVE LIGHTED

IT WOULD BE
WORSE
no
more nor less
indifferently but as much CHANCE
Falls
the plume
rhythmic suspense of the disaster
to bury itself
in the original foam

from which its delirium formerly leapt to the summit
faded
by the same neutrality of abyss

NOTHING
of the memorable crisis
where the event
matured accomplished in sight of all non-existent
human outcomes
WILL HAVE TAKEN PLACE
a commonplace elevation pours out absence
BUT THE PLACE
some lapping below as if to scatter the empty act
abruptly that otherwise
by its falsity
would have plumbed
perdition
in this region
of waves
in which all reality dissolves

EXCEPT
at the altitude
PERHAPS
as far as a place fuses with beyond
outside the interest
signalled regarding it
in general
in accord with such obliquity through such declination
of fire
towards
what must be
the Wain also North
A CONSTELLATION
cold with neglect and desuetude
not so much though
that it fails to enumerate
on some vacant and superior surface
the consecutive clash
sidereally
of a final account in formation
attending
doubting
rolling
shining and meditating
before stopping
at some last point that crowns it

All Thought expresses a Throw of the Dice

CHAPTER 6

HOW THE BIBLICAL GOD "GOES AT RANDOM" IN HEBREW, BUT NOT IN TRANSLATION

While engaged in a theoretical study of the role of chance, in the form of random perturbations of external or internal origin in self-organizing physical and biological mechanisms, I happened across two verses from Leviticus (26:40–41) that seemed to be suitable for the epigram to the article and then for an entire volume on the subject.[1] In these verses God announces how he will react to the future transgressions of the people. I read them in the original Hebrew and cited them in my own translation (seeing no need to consult any standard version): "Because they acted with me in a random way, I too will act with them in a random way." When well-intentioned readers decided to look at the verses in context, they turned up two surprises—for themselves and for me. First surprise: not a single translation of the Bible rendered the word *ḳeri* in the verses as "chance" or "randomness." Second surprise: those readers who were fluent in Hebrew agreed that there was nothing wrong with my literal translation; but they could not accept it in the context, where it is applied not only to the conduct of human beings but also, and especially, to God's. The notion that God might say that he would behave in a random or arbitrary fashion was quite incomprehensible to them. In their eyes, the conceptual difficulty justified that fact that all translations of the Bible make do with approximations, some more and some less successful.

Inferior translations[2] employ two different terms for the same Hebrew word *be-ḳeri,* depending on whether it is referred to human beings or to God. Better ones (in French, English, and German) settle on a single term, but it is never "at random." For example, both the Bible published by the French rabbinate and the French translation by André Chouraqui have "avec hostilité" (with hostility); the New Jewish Publication Society translation adopts the same word ("hostile to Me/them"). The Dhorme translation cleverly employs "à l'encontre," which suggests both an encounter or meeting (one of the meanings of *miḳreh,* from the same root as *ḳeri*) as well as the normal

1. Henri Atlan, *L'Organisation biologique*; the article is reprinted in *Entre le cristal et la fumée*, p. 39.

2. <But also the Vulgate, although the semantic difference between *ex adverso mihi* 'contrary to me' and *contra eos* 'against them' is not great.>

sense of the French idiom, "in opposition to" or "against." Thus in verse 27 of the same chapter, *ḥamat ḳeri* "wrath of randomness" can be rendered naturally to yield, "I will walk *against* you in *wrath* and punish you myself sevenfold for your sins."

What complicates things is that the word *ḳeri* is found in the Bible only in this single context and in this one chapter of Leviticus, which enumerates the afflictions that will strike the Israelites when they abandon the law and covenant of their god, in the two verses already cited as well as in verses 21, 23, 24, and 27–28.

It is only later, in the Hebrew of the Mishnah,[3] that the word *ḳeri* is used with the sense of a nocturnal emission, which takes place "by chance"—unintentionally. Here *ḳeri* has taken on the meaning of the biblical *miḳreh laylah,* an "incident at night" or "nocturnal accident" (Deut. 23:11), in which the word *miḳreh*, from the root *ḳ.r.h*, has the familiar sense of an event that takes place "at random" or "by chance"; that is, accidentally, irregularly, and unintentionally.

I translated these verses as I understood them spontaneously, in what was, I had no doubt, their literal meaning. Whether or not natural events have a meaning depends on how human beings conduct themselves vis-à-vis nature and on their ability to domesticate random chance by assigning it a meaning. If human beings act at random—that is, if they give no meaning to their existence—it really is senseless. Events take place at random, for God acts with them at random. This notion that God's behavior is determined by what human beings do—compatible with the talmudic principle of *middah ke-neged middah* or "measure for measure," extensively developed in midrash and kabbalah—conflicts with such deeply rooted theological ideas that it is dismissed as quite incomprehensible and unimaginable. But since I had no theological a priori, I understood and spontaneously translated this text with its literal meaning, finding in it, what is more, a certain Nietzschean accent, to the point of associating it with a passage from *Zarathustra*:

> I am Zarathustra, the Godless: I still cook up every chance event in *my* pot. And only when it is quite cooked do I bid it welcome, as *my* food.
>
> And verily, many a chance event came to me imperiously: but even more imperiously did my *will* speak back to it—then it went down imploringly on its knees—
>
> Imploring that it might find shelter and heart with me, and urging me flatteringly: "But see, O Zarathustra, how only a friend comes to a friend!"—[4]

Nevertheless, surprised after the fact by the thoroughly unwitting originality I had demonstrated, I began examining the traditional commentators for confirmation of what I was certainly not the only person to find in the Hebrew text. Sure enough, Rashi's commentary on Leviticus 26:21 associates *ḳeri* with *miḳreh* and *'ar'ai*, in the sense of "impermanent," "occasional," or "irregular."

Applied to human behavior, action "at random" is understood as intermittent and irregular observance of the law, at random. What remains is to understand what this phrase can mean when applied to God. The eighteenth-century Moroccan kabbalist and commentator Rabbi Ḥayyim ben Attar answered this question in his exposition of verses 23 and 24. The curses triggered by abandon-

3. M *Berakhot* 3:3; see also B *Berakhot* 20b, 22b, and 26a.

4. Friedrich Nietzsche, *Thus Spoke Zarathustra: A Book for Everyone and Nobody*, trans. Graham Parkes (Oxford University Press, 2005), p. 148.

ment of the covenant are the normal outcome of transgressions. In the first stage of distancing oneself from the law, the retribution for the sin is still related to the sin and can be clearly perceived as punishment for that sin and no other. But when human beings persist in their backsliding, to the point that their relationship with the law is only episodic, accidental, and random, the retribution also seems to be a random blow, with no meaning, no relationship to any sin whatsoever.

Beyond this, however, I could not find unequivocal confirmation, certainly not from traditional readers of the Bible in Hebrew, for what I had taken to be the obvious meaning of this passage.

Maimonides refers to this crux in his *Guide of the Perplexed*.[5] He translates *ḳeri* as "chance" in all cases and manages to integrate the notion of God's "chance conduct" into his theology of a transcendent God and judge who punishes the wicked and rewards the righteous. Hence it is quite astonishing that no translators seem to have taken their cue for this difficult passage from him. It is true, however, that in order to reach this reading Maimonides had to dilute the literal meaning, referring to "accidents *as you call them*."

It was only some years later that I encountered a text that had not been reprinted for many centuries and was, consequently, little known: the thirteenth-century kabbalistic treatise *Ginnat 'egoz* by Joseph Gikatilla:

> The word *ḳeri* indicates a category from which providence has been removed, in the sense of abandonment (*hefḳer*) and of events taking place with no intention, leaving only what is produced by chance (*ḳeri*), with neither consciousness nor intention and with absolutely no element of providence or reflection. And it is that chance (*ḳeri*) that is referred to by the Torah, "if you go with me [in] *ḳeri*" (Lev. 26:21), as if to say: "if you deny my providence and say that everything good and bad that happens to you happens by chance (*derekh ḳeri*), with no element of providence and no intention of reward and punishment, then I will remove my providence totally and you will be completely abandoned *(hefḳer)* to what is produced by the category of chance (*yithavveh me-'inyan ḳeri*)." This is the meaning of v. 24: "I myself will go with you at random (*be-ḳeri*)"—that is, the total removal of his providence, may he be blessed.[6]

The text from which this passage comes merits a closer look. Gikatilla sets out to demonstrate that, despite their common root, *ḳeri* and *miḳreh* have different senses. This difference is not without importance, because it explains the unusual use of the word *ḳeri* in Leviticus 26, which would not be justified were it synonymous with *miḳreh*. For Gikatilla, the difference has to do precisely with the fact that *ḳeri* designates a random event, something that is *hefḳer* or "abandoned," in the sense of ownerless, in no one's possession, not reflecting intention. *Miḳreh*, by contrast, desig-

5. *Guide* 3:36. The topic there is the meaning of Maimonides' first class of biblical commandments, whose utility, in his system, is that they reinforce belief in the greatness of God and

> establish firmly the true principle that God takes notice of our ways, that he can make them successful if we worship him, or disastrous if we disobey him, that [success and failure] are not the result of chance or accident. In this sense we must understand the passage, "If you walk with me by chance" (*be-ḳeri,* Lev. 26:21): i.e., if I bring troubles upon you for punishment, and you consider them as mere accidents, I will again send you some of these accidents as you call them, but of a more serious and troublesome character. This is expressed in the words: "If you walk with me by chance: then I will walk with you also in the fury of chance" (ibid. 27–28). For the belief of the people that their troubles are mere accidents causes them to continue in their evil principles and their wrong actions, and prevents them from abandoning their evil ways. [As we read:] "You have stricken them, but they have not grieved" (Jer. 5:3). For this reason God commanded us to pray to him, to entreat him, and to cry before him in time of trouble. (trans. Friedländer, slightly modified, pp. 331–332)

6. Joseph Gikatilla, *Ginnat 'egoz* (Jerusalem, 1989), p. 326.

nates the *occurrence* of any event, even those that are the product of some intentional action that may give them meaning. So in this chapter of Leviticus, for Gikatilla, the conduct of the Hebrews and of their God is explicitly presented as devoid of meaning, as the random and disordered conduct of human beings divorced from the law and guidance of God/Nature, with no meaning for human beings, without providence. The phrase *ḥamat ḳeri* 'fury of chance' evokes for us the sound and fury of an absurd and violent story in which the human beings who have been abandoned make war on one another.

Another passage of Gikatilla's commentary sheds remarkable light on the relationship between "chance" and natural law, between the possible meaning of human existence and determining causes in nature. In verses 21 and 27 of Leviticus 26, the phrase "I will go with you *be-ḳeri*" is followed by the threat that the Israelites will be chastised sevenfold for their sins. Gikatilla reads this as an allusion to the seven planets, which, according to the scientific astrology of antiquity and the Middle Ages, were the main channels through which nature wielded its deterministic influence. Thus the punishment of his people by the god of Israel, by means of chance events that are themselves the chance outcomes of human actions, coexists with the rigid order of natural determinism. This "chance" is not the absence of order, but only the absence of meaning on the level of human meanings, that is, the lack of an intentional and perhaps moral sense. Are we to understand, conversely, that providential divine behavior—the sequence of events that bear meaning in the history of human beings, both individuals and collectives—is immune to natural determinism? The talmudic dictum *'ein mazzal le-yisra'el*, "Israel is not subject to astral determinism," might lead us to think so. But the Talmud itself gainsays this notion, forcing us to interpret it in the strictly moral context in which it is introduced.[7] That is, we must read it in association with another maxim, "all is in the hands of heaven [that is, determined by the stars] except for the fear of heaven."[8] This is the view of Rabbi Ḥanina, who also maintains that "Israel is subject to astral determinism."[9]

Gikatilla develops an ontology that rejects the hypothesis that a meaningful world is immune to natural determinism. On the contrary, the determining causes acquire their "truth" and intelligibility from formal structures that express the eternal essences of existents. All of these "essences" (*havayot*) derive from the Tetragrammaton—the *shem havayah* 'name of being' (or 'name of essence'),[10] by means of various combinations, permutations, and repetitions of the letters that compose it. To describe the relationship between these essences and nature, he borrows the scholastic terms *natura naturata* and *natura naturans*, translating them into Hebrew and associating them with different names of God.[11] Thus, as I spontaneously imagined when reading Leviticus in Hebrew, the role of *ḳeri*, as the random element in nature corresponding to the random element in human behavior, can easily be fit into an ontology in which the character of the actions

7. B *Shabbat* 156a.

8. B *Berakhot* 33b.

9. On this talmudic discussion see Chapter 3, nn. 43–45; and Atlan, *Enlightenment*, pp. 268–269.

10. <The word *havayah* is an anagram of the Tetragrammaton, which can also be read as a conflation of the three tenses of the verb *to be* (see above, p. 185). Nor should we forget that the word *essence* is derived from the Latin *esse* 'to be'.>

11. Spinoza, of course, took up these terms and incorporated them into his natural philosophy. As compared to transcendental theologies of Aristotelian inspiration, Spinoza seems to have carried the process of naturalizing the notion of essence in the context of immanent thought to its limit. This process, which brings to mind Stoic theology as well as some aspects of the syncretistic Neoplatonism of the Renaissance, seems to have been begun by Gikatilla.

of God/Nature is intelligible, conveying either meaning or absurdity, depending on how human beings represent it to themselves and how they behave in consequence.

So far as I know, these passages from *Ginnat 'egoz* and from Ben Attar's commentary (also of kabbalistic inspiration) are the only evidence that past Hebrew readers understood the word *ḳeri* in Leviticus 26 in the sense of "chance" even when applied to God's behavior.[12]

We learn several things from this adventure in translation. On the one hand, it demonstrates the sometimes baneful influence of theological prejudices on attempts to perceive the literal meaning of the Bible. On the other hand, it is evidence of the liberty that some kabbalists have taken with regard to these preconceived notions, probably because the kabbalistic tradition is more cosmological than theological and seeks knowledge of the world (including its divine aspects) more than of God, even if the former may be seen as the activity of the latter.

Finally, a talmudic dialogue about the significance or insignificance of the suffering caused by illness[13] takes on an interesting aspect in the light of this "mechanistic" notion of meaning that has objective existence only to the precise extent that human beings make it exist. In this discussion, the pains inflicted by nature are interpreted either as signs of an unconscious transgression to be uncovered,[14] or, if the afflicted person searches and does not find one, as trials imposed by God's love. We can then accept them as such, as part of an asceticism in which the trial endured is considered to provide new strength and is ultimately beneficial; or we can reject them, "them and their rewards,"[15] demurring at the benefits they might yield later, in the world to come. If so, the suffering has no meaning and we can ascribe it only to the "natural" activity of parasites and evil "demons" that we must combat and overcome.

This brings us to the possible superposition of two forms of determinism: the determinism of nature, devoid of human meaning, and the determinism of specifically human intelligibility, which is essentially moral. It is the latter that Job complains he cannot perceive in his misfortunes. But the frame tale of the book of Job indicates that he and his friends erred by looking for the sense of his calamities in the simplistic logic of a childish morality in which virtue is automatically rewarded by natural benefits and crimes are punished by catastrophes that are equally natural. In fact, however, these two forms of determinism are identical. It is the same chain of cause-and-effect, which is absurd or meaningful as a function of the world of meanings, respectively impersonal or personal, that human beings invent as speaking and thinking beings, when they are caught in the web of their own actions and of the natural events that intersect them.

Some may ask if this is really what the original text "means." But does their question have a meaning? If the meaning of one text has to be restated in another text, the odds are that the latter really says something else. Translation, from this point of view, can augment a text and

12. As noted, Rashi, in his commentary on verse 21, laconic as usual, understands it in this way with regard to the behavior of human beings, leaving his readers to puzzle out what it might mean with regard to God's actions. But he also cites other possible meanings, such as that conveyed by the Aramaic translation of Onqelos (see below).

13. B *Berakhot* 5a.

14. I would like to thank Irun R. Cohen for drawing my attention to the significance of Rashi's gloss on this passage. <The text of the Talmud is "let him examine his conduct . . . if he examines [it] and finds nothing, . . ."; it is Rashi who glosses this to mean, "if he fails to discover a transgression on account of which it is appropriate for these pains to visit him.">

15. B *Berakhot* 5b.

not just detract from it. When the original text is several thousand years old, any reading, even in the original language, is necessarily a translation. Hence explicit translations are the only way to understand these texts, even if no translation can ever provide more than a partial representation of the original. The combination of all translations in a potentially infinite set makes it possible to construct, post factum, a meaning that is attached, in some fashion, to the underlying text. Can we reasonably imagine that the words "mean" the same thing as they did, after the lapse of thousands of years? The word *ṭeva'* 'nature' does not exist in biblical Hebrew; it is a medieval coinage, derived from the biblical verb *ṭava'* 'sink' or 'impress', probably through mishnaic Hebrew,[16] in which the verb is used for impressing a human form ("nature") by means of a signet ring (*ṭabba'at*, from the same root).[17] A fortiori we may doubt the relevance of applying the notion of chance in nature to the biblical text, with the modern sense of "random" or "unintentional," or, even more so, the "intersection of two independent causal series," to employ Cournot's classic definition. But the same applies to all ideas, including and especially that of God, which has been burdened over the centuries by a theology expressed in Greek, Latin, and Arabic. The Septuagint of Leviticus renders *ḳeri* by *plagios* 'perverse', rather than by *kairos* or *automaton*, both of which mean "chance," though in a sense quite remote from the modern naturalist notion of "random event." Nevertheless, perhaps *plagios* is closer to the ancient concept of *ḳeri*, in which chance is disorder, a perversion of the order of straight thinking. But cunning reason, the *metis* of the Greeks, can employ it, just as it can turn the unexpected to its profit.[18] Cunning against cunning. We find, by this detour through the Greek version, the measure for measure imagined at the outset, *ḳeri* for *ḳeri*. Could classical Greek, with its historical proximity to the Hebrew of the Bible and Mishnah, be the closest semantically of all of the languages in which classical theology has been written, at least with regard to our story of "chance"?

Perhaps; but the Aramaic translations of these same verses show that it is not so simple. Aramaic is probably the closest linguistic relative of biblical Hebrew, both in time and in structure. Nevertheless, the two Aramaic translations of Leviticus, pseudo-Jonathan and Onqelos, render *ḳeri* differently. The first employs *be-'ar'ai* (Rashi's first gloss), which means "accidentally" or "by chance, at random." But pseudo-Jonathan is unable to preserve the symmetry of the Hebrew, which has the same verb in both contexts; although it uses the same adverb for God and human beings, it employs different verbs: "go by accident," applied to human beings, but "act by mere accident" (or, more literally, "by accident in the world"), when applied to God. Thus verses 23 and 24 become: "If you go with me by accident, I will act with you by mere accident." Onqelos renders *be-ḳeri* in both cases as *be-ḳashyu* 'rigidly', stubborn and resistant, as in the biblical idiom *'am ḳeshei 'oref* 'stiff-necked [*lit.* hard-naped] people'. Here it is the element of *opposition* that is retained in the *encounter* with a fortuitous event (the *miḳreh*), which takes place without being summoned, which we encounter in our path, which collides with us. This is reminiscent of Dhorme's translation.[19]

16. M *Sanhedrin* 4:5; B *Sanhedrin* 37a.

17. See Chapter 1, n. 107.

18. See: Marcel Detienne and Jean-Pierre Vernant, *Cunning Intelligence in Greek Culture and Society*, trans. Janet Lloyd (Hassocks, Sussex: Harvester Press, 1978); Atlan, *Enlightenment to Enlightenment*, pp. 122–128.

19. <The Hebrew/Aramaic *'ara'* (ארע), whose semantic field includes "strike," "befall," and "encounter," is a byform of *'ara'* (ערע), the root of *'ar'ai* 'random'.>

The two aspects of the unexpected and the opposing, both inherent in *ḳeri,* are sundered in the two ancient Aramaic translations. But it is precisely this dichotomy that makes it possible to uncover them—and that is why translation is always important.[20]

20. We should note, in conclusion, that all meanings may sometimes be combined in paradoxical and contradictory commentaries that are hard to understand if we are not aware of the polysemy of biblical terms. We find this in Rashi and the supercommentary *Siftei ḥakhamim* on Exodus 3:18, where God appears to Moses and commands him to go to Pharaoh, along with the elders of Israel, and speak to him as follows: "YHWH, the god of the Hebrews, has met (*niḳrah*) with us." All of the multiple senses of the root *ḳ.r.h* are at play here: encounter, hardness, impermanence, fortuitous event, impurity of a nocturnal emission. Noting the deponent form *niḳrah* here, Rashi refers to the identical construction *va-yiḳḳar* (a perfect rather than the imperfect employed here) in Numbers 23:4, where God "meets" Balaam. In his commentary on that verse he cites all of the pejorative senses of the verb, as they apply to Balaam's impure prophecy: it is "an expression [ordinarily used to denote events] of a casual character, an expression for something shameful, an expression for an unclean happening" (Rashi on Num. 23:4; trans. Silberman, vol. 4, p. 113; see Chapter 5, at nn. 22 and 23). By contrast, Abraham Ibn Ezra, commenting on the same verse in Exodus, affirms that the verbs *ḳ.r.h* (קרה), normally "meet, happen," and *ḳ.r.ʿ* (קרא), normally "call," may sometimes be used interchangeably; thus he rejects the contrast, emphasized by Rashi and the midrash quoted above (Chapter 3, n. 66, on Lev. 1:1), between God's summons to Moses and the occasional manifestations of prophecy to Balaam.

INDEX

Abel, 35, 75n31
absurd, and meaning, 305n32
Abulafia, Abraham, 197, 271n216
Achan, 295n10, 296–99
acosmism, versus Spinoza, 287
Adam, 1, 35–36, 37, 99
Adam and Eve, story of, 6
Adam Ḳadmon (primordial man), 5, 32n32, 105–9, 277
agriculture, invention of, 21
Akedah. *See* Isaac, binding of
Akiva son of Joseph, 127
Albertus Magnus, 32n32
alchemical imagery, in modern science, 150
alchemy, 44n59, 169
'alilot (deeds; pretexts), 88n61, 97, 98, 101n83
allegorical interpretation, 237, 241
allocation of land, 298
alphabet of nature, 270, 271n216
alterity, 119, 121–23
Althusser, Louis, 216n83
ambiguity: of biotechnology, 90; of chance, 302n29; of cognitive experiences, 82; compounded, 57; of cunning knowledge, 76, 80, 88; of the divine, 301; of hidden causes, 142; of human nature, 30n25; of knowledge, 39n51, 43, 72–73, 82; of moral consequences of biological discoveries, 91; permanence of our, 88; and polysemy, 47–49; of the possible, 82; of quest for power through knowledge, 30; of relating to demons, 157; of serpent's nature, 81; and the sparks of randomness, 34–37; and the sexual act, 45; of the term "golem," 32n32
American dream, 258n172
Amsterdam, and New Amsterdam, 258n172
ancient texts, value of, 209–10
angelology, 149–50, 152
angels, 129–30, 135–36, 159, 162, 166, 182–84. *See also* cherubim; spirits
anima, 25
animals, for food, symbolism of, 174
animism, 261
'ani YHWH, 108n98
anthropoid, 32n32
anthropomorphism, 181, 206
antisemitism, European, 231
apoptosis, 92n68
Aramaic, 182, 315
arbitrariness, 304
argument from design, for God's existence, 264
Aristotle, 181n122, 290
'arum (naked, shrewd), 47
ascesis, 72, 121
Ashlag, Yehuda, 169–72, 183, 303n29
Ashmedai, 175; and Solomon, 36–37, 39–43, 156–57, 163n72, 175n115; in the *Zohar*, 176n115
Askénazi, Jöel, 211n67
astrology, 154–55, 169
astronomy, 155n45
atoms, theory of, 181n122
Augustine, and original sin, 4
authority, submission to, 257
automaton (chance), 290
Avicebron. *See* Ibn Gabirol
axiological domain, versus cognitive domain, 175
Azazel, 302
'aẓilut, 43n57, 48n68, 51n78, 81, 103n88, 135n8
Azulai, Abraham, 184n138

Babel, story of, 174. *See also* generation of the Tower of Babel; Tower of Babel, myth of
Bacharach, Naphtali, 127, 106n92, 180n122, 269n208
Balaam, 52, 58, 98, 160, 161n66; and chance prophecy, 300, 316n20; as Moses' foil, 56
balance, metaphor of the, 173
Bar Yoḥai, Simeon. *See* Simeon bar Yoḥai
bat ḳol, 162

Bataille, Georges, 84
Bayle, Pierre, 259–60
Becco, Anne, 270
Behemoth, and Leviathan, 86–88
Benaiahu ben Jehoiada, 40–41
Benjamin, Walter, 85–86
Bergson, Henri, 211n65
Bernier, 217n88
Bible, and theology, 314
Bible codes, 17
biological revolution, 9–10
biologization of daily life, 9
biology, 10, 91n68
biophysics of organized systems, 93n69
birth, 19
blasphemy, 37
Blyenbergh, William De, 256n169
Böhme, Jakob, 270n216
Boaz, chance meeting with Ruth, 299
body, 66–72, 93–95, 88–90, 93–95; versus mind, 1, 85, 87
Boss, Gilbert, 205
Bouveresse, Renée, 261n180
Boxel, Hugo, 146–48
breath, 122
Bréhier, Emile, 216–20
Brykman, Geneviève, 230
Buddhism, in European thought, 213

Cain, 35, 75n31
calculus, 261, 266
Canaanite, 22n2
Canguilhem, Georges, 25, 27
capital punishment, 297n16
casting lots, 48n69, 289–94
Catholicism: modus vivendi with science, 271n218
causal explanations, 142, 143, 144n6
causal relationships, 143, 156
causality, 148, 263
causation, and statistics, 18–19
cause and effect, commensurability of, 203n39, 265
causes, 143, 145, 265, 290; and animation of subjects, 149; hidden, 142–46, 153, 166; ignorance of, 292; and reasons, 145, 193
celibacy, 6
certainty, 198, 202n38, 237n129
chaining, 40–41
chance, 3, 19, 44; as the absence of meaning, 314; ambivalence toward, 290; as bearer of hope, 54; and casting lots, 289–94; and conception of a child, 34; desacralized, 290, 302–4; and destiny, 299–301; as divine revelation, 300; experienced as ignorance, 45; as mode of knowledge, 302n29; and natural law, 314; oracular role of, 302; recourse to, 298; sacred versus destructive, 293
chaos, 101–9, 292
cherubim, in the Holy of Holies, 131
children, and ambiguity, 45
choice, to play at being free, 131
chosenness, by deity, 276n225
Christ, versus Moses, 273
Christ and apostles, as philosophers and political leaders, 259
Christianity, 246, 258n171
church, view of science and philosophy by, 268–73
circumcision, 47
cloning, 24n6, 63–64
clothing, 88–89
cognition, replacing prophecy and divination, 243
cognitive domain, versus axiological domain, 175
Cohen, David: on Spinoza, 285–86
Cohen, Hermann, 222
combinatorics, 271n216
compromise, 258
concept, and life, 25–28
confession, 294, 296
conscience, as inferior form of knowledge, 72–74
consciousness, history of, 93, 195n18
Constitution, U. S., Spinoza's influence on, 258
contradictions, intentional, 238–39
control, 22–23, 55, 64
Copernican revolution, 227n113
Cordovero, Moses, 125–26, 128, 171, 196–97; allegory of play, 180–81; concept of an immanent deity, 272n219; on the dangers of esoteric wisdom, 231n120; on God, 69, 272n219; on *Keter*, 287n261; pantheism and panentheism in, 287n261; parable in *Shi'ur komah*, 129; on "rose into thought," 280n239
Cornford, Francis, 189n1
corporeality, of God, and human body's shape, 66–72
correlation, statistical, 18–19
cosmogony, kabbalistic, angels and spirits in, 150
Coudert, Allison, 278n233
Creation, mythic events prior to, 101–9
Crescas, Hasdai, 196n19, 197n24, 286, 288
critical inquiry, 197
critical thinking, about thinking, 15
crowns, on Hebrew letters, 127
cunning, ambiguity of, 88
curiosity, as aspect of Adam's sin, 74
curses, of Adam and Eve, 6–7

da Fano, Menahem Azariah: on God's corporeality, 69n12
daimon, 176–77
Darwin, Charles, 264n195
David, Catherine, 90
Day of Atonement, 88, 302–4. *See also* Purim lottery; Yom Kippur
de Careil, Foucher, 220n95
de Leon, Moses: on Balaam, 160n66
de Prado, Juan, 221
death, as integral to life, 168
delator, 296n14, 302n29
Delmedigo, Joseph: as precursor for Spinoza, 211n67, 284n254
Democritus, 181n122
demon: graveyard, 159; Jewish, 51n78; merger with human (soul), 163–64
demonic power, transformation into divine power, 164
demonology, 149, 152, 154, 158n56, 166; talmudic, 155–57
demons (*shedim*), 1, 35, 42, 175; as created by human actions, 166; as creatures of the devil, 148n15; distinguishing from humans, 162–64; divination involving, 160; in the generation of Babel, 182–84; as hazardous, 49; as hidden causes, 153; and the name YHWH, 188; no longer personified, 176; in 17th-century science, 148n15; as sons

of Adam, 36; in the Talmud, 11; usefulness of, 183
demons and spirits, 142, 155, 156
depersonalization, of technoscience, 111
desacralization: of chance, 19; of the universe, 194
Descartes, René, 146; mistake of, 262–63
design, as argument for God's existence, 264
desire, 84, 171, 177, 253
despair, and hope, 305
destiny, and chance, 290, 299–301
determination, modes of, 95
determining causes of nature, 153–55
determinism, 96, 115, 167, 314; and free will, 117, 292; in kabbalah and in physiological sciences, 109–10; non-theological, 100; partial awareness of, 118; in procreation, 46; of the rabbis, 113n106; in Spinozism, 287; Talmudic, 100–101; versus fatalism, 158n56
deterministic chaotic states, 292
dice, 304
differentiation of humans, 62, 76n32
din (rigor), origin of, 170
Diogenes Laertius, 220n94
discourse, philosophical versus mythological, 196
discourse style, for ethics, 2
disjunction, between inside and outside, 82
divination, 158–62, 184n138, 289
divine, ambiguity of the, 301
divine image (*ẓelem 'elohim*), as reproduction of human image, 166–68
divine law, 251
divine wisdom, and demons, 157
divine world, 193, 195
DNA, as inert, 23, 24n6
Donollo, Shabbetai, 225
door, as symbol, 171
double philosophical language, 200
doubt. *See* certainty
dream interpretation, 152
dreams, 162
Droit, Roger-Pol, 199n30, 213
dualism, versus monism, 156
duty to learn, 159n56
dynamics, 261

East and West: Hebrew versus Egyptian, 210–16, 229n117
Ecclesiastes, 41–42, 116n112, 121–22, 140–41
eclipses: and moral behavior, 108n98
ecologism, 122
ecology, deep, 122
Edel, Suzanne, 270
Edels, Samuel (Maharsha), 298
Eden, nonexistence of evil in, 73
ego, 176–81
'Ein Sof (the Infinite), 185. *See also* Infinite
Eleazar ben Simeon: in the *Zohar*, 58
elements, four basic, 163n73
Eliaschow, Shlomo, 74–79, 97, 104, 106n92, 280n240; dialectic of determinism and choice, 132–40; on free choice and determination, 114; and random drops, 108
Eliezer (Abraham's servant), 161n69
Elijah the Gaon of Vilna, 60n102
'elohim (God, god[s]), 185–86
emanation, 43n57, 48n68
emergence, 91n68
emotions, and statistical tests, 145
enchantment, the need to restore, 153–54
End of Days, 86n57
enrootedness, 120
entelechy, 263n188
entheogens. *See* hallucinogens
Epicurus, 93–94, 191–92
epigenetic level, 64
equality of cause and effect, 265
Eros, 45, 84–86, 131n153, 250, 301; and Thanatos, 8, 79
erotica, sacred, 84
errors, in the pursuit of causes, 145
essence (*'aẓmut*), of the Infinite, 287n261
essence (*havayah*), 314
essence, of the human condition, 30; of life, 25; of things, 133
essences, 287n261, 314
Essenes, 234–35
Esther, book of, 246–51
eternal return, 113
ethical experiences, ontological role of, 118
Ethics (Spinoza), 13, 93n69, 95n71, 110, 205, 206n52; addressees of, 255; and the cult of nothingness, 215; dogmatic character of, 205; God in, 196, 236n128; interpretations of, 216n83; and kabbalah, 279; and rabbinic literature, 288; and salvation, 259; social aspect of, 253
ethics, in the absence of free will, 112; call of, 119–24; levels of, 253
Europe, churches of, view of science and philosophy, 268–73
evil, 6, 8, 49–53; averting, 268; causes of, 302n29; elimination of, 107; origin of, 101, 102, 170, 260, 268, 281n241; at source, 161n66, 168
evolution, 193, 255
exegesis, combined meanings in, 284, 316n20. *See also* interpretation
experience of the one and the many, rational doctrine of, 63n109
experience of the sacred, 45–46
experimental method, 60
explanations for things and events. *See* causes, hidden
Extension, and Thought, 281n241
extermination. *See* proscription (*ḥerem*)
'eẓel (in, alongside), 48n68

faces (*parẓufim*), 277n227
fairness, 298
fall of Adam, 74, 183
farmers, versus nomads, 22
fate, 3, 113. *See also* chance; destiny
fear of YHWH, 160n63
fertilization, 45–46
Ficino, Marsilio, 206n51
fifty gates of understanding, 54n87
finalism, 265
Flood, myth of the, 11, 55–57. *See also* generation of the Flood
Fludd, Robert, 269
food, symbolism of, 174
fool, versus wise, 253
forbidden, and permitted, 183–84, 301
forgetting, 1
formulas, 151, 183n135
Francès, Madeleine, 274n223
Fredriksen, Paula, 234

free choice, 114, 133, 135, 140–41
free necessity, 123
free will, 11, 118–19, 264, 287, 291, 292
freedom, 44, 53, 109–19, 124n129, 130
Friedmann, Georges, 220n95, 278
functional explanations, 191
fundamentalist doctrine, 268n206
fusion, with the infinite, 180

Gamaliel, Rabban, 252n161
game, 111; as allegory, 96, 180–81, 196; played by nature, 76n32, 124
games of chance, 293
games of the Infinite, 124–30
garment, 166, 180–81, 287n261
gender, 44
gene, 23, 24n6, 156
generation of the Wilderness, 37–38
generation of the Flood, 37–38
generation of the Tower of Babel, 37–38, 86n52, 175. *See also* Babel, story of; Tower of Babel, myth of
generosity, versus law, 104n88
Genesis, opening narratives of, 3, 5, 21–22
genetic code, analogy to language, 26–28
genetic discourses, ambiguity in, 24n6
genetic indeterminacy, 64n111
genetics, 23, 24n6
gevurah (contraction), and *ḥesed* (expansion), 102
Gikatilla, Joseph, 82n45, 185n143, 187, 197n24, 291n3; on golem, 32n32; ontology of, 312–13
Girard, René, 294
Glanvill, Joseph, 147n15
Gnostic interpretation, 4
Gnosticism, 149–50
Gobineau, Arthur de, 215
God, 252n163; as abusive, 99; anthropomorphistic depiction, 103n88; as Author of the Bible, 276n225; as creator of the evil impulse, 100; definition of, 260n179; as explanation for events, 250; going at random, 310–16; hidden, 249; idea of, 206; incorporeality of, 272n219; intelligibility of actions of, 313–14; knowledge of true nature of, 274n223; proof of existence of, 205, 264; sufficiency of, 264. *See also* Name (of God)
god of persons, 122
Golden Calf, 38
Golden Rule, 251, 253
Goldschmidt, Richard, 279
golem, 6, 28–34, 169n90
good and evil, 48, 52, 53, 72–74
Gordin, Jacob, 222–24
greatness of soul, 118n115
Greek philosophy, history of, 189, 194, 216–20
ground, 198, 207; of knowledge, 202n38, 204–5
Guéroult, Martial, 73n22, 148n15, 203, 204, 206n52, 267, 284
guilt, 88n61, 90, 97

Hadot, Pierre, 118n115
Hai Gaon, 184n141
Halevi, Judah, 272n219
hallucinations, activation of, 48, 85n52, 172, 178, 182n129
hallucinogens: in the Bible, 195n18; in prophecy, 242n142
Ḥanina, Rabbi, 158n56
Hanukkah, 248n152
harmony, ad hoc versus universal, 117n112
Ḥayyim ben Attar, 311
Ḥayyim Volozhiner, 272n219
head that is not known, 53–54
Hebrew, use of, 182
Hebrews, mental universe of, 195n18
Hegel, G. W. F., 211–14
Heidegger, Martin, 120
Herrera, Abraham Cohen, 151, 196, 211n67, 280n239, 284n254
ḥesed (expansion). See *gevurah*
ḥesed (favor; disgrace), 78n36
heteronomy of the law, transformation into autonomy, 123
Hillel, 251
historical-critical method, 282n242
history of the Jewish people, 243
history of philosophy, 220
hoopoe, 41n54
hope, and despair, 305
Horowitz, Isaiah, 31, 50n76, 51n78, 69n12, 81, 283n247
human beings, 84n47, 99. *See also* man
human nature, 23, 94n70
Hume, David, 156, 203, 264
humor, 303n29

I, 55, 121–22
Ibn al-Arabi, 174n108
Ibn Ezra, Abraham, 185, 282n242; on Azazel, 302n29; concept of an immanent deity, 272n219; exegetical method, 227n113; on Num. 23:4, 316n20; on Pentateuchal authorship, 226
Ibn Gabbai, Meir, 286n261
Ibn Gabirol, Solomon, 50, 196n19, 288
Idel, Moshe, 29, 31, 166
idolatry, 58n99, 131, 158n56, 166, 169
ignorance, 54, 67, 118, 119. *See also* knowledge; unlettered
image (*ẓelem*), 62
image of God, bidirectionality of, 72, 96, 122
imagination, mythological, 192
imitatio Dei, 30n25, 31, 34
immanence, 70, 113, 286
impersonalization, of technology, 108n98
Incarnation, 273
incest, 78n36
incorporations, 130–32
incorporeality, of God, 67, 68–72
Indian East, in the history of Greek philosophy, 216–20
individual welfare, versus public good, 241
individuation, origin of, 102
Ineffable Name. See *shem ha-meforash*
infinite, mathematical concepts of, 202n38; stakes of the, 180–81
Infinite (*'ein sof*), 287n261. See also *'Ein Sof*
infinite chain of being, 96
infinity, of Lurianic kabbalah, 270
inquiry, conditioned nature of, 198
inside, like the outside, 82
intellectual love of God, 124n129
intelligence: analytical (*binah*), 51n78, 170; theoretical, 172
intelligent design, 264
intelligibility, temporal and timeless levels of, 140
intention (*kavvanah*), in prayer versus in Torah study, 108n98

interpretation, 4–6, 14. *See also* exegesis
intersubjectivity, 12
intrigues, divine (*'alilot*), parable of, 96–101
Isaac, binding of, 171, 174, 179n119
Israel (the Jewish people), and astral determinism, 154–55
Israel, Nicolas, 254n166

Jacob, François, 27
Jaynes, Julian, 93, 192
Jeremiah: as creator of golem, 33–34; metaphor of potter's clay, 109
Jews, as free from astral determinism, 154–55
Job, interpretation of last chapter, 88n61
Johanan, Rabbi, 300
Joseph of Hamadan, 260n174, 286n261
Judah Loew ben Bezalel, 29. *See also* Maharal
Judah's Offering—The Spirits Tell (Petaya), 152–53
Judaic studies, postwar, 224
Judaism, relationship to India, 225–26; versus Christianity, 246
Judeo-Christian tradition, 260
just, versus wicked, 256n169
justice, and lotteries, 297

kabbalah, 283; association with religion, 288n266; Christian, 130; ecstatic versus theurgic, 16; irrational currents in, 282; limited, Spinozism as, 279–88; Lurianic, 270; mythical character of, 196–97; notions of, in Europe, 288n266; on evil, 49–52; practical, 17; and Sabbateanism, 222n100; as science, 169; as science and natural philosophy, 151; speculative versus practical, 16; and Spinozism, 212n68; theurgic, 125n129; use of theistic figures of speech, 107
Kant, Immanuel, 200–201, 213, 265
Kepel, Gilles, 216n84
ḳere-ketib, as exegetical technique, 98n79
ḳeri (randomness), 37–38; in Leviticus 26, 310–16; in the Mishnah, 311
Kimḥi, David, 295n11
kinetic energy, conservation of, 261
king, functions of, 191. *See also* sovereign
Knorr von Rosenroth, Christian, 212n68, 269
knowledge, 1; of abominable practices of other nations, 159n56; of absolute determinism of nature, 117; access to supernal, 161n66; as appropriation, 74–75; biblical, 43; of the body, 66; capacity for inner, 83; by causes, 143, 153; and control, 59; and demons, 157; desire for, 177; development of objective, 90; ethics of, 174; externalized, as sin, 78; and fertilization, 28; of the first kind, 255; of future, 159; of God's true nature, 274n223; as good, 38; good and bad, 48; of good and evil, 72, 119; ground of (*see* ground of knowledge); hidden, 304; of ignorance, 304; inadequate, 45; of the intimate, 84; of the left side, 183; limitations of, 34; moral, inferiority of, 72–74; and mystical experience, 45; of nakedness, 76–77; not divided from desire, 84; objective, 84; of the objects of nature, 206; paradoxical experiences of, 72; permitted oracular, 159n58; prescientific, value of, 60; and randomness, 44; of the second kind, 254, 255; of self, 78, 79n37, 89, 117; serpentine, 161, 164; and sexuality, 44–46, 75–76, 161n66; subjective versus objective, 87; temporal and timeless levels of, 140; of the third kind, 206n52, 239, 240, 255; Torah as source of efficacious, 160; two types of, 235; via rational thought, 178; without conscience, 168–69; in the *Zohar*, 51n78
Kook, Abraham Isaac, 52n83, 60n103, 86, 284–85
Kordiakos, 156–57, 163n72
Kortholt, Christian, 260

labor (childbirth), 7
labor (toil), 6
Lagrée, Jacqueline, 217n88
Land of Israel, 252n163
language, 56, 85n52, 177, 198; esoteric, 251; misuse of, 202n38
lashon nofel 'al lashon (linguistic correspondence), 81n40
laughter, 303n29
Lautman, Albert, 265
law, versus generosity, 104n88
law of large numbers, 291
law of levirate marriage, 301
law of the community, obedience to, 242
left side. *See* wisdom, of the left side; knowledge, of the left side
Leibniz, Gottfried Wilhelm, 262, 266–70; and alphabet of nature, 270n216; annotations of Wachter's book on Spinoza, 212n68, 220n95; as ghost writer for van Helmont, 270n211, 271n216; and kabbalah, 259–68; on metempsychosis, 240; and Spinoza, 212n68, 260, 268–73, 278; on universal soul, 217n88
Leibowitz, Yeshayahu, 285n256
Leviathan, and Behemoth, 86–88
leviathan, as serpentine, 87
Levinas, Emmanuel, 119–21, 123
Lewontin, R. C., 156n47
liberation, via acceptance of the Torah, 95, 124n129, 179
Libet, Benjamin, 111n101
libido, 8
Liebes, Yehuda, 33n32, 126
Life, 26, 92n68
life sciences, 28, 91, 92n68
light, direct versus reflected, 171
Lipsius, Justus, 217n88
literal meaning, versus symbolic interpretation, 5
live forces, theorem of, 261, 263
logical constraints, versus ontological constraints, 103n88
logos spermatikos. *See* seminal reason
lots, casting of. *See* casting lots
lottery, 290, 293; on the Day of Atonement, 303; in judicial proceedings, 294–99
love, 79, 123, 125n129; of God, 124n129; platonic, 85–86

Lucretius, 191–92
Lully, Raymond, 271n216
Luria, Isaac, 196

macranthropos, 32n32
Macherey, Pierre, 215, 216n83
machines, 92n68
maggid, angel or demon as, 162
magia naturalis, 270
magic, forbidden versus permitted, 173n107; natural, 44n59, 60–61; talmudic types of, 158n56; time of, 57–61
Maharal, 54, 69, 70, 149, 197n24; attribution of a golem to, 29–30. *See also* Judah Loew
Maharsha. *See* Edels
Maimon, Salomon, 200–207, 267; on kabbalistic interpretation of Scripture, 282n242; methodology of, 281n242, 284n253; on Spinozism, 279
Maimonides, Moses: on Adam's fall, 73–74; critiques of, 14; exegesis, 227n113; on God's chance conduct, 312; influence on kabbalah, 197; on mystery of God's omniscience, 97; on sacrificial ritual, 245; and Spinoza, 250; Spinoza's refutation of, 237–38; status assigned to Moses, 237–38; use of internal contradictions, 238–39
makeshift, 290, 293, 299
Mallarmé, Stéphane, 304–5
man, 76n32, 122. *See also* human being
manna, 163n75
Marcus Aurelius, 118n115
mastery, by the body, 89
mathematical infinite, 266n201
mathematical modeling, 144
mathematical physics, and philosophy, 261–63
mathematics, 204, 206, 261
mathematization of biology, 204n44
Matheron, Alexandre, 236, 238–40, 255n167, 274n223, 281n241
meaning, and the absurd, 305n32; and chance, 314; human being as carrier of, 46; of a text, 314–16
mechanics, 143n2, 290
mechanism, and responsibility, 90–96
mechanists, versus vitalists, 91
Méchoulan, Henry, 221, 228, 230
medical lore, 153n38, 156
Meir, Rabbi, 36
memory, 61, 119
mental mutation, 190
Mersenne, Father, 271n218
Merzbach, Ely, 291n3
Metatron, 41n55, 51n78, 81
metempsychosis, 240. *See also* reincarnation; transmigration of souls
metis, 39, 315
Metis and Themis, 176–77, 179
middah ke-neged middah (measure for measure), 311
miḳreh, 312
mind, and body, 1, 85, 87
miracles, 117n112
Mirandola. *See* Pico della Mirandola
Mishnah, 244, 246, 296
Misrahi, Robert, 115, 274n223, 277n227, 279
mixed multitude, 53–54
Moabites, 300
modeling, mathematical, 144
modernism, end of, 13
modes, 280–81
monads, 270
monism, 10, 260, 278; of Spinoza, 72n22; versus dualism, 156, 205
monotheism, 195n18
months, as succession of births and deaths, 116n112
moon, 116n112
moral code, provisional, 104n88, 251–59
moral progress, 7
moral responsibility, and free will, 11
moral subject, 11, 178
morality, versus science, 175; versus truth, 73
More, Henry, 147n15, 269, 280n239
mortification of the flesh, 6
Mosaic law, as a means of salvation, 249. *See also* Torah
Moses, 54, 75n31, 105n89, 164, 273. *See also* rod of Moses; staff of Moses
Moses de León, 71n18
Moses Germanus, 218n88, 259
mutation, mental, 190, 192
mystical experience, 45, 180
myth: of decadent West, 216n84; and empirical skepticism, 208; and explanatory answers, 207; of origins, 5, 125; and philosophy, 12, 198; questions addressed by, 2; rationalized, 190; and relative relativism, 208; severed from reason, 193; transition to philosophy, 94n70
mythology, as mode of discourse, 196; replacement by philosophy, 191

Naḥman of Bratslav, 131
Naḥmanides, Moses, 58n99, 158–59, 187; on Azazel, 302n29; on demons, 163n73; exegesis, 227n113; on levirate marriage, 301; objections to Maimonides, 238n133
nakedness, 76–77, 78n36
Name (of God), 58, 184–88
natura naturans and *natura naturata*, 62; terms as used by Gikatilla, 70, 123, 185, 314; terms as used by Spinoza, 236n128, 261n179, 287
natural law, and chance, 314
natural light (of reason), 195, 235, 237n129, 244, 258n171, 274n223, 276n225
natural magic, 147n15, 149–52; ancient antecedents of, 210–11
natural selection, 264n195
natural theology, 265
nature, 3, 18
necessity, 263, 265
nefesh ḥayyah (living soul), 96
Negri, Antonio, 216n83
Neoplatonism, 149–50
neshamah (intellectual soul), 96
neuter, 120n117, 121
974 lost generations, 101–9, 134
nomads, versus farmers, 22
nonculpability for sin, 90
nondualism, victory over dualism and idealism, 93n69
nondualist philosophies, 93
nora', 102n83
nora' 'alilah, 99
novelty, experience of, 115–16
number of years, symbolic value of, 103n88

objectivity, versus subjectivity, 176
obligations, deriving from human differentiation, 62
Oldenburg, Henry, 110n100
one, 218n88, 261n179, 280n239
oneness, of YHWH and his name, 184–87
ontological constraints, versus logical constraints, 103n88
oracle, via lottery, 301
Oral Law, 51n78, 81, 233–34, 243, 248
Orient, 212, 218n88, 225n111
oriental phrases, 273n222
original sin, 4, 6
Ornstein, Jacob Meshullam, 122n122
Other and Self, 121
Other, the, 63n109
outside, picturing oneself from the, 84n47
Ozanam, Frédéric, 214, 215n78

pain, and curses of Adam and Eve, 7, 53, 253. *See also* suffering
Paley, William, 264n195
Panaccio, Claude, 209
panentheism, 286, 287n261
pantheism, 286–87
parallelism, of body and mind, 261n180
peace, 39
Peretz, Rabbenu, 284n252
permitted, and forbidden, 183–84, 301
personal god, 108n98, 181
perspective, of moral subject and knowing subject, 174
Petaya, Yehudah, 152–53, 182n129
Pharisees, 232–38, 242–46, 259
philosopher, versus unlettered, 255, 257
philosophical hypotheses, and mythical narratives, 198
philosophical kabbalists, 196
philosophical language, bivalence of, 200
philosophical quest, 235
philosophical questions, mythical answers to, 189–200
philosophical religion, 274n223
philosophical systems, 199, 208–9, 268
philosophizing, 208–9
philosophoi (lovers of wisdom), 194
philosophy: as an activity of thinking, 268; ancient, 3, 94n70; break from theology, 210; as critical inquiry, 208; death of, 208; distance from theology and mythology, 196; escape from myth, 207–8; incorporation of myth in, 196–200, 207–8; and language, 198; and myth, 12; origin of, 189, 220n94; perennial, 277n226; replacing mythology, 191–96; transcendental, 202; versus science, 200; and wonder, 155n43
physicalist tradition, 93
Pichot, André, 84n47
Pico della Mirandola, Giovanni, 212n68, 229n117, 269, 271n216
Pines, Shlomo, 238
Pirsig, Robert, 90
placebo effect, 153
planning, total, of procreation, 64
play, 131–32, 180–81, 303n29
pleasure, 35–36, 42, 51, 253. See also *sha'ashua'*
Plotinus, 216–19
polytheism, 131, 195n18
possession, by a spirit, 159
power, legitimacy of, 22
prayer, 109n98, 256
prediction, theoretical, 115–16
preformationism, 156
prelapsarian knowledge, 77–78
prelapsarian state, 52
principle of sufficient reason, 143n2, 262, 263–65, 290
principle of contradiction, 262
prisca sapientia, 277
prisca theologia, 277
probabilities, misunderstanding of, 18
progress, moral and technical, 7–8
proof, unity of substance, 260n179
prophecy, 161–62, 235; end of, 195, 242–44, 152; Spinoza's theory of, 258n171; versus divination, 159–60; versus wisdom, 276n225
prophetic religion, 274n223
prophetic Scriptures, 241, 244
proprioception, 84n47
proscription (*ḥerem*), 295
providence, divine, 108n98
pseudo-concepts, for describing reality, 206
psyche, 94n70
psychedelics. *See* hallucinogens
psychologization of kabbalistic motifs, 152
psychotherapy, 153n38
public good, versus individual welfare, 241
Purim, 251, 302n29
Purim lottery, as twin of Day of Atonement lottery, 251
Pygmalion, 33n32

questions, and answers, 198
quid facti, and *quid juris*, 201, 202n38, 203n41

rabbinic Judaism, as historical break, 243
rabbinic teachings, as source of inspiration, 3
rabbis, and prophecy, 161
rain, and moral behavior, 108n98
random, mastery of the, 293
random drops, 108, 109n99
random lottery, and desacralized randomness, 290
randomness, 48, 55, 64n111, 291; of birth, 3, 35, 49–53, 59, 64n111; of God, 310–16
random-number generator, 293
Rashi, on Azazel, 302n29; on *'elohim*, 195n18; on Josh. 7:1–19, 295n11; on Num. 23:4, 316n20
rational thought, 178
rationalism, and skepticism, 205
Rava, 31, 32n32
Ravina, 297
reality, dichotomies in, 170; experienced as divided, 87; and shadow, 165
reason, multiple forms of, 190
reasons, change in nature of, 193; versus causes, 145
Recanati, Menaḥem, 174
reciprocity, of god and man, 165
reconstruction, 183, 209. *See also* Spinoza, reconstruction of
Redfield, James, 94n70
reflexivity, 16, 84n47
reincarnation, 255. *See also* metempsychosis; transmigration of souls

relative relativism, 208
religion, 83; common, 276n225; philosophical, 274n223; prophetic, 274n223; true, 276n225; wars of, 268n206
rendering the hands impure, 247
renewal (*ḥiddush*), 116n112
repair (*tikkun*), 38, 137
reproducibility, 145
reproduction, human, 7–8, 21–23
reshut, 131n154
responsibility, 46n63, 123; due to status as autonomous beings, 135; as the ground of free will, 119; to play, 132; and sense of time, 140–41; source of, 114; as trick played by creation, 97
Révah, I. S., 221
revelation, status of, 276
right side. *See* wisdom, of left side and right side
righteous, and wicked, difference of principle, 139
rituals, utility of, 256n169, 257
rod of Moses, 51n78, 81. *See also* staff of Moses
roles, confusion of, 259
"rose into thought" (Hebrew idiom), 105–6, 106n92, 127, 180, 280n239
Rosh Hashanah, 172
ruaḥ (spirit), 96
Ruth, meeting Boaz by chance, 299

Saadia Gaon, 286
Sabbateanism, and Lurianic kabbalah, 222n100
Sabbath, carrying on the, 131n155
sacred, 84, 94n70
Sacred Scripture, 247–48
Sacred Writ, invented by Pharisees, 243
sacrificial ritual, 245
Sadducees, 233–35, 249
Saint-Hilaire, Geoffroy, 156
salvation, 235, 237n129, 246; for all, 251–59; through knowledge, 251; and knowledge of the third kind, 240; and Mosaic law, 249; prerequisite for, 256; road to, 241; of the unlettered, 238–42, 242–46, 253
śar (tutelary angel), 159
Sarug, Israel, 151, 196, 211n67
scale, metaphor of the, 173
scapegoat, 294, 302–4
Schlanger, Jacques, 199
Schmidt, Francis, 244, 245n146
scholar, true, 82–83n46
scholasticism, "genetic" causes of, 155–56
Scholem, Gershom, 29, 107, 181n122, 272n219, 280n239
Schopenhauer, Arthur, 220n95
science, 2, 9, 90, 113; break from magic, 144, 210; in bygone times, 149–52, 169; and magical pre-science, 270; modern, 144, 150; modus vivendi with Catholic dogma, 271n218; objectivity of, 176; of non-Israelite nations, 159; versus morality, 175
scientific literature, flaws in, 18
scientific method, 200
scientific progress, distrust of, 112
Scripture, authority of, 274n223; interpretation of, 227n113, 242, 282n242; prophetic, 241; in shift from prophecy to philosophy, 235–36; status of, 268n206
seal (*ḥotam*) of the individual, 62–65
secret of gestation (*sod ha-'ibbur*), 23, 130
seduction, ambiguity of, 88
Sefer Yeẓirah, as known to Jacob and his sons, 31n30
sefirot, 67n2, 71n18, 122, 185, 195n18, 280n240; sides of, 170
sefirotic tree, 281n241
śeḥok (amusement), 125
self, 55, 119
Self and Other, 121
self-awareness, conjunction with ignorance of true causes, 119
self-consciousness, 77, 193n15
self-organization, 91n68
semen, 1, 29, 46
seminal reason, 10, 28, 44
seminal reason, and genetic determinism, 23–25
Semitic East, in the history of Greek philosophy, 216–20
serpent (*naḥash*), 47, 75–76, 78; ambiguity of, 79–80; brazen, 80; and cunning knowledge, 164; cunning and seductive knowledge of, 89; knowledge possessed by, 78; prelapsarian, 133; and Sammael, 75n32
Seth, 35, 79
sex, 301
sexual knowledge, change in, 78
sexuality, prelapsarian, 89
sha'ashua' (pleasure), 125, 181. *See also* pleasure
Shabbetai Ẓevi, 222, 228
shadow: of a shadow, 162–65; and subject, 166–68
shamanism, disappearance of, 194
shame, 84
shamir, 39–41, 175
shem ha-meforash (the explicit name), 58. *See also* Name (of God)
shiddah, 42–43
Shneor Zalman of Liadi, 272n219
sides, left and right, reintegration of, 172
significance, statistical, 18–19
signs (*'otot*), 117n112
Simeon bar Yoḥai, 58, 170
sin, 9, 102, 173n107; of Adam and Eve, 57, 73–79
singularity of the individual, 63
613 organs and sinews, 167
skeptical philosophers, 199
skepticism, 203
slavery, 64n111
social cohesion, and the scapegoat mechanism, 294
Socrates, 94n70
Solomon, 77n33, 140, 252n61; and Ashmedai, 36–37, 39–43, 156–57, 163n72, 175n115; as wise, 225n111, 233n122, 276
sophoi (sages or wise men), 194
sorcery, versus recourse to demons, 158n56
soul, 25, 28, 68, 93, 96
souls, choice by, 139
sovereign, responsibility of, 241, 253. *See also* king
sovereignty, loss of Jewish, 252
sparks of randomness (*niẓoẓot shel keri*), 1, 9, 34–37, 74, 82
speech, and subject, 181
spells, 60, 183
spermatic knowledge, 10, 29, 35, 155
Spinoza, Baruch, 13–15, 14, 17, 26, 139, 223; access to pre-Christian wisdom, 225; on

Adam's fall, 73–74; and the analytical method, 207; "anti-Judaism" of, 221–31; attacks on kabbalists, 227–28; attacks on Pharisees, 227; attacks on Talmud, 227–28; on the body's limits, 66; as both Oriental and Western, 225; challenge to God's incorporeality, 68–69; on Christ, 105n89, 237n129, 238, 239, 272, 275; "Christianity" of, 273–78; coherence of his thought, 268; comparison with Descartes, 274n223; confirmation of, 112n104; critical method of reading the Bible, 237–38; at defining moment of modern philosophy, 211; dissociation from Christianity, 273; on emotions' effect on human conduct, 110n100; endeavor of, 276; on essence, 314; evaluation of philosophy of, 268; on evil, 50n77; excommunication, 221–22; figure of Christ in, 274–275; first agents of, 231; on freedom, 53n85; goal of, 278; ground for knowledge in, 204–5; his idea of God, 109n100; and Indian universal soul, 218; on infinite understanding, 277n227; influence of Neoplatonism on, 272n219; interpretations of, 216n83; Jewish and Hebrew sources, 211–12; knowledge of Talmud, 232, 234; on knowledge of unity versus parts, 87n60; on literal meaning of Scripture, 226; on love of God, 276n225; and Maimonides, 237–38, 250, 272n219; "Marranism" of, 230; and Mennonites, 223; on mind and body, 11; on modes of revelation, 195–96; monism of, 72n22, 205; mythical element in, 205; at nexus of philosophy and myth, 189; pantheism of, 259; on piety and religion, 254; and Plotinus, 218; provisional moral code and ontology and epistemology, 254n166; quotations of kabbalistic texts, 271; on reading Bible as a philosophical text, 238; reasons for writing the Tractatus, 229; reconstruction of, 206–7; rediscovery of physicalist tradition, 93; rejection of spirits, 146–48; relevance of, 268, 288n266; resistance to notion of divine providence, 219; sources for, 224, 229; sources as exegeses of, 282, 284; on status of Apostles, 236n127, 238; on status of biblical revelation, 276; on submission to the sovereign, 253; terminology of, 274; on three kinds of knowledge, 95n71, 125n129; and theological discourse, 237; theory of knowledge of, 277n226; and thinking subject, 178; and traditional Judaism, 222; unique social position, 271; versus Leibniz, 278; view of ancient Hebrew republic, 232n122; view of Pharisees, 232–238, 250; and Wittgenstein, 207

Spinozism: ambivalence toward, 278n234; and democracy, 241; determinism and free will in, 219, 287; expanded, 279–88; history of, 259; and Jewish-Christian doctrinal rapprochement, 105n89; and kabbalah, 212n68, 282; as pejorative, 268; salvation in, 255–56; timelessness of, 215–16; versus acosmism, 287; versus pantheism, 287

Spinozist, Bayle's definition of, 259

spirits, 35, 48; divination involving, 160; existence of, 146; as hazardous, 49; as hidden causes, 147n15, 153; hierarchy of, 84; and human soul, 163; in 17th-century science, 148n15. *See also* angels; demons and spirits

spiritual automaton, 261n180

spiritual exercises, 130

spirituality, 87

staff of Moses, 47, 51n78, 81. *See also* rod of Moses

statistics, 18–19, 56, 145, 291

stichomythia, 299

stock-breeding, invention of, 21

Stoicism, 220n94

Stoics, 93–94, 118n115, 121

stones, hewn versus unhewn, 175

Strauss, Leo, 238

subject, 149, 176–78, 214n77; and shadow, 166–68

subjectivity, 149, 176

substance, of the Infinite, 287n261

suffering, 266n202, 314. *See also* pain

sufficient reason, principle of, 263, 265

supernatural agents, exclusion of, 193

superstition, 153

Szent-Györgi, Albert, 92n68

Talmud, 3, 13, 109n100, 153n38; on Adam's offspring, 35–36; on culpability for evil, 100; on demons, 155–57; on determinism, 46; on games of chance, 293; on golems, 31, 32n32; on individual uniqueness, 62–63; on Joshua 7:1–19, 294–99; on prophecy versus wisdom, 276n225; on Solomon and Ashmedai, 39–41; on speculation's dangers, 17

talmudic legends, 41

Tanna de-vei Eliyahu, 121–22

Tardieu, M., 275n224

technological advances, 168

technology, demonization of, 113; as magic, 61; time of, 57–61

technoscience, 108n98

teleological judgment, legitimacy of, 265

Temple (Jerusalem), and Ashmedai, 39; legacy of, 244–45; in Lurianic kabbalah, 38; rebuilding of, 245n146

ten, counting by, 67

Ten Commandments, 185–87

teshuvah, 173n107

theistic personifications, by kabbalists, 132, 138

Themis and Metis, 176–77, 179

theology, 143n2, 147n15, 181, 314

thermodynamics, first law of, 262n183

thinking, 178, 209, 280n240

thought, 190

Thought, and Extension, 281n241

thousandfold blessing for Israel, 105n89

throne of glory, 187
tikkun (repair), 137
time, 140–41; divorced from animate beings, 23; experience of, 116; perceptions of, 116n112; and randomness, 34; and reproduction, 61; reversibility of, 34, 65
toil. *See* labor (toil)
Torah, 51n78, 81, 125; and association with demons, 183; of *'azilut*, 73n24, 129, 211n67; conditioned nature of, 54; as drug of life versus drug of death, 175; and freedom, 130; as garment of the Infinite, 127, 130; meanings of, 128, 129; primordial, 211n67; role of, 126; as source of knowledge and holiness, 6; study of, 109n98, 245; transforming sin into obligation, 173n107. *See also* Mosaic law; Oral Law
torment (*nega'*), transformation from pleasure (*'oneg*), 35–36, 42, 51
Tower of Babel, myth of, 11, 55–57, 59n99. *See also* Babel, story of; generation of the Tower of Babel
Tractatus Theologicus-Politicus (Baruch Spinoza), 226, 238, 241, 274n223; anti-Judaism in, 222–23, 228; Christian perspective in, 235; context of, 258; objective of, 234, 241, 259, 274n223; obscure passages in, 277n228
training the body, via indirection, 89
transcendence, and immanence, 70, 286
transformations, mental, 190
transgression, transformed into obligation, 173n107
translation, 77n33, 314–16
transmigration of souls, 240, 255. *See also* metempsychosis; reincarnation
Tree of Knowledge, 43–47; Eliaschow's interpretation, 74–79; and the libido, 8; and Tree of Life, 7–8, 34, 47, 57–58, 138
Tree of Life, as remedy, 9. *See also* Tree of Knowledge
truth, 83, 184, 206
Tschirnhaus, Ehrenfried W. von, 236n128
231 gates, 127
tyranny, obedience to, 257, 258n171

unification, of YHWH, 187
union, of cognitive and ethical, 169; of God and the world, 71; of soul and body, 10, 70, 71
universal religion, 240–41, 254, 257
universalism, 217n88
universality, European claim to, 276n225
unlettered, 255. *See also* salvation of the unlettered
Urim ve-Tummim, 159–60

validity, of statistical hypothesis, 19
van Helmont, François-Mercure, 212n68, 262, 269, 270, 271n216
velocity, 262
Vernant, Jean-Pierre, 190–92, 194
Veyne, Paul, 190
Vidal-Naquet, Pierre, 189n1
violence, 40n51, 79, 304
Vital, Ḥayyim, 37–38, 42, 49, 167
vitalists, versus mechanists, 91
volition, as illusory, 111n101
Volozhiner, Ḥayyim, 68, 71, 165, 181n122, 286
vowel points, Hebrew, 70

Wachter, Johann Georg, 212n68, 286n261, 282
wars of religion, 268n206
Webster, John, 147n15
Western philosophy, versus Oriental philosophy, 219
white space, 172
why, the real question behind, 62n108
wicked, 138, 256n169
wisdom, 47n68, 173, 211n67, 243, 276n225; ancient, 169; intuitive (*ḥokhmah*), in the *Zohar*, 51n78; of the left side, 39, 40n51, 174, 303n29; of left side and right side, 40n51, 169–72, 175, 179
Wisdom (*Ḥokhmah*), 48n68, 125
wise, 253–54
Wittgenstein, Ludwig, 198n26, 202n38
Wolfson, Harry A., 211n67, 280n239
wonder, as source for philosophical thought, 155n43
word, divine, 298
work, 6–7. *See also* labor (toil)
world of meaning, and world of nature, 117n112
world to come, 34

yearning, 83, 175; for yearning, 84–86
Yeḥiel, Rabbenu, 284n252
yezer ha-ra (evil impulse), as created by God, 100
yezer tov and *yezer ra'* (creative impulses), 166–67
YHWH, 184–88
yiḥudim (acts of unification), 170, 184
Yom Kippur, liturgy, 97

Zarathustra, 311
Zealots, 234–35
Zeira, Rabbi, 31
zelem (divine image), 35, 79, 176n115
zimzum, 125n131
Zohar: on Azazel, 302n29; on Creation, 227n113; on demons, 163; on evil, 51; on plain-sense reading of Scripture, 231n120; on Tree of Life, as Torah of *'azilut*, 73n24; on union of soul and body, 70

Cultural Memory in the Present

Rebecca Comay, *Mourning Sickness: Hegel and the French Revolution*

Djelal Kadir, *Memos from the Besieged City, Lifelines for Cultural Sustainability*

Stanley Cavell, *Little Did I Know: Excerpts from Memory*

Jeffrey Mehlman, *Adventures in the French Trade: Fragments Toward a Life*

Jacob Rogozinski, *The Ego and the Flesh: An Introduction to Egoanalysis*

Marcel Hénaff, *The Price of Truth: Gift, Money, and Philosophy*

Paul Patton, *Deleuzian Concepts: Philosophy, Colonialization, Politics*

Michael Fagenblat, *A Covenant of Creatures: Levinas's Philosophy of Judaism*

Stefanos Geroulanos, *An Atheism that Is Not Humanist Emerges in French Thought*

Andrew Herscher, *Violence Taking Place: The Architecture of the Kosovo Conflict*

Hans-Jörg Rheinberger, *On Historicizing Epistemology: An Essay*

Jacob Taubes, *From Cult to Culture*, edited by Charlotte Fonrobert and Amir Engel

Peter Hitchcock, *The Long Space: Transnationalism and Postcolonial Form*

Lambert Wiesing, *Artificial Presence: Philosophical Studies in Image Theory*

Jacob Taubes, *Occidental Eschatology*

Freddie Rokem, *Philosophers and Thespians: Thinking Performance*

Roberto Esposito, *Communitas: The Origin and Destiny of Community*

Vilashini Cooppan, *Worlds Within: National Narratives and Global Connections in Postcolonial Writing*

Josef Früchtl, *The Impertinent Self: A Heroic History of Modernity*

Frank Ankersmit, Ewa Domanska, and Hans Kellner, eds., *Re-Figuring Hayden White*

Michael Rothberg, *Multidirectional Memory: Remembering the Holocaust in the Age of Decolonization*

Jean-François Lyotard, *Enthusiasm: The Kantian Critique of History*

Ernst van Alphen, Mieke Bal, and Carel Smith, eds., *The Rhetoric of Sincerity*

Stéphane Mosès, *The Angel of History: Rosenzweig, Benjamin, Scholem*

Pierre Hadot, *The Present Alone Is Our Happiness: Conversations with Jeannie Carlier and Arnold I. Davidson*

Alexandre Lefebvre, *The Image of the Law: Deleuze, Bergson, Spinoza*

Samira Haj, *Reconfiguring Islamic Tradition: Reform, Rationality, and Modernity*

Diane Perpich, *The Ethics of Emmanuel Levinas*

Marcel Detienne, *Comparing the Incomparable*

François Delaporte, *Anatomy of the Passions*

René Girard, *Mimesis and Theory: Essays on Literature and Criticism, 1959–2005*

Richard Baxstrom, *Houses in Motion: The Experience of Place and the Problem of Belief in Urban Malaysia*

Jennifer L. Culbert, *Dead Certainty: The Death Penalty and the Problem of Judgment*

Samantha Frost, *Lessons from a Materialist Thinker: Hobbesian Reflections on Ethics and Politics*

Regina Mara Schwartz, *Sacramental Poetics at the Dawn of Secularism: When God Left the World*

Gil Anidjar, *Semites: Race, Religion, Literature*

Ranjana Khanna, *Algeria Cuts: Women and Representation, 1830 to the Present*

Esther Peeren, *Intersubjectivities and Popular Culture: Bakhtin and Beyond*

Eyal Peretz, *Becoming Visionary: Brian De Palma's Cinematic Education of the Senses*

Diana Sorensen, *A Turbulent Decade Remembered: Scenes from the Latin American Sixties*

Hubert Damisch, *A Childhood Memory by Piero della Francesca*

José van Dijck, *Mediated Memories in the Digital Age*

Dana Hollander, *Exemplarity and Chosenness: Rosenzweig and Derrida on the Nation of Philosophy*

Asja Szafraniec, *Beckett, Derrida, and the Event of Literature*

Sara Guyer, *Romanticism After Auschwitz*

Alison Ross, *The Aesthetic Paths of Philosophy: Presentation in Kant, Heidegger, Lacoue-Labarthe, and Nancy*

Gerhard Richter, *Thought-Images: Frankfurt School Writers' Reflections from Damaged Life*

Bella Brodzki, *Can These Bones Live? Translation, Survival, and Cultural Memory*

Rodolphe Gasché, *The Honor of Thinking: Critique, Theory, Philosophy*

Brigitte Peucker, *The Material Image: Art and the Real in Film*

Natalie Melas, *All the Difference in the World: Postcoloniality and the Ends of Comparison*

Jonathan Culler, *The Literary in Theory*

Michael G. Levine, *The Belated Witness: Literature, Testimony, and the Question of Holocaust Survival*

Jennifer A. Jordan, *Structures of Memory: Understanding German Change in Berlin and Beyond*

Christoph Menke, *Reflections of Equality*

Marlène Zarader, *The Unthought Debt: Heidegger and the Hebraic Heritage*

Jan Assmann, *Religion and Cultural Memory: Ten Studies*

David Scott and Charles Hirschkind, *Powers of the Secular Modern: Talal Asad and His Interlocutors*

Gyanendra Pandey, *Routine Violence: Nations, Fragments, Histories*

James Siegel, *Naming the Witch*

J. M. Bernstein, *Against Voluptuous Bodies: Late Modernism and the Meaning of Painting*

Theodore W. Jennings, Jr., *Reading Derrida / Thinking Paul: On Justice*

Richard Rorty and Eduardo Mendieta, *Take Care of Freedom and Truth Will Take Care of Itself: Interviews with Richard Rorty*

Jacques Derrida, *Paper Machine*

Renaud Barbaras, *Desire and Distance: Introduction to a Phenomenology of Perception*

Jill Bennett, *Empathic Vision: Affect, Trauma, and Contemporary Art*

Ban Wang, *Illuminations from the Past: Trauma, Memory, and History in Modern China*

James Phillips, *Heidegger's* Volk*: Between National Socialism and Poetry*

Frank Ankersmit, *Sublime Historical Experience*

István Rév, *Retroactive Justice: Prehistory of Post-Communism*

Paola Marrati, *Genesis and Trace: Derrida Reading Husserl and Heidegger*

Krzysztof Ziarek, *The Force of Art*

Marie-José Mondzain, *Image, Icon, Economy: The Byzantine Origins of the Contemporary Imaginary*

Cecilia Sjöholm, *The Antigone Complex: Ethics and the Invention of Feminine Desire*

Jacques Derrida and Elisabeth Roudinesco, *For What Tomorrow . . . : A Dialogue*

Elisabeth Weber, *Questioning Judaism: Interviews by Elisabeth Weber*

Jacques Derrida and Catherine Malabou, *Counterpath: Traveling with Jacques Derrida*

Martin Seel, *Aesthetics of Appearing*

Nanette Salomon, *Shifting Priorities: Gender and Genre in Seventeenth-Century Dutch Painting*

Jacob Taubes, *The Political Theology of Paul*

Jean-Luc Marion, *The Crossing of the Visible*

Eric Michaud, *The Cult of Art in Nazi Germany*

Anne Freadman, *The Machinery of Talk: Charles Peirce and the Sign Hypothesis*

Stanley Cavell, *Emerson's Transcendental Etudes*

Stuart McLean, *The Event and Its Terrors: Ireland, Famine, Modernity*

Beate Rössler, ed., *Privacies: Philosophical Evaluations*

Bernard Faure, *Double Exposure: Cutting Across Buddhist and Western Discourses*

Alessia Ricciardi, *The Ends of Mourning: Psychoanalysis, Literature, Film*

Alain Badiou, *Saint Paul: The Foundation of Universalism*

Gil Anidjar, *The Jew, the Arab: A History of the Enemy*

Jonathan Culler and Kevin Lamb, eds., *Just Being Difficult? Academic Writing in the Public Arena*

Jean-Luc Nancy, *A Finite Thinking*, edited by Simon Sparks

Theodor W. Adorno, *Can One Live after Auschwitz? A Philosophical Reader*, edited by Rolf Tiedemann

Patricia Pisters, *The Matrix of Visual Culture: Working with Deleuze in Film Theory*

Andreas Huyssen, *Present Pasts: Urban Palimpsests and the Politics of Memory*

Talal Asad, *Formations of the Secular: Christianity, Islam, Modernity*

Dorothea von Mücke, *The Rise of the Fantastic Tale*

Marc Redfield, *The Politics of Aesthetics: Nationalism, Gender, Romanticism*

Emmanuel Levinas, *On Escape*

Dan Zahavi, *Husserl's Phenomenology*

Rodolphe Gasché, *The Idea of Form: Rethinking Kant's Aesthetics*

Michael Naas, *Taking on the Tradition: Jacques Derrida and the Legacies of Deconstruction*

Herlinde Pauer-Studer, ed., *Constructions of Practical Reason: Interviews on Moral and Political Philosophy*

Jean-Luc Marion, *Being Given That: Toward a Phenomenology of Givenness*

Theodor W. Adorno and Max Horkheimer, *Dialectic of Enlightenment*

Ian Balfour, *The Rhetoric of Romantic Prophecy*

Martin Stokhof, *World and Life as One: Ethics and Ontology in Wittgenstein's Early Thought*

Gianni Vattimo, *Nietzsche: An Introduction*

Jacques Derrida, *Negotiations: Interventions and Interviews, 1971–1998*, ed. Elizabeth Rottenberg

Brett Levinson, *The Ends of Literature: The Latin American "Boom" in the Neoliberal Marketplace*

Timothy J. Reiss, *Against Autonomy: Cultural Instruments, Mutualities, and the Fictive Imagination*

Hent de Vries and Samuel Weber, eds., *Religion and Media*

Niklas Luhmann, *Theories of Distinction: Re-Describing the Descriptions of Modernity*, ed. and introd. William Rasch

Johannes Fabian, *Anthropology with an Attitude: Critical Essays*

Michel Henry, *I Am the Truth: Toward a Philosophy of Christianity*

Gil Anidjar, *"Our Place in Al-Andalus": Kabbalah, Philosophy, Literature in Arab-Jewish Letters*

Hélène Cixous and Jacques Derrida, *Veils*

F. R. Ankersmit, *Historical Representation*

F. R. Ankersmit, *Political Representation*

Elissa Marder, *Dead Time: Temporal Disorders in the Wake of Modernity (Baudelaire and Flaubert)*

Reinhart Koselleck, *The Practice of Conceptual History: Timing History, Spacing Concepts*

Niklas Luhmann, *The Reality of the Mass Media*

Hubert Damisch, *A Theory of /Cloud/: Toward a History of Painting*

Jean-Luc Nancy, *The Speculative Remark: (One of Hegel's bon mots)*

Jean-François Lyotard, *Soundproof Room: Malraux's Anti-Aesthetics*

Jan Patočka, *Plato and Europe*

Hubert Damisch, *Skyline: The Narcissistic City*

Isabel Hoving, *In Praise of New Travelers: Reading Caribbean Migrant Women Writers*

Richard Rand, ed., *Futures: Of Jacques Derrida*

William Rasch, *Niklas Luhmann's Modernity: The Paradoxes of Differentiation*

Jacques Derrida and Anne Dufourmantelle, *Of Hospitality*

Jean-François Lyotard, *The Confession of Augustine*

Kaja Silverman, *World Spectators*

Samuel Weber, *Institution and Interpretation: Expanded Edition*

Jeffrey S. Librett, *The Rhetoric of Cultural Dialogue: Jews and Germans in the Epoch of Emancipation*

Ulrich Baer, *Remnants of Song: Trauma and the Experience of Modernity in Charles Baudelaire and Paul Celan*

Samuel C. Wheeler III, *Deconstruction as Analytic Philosophy*

David S. Ferris, *Silent Urns: Romanticism, Hellenism, Modernity*

Rodolphe Gasché, *Of Minimal Things: Studies on the Notion of Relation*

Sarah Winter, *Freud and the Institution of Psychoanalytic Knowledge*

Samuel Weber, *The Legend of Freud: Expanded Edition*

Aris Fioretos, ed., *The Solid Letter: Readings of Friedrich Hölderlin*

J. Hillis Miller / Manuel Asensi, *Black Holes / J. Hillis Miller; or, Boustrophedonic Reading*

Miryam Sas, *Fault Lines: Cultural Memory and Japanese Surrealism*

Peter Schwenger, *Fantasm and Fiction: On Textual Envisioning*

Didier Maleuvre, *Museum Memories: History, Technology, Art*

Jacques Derrida, *Monolingualism of the Other; or, The Prosthesis of Origin*

Andrew Baruch Wachtel, *Making a Nation, Breaking a Nation: Literature and Cultural Politics in Yugoslavia*

Niklas Luhmann, *Love as Passion: The Codification of Intimacy*

Mieke Bal, ed., *The Practice of Cultural Analysis: Exposing Interdisciplinary Interpretation*

Jacques Derrida and Gianni Vattimo, eds., *Religion*

Printed and bound by CPI Group (UK) Ltd, Croydon, CR0 4YY

09/06/2025

14685894-0001